Electrons at the Fermi surface

David Shoenberg

Electrons at the Fermi surface

EDITED BY

M. SPRINGFORD

Reader in Experimental Physics
University of Sussex

CAMBRIDGE UNIVERSITY PRESS

CAMBRIDGE

LONDON NEW YORK NEW ROCHELLE

MELBOURNE SYDNEY

CAMBRIDGE UNIVERSITY PRESS
Cambridge, New York, Melbourne, Madrid, Cape Town,
Singapore, São Paulo, Delhi, Tokyo, Mexico City

Cambridge University Press
The Edinburgh Building, Cambridge CB2 8RU, UK

Published in the United States of America by Cambridge University Press, New York

www.cambridge.org
Information on this title: www.cambridge.org/9780521175067

First published 1980
First paperback edition 2011

A catalogue record for this publication is available from the British Library

Library of Congress Cataloguing in Publication data

Main entry under title:
Electrons at the Fermi surface.

'A Festschrift to honour Professor David Shoenberg.'
Bibliography: p.
Includes indexes.
CONTENTS: Lifshitz, I. M. and Kaganov, M. I. Geometric concepts in the electron theory of metals.—Wilkins, J. W. Understanding quasi-particles.—Chambers, R. G. The generalized path-integral approach to transport problems, [etc.]
1. Fermi surfaces. 2. Electronic structure. 3. Free electron theory of metals. 4. Shoenberg, David. I. Springford, M. II. Shoenberg, David.
QC176.8.F4E43 530.4'1 79-50509

ISBN 978-0-521-22337-9 Hardback
ISBN 978-0-521-17506-7 Paperback

Contents

Contributors

Professor O. K. Andersen
Max Planck Institut für Festkörperforschung, Stuttgart, W. Germany

Professor R. G. Chambers
University of Bristol, H. H. Wills Physics Laboratory, Royal Fort, Tyndall Avenue, Bristol BS8 1TL, UK

Dr P. T. Coleridge
Division of Physics, National Research Council of Canada, Ottawa, Canada K1A 0R6

Professor E. Fawcett
Department of Physics, University of Toronto, Toronto, Ontario, Canada M5S 1A7

Professor R. Griessen
Natuurkundig Laboratorium, Vrije Universiteit, De Boelelaan 1081, Amsterdam, The Netherlands

Professor R. J. Higgins
Department of Physics, University of Oregon, Eugene, Oregon, 97403, USA

Dr W. Joss
Laboratorium für Festkörperphysik, Eidgenössishe Technische Hoschule, Zürich, Switzerland

Academician M. I. Kaganov
Institute for Physical Problems, 2 Vorobyevskoye Shosse, Moscow, USSR

Professor M. J. G. Lee
Department of Physics, University of Toronto, Toronto, Ontario, Canada M5S 1A7

Academician I. M. Lifshitz
Institute for Physical Problems, 2 Vorobyevskoye Shosse, Moscow, USSR

Dr G. G. Lonzarich
Cavendish Laboratory, Madingley Road, Cambridge CB3 0HE, UK

Professor D. H. Lowndes
Solid State Division, Oak Ridge National Laboratory, Tennessee, 37830, USA

Professor A. R. Mackintosh
Physics Laboratory I, University of Copenhagen, H C Ørsted Institute, Universitetsparken 5, Copenhagen, Dk-2100, Denmark

Professor J. M. Perz
Department of Physics, University of Toronto, Toronto, Ontario, Canada M5S 1A7

Professor Sir Brian Pippard
Cavendish Laboratory, Madingley Road, Cambridge CB3 0HE, UK

Dr M. Springford
School of Mathematical and Physical Sciences, University of Sussex, Falmer, Brighton BN1 9QH, Sussex, UK

Dr I. M. Templeton
Division of Physics, National Research Council of Canada, Ottawa, Ontario, Canada K1A 0R6

Professor J. W. Wilkins
Laboratory of Atomic and Solid State Physics, Cornell University, Clark Hall, Ithaca, New York 14853, USA

Preface

This book has been conceived as a *Festschrift* to honour Professor David Shoenberg, FRS, on the occasion of his retirement from the Cavendish Laboratory, Cambridge, in 1978. Written by some of his students, colleagues and friends, it has been assembled around the theme of electrons at the Fermi surface.

Shoenberg began his research work as a student of Kapitza in the Royal Society Mond Laboratory at Cambridge in 1932. When shortly afterwards Kapitza was unable to return to England after a visit to Russia, Shoenberg joined him for a year in Moscow. Having become intrigued by the curious oscillations in the magnetism of bismuth at low temperatures – an effect discovered only a few years earlier by de Haas and van Alphen working in Leiden – he developed a torque method for studying the effect, and in 1937 made what was effectively the first experimental determination of a Fermi surface. Interestingly, Landau had at that time just produced an ingenious theory of the de Haas–van Alphen effect which Shoenberg was able to test, although it was some years before a detailed general theory of the effect was to appear. Returning to Cambridge after the war, he became involved with the problem of studying the de Haas–van Alphen effect in other metals. This phase of his work culminated with his beautiful series of experiments in pulsed magnetic fields and his observation in 1958 of de Haas–van Alphen effect oscillations in copper. His pioneering work undoubtedly triggered off the fermiological revolution and during the next fifteen or so years the Fermi surfaces of a great many metals were mapped in detail. Whilst many other physical phenomena have contributed richly to our understanding of the electronic structures of metals, the de Haas–van Alphen effect is uniquely powerful in this respect. Shoenberg has been intimately involved over a period of more than forty years in these developments, first through measurements of de Haas–van Alphen effect frequencies in the investigation of Fermi surfaces, then in pointing out the possibilities of amplitude measurements for studies of the dynamical properties of conduction electrons and of electron scattering and, finally, in the detailed unfolding of waveshapes to yield information on *g*-factors, magnetic interaction and Fermi liquid parameters. The result has been the development of a powerful spectroscopic tool which gives a microscopic picture of conduction electrons and their interactions with the crystal lattice, external fields, impurities, lattice imperfections and each other.

In this volume we have tried to review in a coherent way some of the main themes, both past and present, which have been influenced by Shoenberg's teaching and leadership. The book is arranged in three parts of which the first is concerned with certain general and fundamental issues. First, the history of the familiar geometrical concepts, which are used to describe the electronic properties of metals, is traced. The influence of many-body effects is then considered and, within the framework of Fermi liquid theory, the links between metals and ^{3}He are emphasized. A generalized path integral approach to transport problems is presented as an alternative to the more familiar Boltzmann equation and consideration is given to magnetic interaction in metals, an effect first revealed by Shoenberg's experiments on the de Haas–van Alphen effect and which now conventionally bears his name. Part II relates more specifically to electronic structures and to Fermi surfaces, not simply to their important topological features and intriguing shapes, which give much delight, but rather to the information on the related ground state and elementary excitations of the system to which their study gives access. Transition metals are prominently featured and magnetic excitations in the itinerant ferromagnets iron and nickel are discussed. Included here also is a review of the effect of strain on the Fermi surface, the focus being on three representative metals: lead, copper and tungsten. Given the genesis of the book the underlying theme of the role of quantum oscillations in the study of electrons in metals is an important one, and the final section is concerned exclusively with certain aspects of the de Haas–van Alphen effect. The theory of the de Haas–van Alphen effect in dilute alloys is developed and the complementary measurements of amplitudes and frequencies are shown to yield the variation over the Fermi surface of dynamical properties such as electron quasi-particle lifetimes and *g*-factors. Wherever possible the emphasis throughout is on results and a critical evaluation of contemporary research.

Concerning our choice of units – the c.g.s. system – we should perhaps apologize to present-day students. It does, however, conform to the way in which these fields (and we!) have developed and, by way of rationalization, to Shoenberg's own preference in these matters.

To those many friends of Shoenberg who could have contributed to this volume had space permitted, the editor apologizes. In a sense, however, they are represented since so much of their work is to be found in the bibliography. Finally, the editor is deeply indebted to all the contributors who, in spite of many other commitments, were able to complete their essays so promptly. It is their collective decision not to accept any royalties for their labours but to establish a trust fund which, in accordance with the wishes of David Shoenberg, will be used to promote and support the travels of younger physicists of all nationalities.

M.S.

Part I General principles

1

Geometric concepts in the electron theory of metals

I. M. LIFSHITZ AND M. I. KAGANOV
TRANSLATED BY M. G. PRIESTLEY

What will History say?
History, sir, will tell lies, as usual.
G. B. Shaw, *The Devil's Disciple*

1.1 Introduction

It is generally acknowledged that the last two or three decades have been a period of great success for the quantum physics of solids and, in particular, for the quantum theory of metals. Even if we do not concern ourselves with the development of the microscopic theory of superconductivity (a topic which lies outside the scope of the present review), the achievements have been considerable.

Comparatively long ago a considerable variety of experimental results had been collected which showed the wide difference between one metal and another: it was found that at low temperatures the resistance of some metals increased many times in high magnetic fields, whilst that of others changed by relatively small amounts; it was discovered that the scale of the periods of magnetic oscillations in, say, bismuth and gold differed by several orders of magnitude. Such examples can be continued indefinitely. On the other hand, theory at that time concerned itself with a 'faceless' metal whose properties were generally described by a degenerate Fermi gas. The major part of the work was devoted (at least in expositions of the electron theory of metals)† to those topics that the theory was able to interpret, i.e. the temperature dependences of the resistivity, thermal conductivity, specific heat, etc.

One of the main achievements of the modern electron theory of metals is probably the removal of this contradiction between theory and

† For instance, in the textbooks on the electron theory of metals by Peierls (1955), Sommerfeld & Bethe (1933), and Wilson (1936).

experiment. Each metal acquired its own 'face'. It was understood in what respects the electrons in copper differed from those in zinc. The 'face' of a metal became its Fermi surface, a 'visiting card' describing the constant-energy surface which at zero temperature separates the occupied from the empty states in quasi-momentum space. Fermi surfaces are so diverse in form that one might think that they are the fantasies of a modern artist, rather than a convenient way of describing the properties of conduction electrons.

A specific description of the properties of electrons in different metals or, in other words, the creation of the electron theory of metals in its present form, was made possible by the introduction of geometric examples into the theory of metals and by the use of geometric language. Contemporary work on the physics of metals is embellished with terms such as 'Gaussian curvature', 'external section', 'reference plane', etc.

1.2 Fermi surface: energy spectrum of metals

Although there is no rigorous proof of the following assertion, everyone is sure that:

> in the neighbourhood of the ground state the energy spectrum of any crystal can be described in terms of quasi-particles, and the state of a quasi-particle is determined by its quasi-momentum **p** in periodic **p**-space with a structure given by the crystal geometry. All quasi-particles belong to one of two classes – bosons or fermions; there are many types of bosons (phonons, magnons, excitons, etc.), but the only fermions are electrons.

Boson quasi-particles form an almost ideal gas and the interaction between them becomes less and less as the temperature T is reduced – simply because the number of boson quasi-particles decreases with decreasing temperature (the number of phonons $\propto T^3$, that of magnons in ferromagnets $\propto T^{\frac{3}{2}}$, etc.). The temperature dependence of the number of boson quasi-particles, and hence the temperature variation of many physical properties of crystals, is related to the dispersion law of the quasi-particles, i.e. to the dependence of the quasi-particle energy on quasi-momentum

$$\mathscr{E}_{\mathbf{p}} = \mathscr{E}. \tag{1}$$

In principle this makes it possible to use the experimental temperature dependences of the properties of a solid to establish the boson branches of its spectrum (more specifically, their density of states). Although this method has been given a mathematical basis (Lifshitz 1954), methods using the interaction of penetrating radiation with bosons have been more productive. Since bosons can be created singly, and the probability of the creation of a single boson, if not forbidden by selection rules, is greater than that of multi-quasi-particle creation, then inelastic scattering (of neutrons or phonons) or resonant absorption (of photons or phonons) can be used to determine the dispersion laws of boson quasi-particles.†

The density of conduction electrons in a metal is constant, so that reducing the temperature does not reduce the relative interaction between electrons, which remains of the same order as that with the lattice ions. The conduction electrons in a metal form an electron liquid, and the construction of a consistent theory relies (in all cases of practical interest) on the proximity to the ground state ($T \ll \mathscr{E}_F$, where $\mathscr{E}_F$ is the Fermi energy). According to the Landau theory of a Fermi liquid (Landau 1969) the basic characteristics of the electron subsystem in a metal are as follows.

The dependence $\mathscr{E} = \mathscr{E}_{\mathbf{p}}$ of quasi-particle energy on quasi-momentum is given by the variational derivative of the total electron energy E with respect to the distribution function $n_s(\mathbf{p})$ in the sth band‡

$$\delta E/\delta n(\mathbf{p}) = \mathscr{E}_{\mathbf{p}} \tag{2}$$

and the correlation function $f(\mathbf{p}, \mathbf{p}')$ (the Landau function) determines the change in quasi-particle energy $\mathscr{E}_{\mathbf{p}}$ due to a change in the distribution of electrons in $\mathbf{p}$ space

$$\delta\mathscr{E}_{\mathbf{p}} = \int f(\mathbf{p}, \mathbf{p}')\delta n(\mathbf{p}')\,\mathrm{d}\tau'. \tag{3}$$

When we include the exchange interaction between electrons with spins $\boldsymbol{\sigma}$ and $\boldsymbol{\sigma}'$, we have

$$f(\mathbf{p}, \boldsymbol{\sigma}; \mathbf{p}', \boldsymbol{\sigma}') = f(\mathbf{p}, \mathbf{p}') + \xi(\mathbf{p}, \mathbf{p}')\boldsymbol{\sigma} \cdot \boldsymbol{\sigma}'. \tag{4}$$

In the ground state at $T = 0$ the quasi-particles (electrons) occupy all states with energy less than the Fermi energy $\mathscr{E}_F$, whose value is obtained from the normalization condition: the number of quasi-particles equals the number of electrons in partially filled bands (Landau & Luttinger 1963). For weak excitation there is a redistribution of the electrons near the Fermi surface

$$\mathscr{E}_{\mathbf{p}} = \mathscr{E}_F. \tag{5}$$

† In quasi-particle language, resonant absorption is the transformation of one quasi-particle into another (e.g. a photon into an optical phonon).

‡ In future we shall omit the band index wherever this will not lead to confusion.

Since $\delta\mathscr{E}_{\mathbf{p}} = \mathbf{v}\cdot\delta\mathbf{p}$, where $\mathbf{v}$ is the velocity of an electron at the Fermi surface, the determination of the electron energy spectrum within the framework of the Landau theory of a Fermi liquid reduces to the determination of the shape of the Fermi surface in (5) and of the velocities $\mathbf{v}$ of electrons on it. A complete description of the conduction electrons also includes a knowledge of the Landau function (more exactly a matrix) in (4).

The problem of the decipherment of the electron energy spectrum of metals from the experimental results (mainly from their properties at comparatively high magnetic fields) was formulated *before* the development of a theory of the Fermi liquid. Naturally this used the 'gas' language. It is important to emphasize that the overwhelming majority of the methods which were used to determine (and can in principle be used to determine) the different geometric features of the Fermi surface (see below) remain invariant in the 'translation' from the 'gas' to the 'liquid' language. This has the following consequences: first, the features determined are specifically those of the surface in (5), where $\mathscr{E}_{\mathbf{p}}$ is the dispersion law of the *quasi-particles* (Landau 1969), and secondly the Landau function in (4) does not normally appear in the expressions relating the experimentally measured quantities to the characteristics of the dispersion law for electrons.

On the other hand, the complete elimination of the Landau function from the electron theory of metals (rewriting the theory in the 'gas' language) is, of course, impossible, since $f(\mathbf{p}, \boldsymbol{\sigma}; \mathbf{p}', \boldsymbol{\sigma}')$ appears directly in high-frequency, non-linear, and other properties of metals. The determination of its features is a separate and complex problem, in the solution of which there has been much less success than in the reconstruction of Fermi surfaces.

The determination of the energy spectrum of macroscopic bodies is one stage in the calculation of their properties. Ideally the programme for the theoretical understanding of the physical properties of solids has the following schematic form: first the required characteristics of the quasi-particles in a given body are elucidated; then these are used to calculate any desired macroscopic properties. To calculate transport coefficients one needs to know the scattering cross-sections or mean-free-paths of quasi-particles. If we limit ourselves to the linear response of the metal, the range of problems for which the 'gas' description suffices is still very wide: first, it includes all quasi-static problems ($\omega\tau \ll 1$, where ω is the frequency of the external perturbation, τ is the relaxation time (Lifshitz, Azbel' & Kaganov 1973)) and, secondly, problems such as the anomalous skin effect (London 1940, Pippard 1949, Reuter & Sondheimer 1948) in which the electron distribution function has a peak in a small region of **p**-space. Since the inclusion of Fermi-liquid effects is associated with the addition to the energy of a term containing an integration over the quasi-momentum, the role of Fermi-liquid effects in problems of this type is second order.

Finally, we should emphasize a further important simplifying feature. Electrons in metals are essentially quantum objects. The band nature of the energy spectrum, the concepts of 'quasi-momentum', 'quasi-particle', 'degeneracy', 'Fermi energy', etc., are all consequences of the application to electrons of the laws of quantum mechanics. However, the motion of quasi-particles with quasi-momentum **p** and energy $\mathscr{E} = \mathscr{E}_{\mathbf{p}}$ in external fields is in most cases quasi-classical. In essence, this is because the external fields are relatively small (compared to intra-atomic fields) and they vary over distances large compared to the size of atoms. The main consequence of the (quasi-) classical motion in external fields is that the quasi-momentum can be regarded as the real momentum, while $\mathscr{E}_{\mathbf{p}}$ is the Hamiltonian of a quasi-particle. This makes possible a detailed analysis of the motion of quasi-particles in external fields (Lifshitz & Kaganov 1959), and enables the results of the analysis to be used to calculate the characteristics of the metal.

1.3 Plane sections of the Fermi surface: de Haas–van Alphen effect, galvanomagnetic phenomena

The celebrated work of D. Shoenberg (1939)† on the investigation of the de Haas–van Alphen (dHvA) effect in Bi acted as a stimulus to the formulation of the problem of the reconstruction of the electron energy spectrum of a metal from experimental data. This apparently demonstrated the unlimited scope for using magnetic quantum oscillations to determine the parameters describing the electron gas in a metal. It is true that there were then no formulae relating the dependence of the magnetic moment **M** on magnetic field **H** to these parameters (in the general case); in fact it was not even clear which features of the dHvA effect were the most important for the solution of the spectroscopic problem. It is difficult now to imagine how much labour was expended on, for instance, the description of the envelope of the function $\tilde{\mathbf{M}} = \tilde{\mathbf{M}}(\mathbf{H})$, or how many technical terms were introduced in order to systematize the rapidly

† The appendix to this article gives Landau's theory of the dHvA effect.

accumulating material on the dHvA effect. We are speaking here about the 1950s. It was at this time that there appeared in the theory the term 'an electron with an arbitrary dispersion law', i.e. with an unspecified relation between energy and momentum. It turned out that in many calculations there was no need to be more specific. The foundation for this was laid by Lifshitz & Kosevich (1954, 1955), and the structure of this paper can also be regarded as an example of a definite 'Khar'kov' style, which was later used to investigate a multitude of electronic effects in metals:

(i) The investigation of the classical motion in a magnetic field of an electron with an arbitrary dispersion law, based on the equation

$$\frac{d\mathbf{p}}{dt} = \frac{e}{c}\mathbf{v} \wedge \mathbf{H}; \qquad \mathbf{v} = \frac{\partial \mathscr{E}}{\partial \mathbf{p}}. \tag{6}$$

(ii) The quasi-classical quantization (when necessary) of the classical motion investigated in (i).

(iii) The calculation of the appropriate experimentally observed feature of the effect being studied, using the results of (i) and (ii) and including the degeneracy of the electron gas.

(iv) The explanation and assessment of the spectroscopic possibilities of the effect in question, i.e. the elucidation of just which characteristics of the electrons at the Fermi surface it could be used to determine.

The ideal (but practically unattainable) sum total of the work in this style was seen as a complete detailed determination of the Fermi surface and of the velocities of the electrons over it.

Many Fermi surfaces were named 'monsters'. The electron velocities – being normal to the surface – are its 'bristles'. Thus the ideal outcome of the work is an unshaven monster, especially frightening for a hole surface (where the bristles are directed inwards)!

The results of Lifshitz & Kosevich (1954, 1955) who originated this style can be summarized as follows.

(i) In agreement with (6) the electron trajectory in **p**-space is given by the intersection of its constant-energy surface with the plane $p_z = p_{z0}$ (with $H_x = H_y = 0$; $H_z = H$)

$$\mathscr{E}_{p_x p_y p_z} = \mathscr{E}_0; \qquad p_z = p_{z0}. \tag{7}$$

(ii) If the trajectory is closed, the motion in the plane perpendicular to the magnetic field is quantized, and the dependence of energy on p_z in the nth Landau sub-band is governed by the quasi-classical quantization condition (Lifshitz 1950; Onsager 1952),

$$\mathscr{A}(\mathscr{E}, p_z) = \frac{2\pi e\hbar H}{c}(n + \tfrac{1}{2}),\ n \gg 1, \tag{8}$$

where $\mathscr{A}(\mathscr{E}, p_z)$ is the area included in the orbit in (7).

(iii) The presence of square-root singularities at the boundaries of the sub-bands (with given n) picks out the extremal sections (as a function of p_z) $\mathscr{A}_{\text{ext}}(\mathscr{E})$, whilst the degeneracy fixes the energy $\mathscr{E} = \mathscr{E}_{\text{F}}$. Use of the Poisson summation formula enables the thermodynamic potential Ω, and hence the magnetic moment $\tilde{\mathbf{M}}$, to be written as a sum of oscillatory terms, each of which is related to an extremal section of the Fermi surface with respect to p_z (Fig. 1.1), and moreover the period $\Delta(1/H)$ of the oscillations is inversely proportional to the area of the extremal section:

$$\Delta(1/H) = \frac{2\pi e\hbar}{c} \cdot \frac{1}{\mathscr{A}_{\text{ext}}(\mathscr{E}_{\text{F}})}. \tag{9}$$

The function describing the temperature dependence of the amplitude contains the value of the effective mass

$$m_{\text{c}} = \frac{1}{2\pi} \left.\frac{\partial \mathscr{A}(\mathscr{E}, p_z)}{\partial \mathscr{E}}\right|_{\mathscr{E} = \mathscr{E}_{\text{F}}, p_z = p_z^{\text{ext}}}.$$

(iv) By changing the direction of the magnetic field, measuring the periods and amplitudes, and then comparing the experimental results with the above expressions, one can determine both the shape of the Fermi surface and the velocity of the electrons at the surface (Fermi electrons). This ideal programme was completed by the geometrical insight of Lifshitz & Pogorelov (1954) who showed that, if the surface is convex and has a centre of symmetry, there exists a certain analytical procedure for calculating its shape from the oscillation periods, and the velocities from the effective masses (see §§ 17.7 and 17.8 of Lifshitz *et al.* 1973). Despite all the difficulties, this programme was actually used to

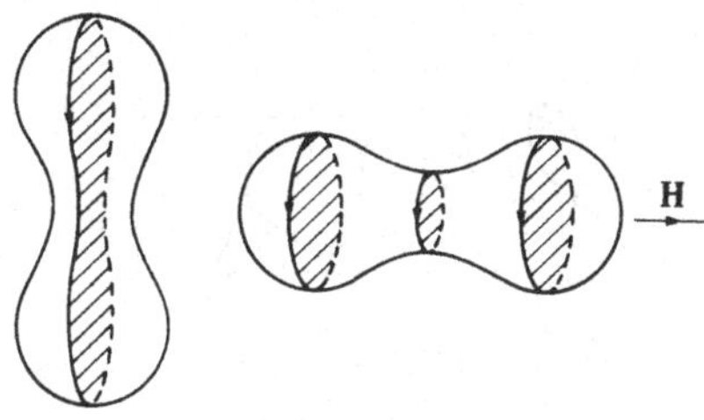

Fig. 1.1. Plane sections of the Fermi surface with extremal areas.

decipher the structure of the Fermi surface in Al and Pb (Gunnersen 1956; Gold 1958).

The calculation of the oscillations does not need a knowledge of the motion in real space of an electron with an arbitrary dispersion law. However, this is needed for the solution of most transport problems, and here we are helped by a simple geometrical theorem which follows from (6): the projection of the real-space orbit on a plane perpendicular to the magnetic field is derived from the **p**-space orbit by a rotation by $\pi/2$ about the field direction and scaling by a factor $c/|e|H$. The motion along the z-axis is given by

$$v_z(t) \equiv v_z[p_x(t), p_y(t), p_{z0}].$$

In terms of the real-space orbit the quantization condition is especially clear; the trajectory of the electron is wound on the magnetic flux, Φ, in such a way as to include a half-integral number of quanta of magnetic flux (Fig. 1.2). The direct correspondence between the orbits in **p**- and **r**-spaces made it possible to use a number of different size effects to investigate the shapes of Fermi surfaces. In principle these size effects have a simple basis. The formulation of the problem picks out electron orbits of a certain size d, which is known (the plate thickness (Sondheimer 1950), sound wavelength (Pippard 1957a), etc.). The equation

$$cD_0/eH = d,$$

where D_0 is the relevant diameter of the Fermi surface (for instance an extremal one, as in Fig. 1.3), shows that for $H = H_d = cD/ed$ some new feature must be observed. What this is, is predicted by the detailed theory (Kaner & Gantmakher 1968), which also makes it clear why the diameter D_0 is singled out. A measurement of H_d in a size effect experiment is thus a way of directly measuring diameters of the Fermi surface. The anisotropy of the effect (the variation of H_d with field direction) is evidence for the anisotropy of the electron energy spectrum, and its investigation is a way of 'calipering' the Fermi surface.

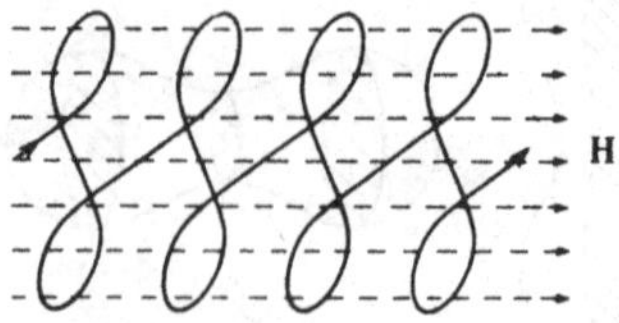

Fig. 1.2. The orbit of an electron includes a half-integral number of flux quanta. $\Phi = (n+\frac{1}{2})\Phi_0$; $\Phi_0 = 2\pi\hbar c/e \approx 4.14\times10^{-7}$ G cm^2.

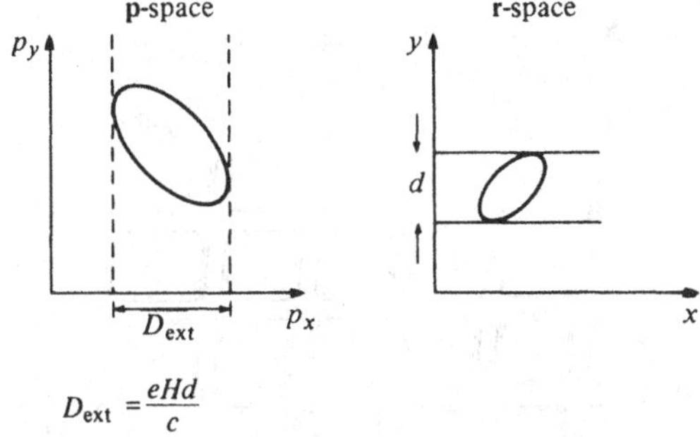

Fig. 1.3. The origin of size effects is the correspondence between orbits in real space and momentum space.

The combination of a size effect and dHvA oscillations is a source of characteristic effects which also have a simple geometric interpretation. Fig. 1.4 shows that for specular reflection from the surface an electron 'jumps across' from one point on the Fermi surface to another. The area $\mathscr{A}_d$ enters into the quantization condition (equation (8)) and this produces a dependence of the period in (9) on plate thickness. Although this effect was predicted in 1953 by Lifshitz and Kosevich, it was observed only comparatively recently, for the 'moustaches' in Sb (Gaĭdukov & Golyamina 1976).

The orbits of electrons in a magnetic field (equation (7)) obey a simple topological classification based on the symmetry properties of the dispersion law. Because of the periodicity of **p**-space there are only two possible cases: for a fixed direction of the magnetic field **H** the orbit in (7) can either break up into closed orbits or go off to infinity along the 'open'

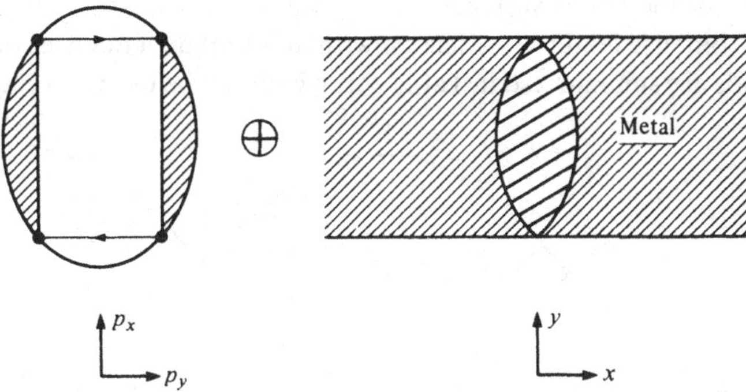

Fig. 1.4. An orbit in **p**-space contains jumps due to reflection from the metal surface.

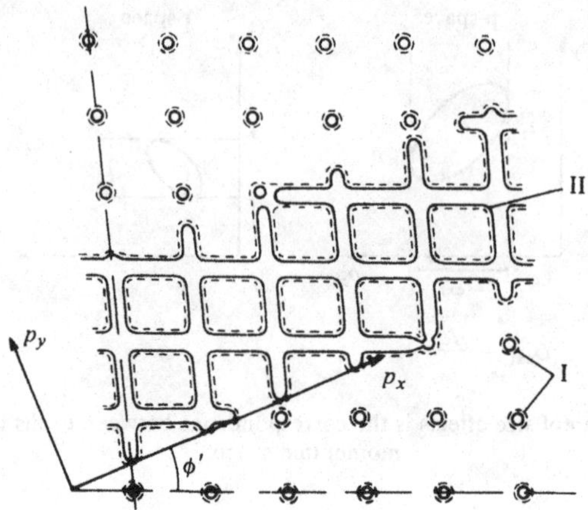

Fig. 1.5. A drawing from Lifshitz & Peschanskiĭ (1958), showing the origin of open orbits in metals with a Fermi surface of the 'monster' type.

direction in Fig. 1.5. It is true that for certain special directions of **H** and special values of p_z it is possible to have self-intersecting orbits (Fig. 1.6) and even two-dimensional networks of open self-intersecting orbits (Fig. 1.7), but for the moment we shall not stop to discuss these comparatively rare and little investigated cases.

The analysis developed by Lifshitz, Azbel' & Kaganov (1956b); Lifshitz & Peschanskiĭ (1958); Alekseevskiĭ, Gaĭdukov, Lifshitz & Peschanskiĭ (1960) showed that galvanomagnetic effects such as the transverse magnetoresistance $\rho_\perp = \rho_\perp(H)$ and the Hall effect make it possible to 'see' the general features of the Fermi surface, i.e. its topology. It turned out that the asymptotic behaviour of the components $\rho_{\alpha\beta}(H)$ $(\alpha, \beta = x, y)$ of the resistivity tensor as $H \to \infty$ is extremely sensitive to the geometry of the Fermi surface.

The geometric features associated with different dependences of $\rho_{\alpha\beta}$ at high fields are summarized in Table 1.1, which can be used as a recipe for

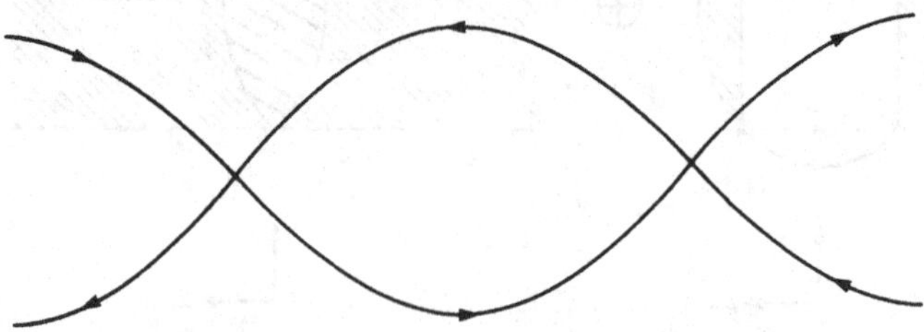

Fig. 1.6. A self-intersecting orbit.

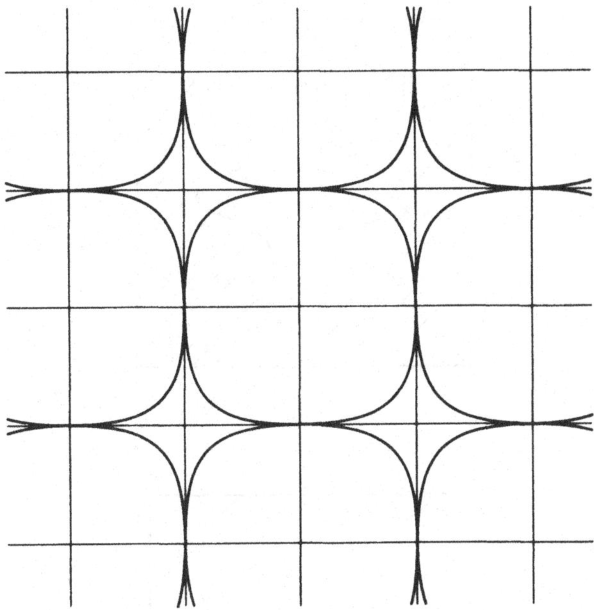

Fig. 1.7. A two-dimensional network of open self-intersecting orbits.

the determination of the topology of the Fermi surface. The overall picture would be clearer if it were not complicated by the different role of electrons and holes (typical electron and hole Fermi surfaces are given in Fig. 1.8). The point is that the value of $\rho_{\alpha\beta}$ is determined not just by the

TABLE 1.1. *The dependence of the metal resistivity on the value of the magnetic field* **H** *and its direction*

Closed Fermi surfaces		Open Fermi surfaces		
$n_1 \neq n_2$	$n_1 = n_2$	$H = \text{const}$	$\theta = \theta_1$	$\theta = \theta_2$
$\Delta\rho$; $\omega_c\tau = 1$; H^2	$\Delta\rho$; $\omega_c\tau = 1$; H^2	$\Delta\rho$; θ_1, θ_2; θ	$\Delta\rho$; $\omega_c\tau = 1$; H^2	$\Delta\rho$; $\omega_c\tau = 1$; H^2

θ is the angle between the magnetic field direction and the crystallographic axis.

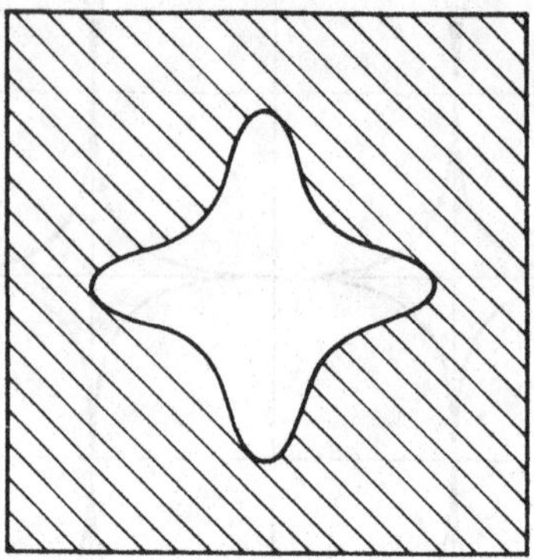

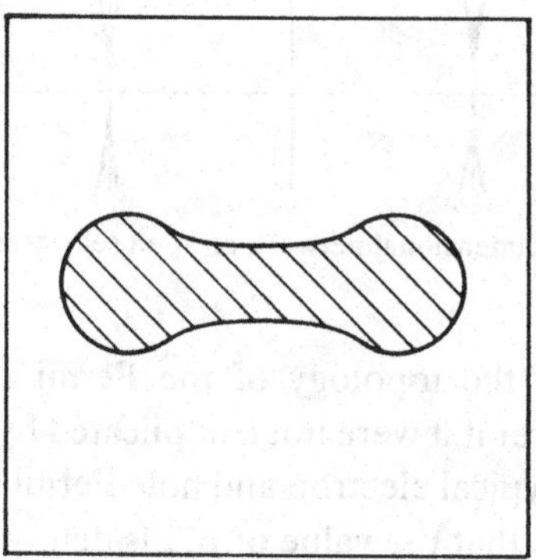

Fig. 1.8. Electron and hole sheets of the Fermi surface.

symmetric part of the conductivity tensor σ_{ik}, but also by the antisymmetric part. The former describes dissipative processes and the latter the Hall effect. In a high magnetic field

$$\sigma_{xy} = \frac{(n_1 - n_2)ec}{H} + \dots, \tag{10}$$

where n_1 is the number of electrons per unit volume, and n_2 is the corresponding number of holes.† The ellipses signify that further terms in the expansion in the reciprocal of the magnetic field have been omitted. It is seen that for $n_1 = n_2$ the expansion starts with terms proportional to $1/H^2$. This is why $\rho_\perp \propto H^2$ for equal numbers of electrons and holes.

† n_1 is the number of *occupied* electron states with $m_c > 0$, and n_2 the number of *empty* electron states with $m_c < 0$.

In order to explain the role of open Fermi surfaces there is no need to resort to calculations using the Boltzmann equation. It is sufficient to use the Einstein relation between the conductivity σ and the diffusion coefficient $\mathscr{D}$, generalized to the case of degenerate statistics,

$$\sigma \approx \frac{e^2 \mathscr{D} n}{\mathscr{E}_F}, \tag{11}$$

and the concepts of kinetic theory are used to calculate the diffusion coefficient. When an electron moves in a closed orbit in a magnetic field, diffusion in the plane perpendicular to the field takes place by jumps by an amount $\sim r_H = cp_F/eH$ (see Fig. 1.9) and with a frequency $\sim 1/\tau$, where τ is the relaxation time. In this case $\mathscr{D} \approx r_H^2/\tau$, and thus

$$\sigma = \sigma_0/(\omega_c \tau)^2; \qquad \omega_c = eH/m_c c. \tag{12}$$

When an electron moves along an open orbit (the open direction is taken to be along p_x), the motion along the y-axis is like that of a free particle and $\mathscr{D}_{yy} \approx v_F^2 \tau$, but along the x-axis it is essentially the same as the motion in a closed orbit and $\mathscr{D}_{xx} \approx r_H^2/\tau$. Hence we have

$$\sigma_{xx} \approx \sigma_0/(\omega_c \tau)^2; \qquad \sigma_{yy} \approx \sigma_0. \tag{13}$$

This considerable difference between the asymptotic forms of $\sigma_{xx}(H)$ and $\sigma_{yy}(H)$ is the basis of the anisotropy of the galvanomagnetic properties of metals with open Fermi surfaces (see §§ 27, 28 of Lifshitz *et al.* 1973).

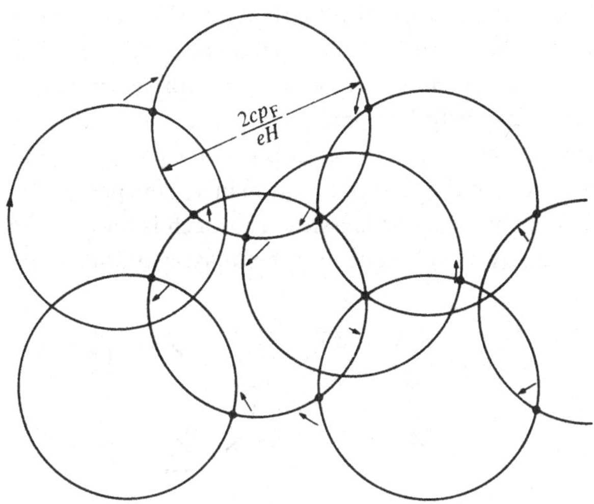

Fig. 1.9. Electron diffusion in a plane perpendicular to the magnetic field.

In a number of cases this understanding (by geometric analysis) of the nature of the electron motion in a magnetic field enables very unusual phenomena to be predicted. The static skin effect is certainly one of these (Azbel' 1963).

The specular reflection of an electron from the surface of a metal renders its motion open parallel to the surface (see Fig. 1.10), and therefore the surface conductivity is greater than the bulk value (see (13)). It is perhaps more surprising that even diffuse scattering of electrons at the boundary does not eliminate this effect. Collisions with the surface occur with frequency $\sim\omega_c$, and the surface conductivity equals $\sigma_0/\omega_c\tau$, which is much greater than $\sigma_0/(\omega_c\tau)^2$. The static skin effect is not, however, simple enough to be amenable to 'back-of-the-envelope' calculation. The derivation of the formulae demands careful analysis of the solution of the Boltzmann equation in the inhomogeneous case (Azbel' & Peschanskiĭ 1965, 1968), and the results show that the best metals for the observation of the static skin effect are those whose bulk resistivity increases with magnetic field.

The change in the type of motion of a conduction electron in a magnetic field, because of the inhomogeneity of the Lorentz force acting on the electron, can easily be understood from the example (Mints 1969) of an electron interacting with a domain wall (a 180° boundary between domains, see Fig. 1.11). The conductivity along the domain wall should be $(\omega_c\tau)^2$ times greater than that across it.

This rapid outline of the theory of galvanomagnetic phenomena should, we feel, be concluded by some appeals for caution. The theory claims to derive the asymptotic behaviour of $\rho_{ik}(H)$, starting from an analysis of the classical motion of electrons with an arbitrary dispersion law and in a sufficiently high magnetic field – such that the Larmor frequency is much greater than the collision frequency ($\omega_c\tau \gg 1$).

The quantization of the electron motion in a plane perpendicular to the magnetic field means that the classical approach is known to be inexact, although it is still quite valid because the quasi-classical condition $\hbar\omega_c \ll$

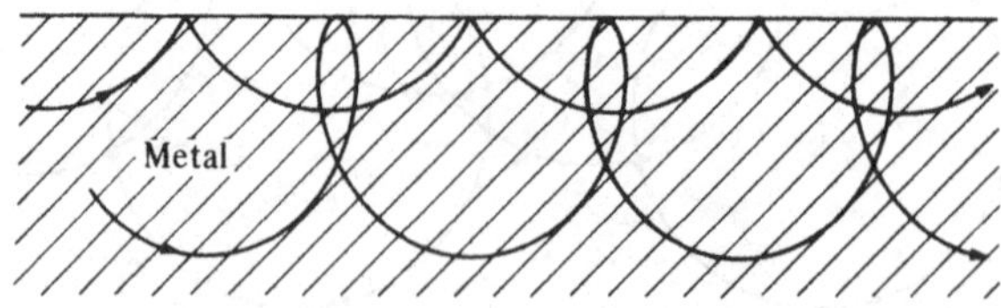

Fig. 1.10. Reflection of an electron at the metal surface can give rise to open orbits.

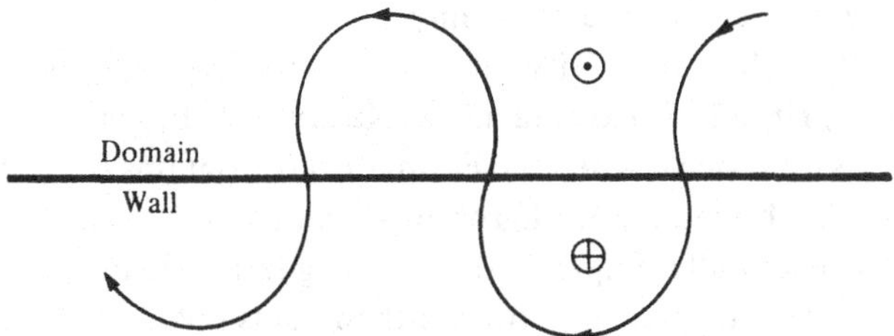

Fig. 1.11. Open orbits can also be created by a change in sign of the Lorentz force at a domain boundary.

$\mathscr{E}_F$ is satisfied right up to enormous magnetic fields, and the oscillatory part of $\rho_{ik}(H)$ – the Shubnikov–de Haas effect (Shubnikov & de Haas 1930; Lazarev, Nakhimovich & Parfenova 1939) – adds 'ripples' to the 'classical' part whose relative amplitude is much less than unity.† The quasi-classical condition ($\hbar\omega_c \ll \mathscr{E}_F$) introduces yet another simplification: to zeroth order in $\hbar\omega_c/\mathscr{E}_F$ the collision operator W is independent of magnetic field. This is because the mean free path of an electron in a crystal containing impurities is of the order of the interatomic spacing a, while the condition $a \ll r_H = cp_F/eH$ becomes $\hbar\omega_c \ll \mathscr{E}_F$. The almost complete absence of a field dependence of W for phonon scattering relies on the comparatively weak condition $\hbar\omega_c \ll (v_F/v_s)T$, where v_s is the velocity of sound. Essentially, the independence of W of H allows us to use the dynamics of conduction electrons to establish the asymptotic form of $\rho_{ik}(H)$.

This comparatively simple theory does not cover all situations. It cannot, of course, deal with the region of arbitrary intermediate fields. The intermediate fields need numerical calculation, which was earlier quite impractical, although over the last few years it has been gradually supplanting analytical methods (we note this with regret). However, there are occasions when an analytic approach is possible at intermediate fields, i.e. when there is in the problem an additional small parameter which has the consequence that the asymptotic behaviour of $\rho_{ik}(H)$ is not reached for $\omega_c\tau \gg 1$, but only at much higher fields. The simplest example of this is when $\Delta n = |n_1 - n_2|$ is small. For such metals (Bi with impurities, for

† The origin of the Shubnikov–de Haas effect is the same as that of the dHvA effect. Moreover, the oscillatory part $\Delta\rho(H)$ of the resistivity can be expressed in terms of the oscillatory part $\Delta\tilde{\mathbf{M}}(\mathbf{H})$ of the magnetic moment (Titeica 1935; Zil'berman 1955; Lifshitz 1957; Lifshitz & Kosevich 1957; Adams & Holstein 1959; Kosevich & Andreev 1960) whilst the dHvA effect is the principal spectroscopic tool for use with metals, the Shubnikov–de Haas effect is commonly used with semi-metals and degenerate semiconductors.

instance) the resistivity saturates only for $\omega_c\tau \gg n/\Delta n \gg 1$, whilst for $n/\Delta n \gg \omega_c\tau \gg 1$ the resistivity increases quadratically with field (Alekseevskiĭ, Brandt & Kostina 1955; Kaganov & Peschanskiĭ 1958). Similar effects must occur when the Fermi surface contains a narrow band of open orbits (Kaganov, Kadigrobov & Slutskin 1967). These two examples are essentially simple, in that the magnetic field effectively does not change the spectrum of electrons or their collisions. It is 'simply' that high fields are needed to single out the electrons responsible for the asymptotic behaviour, since there are relatively few of them.

The situation is more complicated when the magnetic field itself creates the 'responsible' electrons (in essence, by altering the nature of the electron spectrum) or else 'interferes' in the collision process. Both these phenomena allow a geometric interpretation. We shall consider the first under the heading 'magnetic breakdown', but the second is dealt with here. A Fermi surface with narrow necks will demonstrate the 'interference' of the magnetic field in the scattering processes, or, more accurately, the singling out of a particular scattering mechanism. In a sufficiently high magnetic field the resistivity in a direction perpendicular to the neck is determined by the electrons on the neck, but because the neck is small their lifetime in this region of **p**-space differs from the transport lifetime which determines the resistivity at $H = 0$. It can even have a different temperature dependence, for example if the temperature is such that the mean momentum of a thermal phonon is either larger than or of the same order as the neck thickness Δp (see Kaganov *et al.* 1967), i.e. for

$$\frac{T}{\theta_D} \gtrsim \frac{\Delta p}{p_F}, \qquad (14)$$

where θ_D is the Debye temperature.

A somewhat more complicated situation arises when the Fermi surface comes close to its counterpart in a neighbouring unit cell in reciprocal space. An example of this is given in Fig. 1.12. If the separation of the two points A and A' is much less than the size of the unit cell of the reciprocal lattice, then because of small-angle scattering (by phonons or dislocations, for instance) umklapp processes will have an appreciable probability (Pippard 1968). As is well known (Peierls 1955), these play an important part in the temperature dependence of the resistivity. The points A and A' are known as hot spots on the Fermi surface (Young 1968). The effect of a magnetic field is either partially or completely to replace diffusive transport from one hot spot to another by motion along

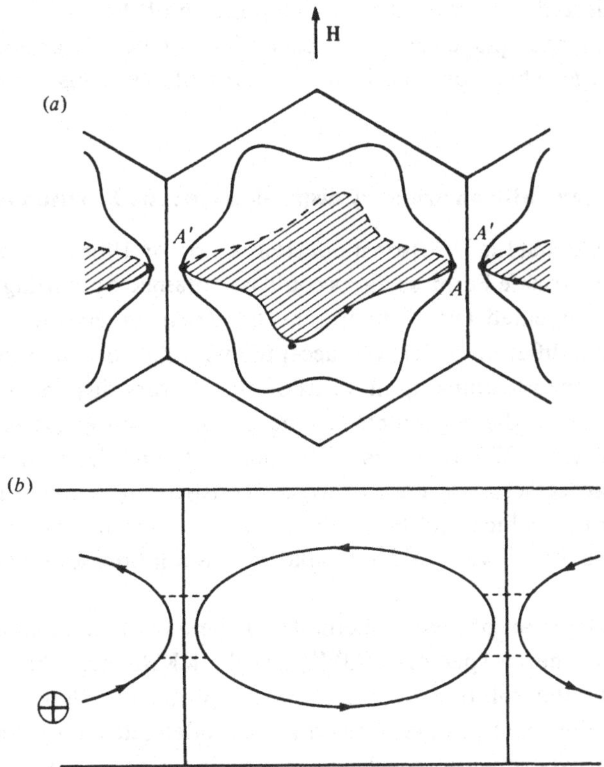

Fig. 1.12. (a) Hot spots on the Fermi surface. (b) An open orbit can be created by scattering.

an orbit (Fig. 1.12), and this naturally increases the role of umklapp processes – the magnetic field 'interferes' in the collision processes. This 'interference' of the magnetic field in the collision processes is especially interesting when, as a result of umklapp processes in the regions where the Fermi surface approaches its counterpart in the next cell, the motion of the electron effectively follows an open orbit (Fig. 1.12(b)) (Young 1968). This effect disappears at sufficiently high fields, since then the electron does not have time to jump onto the neighbouring surface.†

This last example takes us beyond those which refer exclusively to the internal geometry of the Fermi surface. In the phenomena and properties

† We refer readers to the paper by Gurzhi & Kopeliovich (1976) for a discussion of the details of the temperature and field dependence of the transport coefficients of pure metals when these are determined by electron diffusion over the Fermi surface and by jumps between hot spots. This appears to be the first time that the detailed geometry of the Fermi surface (its structure) has been used in a calculation of the temperature dependence of the transport coefficients.

that were described before, the relative position of the pieces of Fermi surface in **p**-space played no part. Each piece of Fermi surface was a separate stage where dramas from the lives of electrons were acted out.

1.4 Electrons with an arbitrary dispersion law: the Harrison model

It must be said that the Harrison model (e.g. Harrison 1966), according to which these intricate Fermi surfaces could be obtained by 'cutting up' the periodically repeated Fermi sphere and then rearranging the spherical segments in a different order, was accepted with difficulty by those who were accustomed to thinking in terms of an arbitrary dispersion law. It was not so much the arguments (Heine 1970; Cohen & Heine 1970; Heine & Weaire 1970) that reconciled them to the model, as that it had numerous successes. The Fermi surfaces constructed in this way were similar to those which had been obtained by comparing experimental results with a theory which did not make any assumptions about the form of $\mathscr{E}_{\mathbf{p}}$.

The development of several methods for the numerical calculation of the electron energy spectrum (OPW, APW, KKR, etc.) changed the approach to the solution of spectroscopic problems. Now the main features of the spectrum are derived from model calculations, and the role of the experimental results is to provide more accurate model parameters (usually components of the pseudopotential). A new ideal scheme was born:

model → comparison with experiment → components of the pseudopotential,

and with their help one can calculate absolutely everything.... However, it must be said that if today we know the Fermi surfaces of most simple metals and many intermetallic compounds, then this is the result of work that has followed exactly this scheme.

In essence, all present-day models for the construction of the electron energy spectrum develop the nearly-free-electron approach (Brillouin 1976). The question which naturally arises is what is a conduction electron? Is it a nearly free electron whose motion is gently perturbed by the periodic lattice field and by external fields, or is it an electron with a complicated dispersion law, moving under the influence of external fields? These questions are sometimes posed directly and sometimes in a veiled form, but one often has to confront them. The answer is not just a matter of taste but is determined, first, by the formulation of the problem and, second, by the ratio of the lattice field to the external fields. Of course, if one wishes to calculate the dispersion law, then one has to use a more or less adequate model, nothing less will do. On the other hand, if some feature of a metal such as its response to an external perturbation is being calculated, then there is a problem of what to take as a basis – a gas of free electrons with well-understood responses to external fields, with subsequent correction of the result to take account of the lattice field, or the opposite approach: first to consider a gas or Fermi liquid of electrons with a complicated dispersion law and then to proceed as described above. The answer to this question is given by comparing a characteristic energy $\mathscr{E}_{ch}$ of the electrons with an energy which is a measure of the interaction of an electron with an external field (let us call this $\Delta\mathscr{E}$). If we are talking about a static magnetic field, then $\Delta\mathscr{E} = \hbar\omega_c$, the separation between magnetic Landau levels, if we are concerned with an electric field, then $\Delta\mathscr{E} = eEl$ is the energy acquired by an electron in a mean free path l, and if we are dealing with a high-frequency field as the origin of transitions, then we have $\Delta\mathscr{E} = \hbar\omega$ (ω is the frequency of the field). When

$$\mathscr{E}_{ch} \gg \Delta\mathscr{E} \tag{15}$$

it is natural to consider the reaction of electrons with an arbitrary dispersion law.

It must be said that it is on just this point that the 'triumphal progress' of the Harrison model should be regarded with some caution. Several decades ago all estimates concurred in the conclusion that $\mathscr{E}_{ch} \approx \mathscr{E}_F$ and that the scale of the Fermi energy $\mathscr{E}_F$ was determined by the electron density $n = z/a^3$, i.e. $\mathscr{E}_F \approx \hbar^2 z^{\frac{2}{3}}/a^2 m \approx 10^{-12}$ erg (z is the number of valence electrons and a the lattice parameter). In other words, so long as the external fields are small compared to the atomic fields there is no need to worry. The minds of those who worked on the quantum theory of solids were dominated by the idea that the energy spectrum of a metal was something rather gross, which was unaffected or almost unaffected by external perturbations. The metals of group V (Bi, Sb, P) were the exception, but they are *semi*metals, i.e. almost semiconductors. It was gradually recognized (and in this learning process an important role was played by experimental investigations – especially of oscillatory effects – as well as by model calculations) that the energy spectrum of a metal is a delicate structure, which is comparatively easily subjected to extensive reconstruction due to external action or to the switching on of new internal forces (e.g. at the phase transition from a paramagnetic to a helicoidal magnetic state (Dzyaloshinskiĭ 1964)). The change in the structure of the electron spectrum is accompanied by a change in the topology of the Fermi surface and is described by a characteristic phase transition which, according to the Ehrenfest classification at $T=0$, should be interpreted as a phase transition of order $2\frac{1}{2}$ (Lifshitz 1960).

The realization of the complexity of the structure of the electron spectrum led to an important change in the estimate of the characteristic energy $\mathscr{E}_{ch}$. The value $\mathscr{E}_{ch}=|\mathscr{E}_F-\mathscr{E}_K|$, where $\mathscr{E}_K$ is the energy at which there is a change in topology of the constant-energy surfaces (Van Hove 1953), and the smallness of $\mathscr{E}_{ch}$ compared to $\mathscr{E}_F$ indicates that singular points in **p**-space lie close to the Fermi surface. This more accurate definition of the inequality in (15) can have widespread consequences; we shall discuss them later (see Section 1.9), but now we return to the concept of 'an electron with an arbitrary dispersion law'. The Harrison model and the theory of the pseudopotential deprived this concept of its mystic aura. It turned out that to understand the origin of the wide variety of Fermi surfaces was not as complicated as we thought twenty years ago. For many years the existence of the pockets of conduction electrons, discovered by the study of oscillatory effects (see (9)), seemed a complete enigma, but it was then explained (in outline) by the fact that in polyvalent metals the starting spheres of the Harrison model repeatedly intersect each other.

If we are not primarily concerned with the origin of the conduction-electron spectrum, then as before we can regard the elementary charge carrier as an electron with a complex dispersion law, whose state is defined by choosing the quasi-momentum **p**. The fact that **p** is the quasi-momentum and not the true momentum is remembered only when the probability of some kind of transition is being calculated – one must remember to include umklapp processes. Otherwise, the quasi-momentum is treated as the true momentum, whilst the energy $\mathscr{E}_{\mathbf{p}}$ is taken to be the kinetic energy of a free electron.

The calculation of collision probabilities demands a knowledge both of the internal geometry of each sheet of Fermi surface and of their relative location in **p**-space. Reference to models (especially to the Harrison model) obviously also helps in this respect. Experimental methods for determining the locations of different pieces of Fermi surface in **p**-space are not very well developed, but there are several 'recipes' in the papers published in the last few years.

1.5 The 'effective zone' on the Fermi surface: anomalous skin effect, ultrasonic absorption, cyclotron resonance

The electrons which make a collision-free interaction with electromagnetic, sound, or spin waves are those whose velocity $\mathbf{v}$ satisfies the Cherenkov condition (see Landau 1969),

$$\mathbf{k}\cdot\mathbf{v}=\omega, \tag{16}$$

where $\mathbf{k}$ and ω are respectively the wave-vector and its frequency. Because of the degeneracy of the electron liquid, the only electrons which take part in the absorption are those at the Fermi energy

$$\mathscr{E}_{\mathbf{p}}=\mathscr{E}_{\mathrm{F}}. \tag{17}$$

Equations (16) and (17) define a line on the Fermi surface, and this is often known as the 'effective zone'. The most interesting case is when $\omega/k \ll v$, i.e. the electron velocity v is much greater than that of the wave.† One can often put $\omega=0$ and then consider that the condition for a collision-free interaction selects those electrons which are moving perpendicular to the wave-vector $\mathbf{k}=k\mathbf{n}$ ($\mathbf{v}\perp\mathbf{k}$). The remaining electrons are ineffective (Pippard 1947, 1954). For a quadratic dispersion law $\mathscr{E}_{\mathbf{p}}=(m^{-1})_{ik}p_ip_k$ and $v_i=(m^{-1})_{ik}p_k$, where $(m^{-1})_{ik}$ is the reciprocal effective mass tensor, and then (16) describes a plane, so that the effective zone is an ellipse whose axes and orientation depend on the values of ω and $\mathbf{k}$. Even a relatively minor change in the Fermi surface leads to a much more complicated form and structure for the effective zone. The effective zones on a dumb-bell-shaped surface for different directions of the wave-vector are shown in Fig. 1.13.

It was in connection with the anomalous skin effect (London 1940; Pippard 1949; Reuter & Sondheimer 1948) that the concept of the effective zone arose in the physics of metals, after it was found that for $\delta \ll l$ (δ is the skin depth) the only electrons which interact with an incident electromagnetic wave are those moving parallel to the metal surface, as shown in Fig. 1.14 (Pippard 1947, 1954; Kaganov & Azbel' 1955). The absorption of ultrasound with wavelength $2\pi/k$ much smaller than the electron mean free path is analogous to the anomalous skin effect (Pippard 1955; Akhiezer, Kaganov & Lyubarskiĭ 1957). Both these processes are characterized by the ineffectiveness of the interaction with all the electrons, except those on the effective zone.

† For a sound wave $\omega/k=v_s\approx 10^5\,\mathrm{cm\,s^{-1}}$, while $v\approx 10^8\,\mathrm{cm\,s^{-1}}$; for an electromagnetic wave $\omega \ll kv$ signifies the priority of spatial dispersion over temporal. It is due to the small size of the skin depth: for $k^{-1}\approx\delta_0\approx c/\omega_{\mathrm{p}}$ (where $\omega_{\mathrm{p}}=(4\pi ne^2/m)^{1/2}$ is the plasma frequency) the condition $\omega\ll(v/c)\omega_{\mathrm{p}}$ is satisfied for frequencies right up to $\sim 10^{-2}\omega_{\mathrm{p}}\approx 10^{13}\,\mathrm{s^{-1}}$.

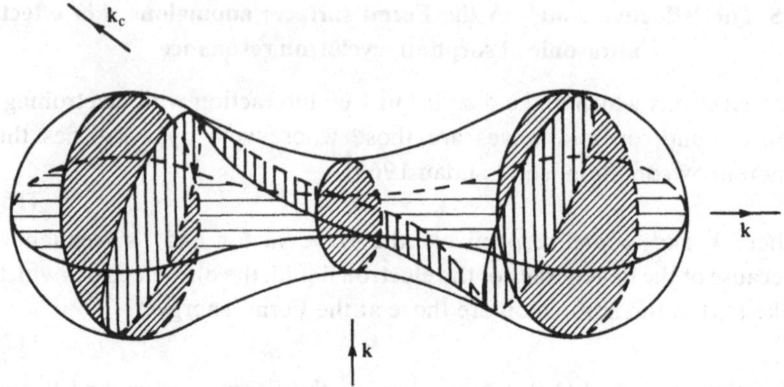

Fig. 1.13. 'Effective zones' on the Fermi surface for different directions of the wave-vector **k**. At $\mathbf{k} = \mathbf{k}_c$ the 'effective zone' changes its topology.

By moving this zone on the Fermi surface and then measuring the resultant values of the surface impedance of the metal, one can investigate the structure of the Fermi surface. Formulae for this are known; they were obtained without any assumptions as to the form of the dispersion law and relate the mean value of the Gaussian curvature of the Fermi surface over the effective zone to the surface impedance of the metal (Pippard 1947, 1954; Kaganov & Azbel' 1955). However, it is difficult to move the effective zone about the Fermi surface. The point is that, because of the high refractive index of the metal, an electromagnetic wave always propagates normal to the metal surface, so that the only possibility is to measure the impedance of different faces of a single crystal as was first done by Pippard (1957b).

The spectroscopic possibilities of the investigation of the electromagnetic properties of metals are extended by the fact that the charge on a conduction electron is equal to that of a free electron, but with ultrasound the position is less favourable. There the coupling with the electrons is

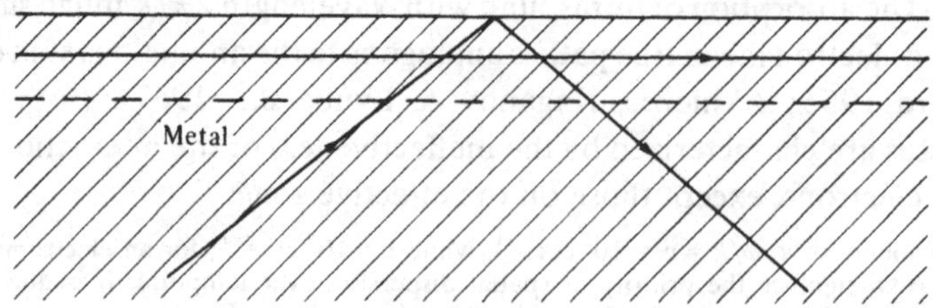

Fig. 1.14. The important electrons in the anomalous skin effect are those which are moving parallel to the metal surface.

described by the deformation potential (Akhiezer 1938; Fröhlich 1950), the determination of whose components is a separate problem. The most obvious method for determining the components of the deformation potential, averaged over the effective zone, for those metals whose electron dispersion law is known, is to use measurements of the electronic part of the ultrasonic absorption coefficient Γ_e. The expression for Γ_e is simple (Pippard 1955; Akhiezer *et al.* 1957),

$$\Gamma_e = \frac{2\omega}{(2\pi\hbar)^3 \rho v_s} \oint_{\mathscr{E}_{p_m} = \mathscr{E}_F} |\Lambda|^2 \frac{dS}{v^2} \delta(\mathbf{n} \cdot \boldsymbol{\nu}); \qquad \boldsymbol{\nu} = \mathbf{v}/v, \tag{18}$$

and besides the components of the deformation potential Λ contains no other unknown quantities (ρ is the density of the metal and v_s the velocity of sound). For some unknown (to us) reason this method of determining the mean values of the deformation potential has received little attention (but see Rayne & Jones 1970).

Effectively, then, the sound wave or the skin effect picks out electrons on the effective zone. The magnetic field forces electrons to move over the Fermi surface in orbits (equation (7)) which in general intersect the effective zone at an arbitrary angle (Fig. 1.15). The investigation of the high-frequency properties of metals in a magnetic field is an everlasting source of information about conduction electrons. Without doubt, the most popular has been cyclotron resonance (or Azbel'–Kaner resonance (Azbel' & Kaner 1956, 1957, 1958)), the nature of which is clear from Fig. 1.16: if for a magnetic field parallel to the surface an electron returns to the effective zone – the skin depth – after a time interval T_H which is a multiple of the field period $T_\omega = 2\pi/\omega$, then the condition for a resonant interaction between the electron and the wave is satisfied; the conductivity must tend to infinity and the impedance, correspondingly, to zero. It is true that the period of revolution $T_H = 2\pi/\omega_c$ (where $\omega_c = eH/m_c c$;

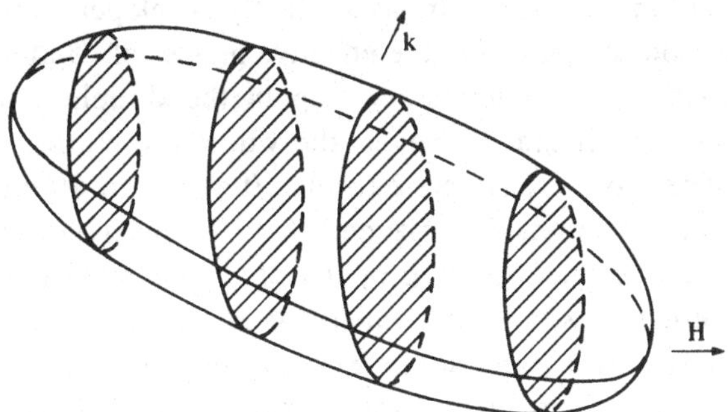

Fig. 1.15. Electron orbits in a magnetic field intersect the effective zone.

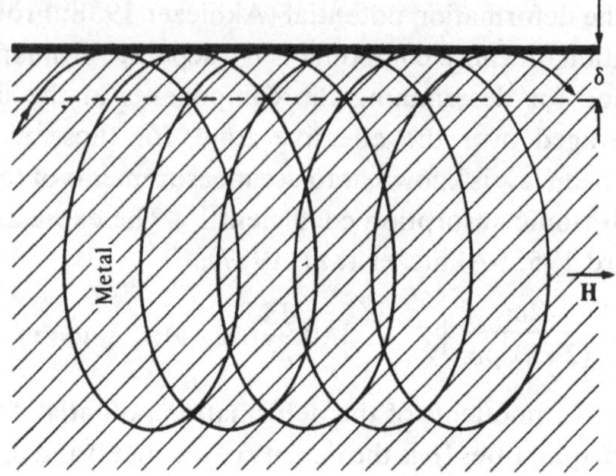

Fig. 1.16. Electrons return to the skin-effect layer when the magnetic field is parallel to the surface.

$m_c = \partial\mathscr{A}/\partial\mathscr{E}$) is different for different electrons. As in most cases, because of the degeneracy the effect is governed by the Fermi electrons, i.e. those electrons with $\mathscr{E} = \mathscr{E}_F$ 'resonate'. The contribution of electrons with different values of p_z depends on the existence of extrema in the dependence of $m_c(\mathscr{E}_F, p_z)$ on p_z. Near the extrema the dependence on the longitudinal component of the momentum is less important than for other values of p_z. The extremal values of the effective mass $m_j \equiv m_c(\mathscr{E}_F, p_{z,j}^{\text{ext}})$ also determine the resonance condition:

$$\omega = n\frac{eH}{m_j c}; \qquad n = 1, 2, 3, \ldots . \tag{19}$$

The investigation of Azbel'–Kaner resonance is thus a method for the experimental determination of the extremal effective masses of conduction electrons. Cyclotron resonance was first observed experimentally by Fawcett (1956) in Sn and Cu. It has gradually developed into a reliable standard method of measuring the effective masses of conduction electrons (Khaĭkin 1962), a necessary step in the decipherment of the electron energy spectrum. However, the value of cyclotron resonance was not confined to these concrete results. It also drew attention to the electrodynamics of a metal in a magnetic field. In fact it had always been considered that the chief optical property of a metal was its reflection of electromagnetic waves, and that when (for $H = 0$) it began to transmit electromagnetic waves it did not differ very much from a dielectric (the dielectric constant of a metal is positive for $\omega > \omega_p$, and absorption is due to the interband internal photoelectric effect). Cyclotron resonance changed our concepts about the interaction of conduction electrons with an electromagnetic field. It turned out that a magnetic field can completely change the nature of the propagation of electromagnetic waves in a metal, to some extent doing away with the skin effect and transforming oscillations which were damped by a factor of $e^{2\pi}$ per wavelength into weakly damped waves with $\mathrm{Im}\, k \ll \mathrm{Re}\, k$. Many different types of weakly damped waves in metals have been discovered, and by no means only on paper (see, for instance, the review articles by Kaner & Skobov (1966, 1968)).

Many of these waves have no direct connection with cyclotron resonance. Neither a discussion of their properties, nor even their enumeration, comes within our present theme. We shall quote only the two simplest examples, to show the effect of the magnetic field.

A *low-frequency elliptically polarized spiral wave* (or *helicon*) (Konstantinov & Perel' 1960; Aigrain 1961) can propagate along a sufficiently high magnetic field H in metals with unequal numbers of electrons and holes ($n_1 \neq n_2$).

The dispersion law of the spiral waves is

$$\omega = \frac{ck^2 H \cos\theta}{4\pi e(n_1 - n_2)}, \quad \omega\tau \ll 1, \tag{20}$$

where θ is the angle between the wave-vector $\mathbf{k}$ and the magnetic field $\mathbf{H}$. The undamped nature of these waves follows from the fact that for metals with $n_1 \neq n_2$ and in a high magnetic field the Hall components of the conductivity are much larger than the dissipative components (transverse, of course). The attenuation length of the spiral wave is determined either by collision processes (i.e. the values of σ_{xx} and σ_{yy}) or by collisionless processes (again the effective zone!).

A *high-frequency magnetoplasma wave* (Buchsbaum & Galt 1961; Kaner & Skobov 1963; Khaĭkin, Fal'kovskiĭ, Edel'man & Mina 1963) propagates for $\omega\tau \gg 1$ and $kr_H \ll 1$ in compensated metals ($n_1 = n_2$) at high fields ($\omega_c \gg \omega$). In the simplest case, the dispersion law of the magnetoplasma wave is the same as that of the well-known Alfven wave

$$\omega = v_a k; \qquad v_a = \frac{H}{(4\pi n\tilde{m})^{\frac{1}{2}}}; \tag{21}$$

v_a is the Alfven velocity ($\tilde{m}$ is a combination of the effective masses of electrons and holes). The dispersion law in (21) can easily be deduced from physical arguments by noting that the frequency ω always appears combined with τ in the form $\tau/(1 - i\omega\tau)$, and the transition from $\omega = 0$ to

$\omega\tau \gg 1$ corresponds to the replacement of τ by i/ω. The effective dielectric constant of the electrons in the metal is $\varepsilon = 4\pi i\sigma/\omega$. For $H = 0$ we have $\sigma = ne^2\tau/m$, and the substitution $\tau \to i/\omega$ leads to a negative dielectric constant $\varepsilon = -\omega_p^2/\omega^2$. For $\omega_c\tau \gg 1$ the dissipative conductivity is $\sigma = (ne^2/m)(1/\omega_c^2\tau)$, and the substitution $\tau \to i/\omega$ leads to a positive dielectric constant.† Because the numbers of electrons and holes are equal, the Hall components play a second-order role.

The influence of a magnetic field on the conduction electrons of course also shows up in the electronic part of the sound attenuation coefficient. The ability of sound to propagate through a metal (in distinction to electromagnetic waves) makes ultrasonic spectroscopic methods especially attractive.

We give below an expression which is the counterpart of (18) and which describes the absorption of sound under the following conditions: $\mathbf{k} \perp \mathbf{H}$, $\omega \ll \omega_c$, $kl \gg 1$, $r_H k \gg 1$. It is easy to see that these four inequalities are not contradictory, provided that the magnetic field is not too high: $1 \ll \omega_c\tau \ll kl$. The 'geometry' of the interaction of an electron with the sound wave gives an especially graphic description (Fig. 1.15): the more frequently the electron intersects the effective zone, the larger is the sound absorption coefficient. The leading term in Γ_e therefore takes a particularly simple form (Gurevich 1959):

$$\Gamma_e(H) \approx \Gamma_e(0) = \frac{2\omega}{(2\pi\hbar)^3 \rho v_s} \oint_{\mathscr{E}_{p_m} = \mathscr{E}_F} 2\pi\omega_c\tau |\Lambda|^2 \frac{dS}{v^2} \delta(\mathbf{n} \cdot \boldsymbol{\nu}), \qquad \mathbf{n} = \mathbf{k}/k; \quad \boldsymbol{\nu} = \mathbf{v}/v. \tag{22}$$

However, it is perhaps the second term in $\Gamma_e(H)$ that is the most interesting, although formally it is kr_H times smaller, because it contains the periodic dependence on the magnetic field

$$\Gamma_e(H) = \Gamma_e^{(0)} + \tilde{\Gamma}_e; \qquad \tilde{\Gamma}_e \approx \Gamma_e^{(0)} \frac{1}{(kr_H)^{\frac{1}{2}}} \sum_m \sin\left(\frac{ck}{eH}\Delta p_{y,m} \pm \frac{\pi}{4}\right). \tag{23}$$

The summation is over all extremal diameters of the Fermi surface (Figs. 1.17 and 1.3). The origin of the periodic dependence is clear, in that the physical situation repeats itself when the number of waves that fit into the orbit diameter changes by unity (Pippard 1957b). The periodic

† This is of course not accidental: for $H = 0$ the frequency ω of the field is greater than the resonance frequency (which in this case is zero!), while for $H \neq 0$ the field frequency ω is less than the resonance frequency ω_c (see above).

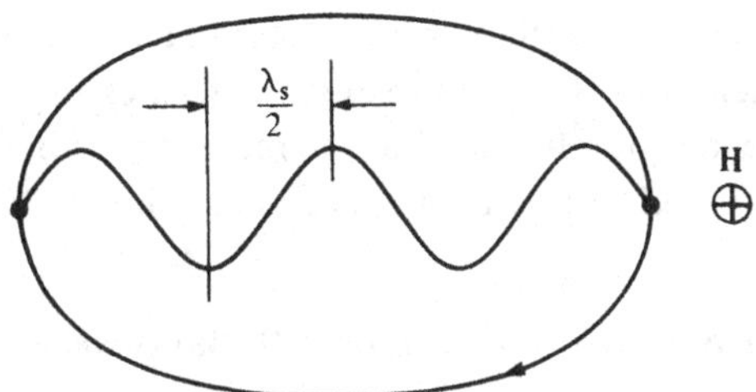

Fig. 1.17. The nature of Pippard oscillations in sound attenuation.

dependence of $\Gamma_e(H)$ became known as Pippard or geometric resonance.† Before finishing this section, let us return once more to (18), i.e. to the absorption of sound at $H = 0$. We do this to draw attention to the factor $1/v^2$ inside the integral. This shows that sound waves interact strongly with slow electrons. In geometric language this statement means that those points in **p**-space where the velocity tends to zero play a special part. It is at these points that new constant-energy surfaces are created or 'necks' disappear (see Fig. 1.18). They are also the origin of the phase

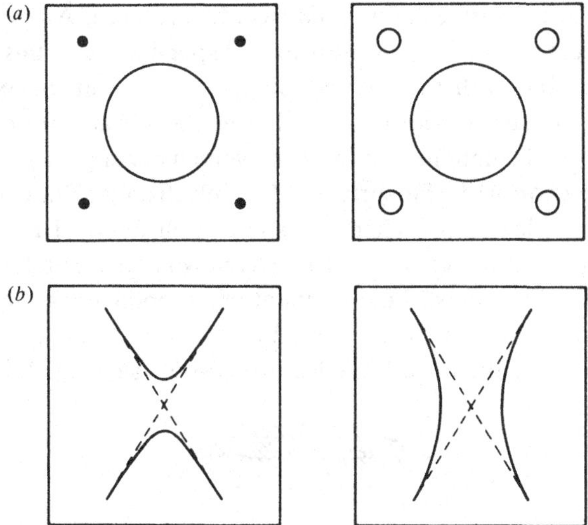

Fig. 1.18. Change in Fermi surface topology at a phase transition of order $2\frac{1}{2}$; (*a*) formation of a new sheet; (*b*) break-up of a neck.

† I remember that in 1957 in the Institute of Physical Problems in Moscow, after one of the sessions of the Conference on Low Temperature Physics, A. B. Pippard told M. Azbel' and myself about his idea of measuring diameters of the Fermi surface from the absorption of sound by metals in a magnetic field (see (23)). At that moment I felt a little sad; I and I. A. Akhiezer had been working on the same problem, but our results were not so complete (comment by M.K.).

transitions of order $2\frac{1}{2}$ (Lifshitz 1960). It follows from (18) that the sound absorption coefficient must be very sensitive to a change in topology of the Fermi surface, despite the fact that the reconstruction of the spectrum takes place at isolated points in **p**-space (Davydov & Kaganov 1972).

1.6 Electron collisions with the surface

The collision of electrons with the sample surface has long attracted the attention of those studying the electron properties of metals. As far back as 1938 Fuchs (1938), who was investigating theoretically the dependence of the resistance of a thin plate on its thickness, formulated his famous boundary conditions describing the partial specular reflection of electrons from the surface. For a long time it seemed that any metal surface appeared rough to a Fermi electron with a wavelength of the order of the interatomic spacing, i.e. it behaved like a clouded mirror. At such a surface an electron must (it was thought) be scattered completely diffusely. Under the pressure of experimental evidence it gradually became clear that this is not so; very often an electron is specularly reflected from the surface of a metal (see, for instance, Andreev 1971). For an electron with an anisotropic dispersion law this has the consequence that both the energy and the component $\mathbf{p}_{\|}=\mathbf{p}-(\mathbf{p}\cdot\mathbf{n})\mathbf{n}$ (where $\mathbf{n}$ is the surface normal) parallel to the surface are conserved, while the normal component $\mathbf{v}\cdot\mathbf{n}$ of the velocity changes sign.

The interaction with the surface, of an electron with a complicated dispersion law, leads to a characteristic complication of a number of familiar properties. For example, it is well known that in an infinitely deep potential well the values of the momentum components are quantized, but for an electron with a complicated dispersion law the chords of the Fermi surface are quantized (Nedorezov 1966) (see Fig. 1.19), which

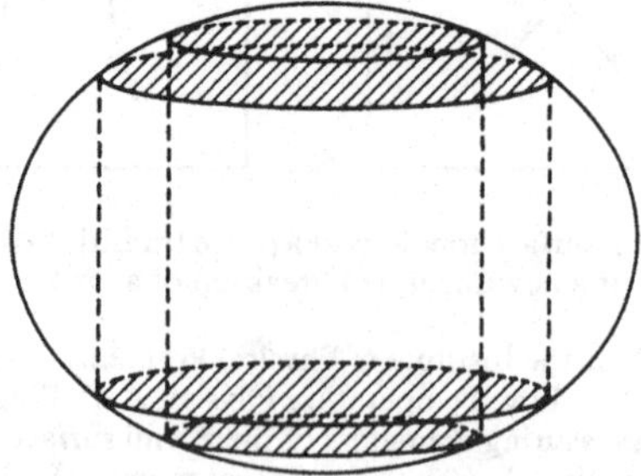

Fig. 1.19. Quantization of the chord picks out electrons on certain lines on the Fermi surface.

makes the picture of the electron energy sub-bands in thin metal plates extremely intricate. With the discovery of the quantum size effect (Ogrin, Lutskiĭ & Elinson 1966; Lutskiĭ, Korneev & Elinson 1966) this phenomenon has ceased to be a piece of theoretical exotica.

The most direct proof of the existence of specular reflection of electrons at the surface is undoubtedly provided by the discovery by Khaĭkin (1960, 1968a, 1968b) of oscillations due to surface levels of the electrons in a magnetic field. If an electron is specularly reflected from the surface in a magnetic field, its motion normal to the metal surface is finite (Fig. 1.10). This motion is quantized, discrete surface energy levels arise, and these give rise to oscillations of the surface impedance as a function of applied magnetic field. A complete theoretical explanation of this effect was given by Nee & Prange (1967).

Khaĭkin oscillations involve only one surface of the sample. When electrons interact with both surfaces of a thin plate (both the plate thickness d and r_H must be much less than the mean free path l), characteristic size-effect resonances can occur and these are easily described in geometric language.

The thin plate picks out from all the orbits the one which just fits into it (Fig. 1.20). This orbit must give rise to a resonance (Lur'e & Peschanskiĭ 1972; Kirichenko, Lur'e & Peschanskiĭ 1976), and this was indeed observed (Volodin, Khaĭkin & Edel'man 1973a, 1973b), yet again in Bi. This enabled non-extremal sections of the Fermi surface to be studied – an opportunity that deserves special attention.

Another mechanism, besides a magnetic field, for returning electrons to the skin-effect layer is reflection from the opposite surface of the sample (Fig. 1.21). If $\delta \ll d$ and $r_H \ll l$, then cyclotron resonance can also be observed from these orbits (Khaĭkin & Edel'man 1964; Peschanskiĭ 1968; Kirichenko, Lur'e & Peschanskiĭ 1976b).

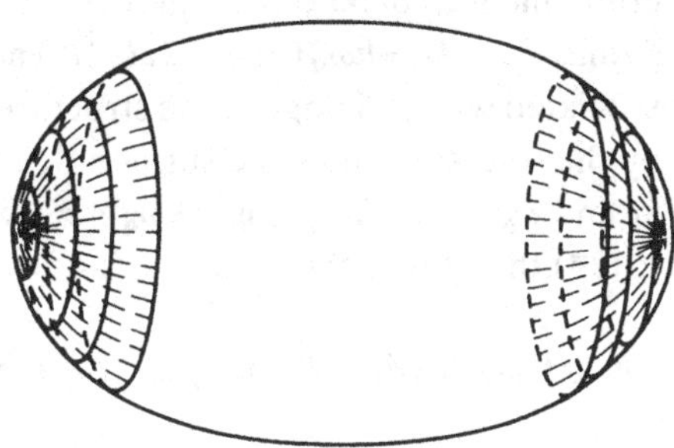

Fig. 1.20. Orbits which can be accommodated inside a thin plate are less than a certain limiting size.

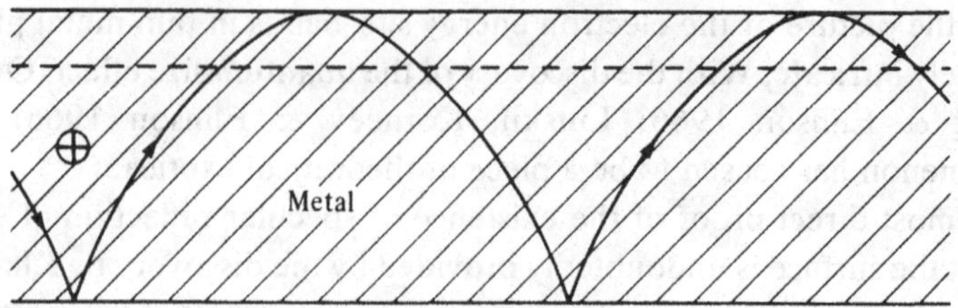

Fig. 1.21. Electrons can return to the skin-effect layer after reflection at the 'back' surface of the sample.

1.7 Intersection of the Fermi surface with its displaced analogue: the Migdal–Kohn anomaly

The anomaly in the dependence of phonon energy on quasi-momentum (the Migdal–Kohn anomaly (Migdal 1958b; Kohn 1959)), which is a consequence of the interaction of phonons with the electrons in a metal, is due to the non-uniform population of electron states in **p**-space and has a clearly formulated geometrical interpretation. Since the Kramers–Kronig relations ensure that anomalies in Im $\omega = \Gamma_{\mathbf{q}}$ lead to anomalies in Re $\omega(q)$ and vice versa, and it is simpler to deal with δ functions than with integrals in the sense of principal values, we shall explain the geometrical location of the special points in phonon **q**-space by using the following expression for $\Gamma_{\mathbf{q}}$, the reciprocal of the phonon lifetime due to electron–phonon collisions:

$$\Gamma_{\mathbf{q}} = \int |M|^2[n_F(\mathscr{E}) - n_F(\mathscr{E} + \hbar\omega_{\mathbf{q}})]\delta(\mathscr{E}_{\mathbf{p}} + \hbar\omega_{\mathbf{q}} - \mathscr{E}_{\mathbf{p}+\mathbf{q}})\,\mathrm{d}^3p. \tag{24}$$

Here $n_F(\mathscr{E})$ is the Fermi function, and the transition matrix element M includes all the other factors. At a temperature of absolute zero the difference in the Fermi functions picks out a region in **p**-space bounded by the surfaces $\mathscr{E}_{\mathbf{p}} = \mathscr{E}_F$ and $\mathscr{E}_{\mathbf{p}} = \mathscr{E}_F - \hbar\omega_{\mathbf{q}}$ (Fig. 1.22(*a*)). The anomaly in the variation of $\Gamma_{\mathbf{q}}$ is associated with a change in the structure of the region of integration given by the intersection of the surface $\mathscr{E}_{\mathbf{p}} + \hbar\omega_{\mathbf{q}} = \mathscr{E}_{\mathbf{p}+\mathbf{q}}$ with the surfaces $\mathscr{E}_{\mathbf{p}} = \mathscr{E}_F$ and $\mathscr{E}_{\mathbf{p}} = \mathscr{E}_F - \hbar\omega_{\mathbf{q}}$. Since $\hbar\omega_{\mathbf{q}} \ll \mathscr{E}_F$, we can expand in powers of $\hbar\omega_{\mathbf{q}}$, and (24) then takes the form

$$\Gamma_{\mathbf{q}} = \hbar\omega_{\mathbf{q}} \int |M|^2 \delta(\mathscr{E}_{\mathbf{p}} - \mathscr{E}_F)\delta(\mathscr{E}_{\mathbf{p}+\mathbf{q}} - \mathscr{E}_F)\,\mathrm{d}^3p. \tag{25}$$

In the δ function describing energy conservation we have left out $\hbar\omega_{\mathbf{q}}$ and replaced $\mathscr{E}_{\mathbf{p}}$ by the Fermi energy $\mathscr{E}_F$. It is seen that the region of

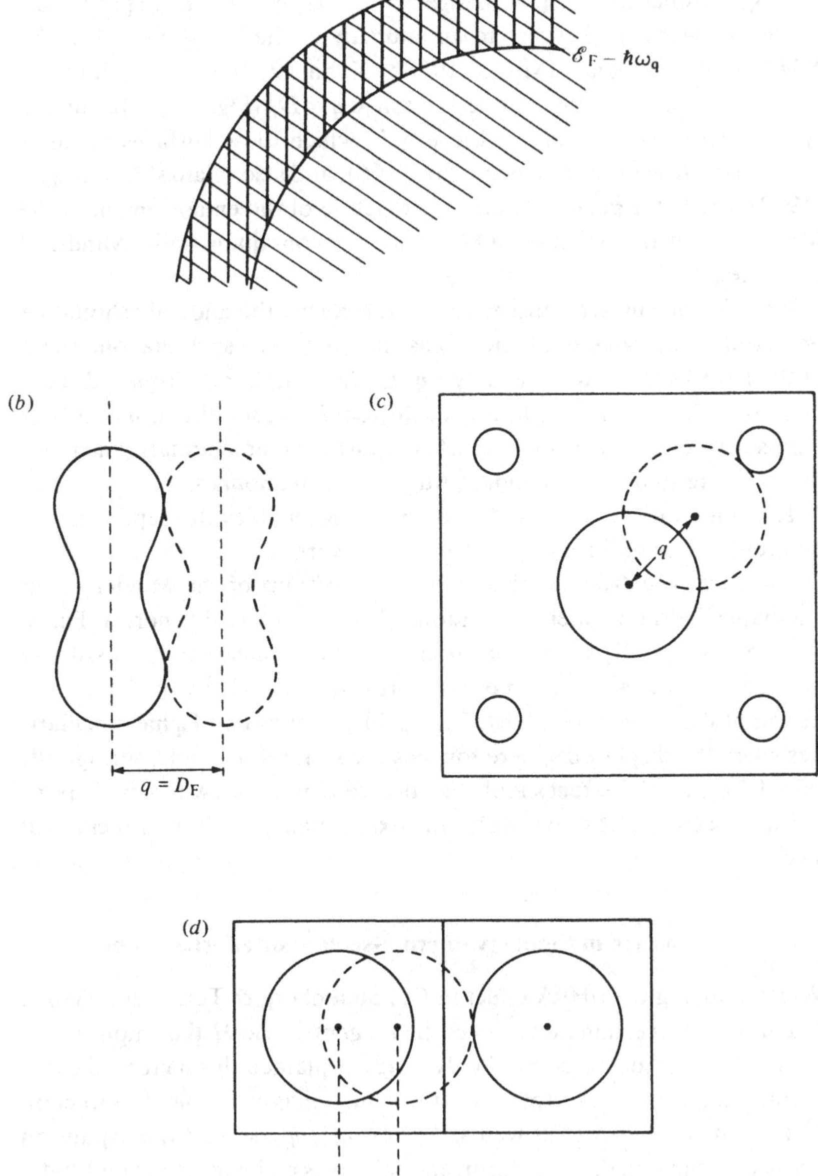

Fig. 1.22. (*a*) Region of integration in equation (24). (*b*) For $q > D_F$ the Fermi surface does not intersect its displaced analogue. (*c*) One sheet intersects another. (*d*) A Fermi surface intersects its analogue in the next cell.

integration in (25) is the line of intersection of the Fermi surface with its analogue displaced by $-\mathbf{q}$. A change in topology of this line (in particular, its disappearance) gives rise to an anomaly in the variation of $\Gamma_{\mathbf{q}}$. The Migdal–Kohn anomaly (Migdal 1958b; Kohn 1959) is the result of the disappearance of this line of intersection at $q = D_F$ (Fig. 1.22(*b*)), and the type of anomaly depends on the way in which these surfaces touch at $q = D_F$ (see Kaganov & Semenenko (1966); and also Afanas'ev & Kagan (1962); the latter paper includes a prediction of the enhancement of the Migdal–Kohn anomalies for a Fermi surface containing finite cylindrical or flat regions).

If the Fermi surface consists of several sheets, the anomaly should be observed at the value of phonon quasi-momentum $\mathbf{q}$ such that one sheet of the Fermi surface displaced by $-\mathbf{q}$ touches another undisplaced sheet (see Fig. 1.22(*c*)). The absorption coefficient *increases* discontinuously as $\mathbf{q}$ passes through this value, and Re ω must have an anomaly which has the opposite sign to the standard Migdal–Kohn anomaly.

The observation of anomalies of this kind enables the separation of different pieces of Fermi surface to be measured.

For a free-electron gas the geometric positions of the Migdal–Kohn anomalies lie on a sphere with radius $2p_F$, but for real spherical Fermi surfaces (Na, K, Rb, Cs) the position is more complicated: the possibility of umklapp processes (or, in other words, the periodicity of $\Gamma_{\mathbf{q}}$ with the period of the reciprocal lattice; $\Gamma_{\mathbf{q}+2\pi\mathbf{b}} = \Gamma_{\mathbf{q}}$) means that $\Gamma_{\mathbf{q}}$ has singularities when the displaced sphere touches the spheres in neighbouring cells (Fig. 1.22(*d*)). This effect should be observed for metals with any shape of Fermi surface and can be useful for fixing their positions in reciprocal space.

1.8 Change in topology of cross-section or effective zone

While studying the dHvA effect in Cu, Shoenberg & Templeton (1973) found that at a certain direction of the magnetic field H the amplitude of the oscillations increased markedly. They explained this increased oscillation amplitude as follows: for this magnetic field direction $\partial^2\mathscr{A}(\mathscr{E}_F, p_z)/\partial p_z^2$ is zero as well as $\partial\mathscr{A}(\mathscr{E}_F, p_z)/\partial p_z$, so that the expansion starts with the fourth-order term, and this gives an increase in amplitude; the small factor $(\hbar\omega_c/\mathscr{E}_F)^{\frac{1}{2}}$ being replaced by the larger factor $(\hbar\omega_c/\mathscr{E}_F)^{\frac{1}{4}}$. It is clear that the topology of the section also changes at this field direction. In Fig. 1.23 we show the different dependences of $\mathscr{A}(\mathscr{E}_F, p_z)$ for different field directions, and also how the topology of the section changes. A change of this kind would be impossible if the surface were everywhere

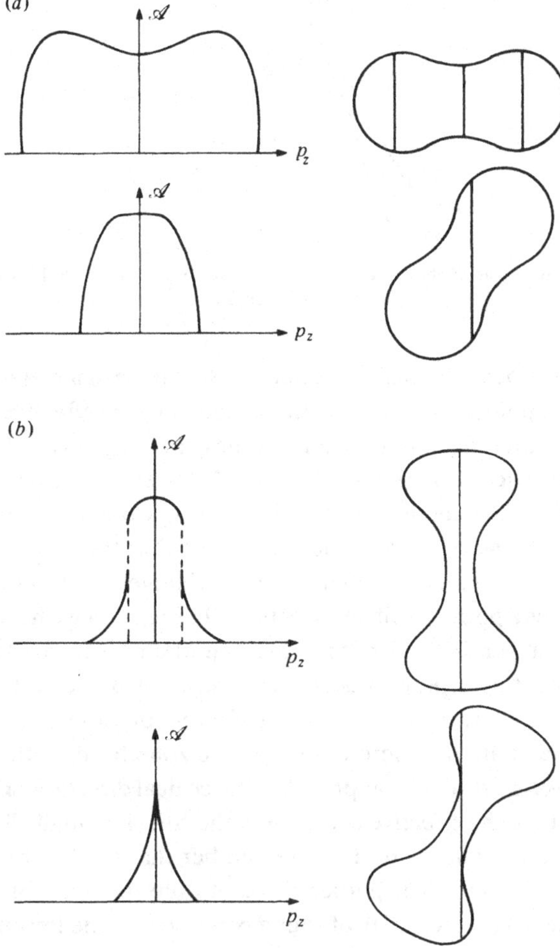

Fig. 1.23. Dependence on p_z of the area of a plane section of the Fermi surface for different directions of magnetic field.

convex. The structure of the section can also change in a way which is not as simple as that shown in Fig. 1.23(*a*). If there exists a field direction for which one of the extremal sections contains a point of self-intersection, then the change in the structure of the sections, and therefore of the spectrum of dHvA oscillations, is more complicated (Fig. 1.23(*b*)), and the amplitude of the critical oscillations (corresponding to a figure-of-eight cross-section) is anomalously small. However, this question has so far received little attention.†

† Azbel' (1960) derived the form of oscillations from non-extremal self-intersecting sections.

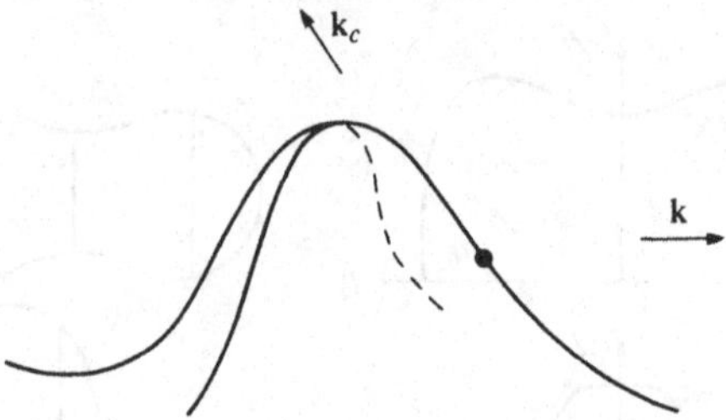

Fig. 1.24. The appearance of an effective zone on a surface similar to a 'hilly region' – at a point of type 0.

The problem of anomalies in the phonon spectrum of a metal which are the result of topological changes in the structure of the effective zones has recently been investigated in detail (Avanesyan, Kaganov & Lisovskaya 1977; Kontorovich & Stepanova 1973, 1978). Let us return to Fig. 1.13. It is seen that the number of effective zones is different for different directions of the wave-vector, and this means that there exists a critical direction $\mathbf{n} = \mathbf{n}_c$ at which the effective zone changes its structure (in Fig. 1.13 it contains a point of self-intersection). In Fig. 1.24 we have sketched part of a Fermi surface which is reminiscent of a hilly region (some time ago we called this type of surface a corrugated plane (Lifshitz *et al.* 1956b)). There is a two-dimensional set of directions for which there is no effective zone at all, but there is an effective zone for directions outside this region. As the direction approaches the critical direction, $\mathbf{n} \to \mathbf{n}_c$, from the region where the effective zone exists, the zone is a small ellipse which contracts to a point at $\mathbf{n} = \mathbf{n}_c$. If we remember that on the effective zone the velocity of electrons is perpendicular to the direction of the wave-vector (we neglect the velocity of sound compared to the Fermi velocity), then the critical points lie on lines of zero Gaussian curvature K on the Fermi surface.† These lines are an essential property of surfaces with dimples, necks, etc. All points on the surface belong to one of three classes:

elliptic ($K > 0$),
parabolic or saddle points ($K < 0$), or
hyperbolic ($K = 0$).

Analysis of the structure of the effective zones (see (16) and (17)) shows that there are two types of hyperbolic points, types 0 and X. At a type 0 point an effective zone is created, while at an X point it is self-intersecting.

† A point of self-intersection of the critical zone on a dumb-bell-shaped surface is shown in Fig. 1.13; a point to which elliptical zones contract is shown in Fig. 1.24.

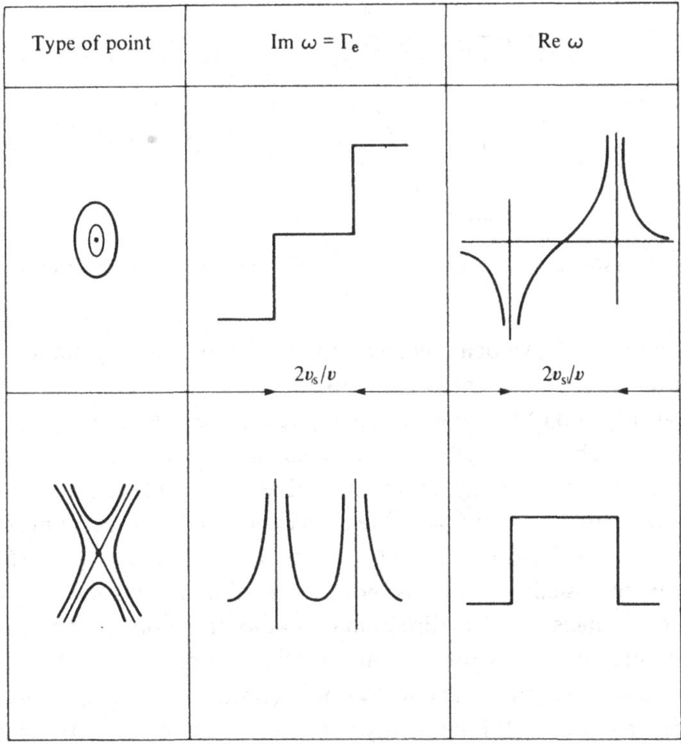

Fig. 1.25. Anomalies in the phonon dispersion law due to points of types 0 and X.

The existence of parabolic points on the Fermi surface, and consequently of critical propagation directions, must lead to anomalies in the angular dependence of the velocity of sound and of other transport coefficients (e.g. the electrical conductivity tensor under conditions of high spatial dispersion).

The phenomenon described represents a rather rare situation where the local geometry of the Fermi surface leads to (or, more accurately, *should* lead to) macroscopically observable effects. It should be emphasized that the structure of (18) is such that the contribution to Γ_e from the neighbourhood of the special point is of the same order as or greater (!) than that from the remainder of the surface.† The structure of the anomalies in $\Gamma_e = \text{Im}\ \omega$ and Re ω for points of type 0 and X is given in Fig. 1.25. The occurrence of a pair of neighbouring anomalies is due to the fact that each special point is accompanied by an 'antipodean' point

† The formula for the components of the electrical conductivity tensor for $kl \gg 1$ is analogous to (18) (see Pippard 1947, 1954; Kaganov & Azbel' 1955).

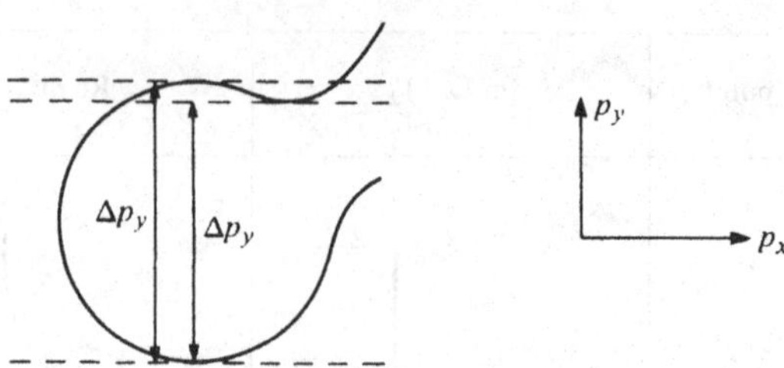

Fig. 1.26. The spectrum of Pippard oscillations is different for different directions of the wave-vector.

with an antiparallel velocity, because the Fermi surface is symmetric with respect to the replacement of $\mathbf{p}$ by $-\mathbf{p}$.

The absorption of sound in intermediate magnetic fields ($1 \ll \omega_c\tau \ll kl$) for $\mathbf{k} \perp \mathbf{H}$, which was described above, also has anomalies associated with a change in the topology of the effective zone. As far as the main (monotonically dependent on H) term is concerned, a direct comparison of (18) and (22) shows that $\Gamma_e^{(0)}(H)$ has the same anomalies as $\Gamma_e(H=0)$. However, the oscillatory part is more interesting, in that the oscillation spectrum changes with the direction of sound propagation. Two similar periods must appear as the critical direction is approached (Fig. 1.26), and the superposition of the corresponding harmonics will be seen as the appearance of beats. It is evidently easiest to observe the increase in the amplitude of the oscillations: if one end of the extremal diameter touches a point of type X, then the amplitude is multiplied by an extra factor $(kr_H)^{\frac{1}{6}}$, while if both ends touch points of type X the factor is $(kr_H)^{\frac{1}{3}}$.

In this article we have dealt hardly at all with scattering processes, on the ground that their consideration takes us far beyond our theme. However, it is difficult to refrain from a description of the elastic scattering of a particle with a complex dispersion law (Lifshitz 1948). As we shall see, the features of this process are also associated with the presence of a line of parabolic points on the constant-energy surfaces.

In elastic scattering, an electron (or any other quasi-particle) 'jumps' from one point on the constant-energy surface to another (energy is conserved). If the surface is everywhere convex, each scattering direction $\boldsymbol{\nu}$ corresponds to a single value of the quasi-momentum $\mathbf{p}_\nu$ of the scattered particle (this is simply determined as the point where a plane perpendicular to $\boldsymbol{\nu}$ touches the constant-energy surface (Fig. 1.27)). If the constant-energy surface has a more complicated form, then there are directions in which scattered particles move for which there are several

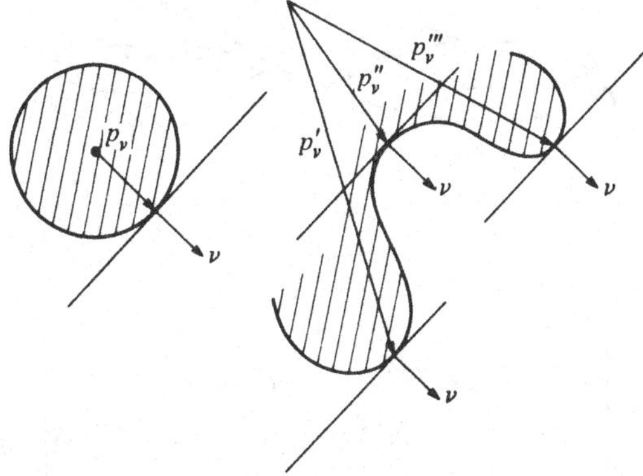

Fig. 1.27. In elastic scattering electrons with different values of quasi-momentum can be moving in the same direction.

different values of quasi-momentum (Fig. 1.27). Moreover, if the scattering direction is parallel to the vector which is normal to the surface at a hyperbolic point ($\boldsymbol{\nu} = \boldsymbol{\nu}_c$, see Fig. 1.24), then two possible values of the momenta of the scattered particles coincide. The character of the scattering thus changes, and this is accompanied by a change in the wave-function of the scattered particle: for $\boldsymbol{\nu} = \boldsymbol{\nu}_c$ it is damped more slowly than for $\boldsymbol{\nu} \neq \boldsymbol{\nu}_c$. It seems to us that this little-studied effect could play an important part in the investigation of the indirect interaction of impurity atoms (via the conduction electrons).

1.9 Electron transitions from one orbit to another: magnetic breakdown

In all the phenomena described above, electrons have either remained at a fixed point in **p**-space on the Fermi surface, 'jumped' from one point to another as the result of scattering, or moved along classical orbits in a magnetic field. The establishment of the fact that most metals have complicated Fermi surfaces coincided with the realization that classical orbits in **p**-space come close to each other, so that there is the possibility of a tunnelling transition from one orbit to another. This phenomenon, which was given the name of magnetic breakdown (Cohen & Falicov 1961), turned out to play an important role in many of the properties of metals (Stark & Falicov 1967).

To understand the nature and importance of magnetic breakdown a distinction must be made between *intra*- and *inter*band transitions. If we consider only the Fermi surface, intraband transitions can occur only over

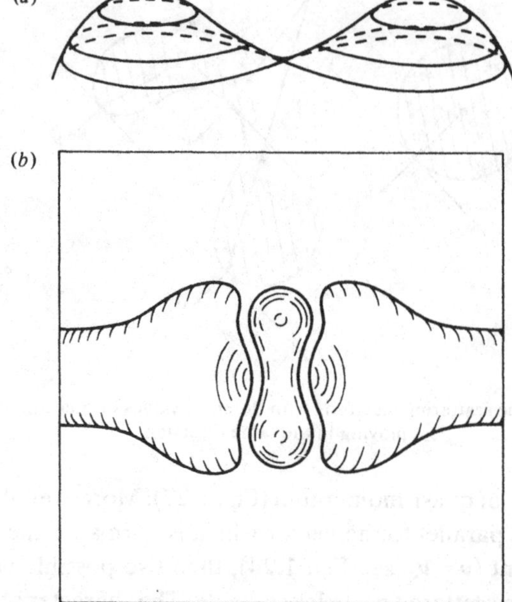

Fig. 1.28. Electrons which take part in intraband magnetic breakdown lie in an infinitely narrow layer (a), but for interband magnetic breakdown they come from a layer of finite thickness (b).

a narrow range of p_z (Fig. 1.28(a)), but interband transitions can take place over a wide range of order $\hbar/a$ (Fig. 28(b)). The range of phenomena associated with interband transitions is known as magnetic breakdown.

Although the aims of this chapter do not include any kind of detailed description of the properties of metals under conditions of magnetic breakdown, we shall point out a few cases which permit a geometrical interpretation.

The region of magnetic breakdown is the region in which the *classical* orbits come close to each other. This means that in the magnetic-breakdown configuration the electron state is a superposition of quasi-classical states. It is for just this reason that it is possible to use a geometric representation to describe magnetic breakdown, while the physical quantities which describe this situation are expressed in terms of the dispersion law and second-rank unitary S-matrices representing the two-channel scattering in the magnetic-breakdown region.† The small

† The square of the modulus of the off-diagonal element of the S-matrix is the breakdown probability W, while the square of the modulus of the diagonal element is the probability of no breakdown, $1-W$ (see Slutskin 1970).

size of the magnetic-breakdown region compared to that of the classical sectors of the magnetic-breakdown configuration makes it possible to calculate the S-matrix for an arbitrary value of the magnetic field without breaking the conditions of quasi-classical motion between the breakdown regions ($\hbar\omega_c \ll \mathscr{E}_F$) (Blount 1962; Slutskin 1967). When the breakdown probability W differs from either zero or unity ($0 < W < 1$), the electron motion has an essentially quantum character, but as the magnetic field H increases $W \to 1$ and the motion again becomes quasi-classical, although the electrons have acquired new properties which were not present at low fields. Because of the spectroscopic character of measurements at a high magnetic field, one should take especial care: the changes can be qualitative and not just quantitative. It is demonstrated in Fig. 1.29 how magnetic breakdown can change the topology of a plane section – it is transformed from an open to a closed section.

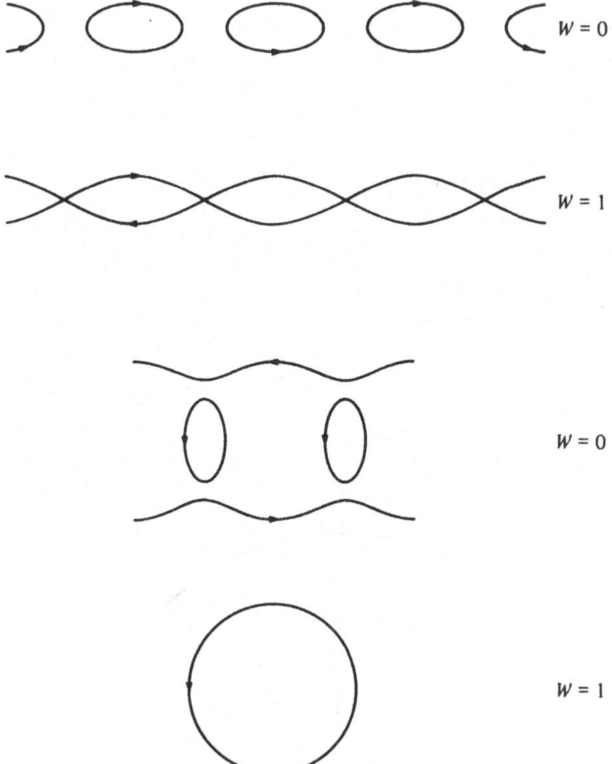

Fig. 1.29. Magnetic breakdown changes the topology of electron orbits.

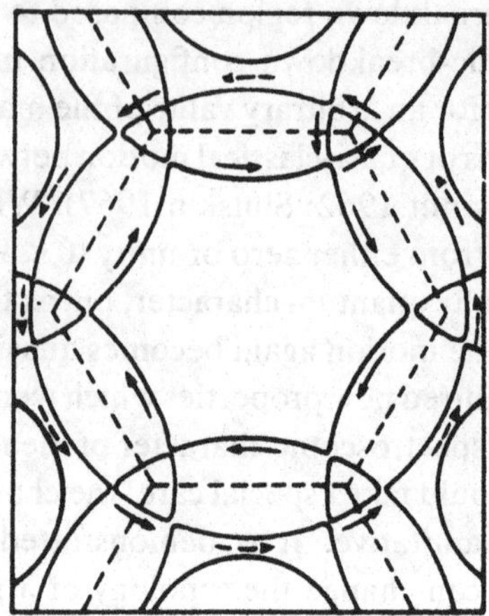

Fig. 1.30. Magnetic-breakdown orbits observed in an investigation of the dHvA effect in Mg.

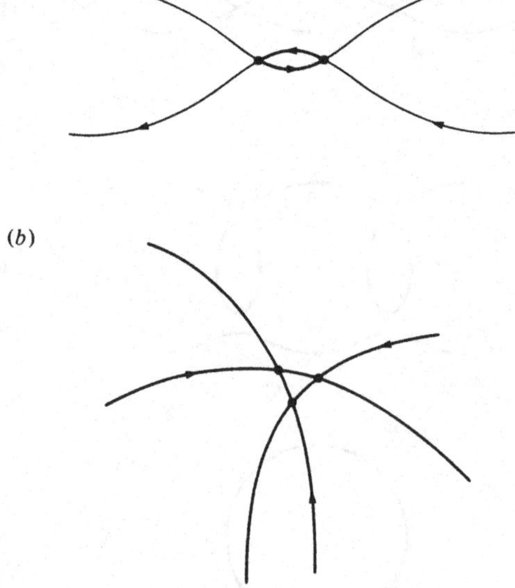

Fig. 1.31. Small orbits which connect large orbits via magnetic breakdown.

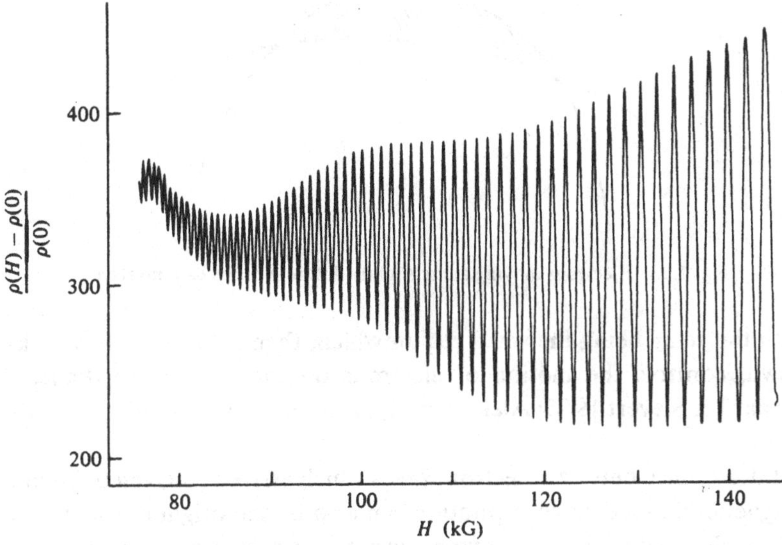

Fig. 1.32. Giant oscillations in the magnetoresistance of Be for **H**||[0001]; measured in the International Laboratory for High Magnetic Fields and Low Temperatures, Wroclaw, Poland, 1972.

The first experimental observation of an effect due to magnetic breakdown was made by Priestley (1963) in a study of the dHvA effect in Mg. Periods (closed sections) were found which were formed by electrons overcoming the barriers separating sectors of classical orbits (Fig. 1.30).

Magnetic breakdown can also be the origin of characteristic oscillatory effects which do not reduce to the dHvA effect. A not unusual situation occurs when two or even three large orbits on the Fermi surface are linked by magnetic breakdown via a small orbit (Figs. 1.31(a) and (b)). One can introduce the concept of an effective breakdown probability W_{eff}, which is the probability that an electron makes a transition from one large sector to the other. This effective probability W_{eff} is a periodic function of reciprocal magnetic field, but although the period is again given by (9) which describes the dHvA effect, the nature of the periodic dependence is rather different. It is similar to the periodic variation of the transmission coefficient of a dielectric plate due to the interference of electromagnetic waves.

The oscillations in W_{eff} are the reason for the giant oscillations in the galvanomagnetic properties of a metal, in which the main groups of electrons participate (Fig. 1.32). We can call this effect the *relay effect,*

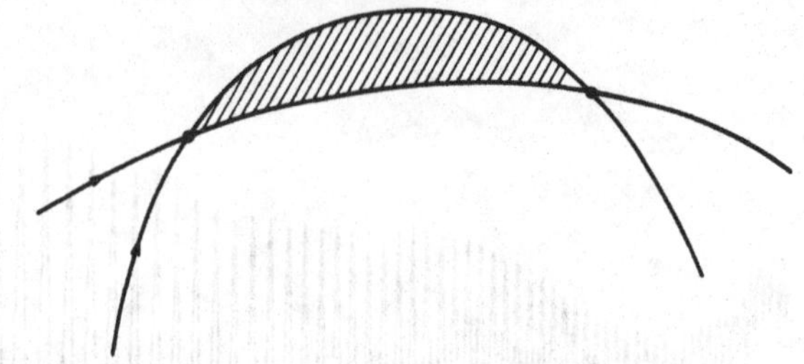

Fig. 1.33. Electrons moving along slightly different orbits can interfere.

with the relay being the small region which, thanks to magnetic breakdown, controls the motion of electrons on the large orbit (Falicov, Pippard & Sievert 1966; Alekseevskiĭ, Slutskin & Egorov 1971; Slutskin 1973).

Interference between electron waves which arrive at a given point in a magnetic-breakdown configuration can also be the origin of oscillatory effects. For this to occur the two paths must differ only slightly from each other (Fig. 1.33). As before, the period of the oscillations is determined by the quantization of an 'area' (Stark & Friedberg 1971, 1972; see also Slutskin 1970). The word 'area' is put in quotation marks in order to emphasize that there is no electron which describes this area. Interference of this kind is observed only in transport effects (Stark & Friedberg 1971, 1972). Only real areas enter into the oscillatory part of the thermodynamic properties.

Magnetic breakdown has been observed in many metals and in a wide variety of properties. Its effects appear in different ways, and the understanding of magnetic-breakdown properties is important in the disentangling of the energy spectrum of a metal, since in the breakdown process an electron finds itself in a region of **p**-space which is inaccessible to classical motion.

1.10 Concluding remarks

To a present-day geometer all our constructions and discussion seem either trivial or else insufficiently rigorous. Physicists do not claim to make any (even minor) discoveries in geometry. The use of geometry in the electron theory of metals is solely to clarify visually many of the deductions and results. Physicists were 'lucky' in that many properties of metals are determined by electrons occupying a surface, and in that sections of this surface together with certain lines and points upon it have a clear physical significance. The understanding of the nature of the metallic state has reached such a level that we have learned to construct these surfaces, to draw lines on them, and to determine the position of relevant points. With the help of these simple geometric models we have also learned to calculate a wide variety of physical properties.

This chapter is not intended as a guide to the questions of priority or to the bibliography. The references that are given may help the reader to become more closely acquainted with the topics mentioned, but we clearly recognize that, had this chapter been written by others, then in a number of cases the references quoted would have differed considerably from those given here.

Acknowledgements

We are grateful to the editor for rendering us the honour of an invitation to take part in this volume in honour of D. Shoenberg, who has played a prominent part in the development of the present physics of the metallic state.

2

Understanding quasi-particles: observable many-body effects in metals and ^{3}He

JOHN W. WILKINS

2.1 Introduction

This chapter attempts to give an overview of our understanding of nearly-independent-particle excitations – quasi-particles – in fermion systems, metals and ^{3}He. To experimentalists it may seem odd to link metals and ^{3}He in a single chapter, but the theory is so intertwined that to talk about one without the other is to miss an opportunity for useful comparisons.

The discussion is arranged in four main sections:
Many-body underpinning for band structure (2.2)
Effective electron–electron interaction (2.3)
Electron–phonon interaction (2.4)
Fermi liquid theory (2.5).

Section 2.2 presents the considerable many-body-physics lore that goes into constructing an effective local band potential that can be used to compute band-structure energies and wave-functions. In addition, this section serves to introduce the concept of regarding the energy as a functional, first of the local density and later of the distribution function for the excitations. This concept is developed in Sections 2.3 and 2.4, reaching its final form in Section 2.5. The intent of Sections 2.3 and 2.4 is to provide specific examples of how the electron–electron and electron–phonon interactions affect various properties. Finally, in Section 2.5, starting with Landau's assumptions (which are a natural extension of earlier arguments) for a Fermi liquid theory, we derive equilibrium properties and, more sketchily, transport properties for ^{3}He and metals.

The central aim of this chapter is to give the reader some feeling for how it is possible to think of the fermion excitations in metals or ^{3}He as nearly independent. As a result the reader should be aware that many concepts will become increasingly sophisticated as the chapter proceeds. The 'distribution factor' $n_{\mathbf{p}s}$ is the best example of this process since it is central to the concept of quasi-particles. Initially $n_{\mathbf{p}}$ is just the distribution function of independent fermions at zero temperature (equation (16)). Later the energy is written as a functional of $n_{\mathbf{p}s}$ which now describes the occupation of excited states (equation (29′)). In Section 2.5 $n_0(\mathbf{p}\sigma)$ is the finite temperature distribution function for interacting quasi-particles (equation (90)) and, furthermore, under the influence of external forces, the distribution function $n(\mathbf{p}\sigma, \mathbf{x}t)$ evolves in space and time (equation (98)).

2.2 Many-body underpinning for band-structure theory

Since the bulk of this book is concerned with experiments which are conventionally interpreted within an independent-particle point of view, it is appropriate to discuss how that approach is developed. This section deals with the understanding of the ground-state energy and the Fermi surface. The reader is also referred to Chapter 5, where these ideas find application in connection with the electronic structure of transition metals.

The long history of understanding the ground-state energies of metals will be completely omitted in order to concentrate on the present method of choice. It is based on a theorem due to Hohenberg & Kohn (1964) who showed that the energy E can be written as a variational functional $E[n]$ of the electron density $n(\mathbf{r})$. In principle, if $E[n]$ is known, then the ground-state energy corresponds to the minimum of $E[n(\mathbf{r})]$.

2.2.1 The energy functional

The total energy functional is written as a sum of terms:

$$E[n] = T[n] + E_{\mathrm{ei}}[n] + E_{\mathrm{es}}[n] + E_{\mathrm{xc}}[n] + E_{\mathrm{ii}}, \tag{1}$$

where all of these are functionals of the electron density $n(\mathbf{r})$ except for the last term E_{ii}, which is due to the interactions between ions. The first term $T[n]$ is the kinetic energy. The second term,

$$E_{\mathrm{ei}}[n(\mathbf{r})] = \int \mathrm{d}\mathbf{r}\, v_{\mathrm{ext}}(\mathbf{r}) n(\mathbf{r}), \tag{2}$$

represents the interactions of the electrons with the ions via an 'external' potential $v_{ext}(\mathbf{r})$. The third term

$$E_{es}[n]=\frac{1}{2}\int d\mathbf{r}\int d\mathbf{r}'\frac{n(\mathbf{r})n(\mathbf{r}')}{|\mathbf{r}-\mathbf{r}'|}$$

is due to the electrostatic (or Coulomb) interaction between the electrons. The remaining electron–electron interactions (due to exchange and correlation, to use the conventional labels) are included in the fourth term

$$E_{xc}[n(\mathbf{r})]=\int d\mathbf{r}\ \varepsilon_{xc}([n(\mathbf{r})];\mathbf{r})n(\mathbf{r}), \qquad (3)$$

in which the extensive character of E_{xc} is utilized to define a local exchange-correlation energy per particle $\varepsilon_{xc}([n(\mathbf{r})];\mathbf{r})$. The main difficulty with using the scheme is specifying v_{ext} and ε_{xc}, a point to be discussed shortly.

Single-particle equations

In order to discuss the ingredients of $E[n]$, it is useful to turn this variational approach into a calculational scheme. To do so one assumes the existence of a complete, orthonormal set of single-particle wave-functions $\psi_{\alpha s}(\mathbf{r})$, where the index α is a set of appropriate quantum numbers which will be connected to energy levels and s is the spin. The electron number density $n(\mathbf{r})$ can then be written as a sum over the occupied levels of the squared amplitude of these wave-functions:

$$n(\mathbf{r})=\sum_{\text{occ}}|\psi_{\alpha s}(\mathbf{r})|^2. \qquad (4)$$

With this identification a variational derivative of $E[n]$ with respect to $n(\mathbf{r})$ corresponds to one with respect to $\psi_{\alpha s}(\mathbf{r})$. Accordingly the minimum of the energy functional (1) is given by the solution of the equations

$$\left\{-\frac{\hbar^2}{2m}\nabla^2+\phi([n(\mathbf{r})];\mathbf{r})+\mu_{xc}([n(\mathbf{r})];\mathbf{r})\right\}\psi_{\alpha s}(\mathbf{r})=\mathscr{E}_\alpha\psi_{\alpha s}(\mathbf{r}), \qquad (5)$$

where $n(\mathbf{r})$ is given by (4), and $\phi+\mu_{xc}$ is the effective potential. The term ϕ is the sum of the 'external' potential of the ions and the Coulomb electrons

$$\phi([n(\mathbf{r})];\mathbf{r})=v_{ext}(\mathbf{r})+e^2\int d\mathbf{r}'\frac{n(\mathbf{r}')}{|\mathbf{r}-\mathbf{r}'|}. \qquad (6)$$

The remaining part of the effective potential, μ_{xc}, turns out to be the exchange-correlation piece of the chemical potential. In any case, it is given formally by

$$\mu_{xc}([n(\mathbf{r})];\mathbf{r}) = \frac{\delta E_{xc}}{\delta n(\mathbf{r})} \tag{7a}$$

$$= \varepsilon_{xc} + n\frac{\delta \varepsilon_{xc}}{\delta n}. \tag{7b}$$

Constructing a reasonable approximation for μ_{xc} is a hard task whose end is not yet in sight.

An anticipatory summary

Up to now, there has been a rather *ad hoc* recitation of 'facts'. An imaginative and anticipatory summary of them may serve to separate the clear features from the unclear.

(i) A variational principle for the total energy has led to a single-particle-like equation, in which the potential is determined by the total density which in turn determines the effective potential. Clearly we have a self-consistency loop which is illustrated in Fig. 2.1. The self-consistent

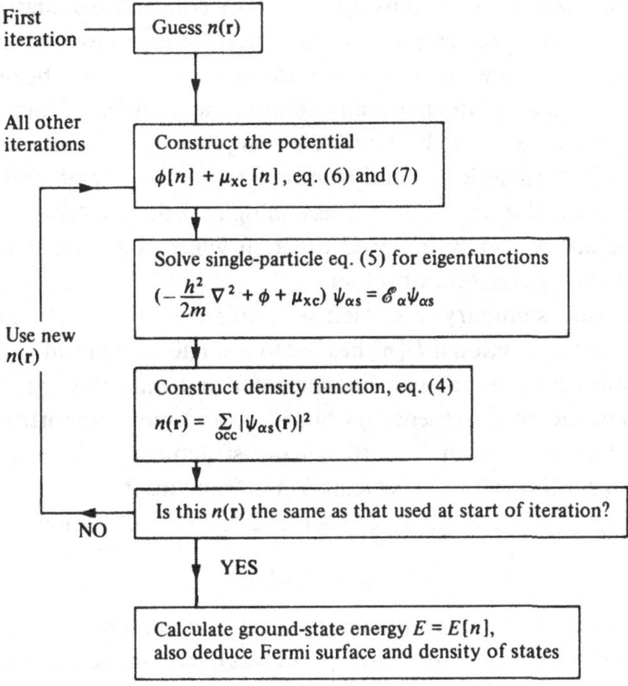

Fig. 2.1. Scheme of self-consistent procedure used in present-day band-structure calculations.

solution of (4) and (5) is equivalent to finding the ground-state energy, i.e. the minimum of $E[n(\mathbf{r})]$.

(ii) The differential equation (5) is of the same form as that solved by conventional band theory. In particular, since the potential has the symmetry of the underlying crystal structure, the solutions of (5) are similar to Bloch functions in that the index α is $(\mathbf{k}, n)$, where $\mathbf{k}$ is a wave-vector in the first Brillouin zone and n is a band index. Thus the surface α_F given by the locus of $(\mathbf{k}, n)$ corresponding to the highest occupied levels is the Fermi surface. The highest occupied levels α_F are determined by noting that they must be consistent with the (fixed) average density n according to

$$n = \frac{1}{V} \sum_{\alpha<\alpha_{F},s} 1 = \frac{1}{V} \int \mathrm{d}\mathbf{r}\, n(\mathbf{r}), \tag{8}$$

where V is the volume of the system. Once α_F is determined by the condition (8), $n(\mathbf{r})$ may be calculated by (4).

(iii) There is one feature of the differential equation (5) that is not easy to describe, namely the eigenvalues $\mathscr{E}_\alpha$. The conventional statement is that they cannot be interpreted as single-particle energies. Certainly the total energy is *not* the sum of them. Furthermore, it has not been shown that $\mathscr{E}_\alpha$ is the energy to create an excitation with quantum numbers α. Nonetheless, it is absolutely standard among band calculators to treat $\mathscr{E}_\alpha$ as the single-particle levels and to calculate the density of states using these $\mathscr{E}_\alpha$. What this amounts to is assuming that the effective potential, $\phi + \mu_{xc}$, is an externally imposed function which does not depend on solving the self-consistency problem.

Finally, this 'summary' can itself be condensed to a brief statement. The variational functional $E[n]$ has led to a single-particle-like differential equation from which can be calculated not only the ground-state energy (i.e. the cohesive energy) but also, and more importantly, the Fermi surface and the single-particle band-structure density of states. In order to proceed with the calculation it remains to specify the ion potential and $E_{xc}[n]$, a task which is faced immediately below.

Electrons and ions

Up to now the terms ions and electrons have been used very loosely. In the original derivation of $E[n]$, the proof was based on the existence of an external potential, i.e. one that did not depend on the electrons. Consequently $v_{ext}(\mathbf{r})$ should be the potential due to the nuclei and $n(\mathbf{r})$ the *total* electron density. From a calculational standpoint this is most impractical; to calculate the band structure of Au, for example, would require the self-consistent calculation of the 68 filled core states. Such an effort, while possible for Au, would be extremely time-consuming and, more to the point, not very informative. The core wave-functions are very insensitive to changes in the potential outside the core. While the core eigenvalues are sensitive to overall shifts in the potential, this effect is only important in the calculation of the total cohesive energy. In view of these considerations it is traditional to include all the core electrons in the construction of the ionic potential $v_{\text{ext}}(\mathbf{r})$ and reserve the electrons in conduction bands for $n(\mathbf{r})$.

In constructing the exchange-correlation energy functional $E_{\text{xc}}[n(\mathbf{r})]$ only the conduction electrons need, at first, to be considered. Later it will be possible to see how some of the results obtained can be used to construct $v_{\text{ext}}(\mathbf{r})$.

Hartree–Fock approximation

Below is a very brief summary of the results of Hartree–Fock (HF) theory adapted to the needs of this chapter. In the next section it will be used to 'derive' $E_{\text{xc}}[n]$ and in later sections it will be used as the 'basis' for Fermi liquid theory. Details of the HF approximation can be found in books by Seitz (1940) and Bethe & Jackiw (1968). In all treatments the Hamiltonian for a system of N fermions which interact by a two-body potential depending only on the distances between them is given by

$$\mathscr{H} = \sum_{i=1}^{N} \frac{\hbar^2 \mathbf{p}_i^2}{2m} + \frac{1}{2} \sum_{i \neq j} V(|\mathbf{r}_i - \mathbf{r}_j|). \tag{9}$$

Here $\mathbf{p}_i$ is the momentum operator for the ith electron. In the case of the electrons, $V(r) = e^2/r$ and one imagines there is a uniform (static) background positive charge density sufficient to achieve overall charge neutrality. (In the case of ^{3}He, $V(r)$ is some potential, such as the 6–12 potential, with an attractive piece at large distance and a repulsive piece at short.) In the HF method a product of single-particle wave-functions satisfying the Pauli principle (a Slater determinant) is used as a trial function ψ_{HF} for minimizing the total energy $E = \langle \psi_{\text{HF}} | \mathscr{H} | \psi_{\text{HF}} \rangle$. To preserve the normalization of the trial wave-functions during the variation requires the following equation to be satisfied

$$\delta(\langle \psi_{\text{HF}} | \mathscr{H} | \psi_{\text{HF}} \rangle - \lambda \langle \psi_{\text{HF}} | \psi_{\text{HF}} \rangle) = 0. \tag{10}$$

The variational calculation results in a set of equations for each of the single-particle wave-functions $\psi_{\alpha s}(\mathbf{r})$ of the form

$$\mathcal{H}_{\text{eff}}\psi_{\alpha s}(\mathbf{r}) = E_{\alpha}\psi_{\alpha s}(\mathbf{r}), \tag{11}$$

where s denotes the spin, and α the other quantum numbers. The effective Hamiltonian $\mathcal{H}_{\text{eff}}$ is really a differential and integral operator given by

$$\mathcal{H}_{\text{eff}}\psi_{\alpha s}(\mathbf{r}) = \left[-\frac{\hbar^2}{2m}\nabla^2 + e^2\int \mathrm{d}\mathbf{r}' \frac{n(\mathbf{r}')}{|\mathbf{r}-\mathbf{r}'|}\right]\psi_{\alpha s}(\mathbf{r})$$
$$-e^2 \sum_{\alpha'<\alpha_{\mathrm{F}}} \int \mathrm{d}\mathbf{r}' \frac{\psi^*_{\alpha' s}(\mathbf{r}')\psi_{\alpha' s}(\mathbf{r})}{|\mathbf{r}-\mathbf{r}'|}\psi_{\alpha s}(\mathbf{r}'). \tag{12}$$

The second term in the square brackets is the so-called Hartree term but is just the Coulomb or electrostatic contribution to the potential due to the electrons (see (6)).

The third term in (12) is the Fock or *exchange* term and may be represented symbolically as $V_{\mathrm{x}}(\mathbf{r})\psi_{\alpha s}(\mathbf{r})$ (the subscript x standing for exchange). This representation is somewhat misleading since $V_{\mathrm{x}}(\mathbf{r})$ is a non-local operator, a very inconvenient fact. Just how inconvenient can be seen by realizing that $\mu_{\mathrm{xc}}([n]; \mathbf{r})$ in (7) must be a non-local operator and hence the beautiful scheme we have been discussing for calculating the ground-state properties of a metal might be rendered intractable by the necessity of solving integral equations for extended systems. Clearly a *local* approximation for $V_{\mathrm{x}}(\mathbf{r})$ is needed.

An argument to construct a reasonable *local* approximation for $V_{\mathrm{x}}(\mathbf{r})$ is as follows: (i) V_{x} is calculated for the free-electron gas; (ii) the dependence of the Fermi wave-vector k_{F} on the density is used as a guess for the dependence of V_{x} on a non-uniform density. In other words the results of many-body calculations for the uniform electron gas are assumed to be valid for a non-uniform case via the connection

$$n(\mathbf{r}) = \frac{k_{\mathrm{F}}^3(\mathbf{r})}{3\pi^2}. \tag{13}$$

In this way E_{xc} is also calculated.

Exchange in a free-electron gas

It turns out that plane waves are a solution of the HF equations. Then the index α is just the wave-vector $\mathbf{p}$ and

$$\psi_{\alpha}(\mathbf{r}) = \psi_{\mathbf{p}}(\mathbf{r}) = \frac{1}{V^{\frac{1}{2}}}\,\mathrm{e}^{\mathrm{i}\mathbf{p}\cdot\mathbf{r}} \tag{14}$$

for a system of volume V. (Note that the system has N electrons so that the average density $n = N/V$. The continuum limit will be taken repeatedly; i.e. N and V going to infinity with n fixed.) The maximum wave-vector for the occupied states is k_F, where $3\pi^2 n = k_F^3$. Then the exchange term becomes

$$V_x\psi_{\mathbf{p}}(\mathbf{r}) = -e^2\frac{1}{V}\sum_{|\mathbf{p}'|<k_F}\int \mathrm{d}r' \frac{\mathrm{e}^{\mathrm{i}(\mathbf{p}'-\mathbf{p})\cdot(\mathbf{r}-\mathbf{r}')}}{|\mathbf{r}-\mathbf{r}'|}\psi_{\mathbf{p}}(\mathbf{r}). \tag{15}$$

The integral has been rewritten so as to show that V_x is local but does depend on wave-vector

$$V_x(\mathbf{p}) = -\frac{1}{V}\sum_{\mathbf{p}'} n^0_{\mathbf{p}'}\frac{4\pi e^2}{|\mathbf{p}-\mathbf{p}'|^2}, \tag{16}$$

where $n^0_{\mathbf{p}'}$ is $\theta(k_F - |\mathbf{p}|)$, the distribution function at zero temperature. In fact the integral in (15) can be done easily to give

$$V_x(p) = -\frac{e^2 k_F}{\pi}\left[1 + \frac{1-(p/k_F)^2}{2p/k_F}\ln\left|\frac{p+k_F}{p-k_F}\right|\right]. \tag{17}$$

Note that $V_x(k_F) = -e^2 k_F/\pi$ which is of the order of -10 eV for metals.

For the purposes of this section the emphasis is on the total energy and the single-particle energies. The only terms of interest are the kinetic and exchange energies. The reason for this is that the electrons are immersed in a uniform background of positive charge cancelling on the average the charge of the electrons. When plane waves are used for $\psi_{\mathbf{p}}$, the electron density $n(\mathbf{x})$ is constant in space. The Hartree term therefore is exactly cancelled by the contribution from the positive background. Accordingly, the single-particle energy can be written

$$E_{\mathbf{p}} = \frac{\hbar^2 p^2}{2m} + V_x(p). \tag{18}$$

On the other hand, since plane waves are exact solutions of the HF equation, if used in calculating $E_{HF} = \langle\psi_{HF}|\mathscr{H}|\psi_{HF}\rangle$ they would give the minimum possible value of the expectation value. To use the language of the treatment of $E[n]$, E_{HF} can be regarded as a functional of the single-particle wave-functions used in ψ_{HF}. The exact solution of the differential equations resulting from $\delta E_{HF}/\delta\psi_\alpha = 0$ (subject to $\langle\psi_{HF}|\psi_{HF}\rangle = 1$) gives the minimum E_{HF}. The result of the calculation gives

$$E_{HF} = \sum_{\mathbf{p}s}\frac{\hbar^2\mathbf{p}^2}{2m}n^0_{\mathbf{p}s} - \tfrac{1}{2}\sum_{\mathbf{p}\mathbf{p}'s} n^0_{\mathbf{p}s}n^0_{\mathbf{p}'s}\frac{4\pi e^2}{|\mathbf{p}-\mathbf{p}'|^2}, \tag{19}$$

where the spin indices have been included for completeness. In this form E_{HF} can be regarded as a functional of $n^0_{\mathbf{p}s}$. Observe that $\delta E_{\mathrm{HF}}/\delta n^0_{\mathbf{p}} = E_{\mathbf{p}}$, a relation that suitably generalized will play a very important role in Fermi liquid theory. In any case it is sufficient for present purposes to note that E_{HF} can be written as the sum of a kinetic energy term and an exchange term E_{x}. Explicit integration of the second term of (19) yields

$$E_{\mathrm{x}} = V\left[-\frac{3}{4}\frac{e^2}{\pi}(3\pi^2 n)^{\frac{1}{3}} n\right]. \tag{20}$$

Comparison of (20) with (3) suggests the obvious definition for the local exchange energy per particle ε_{x}

$$\varepsilon_{\mathrm{x}} = -\frac{3}{4}\frac{e^2}{\pi}(3\pi^2 n)^{\frac{1}{3}}. \tag{21}$$

Clearly, using the scheme (7a), one could calculate the exchange contribution to the chemical potential μ_{x}. From the explicit density dependence of ε_{x} it is trivial to see, via (7b), that

$$\mu_{\mathrm{x}} = \frac{4}{3}\varepsilon_{\mathrm{x}} = -\frac{e^2}{\pi}(3\pi^2 n)^{\frac{1}{3}} = -\frac{e^2 k_{\mathrm{F}}}{\pi}. \tag{22}$$

Exchange-correlation energy for 'real' systems

In summary, the HF equations have been solved exactly for a uniform density electron gas to give a minimum HF energy. From this the density dependence of the exchange energy E_{x} and exchange contribution to the chemical potential μ_{x} have been deduced. What remains to be done? Two things: (i) to include the effect of correlation and (ii) to allow for a non-uniform density.

What is correlation? The only precise definition is that the correlation *energy* is the difference between the exact ground-state energy and the HF energy, $E_{\mathrm{corr}} = E - E_{\mathrm{HF}}$. This is not a very informative definition, given how difficult it is to calculate E_{HF} in practice. Furthermore for almost all other properties the definition, if it can be made, is not so simple. Accordingly practitioners have stopped trying to separate correlation from exchange, and refer to them together; hence the subscript xc. In the case of the chemical potential, Hedin & Lundqvist (1969, 1971) have shown that

$$\mu_{\mathrm{xc}}(n) = \beta(n)\mu_{\mathrm{x}}(n), \tag{23}$$

where $\beta(n)$ is a calculated function of density. Since $\beta(n)$ is of order unity, μ_{x} is a good guide to the size of exchange-correlation effects. As was noted earlier in connection with V_x, μ_x is of order several eV in the range of metallic densities and it is *negative*. Hence exchange-correlation is the primary source of the binding energy in *metals*.

The remaining problem to be disposed of is the fact that in real metals the electronic density is not constant but varies within the Wigner–Seitz cell. Once again the present informal practice is to do the simplest reasonable thing – namely, to rewrite (23) with the local density

$$\mu_{xc}(n(\mathbf{r})) = \beta(n(\mathbf{r}))\mu_x(n(\mathbf{r})). \tag{24}$$

At the time this is being written, no one can quantify how good (or bad) an approximation this is. Using it, or more accurately, using a form of (24) that allows for spin polarization of the electron density, practitioners have calculated the binding energy equilibrium lattice spacing, and bulk modulus of more than 30 metallic elements with encouraging results (Moruzzi, Janak & Williams 1978; see Chapter 6 for some details and a review of the literature).

Finally, there is one point to be tied up: how is the ionic potential $v_{ext}(\mathbf{r})$ computed? It is viewed as a sum of ionic potentials, each of which consists of three terms: an electrostatic piece due to the nucleus, an electrostatic piece due to the frozen core density, and a local exchange-correlation piece calculated also from the frozen core density. Actually this last piece can be done more subtly than this. Since $E[n]$ depends on the total density, one should first compute the sum of the frozen core density and the conduction electron density before computing $\mu_{xc}(n(\mathbf{r}))$. This is quite a sensible procedure, although quite often it is hard to tell from the published band-structure calculation actually what was done.

2.2.2 Summary of many-body effects in band-structure theory or how single-particle-like are the energy bands?

It is worth summarizing the very brief review that has been given of the present understanding of the physical principles underlying 'simple' band-structure calculations.

(i) The standard independent-particle Schrödinger-type equations (5) that are routinely solved can be viewed as equations which hold for the ground state of the fully interacting system at zero temperature. Accordingly, one can calculate the Fermi surface and, with slight hesitation, the density of states.

(ii) The effective local potential $\phi + \mu_{xc}$ in (5) has a simple prescription for its construction, which may, if one guesses badly, require a self-consistent calculation to be performed. But to understand the origin of the prescription requires a full discussion of many-body calculations in the uniform density electron gas. These calculations have been discussed slightly in this section and will be further discussed in the next section, but the full details are beyond the scope of this chapter. There is at present no current review of this field. Furthermore, at the time of writing there is an effort under way to improve the construction of μ_{xc} so as to include better the non-local effects – i.e. variations in the density.

(iii) While one often hears of many-body corrections to band-structure results – indeed such is the aim of this chapter – it should be clear that a great deal of many-body physics has gone into calculating the so-called single-particle energy levels. In fact we shall argue in the course of this chapter that such a good job has been done that the independent-particle picture works extremely well. Most of the corrections we will talk about are relatively small and can be thought of as *multiplicative* corrections to single-particle results. Accordingly in the following sections, the underlying band structure will be treated as free-electron-like because the many-body corrections do turn out to be largely multiplicative. But there is a second reason – namely, that there is almost no work on many-body calculations using realistic energy bands.

Band-structure density of levels and effective masses

There are some notations that simply must be defined in case the reader is not familiar with them. Let $\mathscr{E}_{\mathbf{p}}$ be the single-particle energies resulting from a band-structure calculation; for free electrons $\mathscr{E}^0_{\mathbf{p}} = (\hbar\mathbf{p})^2/2m$. Then any sum over momentum can be turned into an integral over energy by defining a band-structure density of levels per spin, $N_b(\tilde{\mathscr{E}})$, via

$$\frac{1}{V}\sum_{\mathbf{p}} = \int \mathrm{d}\tilde{\mathscr{E}}\, N_b(\tilde{\mathscr{E}}), \tag{25}$$

where $\tilde{\mathscr{E}}$ is an energy measured relative to the Fermi energy μ. In particular

$$N_b(\tilde{\mathscr{E}}) = \frac{1}{2\pi^2}\int_{\mathscr{E}_{\mathbf{p}}} \frac{\mathrm{d}S_{\mathbf{p}}}{4\pi}\frac{1}{|\nabla_{\mathbf{p}}\mathscr{E}_{\mathbf{p}}|}, \tag{26}$$

where the integral is over a surface of constant energy defined by $\mathscr{E}_{\mathbf{p}} = \tilde{\mathscr{E}} + \mu$.

It is possible to define a band-structure effective mass m_b by comparing $N_b(0)$ (i.e. its value at the Fermi surface) with the density of levels for free electrons N_0 according to

$$\frac{m_b}{m} = \frac{N_b(0)}{N_0}. \tag{27}$$

Since

$$N_0 = mk_F/(2\pi^2\hbar^2) \tag{28a}$$

one often sees the formula

$$N_b(0) = \frac{m_b k_F}{2\pi^2\hbar^2}, \tag{28b}$$

but one should keep in mind that k_F is defined by the average density of electrons according to

$$n = \frac{k_F^3}{3\pi^2}$$

and has nothing to do with the shape of the Fermi surface for the real metal. Nonetheless this ratio method suggests a way of defining effective masses to include many-body effects that will be used in the following sections.

2.3 Effective electron–electron interactions in metals

In the next three sections, primary attention will be directed to the low-energy excitations of Fermi systems – metals and liquid ^{3}He – in contrast to Section 2.2 which emphasized ground-state properties. A central concept from that section is the idea of writing the energy of the system as a functional – in the present case, as a functional of the distribution function $n_{\mathbf{p}s}$. Specifically, (19) suggests for low-lying excitations an expression of the form

$$E = \sum_{\mathbf{p}s} \mathscr{E}_{\mathbf{p}} n_{\mathbf{p}s} - \tfrac{1}{2} \sum_{\mathbf{p}\mathbf{p}'s} V_{\mathbf{p},\mathbf{p}'} n_{\mathbf{p}s} n_{\mathbf{p}'s}, \tag{29}$$

where $\mathscr{E}_{\mathbf{p}}$ is a single-particle energy that could be produced from a band-structure calculation, although normally $\mathscr{E}_{\mathbf{p}}$ is taken to have the free-particle form.

Equation (29) represents an intermediate stage in the development of the argument for the Fermi liquid theory. At this point it seems a guess, or at least an approximation. Later it will be asserted that an equation similar to (29), generalized to include spin properly, is *exact.* This assertion has been proven, or nearly so, using many-body theory techniques. In this chapter the assertion must be taken on faith, buttressed by the arguments developed here. The principal differences between (19) and (29) are that (i) the $n_{\mathbf{p}s}$, rather than the equilibrium $n^0_{\mathbf{p}s}$, are used, and (ii) the form of the potential $V_{\mathbf{p},\mathbf{p}'}$ is left open. The use of $n_{\mathbf{p}s}$ means that $E = E[n_{\mathbf{p}s}]$ can be regarded as a functional of possible excitations of the system. About the potential much must be said. In the HF approximation $V_{\mathbf{p},\mathbf{p}'}$ is just the Fourier transform of the Coulomb interaction, and it is a very strong interaction indeed. In fact many-body effects involving the other electrons act to produce an effective potential. This correlation effect known as screening is discussed below.

2.3.1 Screening

To discuss the effective potential between two electrons, consider first the effect of an external negative point charge on a gas of electrons. Clearly the electrons must respond to minimize the energy cost of introducing the test charge. For simplicity, suppose all electrons were expelled from a spherical volume V_0 about the test charge, where V_0 is the volume of uniform background positive charge density that has the magnitude of the test charge. Then outside the volume V_0 the electrons would see no net charge inside V_0 and would be unaffected by the test charge *except* for the fact that their energy would be larger due to their exclusion from the volume V_0. This possible redistribution of the charge is not the most favoured energetically. Instead the electron density increases smoothly with the distance from the test charge, the effective potential becoming increasingly weaker. One says that the external charge is *screened* by the electron gas – i.e. the electron gas adjusts its density so as to make the external charge appear neutral for most of the other electrons. (For a positive test charge the electron gas would adjust *in*ward.) The form of the spatial adjustment is determined both by the condition of chemical equilibrium – i.e. that the chemical potential is a constant, independent of position – and by electrostatics.

Let $U_{\text{ext}}(\mathbf{r})$ be the potential (clearly, Coulombic) generated by the test charge. This $U_{\text{ext}}(\mathbf{r})$ will polarize the electronic density in the vicinity of the test charge, producing a change in the local electronic density $\delta n(\mathbf{r})$ which in turn, via Poisson's equation, generates an induced potential $U_{\text{ind}}(\mathbf{r})$. As a result the effective potential $U_{\text{eff}}(\mathbf{r})$ that the electrons see due to the test charge is

$$U_{\text{eff}}(\mathbf{r}) = U_{\text{ext}}(\mathbf{r}) + U_{\text{ind}}(\mathbf{r}), \tag{30}$$

where

$$4\pi e^2 \delta n(\mathbf{r}) = -\nabla^2 U_{\text{ind}}(\mathbf{r}) = -\nabla^2 (U_{\text{eff}} - U_{\text{ext}}). \tag{31}$$

The more difficult part of the argument is to relate $U_{\text{eff}}(\mathbf{r})$ to $\delta n(\mathbf{r})$ so that the connection between U_{eff} and U_{ext} can be made. The basic idea, stated at the outset, is that the electrochemical potential is constant in equilibrium. Near the test charge where the density is $n + \delta n(\mathbf{r})$, the electrochemical potential, $\mu(n + \delta n(\mathbf{r})) + U_{\text{eff}}(\mathbf{r})$, contains not only the chemical potential μ appropriate to the altered density but also the shift in energy due to the *effective* potential of the test charge. Now far from the test charge these effects are expected to be nil, i.e. U_{eff} and δn should be zero, and the electrochemical potential is just $\mu(n)$. Accordingly, for chemical equilibrium

$$U_{\text{eff}}(\mathbf{r}) + \mu(n + \delta n(\mathbf{r})) = \mu(n). \tag{32}$$

In general this equation will be very difficult to solve. Remember that μ is the sum of the kinetic energy part and the exchange-correlation part, the latter having in general a non-linear (and non-local) dependence on $n(\mathbf{r})$. This is the problem that must usually be solved, but, for simplicity, suppose the test charge is so small that $\delta n \ll n$. Then $\mu(n + \delta n)$ can be expanded in a power series and only the linear term retained. Then it follows that

$$\delta n(\mathbf{r}) = -\left(\frac{\partial n}{\partial \mu}\right) U_{\text{eff}}(\mathbf{r}). \tag{33}$$

The remaining steps of the argument are very short. The Fourier transform of (33) just replaces the variable $\mathbf{r}$ with its transform variable $\mathbf{q}$, while (31) becomes

$$4\pi e^2 \delta n(\mathbf{q}) = q^2 [U_{\text{eff}}(\mathbf{q}) - U_{\text{ext}}(\mathbf{q})]. \tag{34}$$

Accordingly one finds, combining (33) and (34), that

$$U_{\text{eff}}(\mathbf{q}) = \frac{U_{\text{ext}}(\mathbf{q})}{\varepsilon(q)}, \tag{35a}$$

where

$$\varepsilon(q) = 1 + k_s^2/q^2 \tag{35b}$$

and the screening wave-vector

$$k_s^2 = 4\pi e^2 \frac{\partial n}{\partial \mu}. \tag{35c}$$

One says that the external potential is screened by the dielectric function $\varepsilon(q)$.

For an external Coulomb potential $1/r$, whose Fourier transform $4\pi/q^2$ diverges at small q, this divergence is removed by $\varepsilon(q)$ with the result that $U_{\text{eff}}(r) = (e^2/r)\exp(-k_s r)$ is exponentially screened at large distances. In general, the inverse screening length k_s is proportional to the compressibility κ of the electron gas, since $\partial n/\partial\mu = n^2\kappa$ (see Section 2.5). For the free-electron gas one finds $(k_s/2k_F)^2 = 1/\pi k_F a_0$, where a_0 is the Bohr radius. Thus in most metals k_s is comparable to k_F. As a result both $U_{\text{eff}}(\mathbf{r})$ and $\delta n(\mathbf{r})$ differ from zero only in the vicinity – i.e. a few Bohr radii – of the external test charge.

In summary then, the condition of chemical equilibrium leads to the result that any external potential is screened by a dielectric function $\varepsilon(q)$, the long wavelength limit of which, given by (35), is exact even when all many-body corrections are included.

However, the aim of this discussion is to find the effective potential to use in (29). In this case the electrons are interacting not with an external charge, as so far considered, but with other electrons. This fact introduces additional difficulties. In fact, while the long-wavelength limit for the electron–external-charge interaction is known exactly, no such information is available for the electron–electron interaction. However, a reasonable approximation, called the screened Coulomb potential, which will suffice for present discussions, is that

$$V_{\mathbf{p},\mathbf{p}'} = \frac{4\pi e^2}{|\mathbf{p}-\mathbf{p}'|^2\varepsilon(\mathbf{p}-\mathbf{p}')} = \frac{4\pi e^2}{|\mathbf{p}-\mathbf{p}'|^2 + k_s^2} \equiv V_{\mathbf{p}-\mathbf{p}'}, \tag{36}$$

which can be used in the functional energy expression (29),

$$E[n_{\mathbf{p}s}] = \sum_{\mathbf{p}s} \mathscr{E}_{\mathbf{p}} n_{\mathbf{p}s} - \tfrac{1}{2} \sum_{\mathbf{p}\mathbf{p}'s} V_{\mathbf{p}-\mathbf{p}'} n_{\mathbf{p}s} n_{\mathbf{p}'s}. \tag{29'}$$

What then are the consequences of this approximation for the energy functional?

2.3.2 Many-body corrections to specific heat and susceptibility

The discussion at the beginning of this section argued that the energy expression (29) or (29′) should be viewed as a functional of the distribution function for the excitations of the electron gas. This concept can be used to derive the single-particle-like energies including many-body corrections. To do so, suppose that $n^0_{\mathbf{p}}$ is slightly perturbed to $n^0_{\mathbf{p}}+\delta n_{\mathbf{p}}$. Clearly an energy $E_{\mathbf{p}}$ can be identified with this change in the distribution function, according to

$$E[n^0+\delta n]=E[n^0]+\sum_{\mathbf{p}} E_{\mathbf{p}}\delta n_{\mathbf{p}} \tag{37}$$

or more explicitly

$$E_{\mathbf{p}}=\frac{\delta E}{\delta n_{\mathbf{p}}} \tag{38a}$$

$$=\mathscr{E}_{\mathbf{p}}-\sum_{\mathbf{p}'} V_{\mathbf{p}-\mathbf{p}'}n^0_{\mathbf{p}'}, \tag{38b}$$

where (29′) has been used to derive (38b). Although (38b) has the form of an HF energy (c.f. (18)), it is supposed that $V_{\mathbf{p}-\mathbf{p}'}$ includes all correlation effects, even if the construction of it offered here (36) does not. Accordingly (38b) can be used to calculate the change in specific heat, susceptibility, and compressibility due to electron–electron interactions.

Specific heat and the effective mass

The electronic specific heat of a metal at low temperature is given by $\gamma_{\mathrm{e}}T$, where the coefficient γ_{e} is proportional to the density of energy levels per spin at the Fermi surface $N(0)$ according to $\gamma_{\mathrm{e}}(\pi^2/3)k_{\mathrm{B}}^2 2N(0)$. $N(0)$ is proportional to $|\nabla_{\mathbf{p}}E_{\mathbf{p}}|^{-1}$ evaluated at the Fermi surface. Accordingly an effective mass m_{e}^* due to the Coulomb (or electron–electron) interaction can be defined as

$$\frac{m}{m_{\mathrm{e}}^*}=\frac{m}{\hbar^2k_{\mathrm{F}}^2}\mathbf{p}\cdot\nabla_{\mathbf{p}}E_{\mathbf{p}}\bigg|_{|\mathbf{p}|=k_{\mathrm{F}}}. \tag{39}$$

The subsequent calculation is greatly aided, first by the reversible substitution $\mathbf{p}'=\mathbf{p}+\mathbf{q}$ which allows the writing of $E_{\mathbf{p}}=\mathscr{E}_{\mathbf{p}}-\sum_{\mathbf{q}} V_{\mathbf{q}}n_{\mathbf{p}+\mathbf{q}}$, and second by the identity $\nabla_{\mathbf{p}}n^0_{\mathbf{p}+\mathbf{q}}=-\delta(k_{\mathrm{F}}-|\mathbf{p}'|)\hat{\mathbf{p}}'$, where the unit vector $\hat{\mathbf{p}}=\mathbf{p}/|\mathbf{p}|$. Thus

$$\frac{m}{m_{\mathrm{e}}^*}=1+\frac{m}{\hbar^2k_{\mathrm{F}}}\sum_{\mathbf{p}'}\hat{\mathbf{p}}\cdot\hat{\mathbf{p}}'\delta(k_{\mathrm{F}}-|\mathbf{p}'|)V_{\mathbf{p}-\mathbf{p}'}\bigg|_{|\mathbf{p}|=k_{\mathrm{F}}}.$$

Note the effective mass is to be calculated at the Fermi surface, so $\mathbf{p}=k_{\mathrm{F}}\hat{\mathbf{p}}$. The sum is transformed to an integral in the usual manner giving

$$\frac{m}{m_{\mathrm{e}}^*}=1+\frac{mk_{\mathrm{F}}}{2\pi^2\hbar^2}\int\frac{\mathrm{d}\Omega}{4\pi}\cos\theta V(\theta).$$

The angle between the unit vectors $\hat{\mathbf{p}}$ and $\hat{\mathbf{p}}'$ is θ. The factor in front of the integral is just the level density at the Fermi surface N_0, for the non-interacting system. So

$$\frac{m}{m_e^*} - 1 = N_0 \int \frac{d\Omega}{4\pi} \cos\theta V(\theta) = N_0 V_1. \tag{40}$$

The quantity V_1 is the $l = 1$ or 'p-wave' projection of the potential.

For the case of the screened Coulomb potential

$$V(\theta) = \frac{4\pi e^2}{k_s^2 + 2k_F^2(1 - \cos\theta)}. \tag{41}$$

This potential is always positive and increases more strongly with $\cos\theta$ in the range $-1 < \cos\theta < 1$ than $\cos\theta$ itself. V_1 is therefore always a positive quantity and the effective mass m_e^* is smaller than m. A widely used parameter describing the density of the electron gas is the effective interelectronic distance r_s (measured in units of a_0) defined by

$$\frac{1}{\frac{4\pi}{3}(r_s a_0)^3} = n = \frac{k_F^3}{3\pi^2}. \tag{42}$$

In most metals r_s varies between 2 and 5.5. When a screened Coulomb potential is used for $V(\theta)$, the ratio m_e^*/m turns out to be rather insensitive to the value of r_s, for r_s between 2 and 5 (see Table 2.1). The Coulomb interaction is seen to have a negligible influence on the effective mass. This holds true even when more sophisticated many-body calculations are done.

Susceptibility

The specific heat can be thought of as characterizing the response of the electrons to a 'statistical perturbation', the raising of the temperature, which changes $n_\mathbf{p}$. Similarly the susceptibility is the response of the electron gas to another probe or perturbation, the application of an external magnetic field.

In a magnetic field $\mathbf{H}$ (parallel to the z-direction) an electron experiences through its magnetic moment $\boldsymbol{\mu}$ an additional interaction $-\boldsymbol{\mu}\cdot\mathbf{H} = g\mu_B s_z H$. Here g is the gyromagnetic ratio ($g = 2.0023$), μ_B the Bohr magneton $|e|\hbar/(2mc)$ and s_z the component of the spin parallel to the magnetic field ($s_z = \pm\frac{1}{2}$). One might imagine that the field raises the energy of the up-spin electrons and lowers that of the down-spins in the manner shown on Fig. 2.2. The situation (a) is obviously unstable and

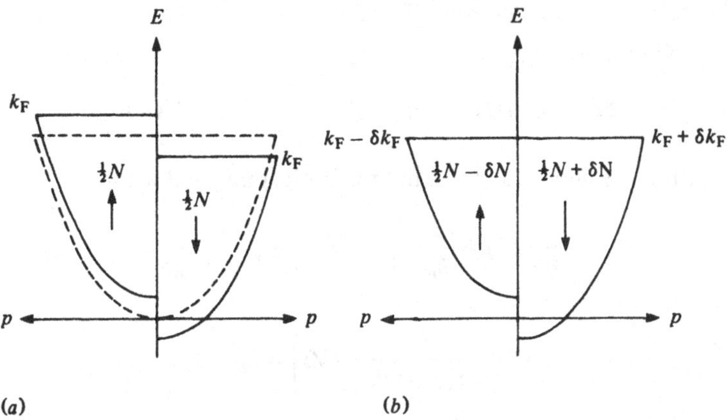

Fig. 2.2. (*a*) Effect of magnetic field on single-particle energies prior to imposing the constancy of the electrochemical potential as in (*b*).

the Fermi surfaces of up- and down-spin electrons adjust as in (*b*), so that the chemical potential of the up- and down-spins are equal. The resulting magnetization can be calculated from the change δN in the number of down-spin electrons (equal in magnitude to the change of spin-up electrons) by

$$\begin{aligned} M &= -g\mu_B(-\tfrac{1}{2})(2\delta N) \\ &= g\mu_B\delta N. \end{aligned}$$

The energy change of a spin-up electron at the Fermi surface – i.e. the change in the up-spin chemical potential – is

$$\delta E_{k_F-\delta k_F,\uparrow}(H) = g\mu_B\tfrac{1}{2}H - \frac{\partial E_p}{\partial p}\delta k_F + \delta N\langle V\rangle. \tag{43a}$$

The first term is the energy change due to the interaction between the magnetic moment of the electron and the external field. The second is the contribution due to the explicit wave-vector dependence of E_p, the change in wave-vector from k_F to $k_F - \delta k_F$ being brought about by the contraction of the Fermi sphere belonging to spin-up electrons. The last term is the change in the interaction term in (29′) due to the decrease δN in the number of spin-up electrons. The angular average $\langle V\rangle = \int (1/4\pi)\, d\Omega\, V(\theta)$ is the $l = 0$ projection or 's-wave' average of the potential at the Fermi surface. The average is evaluated at the undistorted Fermi surface, where $k = k_F$, since $\delta N \ll N$. This last term arises from the explicit dependence of the single-particle energy on the total number of (in this case) spin-up electrons. The corresponding change of down-spin electron energy is

$$\delta E_{k_F+\delta k_F,\uparrow}(H) = -g\mu_B\tfrac{1}{2}H + \frac{\partial E_p}{\partial p}\delta k_F - \delta N\langle V\rangle. \tag{43b}$$

Equating the chemical potential for both spins, and using

$$\frac{\partial E_p}{\partial p}\delta k_F = \frac{\partial E_p}{\partial p}\frac{\delta k_F}{\delta N}\delta N = \frac{\delta N}{(m_e^*/m)N_0}$$

yields

$$2\delta N\left[\frac{1}{(m_e^*/m)N_0} - \langle V\rangle\right] = g\mu_B H.$$

Hence the susceptibility is given by

$$\chi \equiv \frac{M}{H} = \frac{g\mu_B \delta N}{H} = \chi_0 \frac{m_e^*/m}{1-(m_e^*/m)N_0\langle V\rangle}, \tag{44}$$

where the Pauli susceptibility $\chi_0 = (g\mu_B)^2 N_0/2$.

Since in the simplest approximation the potential term comes from the exchange term in the energy, the susceptibility (44) is often referred to as *exchange-enhanced*, although of course $\langle V\rangle$ contains correlations effects. That the susceptibility is enhanced can be seen from the fact that V is positive in (41) (and also in more sophisticated calculations) and the 's-wave' average of V appearing in the denominator of (44) gives rise to an enhancement. A common measure of the enhancement is the so-called Stoner factor, which unfortunately does not have a unique definition but depends on the model being studied. In this case one might choose $S = [1-(m_e^*/m)N_0\langle V\rangle]^{-1}$.

In Table 2.1 are results of numerical calculations of m_e^*/m and χ/χ_0 done (i) with the simple screened Coulomb potential (41), called TF in

TABLE 2.1. *Calculation of m_e^*/m and χ/χ_0 for the electron gas*

r_s		2	3	4	5
$\frac{m_e^*}{m}$	TF	0.95	0.95	0.95	0.96
	Rice (1965)	0.99	1.02	1.06	
$\frac{\chi}{\chi_0}$	TF	1.22	1.28	1.34	1.39
	Rice (1965)	1.26	1.40	1.48	

TF results were calculated using a Thomas–Fermi screened Coulomb interaction (cf. eq. (41), in eqs. (40) and (44)).

Table 2.1, and (ii) by Rice (1965) whose results are characteristic of the more sophisticated many-body treatments. Note that $r_s = 2$ is typical for transition metals and $r_s = 5$ is appropriate for the alkali metal sodium. There are two points to be made about the results. (i) m_e^*/m is nearly unity to within a few per cent. There are no reliable calculations as yet to suggest that including band-structure energies and matrix elements would cause m_e^*/m_b to differ much from unity. Accordingly it is conventional to take $m_e^*/m_b = 1$. (In passing it should be noted that a more realistic calculation such as Rice's gives m_e^*/m, becoming larger than unity as r_s increases. Ashcroft & Wilkins (1965) have shown that the dynamic aspects of screening, omitted from the discussion in this chapter, can explain this effect.) (ii) The enhancement of the susceptibility is more substantial. Furthermore if the level density in (44) were replaced by a large band-structure value the enhancement could be substantial.

How many parameters? and how big?

Two detailed calculations have been presented for the effect of electron–electron interactions on the effective mass (or level density) and on the susceptibility. The results for both properties could be written in terms of two angular averages of an effective potential. Is this true of other properties and how big are these individual angular averages?

There is one important property that has not been calculated – namely the compressibility. With the approximate energy functional (29′) its value, or rather the deviation of that value from the free-electron result, can be deduced without computation for this model. The compressibility κ can be written in the following forms, starting from its definition

$$\frac{1}{\kappa} = -V\left(\frac{\partial P}{\partial V}\right) = V\left(\frac{\partial^2 E}{\partial V^2}\right) = n^2\frac{\partial^2(E/V)}{\partial n^2}. \qquad (45)$$

On the other hand, the susceptibility can be rewritten in a very similar form, using a standard thermodynamic transformation, from $\chi = \partial M/\partial H = -\partial^2(E/V)/\partial H^2$ to

$$\frac{1}{\chi} = \frac{\partial^2(E/V)}{\partial(M/V)^2}. \qquad (46)$$

Now the only term containing the interaction in (29′) is $-\frac{1}{2}\sum_{\mathbf{pp}'s} V_{\mathbf{p}-\mathbf{p}'} n_{\mathbf{p}s} n_{\mathbf{p}'s}$, the spin sum of which can be rewritten

$$\sum_s n_{\mathbf{p}s} n_{\mathbf{p}'s} = \tfrac{1}{2}(n_{\mathbf{p}\uparrow} + n_{\mathbf{p}\downarrow})(n_{\mathbf{p}'\uparrow} + n_{\mathbf{p}'\downarrow}) + \tfrac{1}{2}(n_{\mathbf{p}\uparrow} - n_{\mathbf{p}\downarrow})(n_{\mathbf{p}'\uparrow} - n_{\mathbf{p}'\downarrow}).$$

Hence the density and magnetization enter the interaction term identically. Thus in the approximation of $E[n]$, the changes in the susceptibility and compressibility, according to (45) and (46), must be the same – i.e. $\chi/\chi_0 = \kappa/\kappa_0$. Clearly there is no reason for the many-body corrections to the compressibility and susceptibility to be the same. Thus there is an inadequacy in the model (29′). More precisely, the potential $V_{\mathbf{p}-\mathbf{p}'}$ must have some dependence on spin – a point which will be covered in Section 2.5.

There is another feature of the results that is quite surprising: that the many-body corrections are so small – a few per cent for the mass and some tens of per cent for the susceptibility. How can that be when the Coulomb interaction is so strong? The obvious answer is that the effective interaction is a *screened* Coulomb interaction between electrons which is of relatively short range. Another way of understanding this is to focus on the electron. Its Coulomb potential tends to exclude charge from its vicinity. The remaining positive background screens out the electron's Coulomb potential. Hence the electron and its associated screening cloud form an entity, a *quasi-particle*, which is neutral, obeys Fermi statistics, and interacts weakly with other quasi-particles. The strongly interacting electron gas has been replaced by a set of nearly independent single-particle-like excitations, quasi-particles. This idea of quasi-particles will play a central role in Fermi liquid theory.

Some realistic values

The range of possible many-body corrections due to electron–electron interactions has just been presented. Are they sufficient to explain what is observed? The examples of Pd and Pt given below will show that the answer is resoundingly no. An essential ingredient in making the comparisons is the value of the band-structure level density at the Fermi surface, which Andersen (1970) has calculated for these metals. The expected specific heat coefficient γ_b and susceptibility χ_b due to band-structure effects are compared with the measured values to yield the effective mass m^* and Stoner enhancement factor S according to

$$\frac{\gamma_{\exp}}{\gamma_b} = \frac{m^*}{m_b}, \qquad \frac{\chi_{\exp}}{\chi_b} = S.$$

(For reference, the conversion formulae are $\gamma_b(\mathrm{mJ/mol\,K^2}) = 0.347 N_b$ (states/Ry atom spin) and $\chi_b(\mathrm{cc/mol}) = (4.75\times10^{-6})N_b$ (states/Ry atom spin).)

TABLE 2.2. *Many-body parameters for Pd and Pt*

	$N_{bs}\left(\frac{\text{states}}{\text{Ry atom spin}}\right)$	$\frac{m^*}{m_b}$	$S = \chi_{exp}/\chi_b$	$\langle V \rangle$ (eV)
Pd	16.3	1.66	9.4	0.75
Pt	11.6	1.63	3.8	0.87

See Andersen (1970) and Foner, Dodo & McNiff (1968).

The numbers given in Table 2.2 for Pd and Pt are extraordinary. The mass enhancement is more than 60 per cent over the band-structure value – a result quite beyond anything the present treatment of the electron–electron interaction can explain. On the other hand, the Stoner factor, defined here as $S = (1 - N_b\langle V \rangle)^{-1}$, while quite large, is at least believable. In Table 2.2 the value of $\langle V \rangle$ is seen to be about 1 eV, a number which is consistent with some spin-polarized many-body calculations (Gunnarsson 1976; Janak 1977). (There is some difficulty in comparing the calculated χ with the measured value which includes two diamagnetic contributions: one due to the ion core and one due to the conduction electrons (Landau diamagnetism). It is possible to subtract the ion core contribution. For metals with large N_b, the Landau contribution can probably be neglected since it is expected that $|\chi(\text{Landau})/\chi(\text{Pauli})| \sim (m/m_b)^2$.)

The specific heat presents more significant problems. A simple screened Coulomb interaction can never explain an m_e^*/m_b as large as 1.6 or 1.7. In the next section, the interaction between electrons and phonons will be shown to give a large enhancement of the effective mass. Unfortunately this enhancement is probably not sufficient to explain Pd and Pt, a point that will be briefly discussed in the next two sections.

2.4 Electron–phonon interaction

In this section the effect of the electron–phonon interaction on the specific heat and the Pauli susceptibility of simple metals will be discussed. The effect will be illustrated by a perturbation calculation to second order in the interaction. The results of a more appropriate calculation (using many-body techniques) will be inferred from the perturbation result. A distinction between the density of energy levels and the density of states will be drawn. Finally there will be a series of applications to real systems, including superconductivity and lifetime effects.

At the outset, a review of the electron–phonon interaction is appropriate in order to see what qualitative features are expected. As an ion moves, the bulk of the electrons adiabatically follows its motion because there are no available states to which they could scatter. This Pauli-principle-induced dynamics tends to maintain local charge neutrality and results in a screening of the long-range electron–phonon interaction seen by the few electrons which can scatter, i.e. those lying within a phonon frequency of the Fermi surface. Accordingly the effective electron–phonon interaction will only affect electron energies within a phonon energy of the Fermi energy, a result which is true even when the relative numbers of up- and down-spin electrons are altered by a magnetic field. Accordingly, the effective mass (and hence the specific heat) is affected by the electron–phonon interaction while the susceptibility is not.

For the purpose of the subsequent calculation the electron–phonon interaction or scattering process must be characterized. The electron–phonon matrix $g_{\mathbf{k}'-\mathbf{k}}$ corresponds to a process in which an electron with wave-vector $\mathbf{k}'$ and energy $\mathscr{E}_{\mathbf{k}'}$ scatters to an electron state characterized by ($\mathbf{k}$, $\mathscr{E}_{\mathbf{k}}$) with the emission of a phonon of wave-vector $\mathbf{k}'-\mathbf{k}$ and energy $\omega_{\mathbf{k}'-\mathbf{k}}$. The same matrix element also holds for the process: electron $(\mathbf{k}', \mathscr{E}_{\mathbf{k}'})$ + phonon $(\mathbf{k}-\mathbf{k}', \omega_{\mathbf{k}-\mathbf{k}'}) \rightarrow$ electron $(\mathbf{k}, \mathscr{E}_{\mathbf{k}})$. On the other hand the process: electron $(\mathbf{k}, \mathscr{E}_{\mathbf{k}})$ + phonon $(\mathbf{k}'-\mathbf{k}, \omega_{\mathbf{k}'-\mathbf{k}}) \rightarrow$ electron $(\mathbf{k}', \mathscr{E}_{\mathbf{k}'})$ is described by the matrix element $g_{\mathbf{k}-\mathbf{k}'}$, which is equal to $(g_{\mathbf{k}'-\mathbf{k}})^*$.

2.4.1 Many-body calculation of excitation energy and effective mass

The aim of the calculation below – which will be as close to a real many-body calculation as this chapter will approach – is to deduce the effective mass of the electron due to the electron–phonon interaction. In order to find the momentum derivative of the quasi-particle energy (to use the language of the last section) a definition of the excitation energy for a single-particle-like excitation is required. The following standard prescription is employed: let E_0^N be the ground-state energy of an N-particle system and $E_{\mathbf{p}}^{N+1}$ be the energy of an excited state of the $(N+1)$-particle system with momentum $\hbar\mathbf{p}$. (Among many such excited states the one envisaged here is that which in the absence of interactions would have resulted from simply adding an electron with energy $\mathscr{E}_{\mathbf{p}}$ to the ground state.) Then the excitation energy is defined by

$$E_{\mathbf{p}} = E_{\mathbf{p}}^{N+1} - E_0^N. \tag{47}$$

The calculation of E_0^N to second order in the electron–phonon interaction begins with the ground state, which is a Slater determinant of plane-wave states with all wave-vector states filled up to $|\mathbf{k}| = k_F$. There is no contribution to first order in the interaction since the diagonal matrix elements of $g_{\mathbf{k}'-\mathbf{k}}$ are zero ($|g_\mathbf{q}|^2$ goes to zero as $\mathbf{q}$ does). To second order in the electron–phonon matrix element the ground-state energy of an N-particle system is

$$E_0^N = \sum_{\mathbf{k}s} \mathscr{E}_\mathbf{k} n_{\mathbf{k}s} + \sum_{\mathbf{k}\mathbf{k}'s} \frac{|g_{\mathbf{k}'-\mathbf{k}}|^2 n_{\mathbf{k}'s}(1 - n_{\mathbf{k}s})}{\mathscr{E}_{\mathbf{k}'} - (\mathscr{E}_\mathbf{k} + \omega_{\mathbf{k}'-\mathbf{k}})}. \tag{48}$$

The first term on the right-hand side of (48) is, of course, just the sum of the single-particle unperturbed kinetic energies in the Fermi sea. In the numerator of the second term of the equation the factor $n_{\mathbf{k}'s}$ is present because the single-particle states comprising the unperturbed states are filled to $|\mathbf{k}'| = k_F$, while the factor $(1 - n_{\mathbf{k}s})$ is present because (virtual) transitions are possible only to unfilled states – i.e. those above the Fermi surface. The denominator is, of course, the difference between the unperturbed state and the intermediate state corresponding to the virtual transition in which an electron ($\mathbf{k}'$, $\mathscr{E}_{\mathbf{k}'}$) is promoted above the Fermi surface to the state ($\mathbf{k}$, $\mathscr{E}_\mathbf{k}$) with the spontaneous emission (at zero temperature) of a phonon ($\mathbf{k}' - \mathbf{k}$, $\omega_{\mathbf{k}'-\mathbf{k}}$).

From this discussion one can see how to calculate the energy of an excited state of total momentum $\hbar\mathbf{p}$ of the $(N+1)$-particle system. In the first term of (48) there will be an additional term $\mathscr{E}_\mathbf{p}$ corresponding to the fact that the unperturbed state now includes the single-particle state $\hbar\mathbf{p}$ above the Fermi surface. The second term of (48) is also changed. The sum over occupied states ($\mathbf{k}'$-sum) must be extended to include $\mathbf{p}$ while the sum over empty states ($\mathbf{k}$-sum) to which virtual transitions occur must exclude $\mathbf{p}$. By the prescription above, $E_\mathbf{p}$ is given by

$$E_\mathbf{p} = \mathscr{E}_\mathbf{p} + \sum_\mathbf{k} \frac{|g_{\mathbf{p}-\mathbf{k}}|^2(1 - n_\mathbf{k})}{\mathscr{E}_\mathbf{p} - \mathscr{E}_\mathbf{k} - \omega_{\mathbf{p}-\mathbf{k}}} - \sum_{\mathbf{k}'} \frac{|g_{\mathbf{k}'-\mathbf{p}}|^2 n_{\mathbf{k}'}}{\mathscr{E}_{\mathbf{k}'} - \mathscr{E}_\mathbf{p} - \omega_{\mathbf{k}'-\mathbf{p}}}. \tag{49}$$

(Note: $n_\mathbf{k}$ and $n_{\mathbf{k}'}$ have the same spin as $E_\mathbf{p}$.) Note that the last term in $E_\mathbf{p}$ (which is a characteristic many-body term) arises because there is a term in E_0^N which is *not* in $E_\mathbf{p}^{N+1}$ since the state $\mathbf{p}$ is filled in the unperturbed $(N+1)$-particle state.

The effective mass can be calculated by taking the derivative of (49) with respect to the wave-vector $\mathbf{p}$. However, $E_\mathbf{p}$ is not so much a function of $\mathbf{p}$ as it is of $\mathscr{E}_\mathbf{p}$ or more properly of $\tilde{\mathscr{E}}_\mathbf{p} = \mathscr{E}_\mathbf{p} - \mu$. The matrix element $g_{\mathbf{p}-\mathbf{k}}$ and the phonon frequency are relatively insensitive to the value of $|\mathbf{p}|$

when it is nearly k_F. In the last term of (49), $\mathscr{E}_{\mathbf{k}'}$ is always less than the Fermi energy while $\mathscr{E}_{\mathbf{p}}$ is greater. The denominator of this term is clearly largest for $|\mathscr{E}_{\mathbf{k}'}-\mu|$ and $\mathscr{E}_{\mathbf{p}}-\mu$ of the order of a phonon frequency. The same is true for the first term of (49). So the reasonable thing to do is calculate $E_{\mathbf{p}}$ or $E_{\mathbf{p}}-\mu$ as a function of $\tilde{\mathscr{E}}_{\mathbf{p}}=\mathscr{E}_{\mathbf{p}}-\mu$. For a constant level density in the vicinity of the Fermi surface (this is the assumption of particle–hole symmetry), there is no correction to a single-particle energy for $|\mathbf{p}|=k_F$. Thus the chemical potential is not shifted by the electron–phonon interaction under the assumption of particle–hole symmetry. This is the first evidence of the earlier suggestion that the effects of the electron–phonon interaction are tied to the Fermi surface.

For small $\tilde{\mathscr{E}}_{\mathbf{p}}$, $E_{\mathbf{p}}-\mu$ can be expanded in powers of it. Specifically, as $\tilde{\mathscr{E}}_{\mathbf{p}}$ goes to zero,

$$E_{\mathbf{p}}-\mu=\tilde{\mathscr{E}}_{\mathbf{p}}(1-\lambda). \tag{50}$$

Here the important quantity λ is given by

$$\lambda=N_0\int\frac{\mathrm{d}\Omega_{\mathbf{p}-\mathbf{k}}}{4\pi}\frac{2|g_{\mathbf{p}-\mathbf{k}}|^2}{\omega_{\mathbf{p}-\mathbf{k}}}=N_0\langle 2|g|^2/\omega\rangle, \tag{51}$$

where $|\mathbf{p}|=|\mathbf{k}|=k_F$ (and the angular brackets denote an angular average as in (43a)).

Effective mass

Is this result (50) of second-order perturbation theory reliable? If λ were small compared to unity, then the effective mass due to the electron–phonon interaction m^*_{ep} defined by

$$\frac{m^*_{\mathrm{ep}}}{m}=\frac{\tilde{\mathscr{E}}_{\mathbf{p}}}{E_{\mathbf{p}}-\mu}\bigg|_{|\mathbf{p}|=k_F} \tag{52}$$

could be evaluated. (If (49) were adequate, then m^*_{ep}/m would equal $(1-\lambda)^{-1}$.) The quantity $2g^2/\omega$ can be shown to be equal to the screened Coulomb interaction in a very simple model. Hence λ is at least of the order of $\langle V\rangle$ in the correction to the Pauli susceptibility due to the electron–electron interaction. Generally the range of λ is from a few tenths to more than unity and is clearly not a suitable parameter in which to do perturbation theory. It would appear that since the lowest-order correction is so large it will be necessary to go to quite high order in g^2 to find a believable value of the effective mass m^*_{ep} due to the electron–phonon interaction. There is, however, a form of perturbation theory which (partially) circumvents this objection. In Brillouin–Wigner perturbation theory (Brueckner 1959) one inserts the desired energy ($E_{\mathbf{p}}$ in this case) in the energy denominators instead of the unperturbed energy (e.g. $\mathscr{E}_{\mathbf{p}}$). This procedure clearly sums terms of infinitely high order in g^2, and, if one can solve the resulting integral equation for $E_{\mathbf{p}}$, it may produce results not obtainable by ordinary perturbation theory. In any case, if in (49) $E_{\mathbf{p}}$ is substituted for $\mathscr{E}_{\mathbf{p}}$ in the energy denominators, then since the Fermi energy is not shifted by the electron–phonon interaction an expansion in $E_{\mathbf{p}}-\mu$ is appropriate. For small energies this results in

$$E_{\mathbf{p}}-\mu=\mathscr{E}_{\mathbf{p}}-\mu-\lambda(E_{\mathbf{p}}-\mu). \tag{53}$$

Hence $E_{\mathbf{p}}-\mu=\tilde{\mathscr{E}}_{\mathbf{p}}/(1+\lambda)$ or, via (52),

$$\frac{m^*_{\text{ep}}}{m}=1+\lambda. \tag{54}$$

This result, that the effective mass is always *increased* by the electron–phonon interaction, is quite general.

Characteristic values of λ for a few metals are listed below in Table 2.3. The first row lists calculated values and the second row values deduced from superconducting tunnelling data. Note that there is a wide range in the value of λ. Nonetheless the result for m^*_{ep} (54) has been shown to be correct (Migdal 1958a) up to a term of order $(m/M)^{\frac{1}{2}}$, where M is the mass of the ion. Hence this many-body result is correct to within a few per cent.

The effect of the electron–phonon interaction is concentrated near the Fermi surface. For $E_{\mathbf{p}}-\mu$ much larger than a characteristic phonon energy the effects decrease as $(E_{\mathbf{p}}-\mu)^{-1}$. This can be seen by explicit integration of (49) in which particle–hole symmetry is assumed, the energy dependence of the matrix element and phonon energy are neglected, and $\tilde{E}=E_{\mathbf{p}}-\mu$ is inserted for $\tilde{\mathscr{E}}_{\mathbf{p}}$ in order to reproduce the results of

TABLE 2.3. *Values of the effective electron–phonon coupling constant λ, eq. (51)*

λ (Grimvall 1976)	Pb	Sn	In
Calculated from band structure	1.47	0.89	0.94
Deduced from superconducting tunnelling	1.55	0.72	0.80

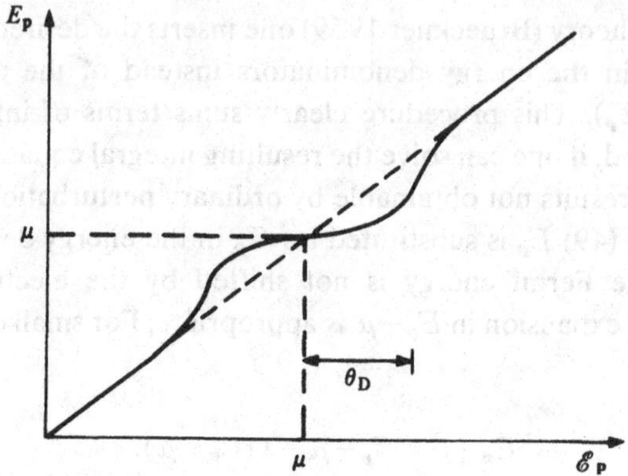

Fig. 2.3. Effect of electron–phonon interaction on the quasi-particle energy $E_{\mathbf{p}}$ as a function of energy $\mathscr{E}_{\mathbf{p}}$ of non-interacting particle.

the more exact calculations (Migdal 1958a). Then the energy integral gives

$$\tilde{E} = \tilde{\mathscr{E}}_{\mathbf{p}} + N_0 \int \frac{d\Omega}{4\pi} g^2(\theta) \ln \left| \frac{\tilde{E} - \omega(\theta)}{\tilde{E} + \omega(\theta)} \right|,$$

where θ is the angle between **p** and **k**. For small $\tilde{E}$ ($\ll$ maximum of $\omega(\theta)$), $\tilde{E} = \tilde{\mathscr{E}}_{\mathbf{p}}/(1+\lambda)$ as in (53). On the other hand, for large $\tilde{E}$ ($\gg$ maximum of $\omega(\theta)$), one finds that $\tilde{E} = \tilde{\mathscr{E}}_{\mathbf{p}} - \omega_0^2/(10\tilde{E})$, where ω_0 is an energy comparable to the Debye energy. This agrees with the anticipated result that the electron–phonon interaction would have relatively little effect on energy levels far from the Fermi surface with respect to the Debye energy. In Fig. 2.3 a schematic plot of $E_{\mathbf{p}}$ versus $\mathscr{E}_{\mathbf{p}}$ indicates the spacing of the energy levels. Since the spacing is compacted near the Fermi level, the low temperature electronic specific heat, which measures the density of energy levels, is enhanced above the free electron value. Other properties which are sensitive to the level density are also affected. The most clear-cut examples are: (i) the mass seen by cyclotron resonance in the anomalous skin regime and (ii) the mass in the temperature-dependent amplitude of the de Haas–van Alphen (dHvA) effect.

Susceptibility

In contrast other properties are essentially unaffected by the electron–phonon interaction or are changed in a quite different way. To parallel the discussion of electron–electron interactions consider the susceptibility. It was argued earlier that the effects due to the electron–phonon interaction are 'tied' to the Fermi surface, and hence the susceptibility will be unchanged from the free-electron value (neglecting the electron–electron interaction for the moment). This follows explicitly from a calculation of the change in the chemical potential for down-spins in a magnetic field **H**, in the presence of which the number of down-spins goes from $N/2$ to $N/2+\delta N$ and the Fermi wave-vector increases to $k_F+\delta k_F$. Then (in parallel with the discussion in the previous section) one expects that $E_{k_F}+\delta k_F$ is different from $E_{k_F}-\frac{1}{2}g\mu_B H$ because (i) $E_\mathbf{p}$ depends on $\mathbf{p}$ (more properly on $\mathscr{E}_\mathbf{p}-\mu$) and so it will increase if $\mathbf{p}$ increases, and (ii) $E_{\mathbf{p}\downarrow}$ formally depends on the number of particles of down-spin according to (49).

$$E_{\mathbf{p}\downarrow}=\mathscr{E}_\mathbf{p}+\sum_\mathbf{k}\left[\frac{|g_{\mathbf{p}-\mathbf{k}}|^2(1-n_{\mathbf{k}\downarrow})}{\mathscr{E}_\mathbf{p}-\mathscr{E}_\mathbf{k}-\omega_{\mathbf{p}-\mathbf{k}}}-\frac{|g_{\mathbf{k}-\mathbf{p}}|^2 n_{\mathbf{k}\downarrow}}{\mathscr{E}_\mathbf{k}-\mathscr{E}_\mathbf{p}-\omega_{\mathbf{k}-\mathbf{p}}}\right]. \qquad [(49)]$$

The effect of (ii) is easily calculated by observing that, as far as the last two terms of (49) are concerned, a small change in the total number of particles (of down-spin) can be represented by replacing $n_\mathbf{k}$ by $n_\mathbf{k}+(\delta N/N_0)\delta(\tilde{\mathscr{E}}_\mathbf{k})$. Of course, $\tilde{\mathscr{E}}_\mathbf{p}$ is also at the Fermi surface so that

$$\delta\mu_\downarrow=-\frac{g}{2}\mu_B H+\left.\frac{\partial E_p}{\partial p}\right|_{p=k_F}+\frac{\delta N}{N_0}\left\{N_0\int\frac{\mathrm{d}\Omega}{4\pi}\frac{2g^2}{\omega}\right\}.$$

The term in the curly brackets is of course λ (see 51), while the second term may be written

$$\left[\left(\frac{\partial E}{\partial\tilde{\mathscr{E}}}\right)\left(\frac{\partial\tilde{\mathscr{E}}}{\partial p}\right)\right]_{\tilde{\mathscr{E}}=0}\frac{\delta k_F}{\delta N}\delta N$$

which (using the result of the perturbation calculation) is $(1-\lambda)\delta N/N_0$. Hence $\delta\mu_\downarrow=-\frac{1}{2}g\mu_B H+(\delta N/N_0)(1-\lambda+\lambda)$ and is unaffected by the electron–phonon interaction. The result holds also when a proper non-perturbative calculation using Green's functions is done.

2.4.2 Density of states versus density of levels

So far attention has been directed to the spacing or density of energy levels. The specific heat is proportional to the density of energy levels – i.e. to

$$\frac{p^2}{2\pi^2|\nabla_\mathbf{p}E_\mathbf{p}|}=N_0\frac{\mathrm{d}\tilde{\mathscr{E}}}{\mathrm{d}\tilde{E}}=N_0(1+\lambda).$$

Any purely 'statistical probe' – one which samples only the spacing of the levels – will see this enhancement, as will any probe which picks out single-particle excitations, such as cyclotron resonance in the anomalous skin regime.

Tunnelling example

Many processes are more complicated in that they depend not only on the energy levels but also on the wave-functions. To illustrate this fact, consider the example of tunnelling from one metal to another through a thin oxide barrier. This process at zero temperature is schematized in Fig. 2.4. A particle with wave-vector **p** in the left-hand metal M_l tunnels through the potential barrier to a state **k** in the right-hand metal M_r. Let T be the matrix element (assumed constant) for the process, and V be the bias of the left-hand metal with respect to the right, i.e. $\mu_l - \mu_r = V$.

Then to lowest order in T the transition rate for tunnelling from M_l to M_r is given by

$$\omega_{r\leftarrow l} = \frac{2\pi}{\hbar}\sum_F \left|\left\langle F\left| T\sum_{\mathbf{k}_\mathbf{p}} C_\mathbf{k}^{(r)+} C_\mathbf{p}^{(l)}\right| I\right\rangle\right|^2 \delta(E_F - E_I),$$

where $C_\mathbf{p}^{(l)}$ ($C_\mathbf{k}^{(r)+}$) is an operator which removes (adds) a particle with wave-vector **p** in M_l (**k** in M_r). The letters F and I refer to the final and initial states, respectively. Of course, they can be written as products of the right- and left-hand states, since in the absence of T they do not interact; in particular the initial state can be written

$$|I\rangle = |0, N_l\rangle|0, N_r\rangle,$$

where, for example, $|0, N_l\rangle$ is the ground state of M_l with N_l particles. A typical final state is

$$|F\rangle = |-\mathbf{p}, N_l - 1\rangle|k, N_r + 1\rangle,$$

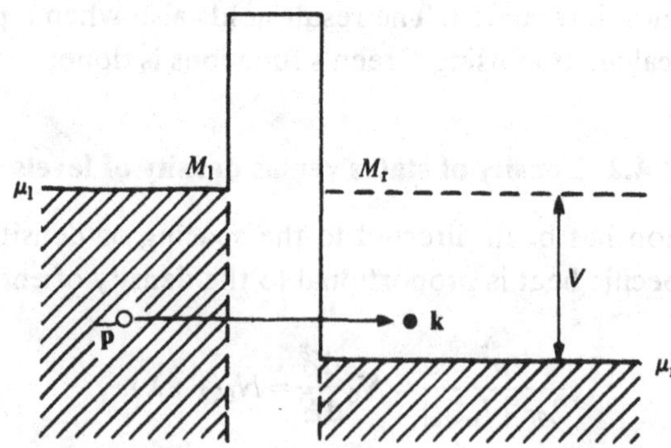

Fig. 2.4. Tunnelling of particle with wave-vector **p** in left-hand metal through a barrier to state **k** in right-hand metal. The chemical potentials are biased by voltage V.

where $|-\mathbf{p}, N_l-1\rangle$ is an excited state with total momentum $-\hbar\mathbf{p}$ of M_l with N_l-1 particles, and $|\mathbf{k}, N_r+1\rangle$ is an excited state with total momentum $\hbar\mathbf{k}$ of M_r with N_r+1 particles. The argument of the energy-conserving delta function can be rewritten

$$\begin{aligned} E_F-E_I &= E_{\mathbf{k}}^{N_r+1}+E_{\mathbf{p}}^{N_l-1}-(E_0^{N_r}+E_0^{N_l}) \\ &= (E_{\mathbf{k}}^{N_r+1}-E_0^{N_r}-\mu_r)+(E_{\mathbf{p}}^{N_l-1}-E_0^{N_l}+\mu_l)-(\mu_l-\mu_r) \\ &= \tilde{E}_{\mathbf{k}}^{(r)}+\tilde{E}_{\mathbf{p}}^{(l)}-V, \end{aligned}$$

where $\tilde{E}_{\mathbf{k}}^{(r)}$ and $\tilde{E}_{\mathbf{p}}^{(l)}$ are single-particle excitation energies of M_r and M_l, respectively, measured relative to μ_r and μ_l, respectively. Finally the identity

$$\delta(E_F-E_I)=\int d\omega\, \delta(\tilde{E}_{\mathbf{k}}^{(r)}-\omega)\delta(\omega-V+\tilde{E}_{\mathbf{p}}^{(l)}) \tag{55}$$

can be used to rewrite the transition rate as

$$\omega_{r\leftarrow l}=\frac{2\pi}{\hbar}|T|^2\int d\omega\, N_T^{(r)}(\omega)N_T^{(l)}(\omega-V),$$

where

$$N_T^{(r)}(\omega)=\sum_{\mathbf{k}}|\langle N_r+1, \mathbf{k}|C_{\mathbf{k}}^{(r)+}|0, N_r\rangle|^2\delta(\omega-\tilde{E}_{\mathbf{k}}^{(r)})$$

and

$$N_T^{(l)}(\omega)=\sum_{\mathbf{p}}|\langle N_l-1, -\tilde{\mathbf{p}}|C_{\mathbf{p}}^{(l)}|0, N_l\rangle|^2\delta(\omega-\tilde{E}_{\mathbf{p}}^{(l)}).$$

The quantity

$$N_T(\omega)=\sum_{\mathbf{k}}|\langle N+1, \mathbf{k}|C_{\mathbf{k}}^{+}|0, N\rangle|^2\delta(\omega-\tilde{E}_{\mathbf{k}}) \tag{56}$$

is called the *density of states*. It is clearly more than a density of energy levels since it also depends on the probability of adding one particle to the system to produce a quasi-particle with energy $\tilde{E}_{\mathbf{k}}$. For non-interacting particles, of course, $\langle N+1, \mathbf{k}|C_{\mathbf{k}}^{+}|0, N\rangle$ is unity and $N_T(\omega)$ is just the density of energy levels. But in the presence of interactions, in particular the electron–phonon interaction, the ground and excited state wave-functions are considerably changed. For example, the wave-function of the ground state is not just a simple Slater determinant but includes also states in which various phonons are virtually excited. Since the total wave-function is normalized, the amplitude of the free-electron Slater determinant is clearly less than one. Likewise the excited states are modified.

Wave-function renormalization

As in the calculation of $E_{\mathbf{p}}$, simple perturbation theory can be used to get the lowest-order term correct. The result of a careful many-body calculation (Wilkins 1968) yields the amplitude appearing in the density of states

$$|\langle N+1, \mathbf{p}|C^+_{\mathbf{p}\uparrow}|0, N\rangle|^2 = (1-\tfrac{1}{2}\lambda)^2 = 1-\lambda \tag{57}$$

to second order in the electron–phonon interaction, and gives

$$=\frac{1}{1+\lambda} \tag{58}$$

exactly. This permits the density of states to be calculated. Since $\tilde{E}_{\mathbf{k}} = \tilde{\mathscr{E}}_{\mathbf{k}}/(1+\lambda)$, (56) yields

$$N_T(\omega) = N_0 \int \mathrm{d}\tilde{\mathscr{E}} \frac{1}{1+\lambda}\,\delta\!\left(\omega - \frac{\tilde{\mathscr{E}}}{1+\lambda}\right) = N_0. \tag{59}$$

The change in the energy level density is exactly compensated by the renormalization of the single-particle amplitude so that the density of states is equal to the free-particle density when the effect of the electron–phonon interaction is properly included.

In principle this result could be checked by a careful tunnelling experiment, but the experiment is very difficult for a variety of reasons. However, a related result for superconductors has been confirmed experimentally. Of course the result is much more general since all transport processes involve the transfer of electrons from one state to another, so that the density of states (and *not* the density of energy levels) is the relevant quantity. Accordingly one finds that for d.c. transport processes (electrical and thermal conductivity, NMR spin–lattice relaxation time, etc.) there is no effective mass enhancement due to the electron–phonon interaction. For example, in $\sigma = ne^2\tau/m$ there is no factor of $(1+\lambda)$, although there are effects due to band structure and electron–electron interactions. Recently it has been shown (Opsal, Thaler & Bass 1976; Lyo 1977; Vilenkin & Taylor 1979) that renormalization effects do enter into the thermopower. Clearly there are a great many properties which may or may not be affected by the electron–phonon interaction. There has been considerable effort (see, e.g., Holstein 1964; Prange & Kadanoff 1964; Prange & Sachs 1967; Enz

1968; Grimvall 1976, 1980) to produce consolidated descriptions of the effects of the electron–phonon and electron–electron interactions on equilibrium and transport properties. Some of the more important results will be summarized in the section on Fermi liquid theory.

2.4.3 Extension to real metals

The treatment in this section up to now has been typical of many-body treatments – almost nothing has been very specific. For example, with respect to the electron–phonon matrix element there has been no mention of longitudinal versus transverse phonons, of normal versus umklapp processes, or of the fact that the initial and final states refer to parts on possibly complicated Fermi surfaces described by Bloch wave-functions. Furthermore, the calculations have been done at zero temperature and for excitations very near the Fermi surface. But typical phonon energies correspond to temperatures (and energies) of a few 100 K. Accordingly there must be important effects occurring as a function of temperature and energy. To name one, many metals and alloys are superconductors and the principal mechanism for this phenomenon is the electron–phonon interaction. Clearly all these topics – other examples include all transport processes – cannot be covered in one chapter. Attention will be directed to those areas which seem broadest and best-defined.

$\alpha^2F(\omega)$ and superconductivity

The electron–phonon enhancement of the effective mass (or specific heat) is given by (54) and (51)

$$\frac{m^*_{\mathrm{ep}}}{m}-1=\lambda=N_0\int\frac{\mathrm{d}\Omega_{\mathbf{p}-\mathbf{k}}}{4\pi}\frac{2|g_{\mathbf{p}-\mathbf{k}}|^2}{\omega_{\mathbf{p}-\mathbf{k}}}. \qquad [(51)]$$

This result, here an s-wave average of g^2/ω, can be generalized to an arbitrary Fermi surface by noting that the band-structure level density (26) is

$$N_{\mathrm{b}}(0)=\frac{1}{2\pi^2}\int\frac{\mathrm{d}S_{\mathbf{p}}}{4\pi}\frac{1}{|\nabla_{\mathbf{p}}\mathscr{E}_{\mathbf{p}}|}=\frac{1}{(2\pi)^3}\int\frac{\mathrm{d}S_{\mathbf{p}}}{|v_{\mathrm{F}}(\mathbf{p})|},$$

where $|v_{\mathrm{F}}(\mathbf{p})|$ is the magnitude of the Fermi velocity at Fermi surface. Then an obvious generalization of λ, one sustained by detailed calculations, is

$$\lambda=2\int\mathrm{d}\omega\frac{\alpha^2F(\omega)}{\omega}, \qquad (60)$$

where

$$\alpha^2F(\omega)=\int\frac{dS_{\mathbf{p}}}{|v_F(\mathbf{p})|}\int\frac{dS_{\mathbf{k}}}{|v_F(\mathbf{k})|}\frac{|g_{\mathbf{p}-\mathbf{k}}|^2}{(2\pi)^3}\delta(\omega-\hbar\omega_{\mathbf{p}-\mathbf{k}})\Big/\int\frac{dS_{\mathbf{p}}}{|v_F(\mathbf{p})|}. \tag{61}$$

Note that for a spherical Fermi surface (61) substituted into (60) reduces to (51). The quantity $\alpha^2F(\omega)$ was originally invented (Scalapino, Schrieffer & Wilkins 1966) for use in simply characterizing the electron–phonon interaction part of equations associated with energy- and temperature-dependent effects in superconductors. Since then it has had a wide application to the properties of normal metals. The reason it has proved so popular, however, is that it can be extracted in a fairly clean way from superconducting tunnelling experiments (McMillan & Rowell 1969). Hence it can then in principle be used to calculate normal properties. (Usually the normal properties emphasize the low-frequency part of $\alpha^2F(\omega)$ while the superconducting ones emphasize the higher frequency parts. As a result the $\alpha^2F(\omega)$ deduced from superconducting tunnelling is not too accurate at low frequencies, where it is needed for normal metals.)

There is one extraordinary connection between λ and the superconducting temperature that simply must be pointed out, even if it is not very germane to the general topic of this chapter. This is that the superconducting temperature T_c is proportional to $\exp[-(1+\lambda)/\lambda]$. There is a more precise relation (McMillan 1968) that works well for a large class of superconductors:

$$\ln\left(\frac{1.45T_c}{\theta_D}\right)=-\frac{1.04(1+\lambda)}{\lambda-(1+0.62\lambda)\mu^*}, \tag{62}$$

where μ^* is a number of order 0.1 which arises from the effects of the Coulomb potential (which acts to prevent superconductivity) and θ_D is the Debye temperature. So there is a very close connection between T_c and λ. For example, if the specific heat enhancement in Pd (see Section 2.3) were all due to the electron–phonon interaction, then λ would be 0.66 and Pd should be a superconductor which it is not. Accordingly there must be another mechanism that tends to increase m^* while decreasing T_c, a mechanism which will be mentioned in Section 2.5.

In Fig. 2.5 plots of $\alpha^2F(\omega)$ for lead and mercury are shown. The two peaks for Pb are for the transverse and longitudinal phonons. For Hg there is a transverse peak but the high-frequency part is relatively structureless. These are rather extreme examples. Often for analytic

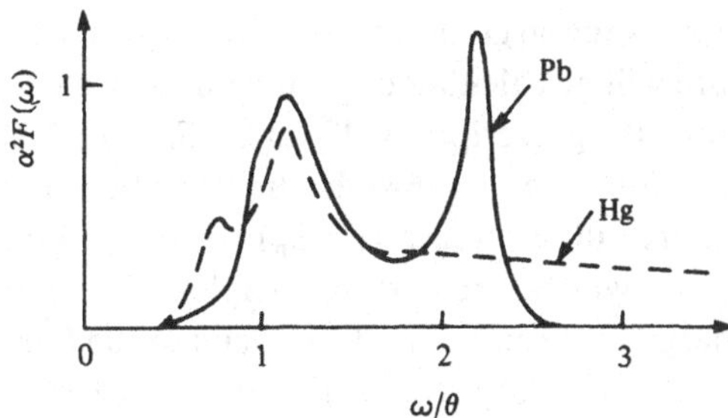

Fig. 2.5. Plots of $\alpha^2F(\omega)$ for lead and mercury versus phonon frequency scaled by the appropriate Debye temperature.

calculations, the assumption

$$\alpha^2F(\omega) = \lambda\,(\omega/\theta_D)^2, \qquad 0<\omega<\theta_D \tag{63}$$

is made; here θ_D is the Debye energy. Clearly (63) is a gross distortion for the plots in Fig. 2.5, which were deduced from superconducting tunnelling experiments, while nonetheless being quite useful for making analytic estimates.

Quasi-particle self-energy and lifetime

The effects of the electron–phonon interaction at finite temperature and frequency are trivial to calculate, provided one is willing to accept several many-body results on faith. The principal point of this subsection is to introduce the concept of a self-energy. In fact this idea has been implicit in the previous discussion. Crudely speaking, the self-energy is the change in the single-particle energy due to interactions. More precisely, one can write (if no expert is looking very closely)

$$\tilde{E}_{\mathbf{p}} = E_{\mathbf{p}} - \mu = \tilde{\mathscr{E}}_{\mathbf{p}} + \Sigma_{\mathbf{p}}(\tilde{E}_{\mathbf{p}}), \tag{64}$$

where $\Sigma_{\mathbf{p}}(E)$ is the *self-energy*. For the case of the electron–electron interaction $\Sigma_{\mathbf{p}}(\tilde{E}_{\mathbf{p}}) = -\Sigma_{\mathbf{p}'} V_{\mathbf{p}-\mathbf{p}'} n_{\mathbf{p}'}$, while for the electron–phonon interaction the self-energy is given by (49) if the $\mathscr{E}_{\mathbf{p}}$s in the denominators are replaced by $E_{\mathbf{p}}$. Both of these examples are incomplete representations of the self-energy. The reason is simple: if an interaction can shift the energy, it can also produce scattering among the excitations – i.e. the quasi-particles will have a finite lifetime; they will decay. Consequently, the actual self-energy must be *complex*, or more to the point, the quasi-particle energy $\tilde{E}_{\mathbf{p}}$ in (64) is complex. It turns out to be relatively easy to build this effect into the calculation of the self-energy and also to include finite-temperature effects.

The main steps in the arguments are: (i) the ground-state energy at finite temperature will be calculated, (ii) the quasi-particle energy will be calculated either by the prescription (47) or by $E_{\mathbf{p}} = \delta E_0^N/\delta n_{\mathbf{p}}$, and (iii) the self-energy will be identified by (64). In the transition from (ii) to (iii) all $\tilde{\mathscr{E}}_{\mathbf{p}}$s in denominators will be replaced by $\tilde{E}_{\mathbf{p}}$ in order to anticipate the exact calculation, and $\tilde{E}_{\mathbf{p}}$ will be treated as complex. The zero-temperature ground-state energy (48) can be used to calculate the finite-temperature energy where now $n_{\mathbf{p}} = (\exp \beta\tilde{\mathscr{E}}_{\mathbf{p}} + 1)^{-1}$ and there are real phonons present, the number for a given energy $\omega_{\mathbf{q}}$ being the Bose factor $N_{\mathbf{q}} = (\exp \beta\omega_{\mathbf{q}} - 1)^{-1}$. Consequently the energy is

$$E_0^N = \sum_{\mathbf{k}s} \mathscr{E}_{\mathbf{k}} n_{\mathbf{k}s} + \sum_{\mathbf{k}\mathbf{k}'s} \frac{|g_{\mathbf{k}'-\mathbf{k}}|^2 n_{\mathbf{k}}(1-n_{\mathbf{k}'})(1+N_{\mathbf{k}-\mathbf{k}'})}{\mathscr{E}_{\mathbf{k}} - \mathscr{E}_{\mathbf{k}'} - \omega_{\mathbf{k}-\mathbf{k}'}} + \sum_{\mathbf{k}\mathbf{k}'s} \frac{|g_{\mathbf{k}-\mathbf{k}'}|^2 n_{\mathbf{k}}(1-n_{\mathbf{k}'})N_{\mathbf{k}-\mathbf{k}'}}{\mathscr{E}_{\mathbf{k}} + \omega_{\mathbf{k}-\mathbf{k}'} - \mathscr{E}_{\mathbf{k}'}}. \tag{65}$$

The second term of (65) is similar to that of (48), the principal difference being the additional factor $1 + N_{\mathbf{k}-\mathbf{k}'}$ which corresponds to the fact that at finite temperature the emission of virtual phonons in the intermediate state is not only spontaneous (the '1') but also stimulated (the '$N_{\mathbf{k}-\mathbf{k}'}$'). The third term in (65), which is of course not present at zero temperature, represents that contribution to the ground-state energy where one of the phonons present in the unperturbed ground state is *absorbed*, so that the intermediate state contains one less phonon.

The quasi-particle energy can be calculated by taking the functional derivative, i.e. $E_{\mathbf{p}} = \delta E_0^N/\delta n_{\mathbf{p}}$. (Note that $\delta n_{\mathbf{k}}/\delta n_{\mathbf{p}} = \delta_{\mathbf{k},\mathbf{p}}$.) All intermediate algebra has been omitted. Furthermore $\mathscr{E}_{\mathbf{p}}$ has been replaced by $E_{\mathbf{p}}$, which would occur in a proper many-body calculation. Finally the angular integral involving $|g_{\mathbf{k}-\mathbf{p}}|^2$ has been replaced by an energy integral over $\alpha^2F(\omega)$. Hence

$$E_{\mathbf{p}} = \mathscr{E}_{\mathbf{p}} + \int \mathrm{d}\tilde{\mathscr{E}}_{\mathbf{k}} \int \mathrm{d}\omega\, \alpha^2F(\omega)\left[\frac{N(\omega)+1-n_{\mathbf{k}}}{\tilde{E}_{\mathbf{p}} - \tilde{\mathscr{E}}_{\mathbf{k}} - \omega} + \frac{N(\omega)+n_{\mathbf{k}}}{\tilde{E}_{\mathbf{p}} - \tilde{\mathscr{E}}_{\mathbf{k}} + \omega}\right], \tag{66}$$

where $N(\omega)$ is the Bose factor for energy ω. The final step in the argument is the identification of the self-energy via (64). In the calculation of the effective mass m^*_{ep} it was noted that the momentum dependence was negligible. Hence the momentum dependence of the self-energy can be dropped for the case of the electron–phonon interaction, i.e.

$$\tilde{E}_{\mathbf{p}} = \mathscr{E}_{\mathbf{p}} + \Sigma(\tilde{E}_{\mathbf{p}}). \tag{64'}$$

Furthermore the dummy variable $\tilde{\mathscr{E}}_{\mathbf{k}}$ is replaced by E'. Thus (66) and (64′) imply

$$\Sigma(E)=\int d\omega\, \alpha^2F(\omega)\int dE'\left[\frac{N(\omega)+n(-E')}{E-E'-\omega}+\frac{N(\omega)+n(E')}{E-E'+\omega}\right], \quad (67)$$

where the Fermi function $n(E)=(\exp\beta E+1)^{-1}$ (Abrikosov, Gor'kov & Dzyaloshinskiĭ 1975).

The important point to remember is that $\tilde{E}_{\mathbf{p}}$ and $\Sigma(\tilde{E}_{\mathbf{p}})$ are complex. If the quasi-particle picture is to be useful the imaginary part must be small compared to the real part. In other words, if one writes

$$\tilde{E}_{\mathbf{p}}=\operatorname{Re}\tilde{E}_{\mathbf{p}}-\frac{i\hbar}{2\tau(\tilde{E}_{\mathbf{p}})} \quad (68)$$

then it is expected that the inverse lifetimes τ can be evaluated at the real value of $\tilde{E}_{\mathbf{p}}$. If $E=\operatorname{Re}\tilde{E}_{\mathbf{p}}$, then $\tau(\tilde{E}_{\mathbf{p}})=\tau(E)$. (The sign of the imaginary part of (68) is chosen by noting that if the states evolve according to $\exp(-i\tilde{E}_{\mathbf{p}}t)$ the choice must correspond to a decaying state.) This observation means that

$$\frac{\hbar}{2\tau(E)}=\operatorname{Im}\Sigma(E+i0^+),$$

where 0^+ is a positive infinitesimal. With these observations out of the way explicit forms for the real and imaginary self-energy can be written down:

$$\operatorname{Re}\Sigma(E,T)=-\tfrac{1}{2}\int d\omega\,\alpha^2F(\omega)\int dE'(E'+\omega)^{-1}$$

$$\times\left\{\tanh\left[\frac{\beta}{2}(E+E')\right]+\tanh\left[\frac{\beta}{2}(E-E')\right]\right\} \quad (69a)$$

and

$$\frac{1}{\tau(E,T)}=\frac{2\pi}{\hbar}\int d\omega\,\alpha^2F(\omega)[2N(\omega)+f(E+\omega)+f(\omega-E)], \quad (69b)$$

where the inverse lifetime result is exactly what a careful Fermi golden rule calculation would yield. The temperature variable has been explicitly exhibited in (69).

There are, of course, numerous applications of (69). Only a few will be mentioned here, and they will all be based on the simple form (63) for $\alpha^2F(\omega)$. Some limiting forms (Migdal 1958a; Wilkins 1968; Grimvall

1969; Allen 1975) for the inverse lifetime are

$$\frac{\hbar}{\tau(E,0)}=\begin{cases}\frac{2\pi}{3}\lambda\theta_D(E/\theta_D)^3, & E<\theta_D\\ \frac{2\pi}{3}\lambda\theta_D, & E>\theta_D\end{cases} \tag{70}$$

and

$$\frac{\hbar}{\tau(0,T)}=\begin{cases}14\pi(1.2)\lambda\theta_D(T/\theta_D)^3, & T<\theta_D\\ 2\pi\lambda T, & T>\theta_D.\end{cases} \tag{71}$$

The temperature-dependent expressions are well known from transport theory for the thermal conductivity and high-temperature electrical conductivity. In addition the T^3 behaviour has been observed in radio-frequency-size-effect (RFSE) experiments on several metals (Wagner 1979).

On the other hand, the energy-dependent lifetime limits the validity of the quasi-particle concept. For excitation energies small compared to the Debye energy the damping of the quasi-particle is quite small. But when E is of order θ_D, $\hbar/\tau$ is of the same size. Hence these states are heavily damped, and the quasi-particle picture breaks down. Nonetheless it has proved possible to perform accurate many-body calculations without assuming the existence of quasi-particles (Scalapino *et al.* 1966; Schrieffer 1964).

Consider briefly the real part of the self-energy. For energies and temperatures large compared to the phonon energy, the real part of Σ is negligible. In the case of specific heat enhancement, one would suppose it to be $1+\lambda(T)$, where a reasonable estimate for $\lambda(T)$ is

$$\lambda(T)=-\left.\frac{\partial\Sigma(E,T)}{\partial E}\right|_{E=0}. \tag{72}$$

(Of course, things are never quite as simple as that, but this formula is a good guide.) Simple calculations yield

$$\lambda(T)=\begin{cases}\lambda\left[1-\frac{2\pi^2}{3}\left(\frac{T}{\theta_D}\right)^2\ln\left(\frac{\theta_D}{T}\right)\right], & T\ll\theta_D\\ 1.05\lambda(\theta_D/\pi T)^2, & T\gg\theta_D.\end{cases} \tag{73}$$

Hence a good measurement of the electronic specific heat at both high and low temperatures would permit a direct determination of $\lambda\ (T=0)$. Such an experiment is more easily suggested than performed, since there are many sources of pseudo-linear-in-temperature specific heat contributions at high temperatures – e.g. thermal expansion, anisotropic phonons, etc.

Temperature-dependent amplitude of the dHvA effect

Given the context of this book, the most obvious question is: can these energy- and temperature-dependent effects be seen in dHvA oscillations? The answer is yes, but the effect is surprisingly dull. It was recognized some time ago (Fowler & Prange 1965; Wilkins & Woo 1965) that at low temperature the temperature-dependent amplitude A of the dHvA has an electron–phonon mass enhancement in the exponential

$$A = \exp\left[-\frac{2\pi^2 k_B T}{\hbar\omega_c}(1+\lambda)\right], \qquad T \ll \theta; \tag{74}$$

and the natural assumption was that at higher temperature λ should be replaced by $\lambda(T)$ (see (73)). Something like a 10 per cent increase in $\lambda(T)$ was expected for $T \sim \theta_D/3$. Of course, there was the disturbing aspect that one knew that the lifetime (à la Dingle) of the quasi-particles might also enter into that amplitude. Nonetheless the hope was that it grew so slowly at low temperatures that it would not affect A sufficiently to prevent $\lambda(T)$ from being observed. The experiment was done for the most likely element, Hg, by Palin (1972). Roughly speaking, the amplitude (74) with the *zero-temperature* λ fit the data. What is the explanation for this surprising result? There is a completely satisfactory many-body calculation (Engelsberg & Simpson 1970; Engelsberg 1978) which finds (74) to be true in general provided that $\omega_c \ll \theta_D$. This point is discussed more fully in Chapter 9, from which it is evident that, for most systems, the dHvA amplitude cannot be used to measure either the temperature-dependent enhancement, $\lambda(T)$, or the quasi-particle lifetimes, $\tau(T)$.

2.5 Fermi liquid theory

This chapter could have ended with the previous section but to have done so would have been to omit a major conceptual framework for metals and ^{3}He. In fact the presentation has been organized to facilitate the discussion of this framework: Landau's (1957a,b) Fermi liquid theory. Many of the ideas introduced so far – energy functional, quasi-particles, angular average of effective potentials – are important elements of Fermi liquid theory. The self-energy concept discussed in the last section is essential to the derivation of Fermi liquid theory (Landau 1959; Pitaevskii 1959;

Luttinger & Nozières 1962). The presentation here will, of necessity, be rather abbreviated, and the reader is directed to the book of Pines & Nozières (1966), which offered a fairly complete discussion at the time it was written. Quite recently Baym & Pethick (1978) have written a review article on ^{3}He. On the other hand, there is no recent comprehensive treatment for metals. There are at least two reasons for this: (i) there is a very large number of transport properties to be catalogued (it is not yet clear what the many-body corrections are for all of them) (ii) the importance of band structure necessitates the coverage of many different elements, a project still to be completed. There is an additional reason for believing that the Fermi liquid approach to metals, now out of fashion, may experience a revival. Tunnelling experiments on superconductors have led to an improved understanding of the electron–phonon interaction – an understanding which stimulated renewed interest in normal metals. Much of the notation and approach from superconductivity, especially $\alpha^2F(\omega)$, is now widely used to calculate and describe normal metal transport phenomena. In the same way, the relatively recent (1972) discovery of superfluidity in ^{3}He has already stimulated a renewed interest in the properties of normal ^{3}He. The resulting improved understanding of Fermi liquid theory can only have a positive impact on experiments and theory for normal metals, especially at ultra-low temperatures where electron–electron scattering will dominate. But that is in the future. For the present section, the assumptions of Fermi liquid theory, which are natural extensions of the previous sections, are used to discuss first equilibrium properties and then transport properties.

2.5.1 Equilibrium properties

The equilibrium properties – specific heat C, spin susceptibility χ, and the compressibility κ of the Fermi liquid ^{3}He – are used to introduce (once again!) the idea of characterizing the system by a few parameters. In particular each property can be related to various angular averages of an effective 'potential', commonly denoted by $f(\mathbf{p}\sigma, \mathbf{p}'\sigma')$. Moreover, the same averages also show up in expressions derived for transport properties, so it is possible to obtain a check on the agreement of theory with experiment.

How many parameters?

Below, very brief 'derivations' of the several properties are given. The point of the derivations is to introduce as few assumptions as possible so that in comparison with experiment it is clear what must be explained. In what follows it is supposed that all calculations are done for low temperature, which means a temperature low compared with the Fermi temperature, $k_B T_F = \mu$. For ^{3}He the appropriate T_F is of order 1 K, while for metals it is $\sim 10^4$ K.

Consider first the specific heat C. The entropy associated with a system of fermions is (Landau & Lifshitz 1958)

$$S = -k_B \sum_{\mathbf{p}\sigma} [n_{\mathbf{p}\sigma} \ln n_{\mathbf{p}\sigma} + (1 - n_{\mathbf{p}\sigma}) \ln (1 - n_{\mathbf{p}\sigma})]. \tag{75}$$

This result depends only on statistics and is independent of the interaction. Here $n_{\mathbf{p}\sigma}$ is the distribution function of a given state $\mathbf{p}\sigma$. In equilibrium $n_{\mathbf{p}\sigma} = \{\exp[\beta(E_{\mathbf{p}\sigma} - \mu)] + 1\}^{-1}$, where $E_{\mathbf{p}\sigma}$ is the quasi-particle energy. (That this is a consequence of the Fermi liquid theory will appear later.) The specific heat is given by

$$\begin{aligned} C &= T\frac{\partial S}{\partial T} \\ &= -k_B T\, 2 \int \mathrm{d}\tilde{E}\, N(\tilde{E}) \frac{\partial n}{\partial T} \ln\left(\frac{n}{1-n}\right), \end{aligned}$$

where $\tilde{E} = E - \mu$ is the energy measured from the chemical potential (the factor 2 comes from summing over spin), and $N(\tilde{E})$ is the level density. For the equilibrium form of n, there is the low-temperature expansion

$$\frac{\partial n}{\partial \tilde{E}} = -\left[\delta(\tilde{E}) + \frac{(\pi k_B T)^2}{6} \delta''(\tilde{E}) + (\text{higher terms in } T)\right].$$

Without further approximation the low-temperature specific heat can be written as

$$C = T^2 \frac{(\pi k_B T)^2}{6} 2 \int \mathrm{d}\tilde{E}\, N(\tilde{E})\, \delta''(\tilde{E}) \frac{1}{2} \frac{\partial}{\partial T} \left(\frac{\tilde{E}}{T}\right)^2$$

and hence

$$C = \frac{\pi^2}{3} k_B^2 2N(0) T. \tag{76}$$

The comparison with the free fermion specific heat C_0 suggests once again that the natural parameter for characterizing the specific heat is the effective mass m^*:

$$C/C_0 = N(0)/N_0 = m^*/m. \tag{77}$$

To calculate the spin susceptibility consider a set of spin-$\frac{1}{2}$ fermions with magnetic moment $\boldsymbol{\mu} = g\gamma_0\mathbf{s} = \gamma_0\boldsymbol{\sigma}$. (The g-value is taken to be 2 so $\boldsymbol{\sigma} = 2\mathbf{s}$.) Here γ_0 may be either an appropriate nuclear magneton (He^3) or the usual Bohr magneton (electrons). The component of the magnetization M in the z-direction is

$$M_z = \gamma_0 \sum_{\mathbf{p}} (n_{\mathbf{p}\uparrow} - n_{\mathbf{p}\downarrow}).$$

In a non-interacting system the single-particle energies $\mathscr{E}_{\mathbf{p}\sigma}$ are changed when the magnetic field is turned on according to $\mathscr{E}_{\mathbf{p}\sigma}(H) = \mathscr{E}_{\mathbf{p}\sigma} - \gamma_0 H\sigma_z$. (The magnetic field defines the z-direction.) For the interacting system consider the possibility that the single-particle energies in the presence of the field are affected by the interactions. In the limit of a small magnetic field ($\gamma_0 H \ll \mu$) suppose

$$E_{\mathbf{p}\sigma}(H) = E_{\mathbf{p}} - \gamma H\sigma_z,$$

where γ will be determined later. Then the magnetization is

$$M_z = \gamma_0 \sum_{\mathbf{p}} [n(E_{\mathbf{p}} - \gamma H) - n(E_{\mathbf{p}} + \gamma H)]$$

$$= 2\gamma H\gamma_0 \sum_{\mathbf{p}} \left(-\frac{\partial n}{\partial E_{\mathbf{p}}}\right) = (2N(0)\gamma\gamma_0)H.$$

The susceptibility $\chi = M_z/H$, or rather its ratio to the susceptibility for the non-interacting system $\chi_0 = 2N_0\gamma_0^2$, defines another parameter γ/γ_0:

$$\frac{\chi}{\chi_0} = \frac{N(0)}{N_0}\frac{\gamma}{\gamma_0}. \tag{78}$$

The final parameter, the compressibility, while having a thermodynamic definition, can be related to the sound velocity. Through thermodynamic identities the compressibility $\kappa^{-1} = -V(\partial P/\partial V)_N$ can be related to

$$\frac{1}{\kappa} = n^2 \frac{\partial \mu}{\partial n}, \tag{79}$$

which can be compared to the non-interacting value $\kappa_0^{-1} = 2n\mu_0/3$ to define a third parameter κ_0/κ. The compressibility can also be related to the velocity of an ordinary or hydrodynamic sound wave propagating through the liquid. The velocity of hydrodynamic sound is given by

$$v_s^2 = \frac{1}{mn\kappa} = \frac{n}{m}\left(\frac{\partial \mu}{\partial n}\right). \tag{80}$$

The result (80) defines a third parameter

$$\frac{\kappa_0}{\kappa}=\frac{v^2}{v_{s0}^2}, \tag{81}$$

where

$$v_{s0}^2=(n/m)(\partial\mu_0/\partial n)=\tfrac{1}{3}(\hbar k_F/m)^2=\tfrac{1}{3}(v_F^0)^2 \tag{82}$$

is the 'sound velocity' for non-interacting fermions. The hydrodynamic mode (80) is possible only for a neutral Fermi system where there are no long-range forces. Such forces as, for example, the Coulomb potential between electrons in a metal, completely change the character of the density wave. The potential enters the dispersion law in an essential way and leads to a high-frequency mode (a plasma oscillation).

In Table 2.4 typical values of parameters C/C_0, χ/χ_0 and κ_0/κ are listed for ^{3}He at three different pressures (or, equivalently, densities). Note that the Fermi wave-vector k_F, given by $(3\pi^2 n)^{\frac{1}{3}}$, is comparable to metallic values. On the other hand since the ^{3}He mass is six thousand times larger than the electron mass the Fermi velocity is much smaller. The striking aspects of the three ratios are how different they are from each other and how much they depend on pressure. Note in particular that the susceptibility increases with pressure while the compressibility decreases. This is in marked contrast to the 'prediction' of the second section where χ and κ both depend in the same way on the effective potential (which there did not depend on spin). These data are the strongest possible evidence that in the theory the 'potential' must have a piece that depends on spin. This table further indicates that the theory must have at least three adjustable parameters (for each density). Such a theory would be useless unless it made other predictions which depended on those same parameters. The beauty of Landau's theory is that it does just this. He predicted a new density oscillation, called zero sound at low temperatures (and high frequencies) whose velocity depends on these parameters (and in a different way from ordinary sound). This mode has been seen and its properties, including its damping, agree with the predictions.

TABLE 2.4. *Equilibrium properties of ^{3}He*

Pressure (bar)	$k_F(10^8\,\mathrm{cm}^{-1})$	$v_F(10^3\,\mathrm{cm\,s}^{-1})$	C/C_0	χ/χ_0	κ_0/κ
0	0.79	5.5	3.01	9.1	3.7
18	0.86	3.7	4.93	18.3	10.7
34.36	0.89	3.0	6.22	23.9	15.3

Basic formulation of Fermi liquid theory

In the following description of Landau's (1957a,b) theory the argument closely follows his, starting with the assumptions and then proceeding to the consequences. Landau imagined that the interacting system arose from the system of free fermions by gradually turning on the interaction. He claimed that it was still possible in the interacting system to speak of well-defined single-particle excitations or 'quasi-particles' in a one-to-one correspondence with the non-interacting system. There are three main assumptions underlying Landau's theory:

(i) Even in the presence of interactions the Fermi surface remains 'sharp', i.e. all states $\mathbf{p}\sigma$ are occupied for $|\mathbf{p}| < k_F$ in the ground state (at $T = 0$ K) and the rest are unoccupied. k_F is determined by the number of particles per unit volume n through $k_F^3 = 3\pi^2 n$ (n is the number density of ^{3}He atoms). The low-lying single-particle excitations, which are well defined near the Fermi surface, possess a given momentum $\mathbf{p}$, spin $\hbar\sigma/2$, energy $E_{\mathbf{p}\sigma}$ and obey Fermi statistics.

(ii) The total energy is a functional of the distribution function $n_{\mathbf{p}\sigma}$ of all the states $\mathbf{p}\sigma$, i.e. $E = E[\{n_{\mathbf{p}\sigma}\}]$. More specifically, by writing for the change in total energy due to changes in the distribution function $\delta n_{\mathbf{p}\sigma}$ at the Fermi surface $\delta E = \sum_{\mathbf{p}\sigma} E_{\mathbf{p}\sigma}\delta n_{\mathbf{p}\sigma}$ or $E_{\mathbf{p}\sigma} = \delta E/\delta n_{\mathbf{p}\sigma}$, Landau defined the 'quasi-particle' energy $E_{\mathbf{p}\sigma}$ which is itself a functional of $n_{\mathbf{p}\sigma}$. He further assumed that this energy could be regarded as the Hamiltonian function of the added particle (in state $\mathbf{p}\sigma$) in the self-consistent field of the other particles and external fields.

(iii) As $E_{\mathbf{p}\sigma}$ is also a functional of the $n_{\mathbf{p}\sigma}$s, one may write

$$\delta E_{\mathbf{p}\sigma} = \sum_{\mathbf{p}'\sigma'} f(\mathbf{p}\sigma, \mathbf{p}'\sigma')\delta n_{\mathbf{p}'\sigma'}, \tag{83}$$

or in terms of the total energy E,

$$\frac{\delta^2 E}{\delta n_{\mathbf{p}\sigma}\delta n_{\mathbf{p}'\sigma'}} = f(\mathbf{p}\sigma, \mathbf{p}'\sigma'). \tag{84}$$

From this second expression it follows that the 'interaction' $f(\mathbf{p}\sigma, \mathbf{p}'\sigma')$ between quasi-particles is symmetric in $\mathbf{p}\sigma$ and $\mathbf{p}'\sigma'$. It is further assumed that one may write f as the sum of a spin-independent part and a spin-dependent part of the exchange type, i.e.

$$f(\mathbf{p}\sigma, \mathbf{p}'\sigma') = f_{\mathbf{p}-\mathbf{p}'} + g_{\mathbf{p}-\mathbf{p}'}\boldsymbol{\sigma}\cdot\boldsymbol{\sigma}'. \tag{85}$$

For translationally invariant systems f and g depend only on $(\mathbf{p}-\mathbf{p}')$, where both $\mathbf{p}$ and $\mathbf{p}'$ are at the Fermi surface ($|\mathbf{p}|=|\mathbf{p}'|=k_F$). So they, in fact, only depend on the angle θ between $\mathbf{p}$ and $\mathbf{p}'$.

In connection with the assumptions and definitions a useful quantity is $E_0(\mathbf{p}\sigma)$ which by definition is related to the quasi-particle energy through

$$E_{\mathbf{p}\sigma} \equiv E(\mathbf{p}\sigma) = E_0(\mathbf{p}\sigma) + \sum_{\mathbf{p}'\sigma'} f(\mathbf{p}\sigma, \mathbf{p}'\sigma')\delta n_{\mathbf{p}'\sigma'}. \tag{86}$$

Thus $E_0(\mathbf{p}\sigma)$ is the increase in the energy of the system when a single quasi-particle $(\mathbf{p}, \sigma)$ is added to an (otherwise) undisturbed system (i.e. $\delta n = 0$). If $|\mathbf{p}| = k_F$ then this energy is just the chemical potential. The effective mass m^* as well as the density of states at the Fermi surface $N(0)$ is defined in terms of E_0 in the customary fashion:

$$N(0) = \frac{p^2}{2\pi^2|\nabla_{\mathbf{p}} E_0(\mathbf{p})|}\bigg|_{|\mathbf{p}|=k_F} = \frac{m^* k_F}{2\pi^2\hbar^2}. \tag{87}$$

Another useful convention, since $f(\mathbf{p}\sigma, \mathbf{p}'\sigma')$ is only needed for momenta at the Fermi surface, is to expand it in Legendre polynomials. In particular,

$$2N(0)f(\mathbf{p}\sigma, \mathbf{p}'\sigma')|_{|\mathbf{p}|=|\mathbf{p}'|=k_F} = \sum_l (2l+1)(F_l + G_l \boldsymbol{\sigma}\cdot\boldsymbol{\sigma}')P_l \cos\theta. \tag{88}$$

Here θ is the angle between $\mathbf{p}$ and $\mathbf{p}'$. These F_ls and G_ls are called Fermi liquid parameters.

Landau's assumptions are sufficient to relate the specific heat, susceptibility, and compressibility to F_0, F_1 and G_0. To start with, the equilibrium distribution law for the quasi-particles must be obtained. Consider the change δS in the entropy due to variations $\delta n_{\mathbf{p}\sigma}$ of the distribution function from the equilibrium one $n_0(\mathbf{p}\sigma)$; from (75) it follows that

$$\delta S = -k_B \sum_{\mathbf{p}\sigma} \ln\left[\frac{n_0(\mathbf{p}\sigma)}{1-n_0(\mathbf{p}\sigma)}\right]\delta n_{\mathbf{p}\sigma}. \tag{89}$$

Since $T\delta S = \delta E - \mu\delta N = \sum_{\mathbf{p}\sigma}(\tilde{E}_{\mathbf{p}\sigma} - \mu)\delta n_{\mathbf{p}\sigma}$, a comparison of the coefficients of the arbitrary variation $\delta n_{\mathbf{p}\sigma}$ yields

$$\frac{E_{\mathbf{p}\sigma}-\mu}{k_B T} = -\ln\left[\frac{n_0(\mathbf{p}\sigma)}{1-n_0(\mathbf{p}\sigma)}\right]$$

or

$$n_0(\mathbf{p}\sigma) = \frac{1}{\exp[\beta(E_{\mathbf{p}\sigma}-\mu)]+1}. \tag{90}$$

With this equilibrium distribution function, which was assumed at the start of this section (see sentences after (75)), the result for the ratio of the specific heat C of the Fermi liquid to that of the non-interacting system C_0 is

$$C/C_0 = N(0)/N_0 = m^*/m. \qquad [(77)]$$

To find the sound velocity v_s of the interacting system requires the calculation of the change $\delta\mu$ of the chemical potential due to a change δn in the density or equivalently to a change δk_F in the Fermi wave-number. (Remember the volume V is held constant.) This amounts to considering the change in occupation number $\delta n_{\mathbf{p}\sigma}$ given by

$$\delta n_{\mathbf{p}\sigma} = \left|\frac{\partial n}{\partial \mathbf{p}}\right|_{|\mathbf{p}|=k_F} \delta k_F = \delta(p-k_F)\delta k_F = \delta(E_0-\mu)\left.\frac{\partial E_0}{\partial \mathbf{p}}\right|_{|\mathbf{p}|=k_F} \delta k_F.$$

The corresponding change in energy at the Fermi surface, i.e. $\delta\mu$, is

$$\delta\mu = \left.\frac{\partial E_0(\mathbf{p}\sigma)}{\partial p}\right|_{|\mathbf{p}|=k_F} \delta k_F + \sum_{p'\sigma'} f(\mathbf{p}\sigma, \mathbf{p}'\sigma')\delta(E_0(\mathbf{p}'\sigma')-\mu)\frac{k_F}{m^*}\delta k_F.$$

The form (88) is used for the interaction. The angular averaging reduces the l-sum to the $l=0$ term and the spin summation removes G_0. Hence

$$\delta\mu = (1+F_0)\frac{\hbar^2 k_F}{m^*}\delta k_F = (1+F_0)\frac{\hbar^2 k_F}{m^*}\left(\frac{k_F}{3n}\delta n\right).$$

The last equality follows from the proportionality of k_F to $n^{\frac{1}{3}}$. Thus

$$v_s^2 = \frac{n}{m}\frac{\delta\mu}{\delta n} = \frac{1}{m}\frac{1}{3m^*}(\hbar k_F)^2(1+F_0) \qquad (91)$$

or, in terms of the sound velocity of the non-interacting system v_{s0},

$$v_s^2/v_{s0}^2 = \frac{1+F_0}{(m^*/m)} = \kappa_0/\kappa.$$

Finally, the susceptibility χ can be related to G_0 from the change of energy $\delta E_{\mathbf{p}\sigma} = -\gamma\boldsymbol{\sigma}\cdot\mathbf{H}$ due to an external magnetic field

$$\delta E_{\mathbf{p}\sigma} = -\gamma_0\boldsymbol{\sigma}\cdot\mathbf{H} + \sum_{\mathbf{p}'\sigma'} f(\mathbf{p}\sigma, \mathbf{p}'\sigma')\delta n_{\mathbf{p}'\sigma'}.$$

The first term is the change of energy due to the interaction between the magnetic moment of the particle and the field; the second is caused by the redistribution $\delta n_{\mathbf{p}'\sigma'}$ of all the other particles due to the influence of the field. Since

$$\delta n_{\mathbf{p}'\sigma'} = \frac{\delta n}{\delta E}\delta E = \frac{\partial n_{\mathbf{p}'\sigma'}}{\partial E_{\mathbf{p}'\sigma'}}(-\gamma\boldsymbol{\sigma}'\cdot\mathbf{H}),$$

it follows that

$$-\gamma\boldsymbol{\sigma}\cdot\mathbf{H} = -\gamma_0\boldsymbol{\sigma}\cdot\mathbf{H} + \tfrac{1}{2}\sum_{\sigma'}\sum_{l}\int\frac{\mathrm{d}\Omega'}{4\pi}(2l+1)(F_l + G_l\boldsymbol{\sigma}\cdot\boldsymbol{\sigma}')P_l(\cos\theta)\gamma\boldsymbol{\sigma}'\cdot\mathbf{H}.$$

Again the angular integral reduces the l-sum to $l = 0$, but in this case the spin sum eliminates F_0. The equation reduces to $-\gamma = -\gamma_0 + \gamma G_0$, i.e.

$$\gamma = \gamma_0/(1+G_0).$$

So

$$\chi = 2N(0)\gamma_0^2/(1+G_0) \tag{92}$$

and

$$\frac{\chi}{\chi_0} = \frac{m^*}{m}\,\frac{1}{1+G_0}. \tag{93}$$

There is one additional relation between equilibrium properties and parameters deduced by Landau (1957a). He showed that for a translationally invariant system such as ^{3}He (but *not* a metal)

$$m^*/m = 1 + F_1 \quad (^3\text{He only}). \tag{94}$$

Accordingly in Table 2.5, the ratios have been listed for the general case and the translationally invariant system separately. That (94) does *not* apply for metals is not really surprising. The prescription (84) $\delta^2E/\delta n_{\mathbf{p}}\delta n_{\mathbf{p}'} = f_{\mathbf{p}-\mathbf{p}'}$ permits the ready identification of the effective interaction for Coulomb (electron–electron) scattering from (29′),

$$(f_{\mathbf{p}-\mathbf{p}'})_{\mathrm{c}} = -V_{\mathbf{p}-\mathbf{p}'} \tag{95a}$$

and for electron–phonon scattering from (49)

$$(f_{\mathbf{p}-\mathbf{p}'})_{\mathrm{ep}} = \frac{2|g_{\mathbf{p}-\mathbf{p}'}|^2}{\omega_{\mathbf{p}-\mathbf{p}'}}. \tag{95b}$$

If only Coulomb scattering were present, then (94) would follow from (40): $m/m_{\mathrm{e}}^* - 1 = N_0\langle V\cos\theta\rangle = -(m/m_{\mathrm{e}}^*)N(0)\langle f_{\mathrm{c}}\cos\theta\rangle = -(m/m_{\mathrm{e}}^*)(F_1)_{\mathrm{c}}$ or $m_{\mathrm{e}}^*/m = 1 + (F_1)_{\mathrm{c}}$. Since $(F_1)_{\mathrm{c}}$ is small (see the discussion in Section 2.2), the primary contribution to the effective mass is from the electron–phonon interaction. According to (51)

$$m_{\mathrm{ep}}^*/m - 1 = N_0\langle 2|g|^2/\omega\rangle = N_0\langle f_{\mathrm{ep}}\rangle \approx (F_0)_{\mathrm{ep}}, \tag{96}$$

where the last equality holds only in the limit of small $(F_0)_{ep}$. Hence the presence of an additional scattering mechanism, associated with the presence of ions (which 'disturb' the translational invariance of the system), destroys the simple relation between m^*/m and F_1. Leggett (1968) has shown for metals that the inequality $m^*/m > (1+F_1)$ holds. Accordingly in metals m^*/m is simply treated as an additional parameter.

Constraints on the Fermi liquid parameters

How big are the Fermi liquid parameters? And are they reasonable values? The first question is easy to answer. For the case of *metals* it was discovered that the mass enhancements (over the band-structure value) were usually no more than a factor of two. (An exception to this may be the rare earth metals, where there is some evidence that enhancement factors as big as five or more are possible.) The largest Stoner factor for a non-magnetic element is nearly ten for Pd, although there are larger values for alloys. Finally, for the compressibility the rough expression $\kappa_0/\kappa \approx 1 - r_s/6$ indicates that the electron gas compressibility (as opposed to the value for the metal) may get quite large for the heavier alkali metals.

The Fermi liquid parameters for ^{3}He are shown in Table 2.6 and have been deduced from the measured ratios in Table 2.4 with the aid of the formulae in Table 2.5. (Note that alternative sets of symbols for the Fermi parameters are also shown in the table.) The most surprising aspect is how large F_0 is and how sensitive it is to pressure. On the other hand, even though G_0 changes little, its closeness to -1 is important. By the way, the

TABLE 2.5. *Summary of many-body corrections to equilibrium properties*

Property	Formula	Ratio of formula to that for non-interacting system	Same ratio specialized to the case of ^{3}He
Specific heat	$C = 2N(0)\dfrac{\pi^2}{3}k_B^2 T$	$\dfrac{C}{C_0} = \dfrac{m^*}{m}$	$\dfrac{C}{C_0} = 1 + F_1$
Compressibility	$\dfrac{1}{mn\kappa} = v_s^2 = \dfrac{(\hbar k_F)^2}{3mm^*}(1+F_0)$	$\dfrac{\kappa_0}{\kappa} = \dfrac{v_s^2}{v_{s0}^2} = \dfrac{1+F_0}{m^*/m}$	$\dfrac{\kappa_0}{\kappa} = \dfrac{1+F_0}{1+F_1}$
Susceptibility	$\chi = 2N(0)\dfrac{\gamma_0^2}{1+G_0}$	$\dfrac{\chi}{\chi_0} = \dfrac{m^*/m}{1+G_0}$	$\dfrac{\chi}{\chi_0} = \dfrac{1+F_1}{1+G_0}$

TABLE 2.6. *Fermi liquid parameters for* ^3He

Symbols used in this chapter	F_0	F_1	G_0
Symbols used in Landau (1957a,b)	F_0	$F_1/3$	$\bar{\psi}_0/4$
Symbols used in Pines & Nozières (1966)	F_0^s	$F_1^s/3$	F_0^a
Symbols used in Platzman & Wolff (1973)	A_0	A_1	B_0
Symbols used in Wheatley (1975)	F_0	$F_1/3$	$Z_0/4$
Pressure (bar)			
0	10.1	2.01	−0.67
18	52.8	3.93	−0.73
34.36	94.1	5.22	−0.74

Note added in proof. Recent measurements of the specific heat by Alvesalo *et al.* (*Phys. Rev. Lett.* **33** (1979) 1509) suggest these values must be significantly altered.

natural definition of the Stoner factor for ^{3}He is $(1+G_0)^{-1} = S$, so that S is of order three to four. This fact, for reasons that are discussed in the next paragraph but one, leads some to suggest that ^{3}He is nearly ferromagnetic. Now if it were ferromagnetic, there would be a low-frequency mode, a magnon or spin wave, which would renormalize the properties of ferromagnetic ^{3}He. Since it is not, these spin waves are only overdamped spin fluctuations or paramagnons, to give them their faddish name, which nonetheless will contribute to the Fermi liquid parameters. Of course the contribution is greater the closer the system is to being ferromagnetic. A discussion of spin fluctuations in ^{3}He may be found in the Baym & Pethick review article (1978) which can be used to find the extensive literature in this area.

The question of how believable the Fermi liquid parameters are is ultimately a very deep question which will not be faced. For the case of metals they are quite believable since the second and third sections were devoted to the ideas involved in calculating them. As for ^{3}He, there have been several efforts to calculate them from first principles, which, while not terribly successful, have yielded semi-reasonable numbers. More to the point of this chapter, Fermi liquid theory can place some constraints on their values. For a stable system the free energy $\mathcal{F}$ $(=E-\mu N-TS)$ must be a local minimum. Now $\delta\mathcal{F}/\delta n(\mathbf{p}\sigma)=0$ was already used to derive the form of the equilibrium distribution function. The expansion of $\mathcal{F}$ to second order in $\delta n(\mathbf{p}\sigma) = (\delta n_0/\delta E_0)\sum_l a_l P_l(\cos\theta)$, coupled with the requirement that $\delta^2\mathcal{F}/\delta a_l^2 > 0$, yields the condition that $F_l > -1$ and $G_l > -1$ for all l. If any given F_l (or G_l) were less than -1 the system would gain energy by distorting its Fermi surface according to that particular l. The most simple to visualize is the case $G_0 < -1$, where the Fermi sphere for one spin grows at the expense of the other and the system goes ferromagnetic.

There is one other constraint, which comes from the microscopic condition, that the forward scattering amplitude for two fermion quasi-particles of the same spin be zero. When this condition is translated from the microscopic theory, it results in the condition that

$$\sum_l (2l+1)\left(\frac{F_l}{F_l+1}+\frac{G_l}{G_l+1}\right)=0. \tag{97}$$

Note that the values of F_0, F_1 and G_0 in Table 2.6 do not by themselves satisfy this condition. A common model is to set $F_l = G_l = 0$ for $l \geq 2$ with G_1 taking up the slack. Note, however, that if one of the parameters is very near -1, this model is not reasonable and a great many F_l and G_l are needed to satisfy (97).

The constraint (97), called a sum rule, is for translationally invariant Fermi liquids. As has been noted earlier, the presence of two kinds of interactions makes the metal situation more complicated. There is an analogous sum rule for metals (Brinkman, Platzman & Rice 1968). Why, the reader may ask, does anyone want to know the other Fermi liquid parameters if they do not affect equilibrium properties? The answer is that they occur in the transport properties. In ^{3}He, the zero sound velocity depends on all the F_ls. Furthermore, there has been a considerable effort to describe the scattering processes in terms of the Fermi liquid parameters. Since all descriptions are approximate, the sum rule (97) provides some guidance. In the case of *metals*, there are experiments that can measure additional Fermi liquid parameters.

2.5.2 Transport properties, the Boltzmann equation

The field of transport properties, including time-dependent phenomena, in ^{3}He and metals is vast. Here there will be space only for a few comments and the statement of a few results. A major concept is that of the Boltzmann equation, which describes the temporal and spatial evolution of excitations in response to external perturbation. (For a discussion of the path integral approach to the solution of transport problems in metals, the reader is referred to Chapter 3.) In general, the total time rate of change, $\mathrm{D}n/\mathrm{D}t$, of the distribution function in the absence of collisions plus the time rate of change, $(\partial n/\partial t)_{\mathrm{coll}}$, due to collisions must be zero. Hence

$$\frac{\mathrm{D}n}{\mathrm{D}t}=\frac{\partial n}{\partial t}+\mathbf{v_p}\cdot\nabla_{\mathbf{x}}n+\dot{\mathbf{p}}\cdot\nabla_{\mathbf{p}}n=I(n)=-\left(\frac{\partial n}{\partial t}\right)_{\mathrm{coll}}, \tag{98}$$

where $I(n)$ is the collision integral. According to Landau, the quasi-particle energy $E(\mathbf{p}\sigma, \mathbf{x}t)$ can be regarded as the Hamiltonian function of the quasi-particle $(\mathbf{p}, \sigma)$ in the self-consistent field of the other particles and the external fields. This rather bold assumption, since proved by many-body methods, means

$$\mathbf{v_p} = \nabla_\mathbf{p} E(\mathbf{p}\sigma, \mathbf{x}t) \tag{99a}$$

and

$$\dot{\mathbf{p}} = -\nabla_\mathbf{x} E(\mathbf{p}\sigma, \mathbf{x}t). \tag{99b}$$

There remains only to give an explicit form for the collision integral $I(n)$. In fact, there is no point in doing so since there is no space to discuss usefully the implications of its form. Instead, when $I(n)$ is considered in the next subsection, it will only be in terms of the relaxation-time approximation.

Equations (98) and (99) represent the Boltzmann equation for a Fermi liquid with short-range forces. For the case of long-range forces, such as the Coulomb interaction in metals, Silin (1958) first pointed out that Landau's approach is also valid if it is understood as follows. (i) The interaction $f(\mathbf{p}\sigma, \mathbf{p}'\sigma')$ does not include the Hartree contribution (see (12)). (This is, of course, no surprise since in the discussion in Section 2.3 it was clear that $V_{\mathbf{p}-\mathbf{p}'}$ included only exchange-correlation effects.) (ii) The Hartree term only contributes in the sense that deviations of the distribution from equilibrium $\delta n_{\mathbf{p}\sigma}$ generate an electric field $\mathbf{E}$ via Poisson's equation. (iii) Of course, the electric field and the magnetic induction $\mathbf{B}$ generated by deviations from equilibrium will exert forces back on the electrons. In summary, (99) is modified explicitly to

$$-\nabla_\mathbf{x} E(\mathbf{p}\sigma, \mathbf{x}t) = -\nabla_\mathbf{x} \sum_{\mathbf{p}'\sigma'} f(\mathbf{p}\sigma, \mathbf{p}'\sigma')\delta n(\mathbf{p}'\sigma', \mathbf{x}t) + e\left(\mathbf{E} + \frac{1}{c}\nabla_\mathbf{p} E_{\mathbf{p}\sigma} \wedge \mathbf{B}\right), \tag{100}$$

where the electric field satisfies

$$\nabla \cdot \mathbf{E} = 4\pi e \sum_{\mathbf{p}\sigma} \delta n(\mathbf{p}\sigma, \mathbf{x}t). \tag{101}$$

Of course the remaining Maxwell's equations must be satisfied: two trivially, $\nabla \wedge \mathbf{E} + (1/c)\,\partial\mathbf{H}/\partial t = 0$ and $\nabla \cdot \mathbf{H} = 0$, and one with more computation,

$$\begin{aligned} \nabla \wedge \mathbf{B} - (1/c)\,\partial\mathbf{E}/\partial t &= 4\pi e \sum_{\mathbf{p}\sigma} \nabla_\mathbf{p} E_{\mathbf{p}\sigma} n(\mathbf{p}\sigma) + 4\pi\gamma_0 \nabla \wedge \sum_{\mathbf{p}\sigma} \boldsymbol{\sigma} n_{\mathbf{p}\sigma} \\ &= \frac{4\pi}{c} e\mathbf{J} + 4\pi \nabla \wedge \mathbf{M}, \end{aligned} \tag{102}$$

which, of course, is just $\nabla \wedge \mathbf{H} - (1/c)\,\partial\mathbf{E}/\partial t = (4\pi e/c)\mathbf{J}$, where $\mathbf{J}$ is the number-current density. The interested reader can consult the copious literature devoted to solutions of the Landau–Silin–Boltzmann equation. (Silin 1959; Pines & Nozières 1966; Platzman & Wolff 1973).

Relaxation time

That the interparticle forces also lead to scattering between the quasi-particles is described in the collision integral. Its effect on transport properties has been studied extensively (Brooker & Sykes 1968; Højgaard Jensen, Smith & Wilkins 1968, 1969; Baym & Pethick 1978). For simplicity, the relaxation-time approximation for $I(n)$ will be used. In it, $I(n)$ is written

$$I(n) = -\frac{n - \tilde{n}}{\tau}, \tag{103}$$

where τ is the relaxation time and $\tilde{n}$ is some *local* equilibrium distribution function which can depend on the nature of the perturbation to the system.

The temperature and energy dependence of the relaxation time can be estimated by a simple golden rule argument. Let $W(\mathbf{p}_1, \mathbf{p}_2, \mathbf{p}_1', \mathbf{p}_2')$ be the collision probability for two quasi-particles with momenta $\mathbf{p}_1$ and $\mathbf{p}_2$ to scatter to $\mathbf{p}_1'$ and $\mathbf{p}_2'$, respectively. Then the rate for a quasi-particle $\mathbf{p}_1$ to decay by this scattering process is

$$\frac{1}{\tau(\mathbf{p}_1)} = \sum_{\mathbf{p}_1'\mathbf{p}_2'} \int d\mathbf{p}_2\, \delta(\mathbf{p}_1 + \mathbf{p}_2 - \mathbf{p}_1' - \mathbf{p}_2')\delta(E_0(\mathbf{p}_1) + E_0(\mathbf{p}_2) - E_0(\mathbf{p}_1') - E(\mathbf{p}_2'))$$

$$\times W(\mathbf{p}_1\mathbf{p}_2, \mathbf{p}_1'\mathbf{p}_2')n_0(\mathbf{p}_2)[1 - n_0(\mathbf{p}_1')][1 - n_0(\mathbf{p}_2')], \tag{104}$$

where the spin indices have been suppressed and the two delta functions conserve momentum and energy. The factors $\{n_0(\mathbf{p}_2)\}$ and $\{(1 - n_0(\mathbf{p}_1'))(1 - n_0(\mathbf{p}_2'))\}$ are present because the process depends on the probability of $\mathbf{p}_2$ being occupied and $\mathbf{p}_1'$ and $\mathbf{p}_2'$ being unoccupied. For $\mathbf{p}_1$ near the Fermi surface, the Fermi factors together with energy and momentum conservation constrain $\mathbf{p}_2$, $\mathbf{p}_1'$ and $\mathbf{p}_2'$ to be near the Fermi surface also. In fact, the integral can be reduced to a product of two integrals: one over the angles, involving only W, and the other over energies, involving only the Fermi factors. The energy and temperature dependence arise entirely from the latter. In the notation that $\tilde{E}_n =$

$E_0(\mathbf{p}_n)-\mu$, and letting $\tilde{E}_2$ go to $-\tilde{E}_2$, τ can be written

$$\frac{1}{\tau(\tilde{E}_1)}=\langle W\rangle\int \mathrm{d}\tilde{E}_2\,\mathrm{d}\tilde{E}'_1\,\mathrm{d}\tilde{E}'_2\,\delta(\tilde{E}_1-\tilde{E}_2-\tilde{E}'_1-\tilde{E}'_2)$$

$$\times[1-n_0(\tilde{E}_2)][1-n_0(\tilde{E}'_1)][1-n_0(\tilde{E}'_2)]$$

$$=\langle W\rangle\tfrac{1}{2}\{(\pi k_\mathrm{B}T)^2+\tilde{E}_1^2\},\tag{105}$$

where the energy integral has been done exactly, and $\langle W\rangle$ is some complicated angular integral. That the result must be proportional to T^2 follows from dimensional analysis: since the Fermi factor is a function of $(\tilde{E}/T)$ alone, the integrals over energy would give T^3 if the delta function did not restrict one energy. The principal result, that the quasi-particle relaxation rate at the Fermi surface is proportional to T^2, is also true for metals, where it has apparently been observed in the low-temperature resistance of some transition metals. This relaxation time is also seen in the transport properties of ^{3}He. While the temperature dependences of the thermal conductivity, spin diffusion and viscosity are reasonably well understood, the magnitudes present more of a problem due to imperfect modelling of $W(\mathbf{p}_1\mathbf{p}_2,\,\mathbf{p}'_1\mathbf{p}'_2)$ in terms of Fermi liquid parameters. That the relaxation rate is proportional to $\tilde{E}^2$ means that the quasi-particle concept must break down for sufficiently large $\tilde{E}$, on the order of μ it turns out. Hence for the particle–particle interaction, and in the case of metals also for the electron–phonon interaction, quasi-particles are well defined only near the Fermi surface.

Zero and first sound

Consider the propagation of a density mode (or sound wave) of frequency ω (and sound velocity v_s). If the relaxation rate is large compared to the frequency ($\omega\tau\ll 1$) there are many quasi-particle collisions during one oscillation of the wave – i.e. the Fermi liquid is in local thermodynamic equilibrium with the oscillation. In this regime, hydrodynamics holds and one knows that the sound mode, also called first sound, travels with a sound velocity v_s given by the local compressibility, $v_\mathrm{s}^2=(n/m)(\partial\mu/\partial n)$ (see (80)), and a damping proportional to $\omega^2\tau$ – i.e. a frequency and temperature dependence of $(\omega/T)^2$ (Landau & Lifshitz 1959).

On the other hand, at sufficiently low temperatures $1/\tau$ can decrease until the relaxation rate is small compared to the frequency ($\omega\tau\gg 1$). In this so-called collisionless regime, the quasi-particles are *not* in local thermodynamic (or hydrodynamic) equilibrium. Once again a sound mode can propagate. This mode, called *zero sound*, has a sound velocity with a different dependence on the Fermi liquid parameters from first sound and an attenuation that is proportional to $1/\tau$ – i.e. an attenuation proportional to T^2 with *no* frequency dependence. This new mode has been seen experimentally (Abel, Anderson & Wheatley 1966) at low temperature, with a transition to ordinary sound with increasing temperature as $\omega\tau$ drops below unity. The results of this experiment, which demonstrated that Landau–Fermi liquid theory is predictive, are shown in Fig. 2.6. The natural log of the attenuation is plotted versus ln T in Fig. 2.6(a). At low temperature (below 10 mK) the slope is two and independent of the two frequencies used in this experiment, which is

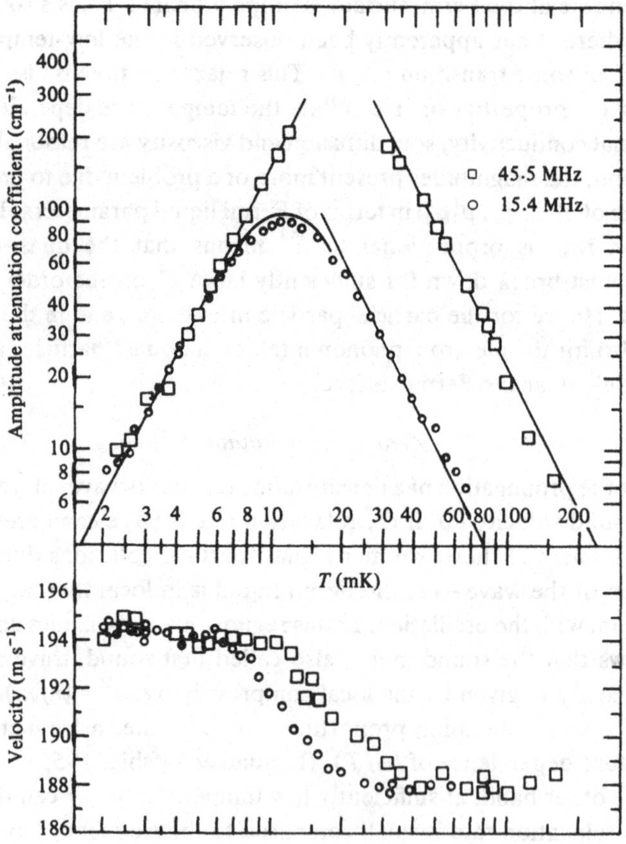

Fig. 2.6. Amplitude attenuation coefficient and sound propagation velocity as a function of temperature in ^{3}He liquid at 0.32 atm. and for frequencies 15.4 and 45.5 MHz (Abel *et al.* 1966).

consistent with the attenuation of zero sound being proportional to τ^{-1} or T^2. The continuously increasing attenuation with temperature reverses itself at higher temperatures and the slope becomes *minus* two. Here the attenuation is proportional to the square of the frequency. This is, of course, consistent with the prediction for ordinary, or first, sound. The change in velocity occurs at the same temperature as for the damping (Fig. 2.6(*b*)). The velocities for zero and ordinary sound do not differ by much even though the formulae determining them are quite different. The measured zero sound and first sound velocities agree well with the predicted values based on the Fermi liquid parameters determined from the compressibility and specific heat.

There is a similarly impressive result for metals. Schultz & Dunifer (1967) observed a spin-wave satellite associated with a conduction electron spin resonance line in Na and K. According to the theory (Platzman & Wolff 1967) this satellite's existence depends on the Fermi liquid parameter G_0 being non-zero. Hence the experiment permitted a fairly direct determination of G_0. In addition there has been considerable work on magnetoplasma modes, which depend less directly on the details of the Landau–Silin–Boltzmann equation (Platzman & Wolff 1973). The interested reader should consult the literature for more details.

In summary, then, this chapter has provided a brief view of how interactions of the electrons with other electrons yield the basic potential by which band states are defined and of how electron–electron and electron–phonon interactions can lead to a simple picture of weakly interacting quasi-particles from which all properties flow.

Fermi liquid calipers

Given the genesis of this book, the immense contribution of David Shoenberg to our understanding of electrons at the Fermi surface, it seems only proper to end with a brief description of the Fermi surface from the many-body point of view. The obvious questions are: (i) Is there really a sharp Fermi surface? (ii) Does the sum over all states inside the Fermi sea equal the number of electrons? and (iii) Do dHvA and other Fermi surface caliper experiments measure the Fermi surface with no many-body corrections? That yes is the answer to all three questions is obvious from the way they were stated. These questions are surprisingly deep, and here there can hardly be more than a listing of what is known, rather than a detailed discussion.

Sharpness of the Fermi surface. That at zero temperature there is a *discontinuous* change in the distribution function, i.e. a sharply defined surface, as the energy passes through the Fermi energy was first shown by Migdal (1957). The proof of these results depends on properties of the self-energy, a concept referred to in Section 2.4. In the course of the proof a disturbing fact emerges: the occupancy of plane-wave states (at zero temperature) is not unity inside the Fermi sea and zero outside. Instead, as shown in Fig. 2.7, the plane-wave occupancy factor differs strongly from the independent-particle distribution function and, from what is equivalent in Fermi liquid theory, the equilibrium quasi-particle distribution function. The reason for this arises from the fact that in order to achieve the lowest ground-state energy, plane-wave momentum states above the Fermi wave-vector are occupied (with a resulting larger kinetic energy) in order to take proper advantage of the attractive exchange-correlation energy. The low-lying excited states – i.e. the quasi-particles – have wave functions which can be thought of as strongly renormalized plane-wave states. As a result, when one calculates matrix elements for scattering between quasi-particles, the resulting matrix elements are greatly reduced from their plane-wave values. That, in a nutshell, is the 'miracle' of the Fermi liquid theory: The process of renormalizing plane-wave states into quasi-particle states of unit occupancy produce weakly interacting quasi-particles, with no terrible mush left over from the renormalization process.

Volume of the Fermi sea. The volume of the Fermi sea is defined by that surface, the Fermi surface, where the quasi-particle energies are equal to the chemical potential of the interacting system. The miracle is that many-body theory can show (Luttinger 1961) that the volume is equal to the volume of the non-interacting Fermi sea. If the resulting Fermi surface is characterized by some $\mathbf{k}_F$ (which may, for a metal, map out a

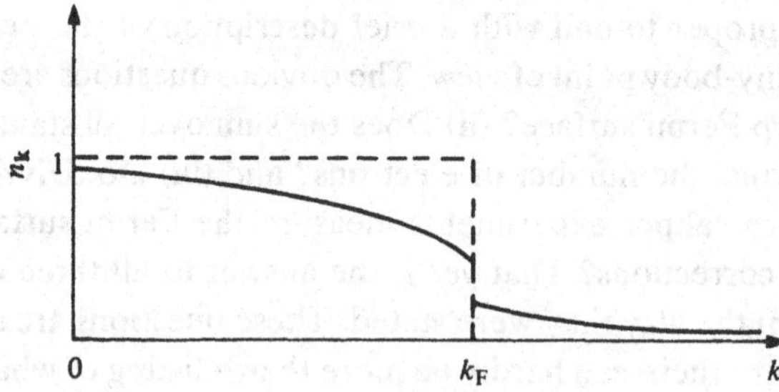

Fig. 2.7. Effect of many-body interactions on plane-wave occupancy factor. Independent-particle model shown with dashed lines. Note that there is still finite discontinuity at the Fermi level.

complicated surface or set of surfaces), then

$$\frac{1}{V}\sum_{s,\mathbf{k}<\mathbf{k}_F} 1 = n,$$

where n is the density of fermions. This result is an essential part of the first of Landau's assumptions and is also essential to constructing the effective potential (see (8)) used to calculate band-structure levels.

Measuring the Fermi surface. Given that there is a sharp Fermi surface containing a well-characterized volume, can it be measured? The clue to this follows from the fact that the wave-vector, by which states are counted, is unaffected by the renormalization process. In the case of dHvA, the magnetic field enters the Hamiltonian via the kinetic energy: $\mathbf{p}$ going into $\mathbf{p}-e\mathbf{A}/c$ as a consequence of gauge invariance. Landau's argument for quantized levels follows directly from this Hamiltonian form, and hence the periods of the oscillation in the magnetization are independent of electron–electron and electron–phonon interactions and serve as a direct caliper of the Fermi surface. This argument was first pointed out by Luttinger (1961). A more detailed microscopic theory was given by Bychkov & Gor'kov (1962). Similar treatments can be given for other caliper experiments. For example, the wave-function-dependent conductivity can be calculated exactly from the Boltzmann equation. Specifically,

$$\sigma(q) = e^2 k_F^2/(4\pi\hbar q)$$

and is free of Fermi liquid effects. But this is the conductivity, needed to calculate the anomalous skin effect (Kittel 1963), that was first used to measure a Fermi surface, that of Cu (Pippard 1957a). How fortunate it is that these two effects used to caliper the Fermi surface survive intact from the cauldron of the interacting Fermi liquid.

Acknowledgements

This chapter is loosely based on some lectures given at Nordita in Copenhagen in 1968. Those notes (Wilkins 1968) were prepared with the considerable assistance of Henrik Smith, now Professor at the University of Copenhagen.

The author must acknowledge that any errors in content, syntax, or spelling that have crept into the originally flawless manuscript are due to the expertise of four Cornell University physics graduate students: Paul Muzikar, Jorge Müller, Roy Richter and David Wood, who read this chapter during its rapid production. The physical preparation was masterminded by Maureen Barnato, who produced order out of chaos.

3

The generalized path-integral approach to transport problems

R. G. CHAMBERS

3.1 Introduction

In the hands of David Shoenberg and others, the de Haas–van Alphen (dHvA) effect has become by far the most powerful method of studying the properties of conduction electrons in pure metals and very dilute alloys. Part of its power derives from the fact that the magnetic susceptibility is not a transport property but a thermodynamic property: to develop the theory of the effect, one does not need to solve a transport equation. Most of the other classic tools used in Fermi surface studies – the anomalous skin effect, cyclotron resonance, the r.f. size effect, magnetoresistance and so on – do, however, involve the solution of a transport equation, usually a fairly complicated one. In tackling such problems, the path-integral method (Chambers 1952) has sometimes proved a useful alternative starting point to the Boltzmann equation. As originally formulated, this method involved the assumption that collisions could be represented by a well-defined relaxation time $\tau(\mathbf{k})$, independent of the nature of the perturbation. More recently, it has become clear that this is a very restrictive assumption, and interest has turned to a more realistic treatment of the scattering process. In this chapter we look at the ways in which the path-integral method can be generalized to take proper account of scattering.

3.2 The Boltzmann equation

The path-integral approach yields simply an integrated form of the Boltzmann equation (Chambers 1950); as in the corresponding neutron transport problem (Davison 1957), it is largely a matter of taste whether one constructs the path-integral expression directly from first principles or by integration of the Boltzmann equation. A first-principles approach has perhaps the advantage that the physical origin of the resulting expression is somewhat clearer, and that any boundary conditions can be included more directly, but the limitations of the path-integral method are in any case precisely the same as those of the Boltzmann equation, and it is worth recalling these limitations at the outset.

First, we need to treat the electrons semi-classically, as localized wave-packets, so that we can treat the distribution function $f(\mathbf{k}, \mathbf{r}, t)$ as a function of position $\mathbf{r}$ as well as of wave-vector $\mathbf{k}$ and time t. Electron–electron collisions are normally neglected, and collisions with phonons and lattice imperfections assumed to be of sufficiently short duration that each collision can be treated as a separate event, independent of other collisions; the assumption that collisions are uncorrelated also involves the 'repeated random phase' approximation. Further, one assumes that collisions are sufficiently infrequent for the energy of the electron between collisions to remain well-defined.

The assumptions involved in using the Boltzmann equation have been well reviewed by Chester (1963) and Kubo (1973). Fortunately, there is good evidence that these assumptions lead to no appreciable error in practice: the assumptions can be avoided by using instead the Kubo–Greenwood density-matrix approach to transport problems, and where this has been done the results are generally in full agreement with those obtained from the Boltzmann equation (see e.g. Chester 1963), at any rate for the reasonably pure crystalline solids that concern us here. We can therefore be confident that these assumptions will not lead us far astray.

In terms of the variables $\mathbf{k}$, $\mathbf{r}$, t, we can write the Boltzmann equation in the form

$$\frac{\partial f_{\mathbf{k}}}{\partial t} = -\frac{\partial f_{\mathbf{k}}}{\partial \mathbf{r}} \cdot \mathbf{v} - \frac{\partial f_{\mathbf{k}}}{\partial \mathbf{k}} \cdot \dot{\mathbf{k}} + \left(\frac{\partial f_{\mathbf{k}}}{\partial t}\right)_{\text{coll}}, \tag{1}$$

where for brevity we have written $f_{\mathbf{k}}$ for $f(\mathbf{k}, \mathbf{r}, t)$. For the collision term, we have as usual

$$\left(\frac{\partial f_{\mathbf{k}}}{\partial t}\right)_{\text{coll}} = -\int d\mathbf{k}' \, [Q_{\mathbf{k}\mathbf{k}'} f_{\mathbf{k}}(1-f_{\mathbf{k}'}) - Q_{\mathbf{k}'\mathbf{k}} f_{\mathbf{k}'}(1-f_{\mathbf{k}})], \tag{2}$$

where $Q_{\mathbf{k}\mathbf{k}'}\, d\mathbf{k}'$ is the probability per unit time that an electron will be scattered out of state $\mathbf{k}$ into a volume element $d\mathbf{k}'$ around state $\mathbf{k}'$ if $\mathbf{k}$ is occupied and $\mathbf{k}'$ is empty; $Q_{\mathbf{k}'\mathbf{k}}\, d\mathbf{k}'$ is the converse probability of scattering from $\mathbf{k}'$ to $\mathbf{k}$ if $\mathbf{k}'$ is occupied and $\mathbf{k}$ is empty. The first term on the right of

(2), integrated over all $\mathbf{k}'$ (and summed over bands, if interband scattering is important) thus gives the total rate of scattering out of $\mathbf{k}$ into other states, and the second term gives the total rate of scattering in from other states.

When (2) is inserted into (1), the resulting integral equation for $f_\mathbf{k}$ is complicated by the presence of the non-linear terms in $f_\mathbf{k} f_{\mathbf{k}'}$, and one usually confines attention to situations where the non-linear terms either vanish or can be neglected. In non-degenerate semiconductors, for example, where $f_\mathbf{k}, f_{\mathbf{k}'} \ll 1$, we can set $(1-f_\mathbf{k})$ and $(1-f_{\mathbf{k}'}) \approx 1$, and (2) reduces to

$$\left(\frac{\partial f_\mathbf{k}}{\partial t}\right)_{\text{coll}} = -\int \mathrm{d}\mathbf{k}' \,(Q_{\mathbf{k}\mathbf{k}'} f_\mathbf{k} - Q_{\mathbf{k}'\mathbf{k}} f_{\mathbf{k}'}), \tag{3}$$

where $Q_{\mathbf{k}\mathbf{k}'}$ and $Q_{\mathbf{k}'\mathbf{k}}$ need not be equal if the scattering is inelastic. In degenerate systems we cannot make this approximation, but now the non-linear terms vanish identically if the scattering is purely elastic, because for elastic scattering $Q_{\mathbf{k}\mathbf{k}'} = Q_{\mathbf{k}'\mathbf{k}}$. This follows from the fact that in thermal equilibrium we must have $(\partial f/\partial t)_{\text{coll}} = 0$, and the principle of detailed balance requires, whether the scattering is elastic or inelastic, that

$$Q_{0\mathbf{k}\mathbf{k}'} f_{0\mathbf{k}}(1 - f_{0\mathbf{k}'}) = Q_{0\mathbf{k}'\mathbf{k}} f_{0\mathbf{k}'}(1 - f_{0\mathbf{k}}) \quad (= P_{\mathbf{k}\mathbf{k}'}, \text{ say}). \tag{4}$$

Here Q_0, f_0 are the values of Q and f in thermal equilibrium, so that $f_{0\mathbf{k}}$, for example, is the Fermi–Dirac function for energy $\mathscr{E}_\mathbf{k}$. If the scattering is elastic, so that the transition probabilities vanish unless $\mathscr{E}_\mathbf{k} = \mathscr{E}_{\mathbf{k}'}$, we have $f_{0\mathbf{k}} = f_{0\mathbf{k}'}$, and it then follows from (4) that $Q_{0\mathbf{k}\mathbf{k}'} = Q_{0\mathbf{k}'\mathbf{k}}$. Since we do not expect departures from equilibrium to affect Q for elastic scattering, (2) thus reduces to

$$\left(\frac{\partial f_\mathbf{k}}{\partial t}\right)_{\text{coll}} = -\int \mathrm{d}\mathbf{k}' \, Q_{\mathbf{k}\mathbf{k}'}(f_\mathbf{k} - f_{\mathbf{k}'}), \tag{5}$$

which can clearly also be written in the form

$$\left(\frac{\partial f_\mathbf{k}}{\partial t}\right)_{\text{coll}} = -\int \mathrm{d}\mathbf{k}' \, Q_{\mathbf{k}\mathbf{k}'}(g_\mathbf{k} - g_{\mathbf{k}'}), \tag{5a}$$

where $g_\mathbf{k} = f_\mathbf{k} - f_{0\mathbf{k}}$.

It is a little more laborious to reduce (2) to linear form in degenerate systems with inelastic scattering, for example phonon scattering. Equation (4) must again hold in thermal equilibrium, but we no longer have $f_{0\mathbf{k}} = f_{0\mathbf{k}'}$ because we no longer have $\mathscr{E}_\mathbf{k} = \mathscr{E}_{\mathbf{k}'}$; instead we have $\mathscr{E}_\mathbf{k} =$

$\mathscr{E}_{\mathbf{k}'} \pm \hbar\omega_{\mathbf{q}}$, where $\hbar\omega_{\mathbf{q}}$ is the energy of the phonon, of wave-vector $\mathbf{q}$, emitted or absorbed in the scattering process. It follows that we no longer have $Q_{0\mathbf{k}\mathbf{k}'} = Q_{0\mathbf{k}'\mathbf{k}}$; instead we must have $Q_{0\mathbf{k}\mathbf{k}'}/Q_{0\mathbf{k}'\mathbf{k}} = \exp(\pm\hbar\omega_{\mathbf{q}}/k_BT)$, the sign again depending on whether the phonon is emitted or absorbed. In fact the detailed theory of phonon scattering shows that $Q_{\mathbf{k}\mathbf{k}'}/Q_{\mathbf{k}'\mathbf{k}} = (N_{\mathbf{q}}+1)/N_{\mathbf{q}}$ if the transition $\mathbf{k}\to\mathbf{k}'$ involves emission of a phonon (i.e. if $\mathscr{E}_{\mathbf{k}} = \mathscr{E}_{\mathbf{k}'} + \hbar\omega_{\mathbf{q}}$), and $Q_{\mathbf{k}\mathbf{k}'}/Q_{\mathbf{k}'\mathbf{k}} = N_{\mathbf{q}}/(N_{\mathbf{q}}+1)$ in the converse case. Here $N_{\mathbf{q}}$ is the phonon distribution function; in thermal equilibrium, $N_{\mathbf{q}} = N_{0\mathbf{q}} = [\exp(\hbar\omega_{\mathbf{q}}/k_BT) - 1]^{-1}$, so that $Q_{0\mathbf{k}\mathbf{k}'}/Q_{0\mathbf{k}'\mathbf{k}}$ is indeed equal to $\exp(\pm\hbar\omega_{\mathbf{q}}/k_BT)$.

But, away from thermal equilibrium, $N_{\mathbf{q}}$ may well differ from $N_{0\mathbf{q}}$, just as $f_{\mathbf{k}}$ and $f_{\mathbf{k}'}$ differ from their equilibrium values $f_{0\mathbf{k}}, f_{0\mathbf{k}'}$, and then

$$Q_{\mathbf{k}\mathbf{k}'}/Q_{0\mathbf{k}\mathbf{k}'} = (N_{\mathbf{q}}+1)/(N_{0\mathbf{q}}+1) \text{ or } N_{\mathbf{q}}/N_{0\mathbf{q}}, \tag{6}$$

depending on whether a phonon is emitted or absorbed, with corresponding expressions for $Q_{\mathbf{k}'\mathbf{k}}/Q_{0\mathbf{k}'\mathbf{k}}$. If we introduce functions $\phi_{\mathbf{k}}$, $\phi_{\mathbf{k}'}$ and $\phi_{\mathbf{q}}$ defined by

$$f_{\mathbf{k}} = f_{0\mathbf{k}} - k_BT\phi_{\mathbf{k}}(\partial f_{0\mathbf{k}}/\partial\mathscr{E}),\ f_{\mathbf{k}'} = f_{0\mathbf{k}'} - k_BT\phi_{\mathbf{k}'}(\partial f_{0\mathbf{k}'}/\partial\mathscr{E}),$$
$$N_{\mathbf{q}} = N_{0\mathbf{q}} - k_BT\phi_{\mathbf{q}}(\partial N_{0\mathbf{q}}/\partial\mathscr{E}),$$

and use $k_BT(\partial f_0/\partial\mathscr{E}) = -f_0(1-f_0)$, $k_BT(\partial N_0/\partial\mathscr{E}) = -N_0(1+N_0)$, we can write

$$f_{\mathbf{k}} = f_{0\mathbf{k}}[1+\phi_{\mathbf{k}}(1-f_{0\mathbf{k}})],\ (1-f_{\mathbf{k}}) = (1-f_{0\mathbf{k}})(1-\phi_{\mathbf{k}}f_{0\mathbf{k}}),$$

with corresponding expressions for $f_{\mathbf{k}'}$, and

$$N_{\mathbf{q}} = N_{0\mathbf{q}}[1+\phi_{\mathbf{q}}(1+N_{0\mathbf{q}})],\ (1+N_{\mathbf{q}}) = (1+N_{0\mathbf{q}})(1+\phi_{\mathbf{q}}N_{0\mathbf{q}}).$$

If we insert these expressions in (6) and (2), we find after a little algebra that

$$\left(\frac{\partial f_{\mathbf{k}}}{\partial t}\right)_{\text{coll}} = -\int d\mathbf{k}'\, P_{\mathbf{k}\mathbf{k}'}[\phi_{\mathbf{k}} - \phi_{\mathbf{k}'} \pm \phi_{\mathbf{q}} + \phi_{\mathbf{k}}\phi_{\mathbf{k}'}(f_{0\mathbf{k}} - f_{0\mathbf{k}'}) + \ldots], \tag{7}$$

where $P_{\mathbf{k}\mathbf{k}'}$ is given by (4), and we take $+\phi_{\mathbf{q}}$ if $\mathscr{E}_{\mathbf{k}'} > \mathscr{E}_{\mathbf{k}}$, $-\phi_{\mathbf{q}}$ if $\mathscr{E}_{\mathbf{k}'} < \mathscr{E}_{\mathbf{k}}$. The remaining terms in the exact expression, omitted from (7), involve $\phi_{\mathbf{k}}\phi_{\mathbf{q}}$, $\phi_{\mathbf{k}'}\phi_{\mathbf{q}}$ and $\phi_{\mathbf{k}}\phi_{\mathbf{k}'}\phi_{\mathbf{q}}$.

In most problems of practical interest, we can safely make the 'Bloch assumption' that the phonon equilibrium is not disturbed appreciably by electric currents (or even by heat currents, which in metals are carried mainly by the electrons), and neglect the term $\pm\phi_{\mathbf{q}}$, together with the higher terms involving $\phi_{\mathbf{q}}$. (The one case where this assumption leads to serious error, of course, is in the calculation of the thermoelectric power at intermediate temperatures, where the non-zero $\phi_{\mathbf{q}}$ produces a large 'phonon drag' contribution.) There remains the non-linear term in $\phi_{\mathbf{k}}\phi_{\mathbf{k}'}$. If this term is large enough to be significant, it will lead to non-linearities in the current/voltage characteristic of the metal, i.e. to departures from Ohm's law. Unless we are interested in such non-linearities this term too can be neglected, leaving the linearized expression.

$$\left(\frac{\partial f_{\mathbf{k}}}{\partial t}\right)_{\text{coll}} = -\int \mathrm{d}\mathbf{k}'\, P_{\mathbf{k}\mathbf{k}'}(\phi_{\mathbf{k}} - \phi_{\mathbf{k}'}) \tag{8}$$

$$= -\int \mathrm{d}\mathbf{k}'\, [R_{\mathbf{k}\mathbf{k}'}(f_{\mathbf{k}} - f_{0\mathbf{k}}) - R_{\mathbf{k}'\mathbf{k}}(f_{\mathbf{k}'} - f_{0\mathbf{k}'})] \tag{9}$$

$$= -\int \mathrm{d}\mathbf{k}'\, (R_{\mathbf{k}\mathbf{k}'}g_{\mathbf{k}} - R_{\mathbf{k}'\mathbf{k}}g_{\mathbf{k}'}), \tag{9a}$$

where

$$R_{\mathbf{k}\mathbf{k}'} = Q_{0\mathbf{k}\mathbf{k}'}(1 - f_{0\mathbf{k}'})/(1 - f_{0\mathbf{k}}),\; R_{\mathbf{k}'\mathbf{k}} = Q_{0\mathbf{k}'\mathbf{k}}(1 - f_{0\mathbf{k}})/(1 - f_{0\mathbf{k}'}), \tag{10}$$

so that

$$R_{\mathbf{k}\mathbf{k}'}(\partial f_{0\mathbf{k}}/\partial \mathscr{E}) = R_{\mathbf{k}'\mathbf{k}}(\partial f_{0\mathbf{k}'}/\partial \mathscr{E}). \tag{10a}$$

We note that (9) can still be used when elastic scattering processes and inelastic processes are both present: for the elastic processes, $f_{0\mathbf{k}} = f_{0\mathbf{k}'}$, so that $R_{\mathbf{k}\mathbf{k}'} = Q_{0\mathbf{k}\mathbf{k}'}$ and $R_{\mathbf{k}'\mathbf{k}} = Q_{0\mathbf{k}'\mathbf{k}}$; since $Q_{0\mathbf{k}\mathbf{k}'} = Q_{0\mathbf{k}'\mathbf{k}}$ for elastic processes, (9) or (9a) then reduces to (5) or (5a).

If we interpret the first term in (5) as representing scattering out of $\mathbf{k}$ into other states, and the second as representing scattering into $\mathbf{k}$ from other states, it appears that the effect of the outward scattering is to cause $f_{\mathbf{k}}$ to relax exponentially to zero with a relaxation time $\tau(\mathbf{k})$ given by

$$1/\tau(\mathbf{k}) = \int \mathrm{d}\mathbf{k}'\, Q_{\mathbf{k}\mathbf{k}'}. \tag{11}$$

On the other hand, if we look at (5a) and choose to interpret the first term of *that* equation as the outward scattering, it appears that outward scattering reduces $f_{\mathbf{k}}$ exponentially to $f_{0\mathbf{k}}$, rather than to zero, with $\tau(\mathbf{k})$ again given by (11). Likewise in (9) or (9a) outward scattering produces an exponential relaxation of $f_{\mathbf{k}}$, with relaxation time $\tau(\mathbf{k})$ given by

$$1/\tau(\mathbf{k}) = \int \mathrm{d}\mathbf{k}'\, R_{\mathbf{k}\mathbf{k}'}, \tag{12}$$

but whether $f_\mathbf{k}$ relaxes to zero or to $f_{0\mathbf{k}}$ depends on whether we take the 'outward scattering' term to be $R_{\mathbf{kk}'}f_\mathbf{k}$ or $R_{\mathbf{kk}'}g_\mathbf{k}$. These differences have no physical significance, of course: they merely depend on what we choose to call the outward scattering part of the collision integral. In the path-integral method, we take the outward scattering term in (9) to be $R_{\mathbf{kk}'}f_\mathbf{k}$, so that all the remaining terms contribute to the inward scattering.

By definition, the rate at which electrons are *actually* scattered out of state **k** is in fact given by the first term of (2), which gives a scattering rate

$$1/\tau^0(\mathbf{k}) = \int \mathrm{d}\mathbf{k}'\; Q_{\mathbf{kk}'}(1-f_{\mathbf{k}'}). \tag{13}$$

Since $1-f_{\mathbf{k}'} \sim \frac{1}{2}$ at the Fermi surface, the 'true' lifetime of an electron in state **k** is thus about twice the effective lifetime defined by (12), because the exclusion principle reduces outward scattering by this factor. But this again has no particular physical significance, because the inward scattering is likewise reduced: for all practical purposes, the relevant parameter is the $\tau(\mathbf{k})$ defined by (12) (which reduces to (11) if scattering is elastic).

In the dHvA effect, where we are concerned with the lifetime broadening of electron energy levels, we might at first sight expect $\tau^0(\mathbf{k})$ to be relevant: we could argue that the outward scattering causes the electron wave-function $\psi_\mathbf{k}$ to decay as $\exp[-t/2\tau^0(\mathbf{k})]$ (remembering that $f \propto |\psi|^2$), and this would lead to a broadening $\Delta\mathscr{E} = \frac{1}{2}\hbar/\tau^0(\mathbf{k})$. (The factor of 2 here corresponds to the factor of 2 introduced by Brailsford (1966) to correct Dingle's original treatment of lifetime broadening.) But this overlooks an additional source of lifetime broadening: the uncertainty in the time at which the electron was scattered *into* state **k**. If state **k** is initially empty ($f_\mathbf{k} = 0$), (2) shows that the inward scattering rate is

$$1/\tau^\mathrm{i}(\mathbf{k}) = \int \mathrm{d}\mathbf{k}'\; Q_{\mathbf{k}'\mathbf{k}}f_{\mathbf{k}'}, \tag{13a}$$

and an exact treatment shows that this produces an additional broadening $\frac{1}{2}\hbar/\tau^\mathrm{i}(\mathbf{k})$: the total broadening is thus $\frac{1}{2}\hbar[1/\tau^0(\mathbf{k}) + 1/\tau^\mathrm{i}(\mathbf{k})]$. If we put $f_{\mathbf{k}'} = f_{0\mathbf{k}'}$ and $Q = Q_0$ in (13), (13a), since we are dealing now with a system in thermal equilibrium, it is easily shown that this reduces to $\Delta\mathscr{E} = \frac{1}{2}\hbar/\tau(\mathbf{k})$ with $\tau(\mathbf{k})$ given by (12): here too, $\tau(\mathbf{k})$ turns out to be the relevant parameter for both elastic and inelastic scattering (Mann & Schmidt 1975). (For inelastic scattering, however, the effect of τ on the dHvA signals is very small, for somewhat subtle reasons: see Chapter 9.)

3.3 The relaxation time approximation

Even when the collision integral (2) has been reduced to the linearized form (9), the integral equation which results on combining this with (1) remains fairly intractable, and for many years it has been customary to simplify the problem by making the 'relaxation time approximation', in which it is assumed that the collision term can be written in the form

$$\left(\frac{\partial f_{\mathbf{k}}}{\partial t}\right)_{\text{coll}} = -\frac{f_{\mathbf{k}} - f_{0\mathbf{k}}}{\tau^{e}(\mathbf{k})} = -\frac{g_{\mathbf{k}}}{\tau^{e}(\mathbf{k})}, \tag{14}$$

where we write $\tau^{e}(\mathbf{k})$ to distinguish this 'effective' relaxation time from the $\tau(\mathbf{k})$ defined by (12). There has in the past been some debate over the validity of (14), but this has been largely a matter of semantics rather than physics. Depending on one's viewpoint, (14) is always true, or never true, or true in a small and highly restricted set of situations (Chambers 1974). It is always possible, in principle at least, to solve the full Boltzmann equation (1), (2) for any problem, and once it has been solved, one can insert the resultant values of $(\partial f_{\mathbf{k}}/\partial t)_{\text{coll}}$ and $f_{\mathbf{k}} - f_{0\mathbf{k}}$ in (14) and simply use (14) to *define* an effective relaxation time $\tau^{e}(\mathbf{k})$ (Chambers 1974; Sorbello 1974a). It is more appropriate to speak of this as the relaxation-time formalism, rather than the relaxation time approximation, since clearly no approximation is involved. But the relaxation time so defined will in general differ from one problem to the next, and for any given problem it may well depend on $\mathbf{r}$ and t as well as on $\mathbf{k}$. A problem-dependent relaxation time of this kind is of little help in *solving* the Boltzmann equation, though it may provide a compact way of summarizing the solution, once it has been obtained.

One can indeed seek for an iterative solution of the full Boltzmann equation by starting with the pair of equations (1), (14) (with some suitable initial choice of $\tau^{e}(\mathbf{k})$), evaluating $g_{\mathbf{k}}$, inserting this $g_{\mathbf{k}}$ in (9a) to evaluate $(\partial f_{\mathbf{k}}/\partial t)_{\text{coll}}$, and then using (14) to derive an improved $\tau^{e}(\mathbf{k})$ for further iteration (Chambers 1968). But for the relaxation-time approximation as such to be really useful, $\tau^{e}(\mathbf{k})$ would need to be a 'universal' relaxation time, independent of the form of perturbation, and it is easy to see that this will never happen in practice (Chambers 1968, 1969). Briefly, one would need the integral over the second term in (9a) to vanish, whatever the form of $g_{\mathbf{k}'}$. Since the only condition on $g_{\mathbf{k}'}$ is that imposed by particle conservation, $\int d\mathbf{k}'\, g_{\mathbf{k}'} = 0$, it follows that we would need $R_{\mathbf{kk}'}$ to be a constant, independent of $\mathbf{k}, \mathbf{k}'$: an electron in state $\mathbf{k}'$ would be scattered with equal probability to any other point on the same energy surface (for elastic scattering), or to any other point in **k**-space (for inelastic scattering). The effect of this 'catastrophic' scattering would be to randomize the electron motion completely at each collision: precisely the assumption made in the original formulation of the path-integral method, which thus implicitly made use of the relaxation-time approximation. Moreover, if the scattering were indeed catastrophic, $\tau^e(\mathbf{k})$ could no longer depend on **k**: for elastic scattering it could still depend on energy, but in the presence of inelastic scattering it would need to be the same for all states **k**. In practice, one often assumes τ^e to be **k**-dependent in using the relaxation-time approximation or in using the path-integral method, and usually this logical inconsistency leads to no obvious difficulties, but it can and does sometimes give trouble with particle non-conservation.

Some confusion has been caused in the past by authors asserting that under certain conditions 'a relaxation time exists', when they mean merely that the same formal relaxation time $\tau^e(\mathbf{k})$ applies to two different problems. If scattering is elastic or quasi-elastic, for example, so that (5) holds, it is simple to show that in the relaxation-time formalism the same value of $\tau^e(\mathbf{k})$ applies to the d.c. electrical conductivity and to the d.c. thermal conductivity, and this is the sense in which Wilson (1953) uses the phrase. More generally, in any d.c. transport process we shall have $g_{\mathbf{k}'} = -g_{-\mathbf{k}'}$, by symmetry (except close to the sample surface), so that if $R_{\mathbf{k}'\mathbf{k}} = R_{-\mathbf{k}'\mathbf{k}}$ the second term of (9a) will vanish on integration, and (9a) reduces to the form (14), with $\tau(\mathbf{k})$ given by (12) (Herring & Vogt 1956). The condition $R_{\mathbf{k}'\mathbf{k}} = R_{-\mathbf{k}'\mathbf{k}}$ may indeed be approximately satisfied in some semiconductors, and also sometimes for impurity scattering in metals (Blaker & Harris 1971), but here again the resultant relaxation time is certainly not a 'universal' one: it is relevant only to the d.c. bulk transport properties, for which $g_{\mathbf{k}'} = -g_{-\mathbf{k}'}$.

3.4 The vector mean free path

In discussing the exact solution of the transport problem for arbitrary form of the scattering probability $R_{\mathbf{k}\mathbf{k}'}$, it is convenient to start with the d.c. transport properties: the behaviour of a bulk sample (so that surface effects can be neglected) under the influence of a uniform steady applied field **E** or temperature gradient $\mathbf{G} = -\nabla T$. It was first shown by Price (1957, 1958) that the perturbation $g_{\mathbf{k}}$ to the distribution function can then be described exactly in terms of a vector mean free path **L**:

$$g_{\mathbf{k}} = -e\mathbf{E} \cdot \mathbf{L}_{\mathbf{k}}(\partial f_{0\mathbf{k}}/\partial \mathscr{E}) \tag{15}$$

for an applied field **E**, and

$$g_{\mathbf{k}} = -k_{\mathrm{B}}\eta_{\mathbf{k}}\mathbf{G}\cdot\mathbf{L}_{T\mathbf{k}}(\partial f_{0\mathbf{k}}/\partial\mathscr{E}), \tag{16}$$

where $\eta_{\mathbf{k}} = (\mathscr{E}_{\mathbf{k}} - \mathscr{E}_{\mathrm{F}})/k_{\mathrm{B}}T$, for an applied temperature gradient **G**. The two free paths **L** and $\mathbf{L}_T$ will not in general be equal, unless the scattering is elastic. Price also showed that the vector mean free path **L** has a simple physical interpretation: it is just the average distance travelled by an electron before passing through a given point in the metal with velocity $\mathbf{v}_{\mathbf{k}}$. Thus if $\langle\mathbf{v}(t)\rangle_{\mathbf{k}}$ is the average velocity at time t of a group of electrons all of which have velocity $\mathbf{v}_{\mathbf{k}}$ at $t = 0$, we can write

$$\mathbf{L}_{\mathbf{k}} = \int_{-\infty}^{0} \langle\mathbf{v}(t)\rangle_{\mathbf{k}}\,\mathrm{d}t. \tag{17}$$

It is perhaps easier to envisage the behaviour of such a group of electrons *after* time $t = 0$: because of collisions, their motion will become increasingly randomized as time goes by, until eventually they lose all 'memory' of their initial velocity $\mathbf{v}_{\mathbf{k}}$. Their mean velocity $\langle\mathbf{v}(t)\rangle_{\mathbf{k}}$ will correspondingly fall to zero for $t \to \infty$, and the mean distance through which they have moved since $t = 0$ tends to the limit

$$\mathbf{L}_{\mathbf{k}}^{+} = \int_{0}^{\infty} \langle\mathbf{v}(t)\rangle_{\mathbf{k}}\,\mathrm{d}t. \tag{18}$$

In just the same way, if we trace the history of this group of electrons backwards in time from $t = 0$, their motion will again become increasingly randomized, so that $\langle\mathbf{v}(t)\rangle_{\mathbf{k}} \to 0$ for $t \to -\infty$, and on the average the distance through which they have moved before $t = 0$ is given by (17). The evolution in space of this group of electrons, before and after $t = 0$, is shown schematically in Fig. 3.1. As we shall see later, $\mathbf{L}_{\mathbf{k}}^{+} = \mathbf{L}_{\mathbf{k}}$ if no magnetic field **B** is applied to the metal; if $\mathbf{B} \neq 0$, we have $\mathbf{L}_{\mathbf{k}}^{+}(\mathbf{B}) = \mathbf{L}_{\mathbf{k}}(-\mathbf{B})$.

Clearly, if individual scattering events only deflect the electron through a small angle, the distance **L** may be substantially greater than the mean distance between individual collisions. The corresponding effect of 'persistence of velocities' has been well known in the kinetic theory of gases for many years (Jeans 1921); in the theory of metals, the idea of an effective free path as a 'forgetting distance' seems to have been first introduced by Supek (1940).

As in the kinetic theory of gases, the effective free path may depend on the phenomenon being studied: as we shall see, $\mathbf{L}_T$ in (16) is given by

$$\eta_{\mathbf{k}}\mathbf{L}_{T\mathbf{k}} = \int_{-\infty}^{0} \langle\eta(t)\mathbf{v}(t)\rangle_{\mathbf{k}}\,\mathrm{d}t, \tag{19}$$

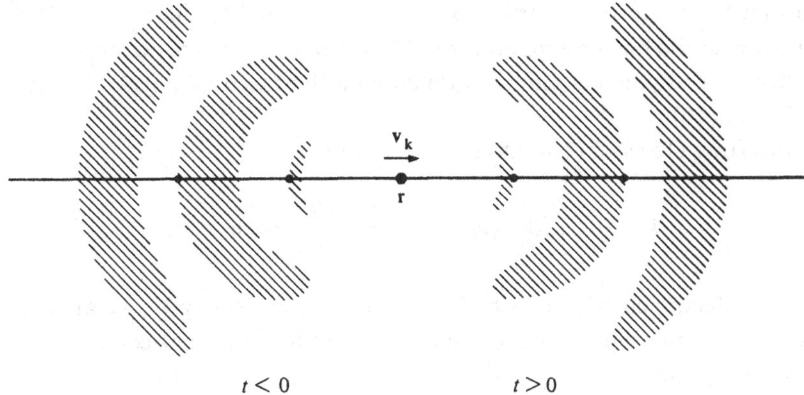

Fig. 3.1. Showing schematically the evolution from $t \ll 0$ to $t \gg 0$ of a group of electrons all of which pass through the point **r** in state **k** at $t=0$. The shaded regions represent scattered electrons, and the dots at the outer edges of the shaded regions represent unscattered electrons, falling in number as $\exp[-|t|/\tau(\mathbf{k})]$. For clarity, the shaded regions are shown as non-overlapping: they would in fact be much larger, extending inwards so that each region overlaps those of smaller $|t|$. For large $|t|$, the shaded regions will be spheres (for free electrons) expanding as $|t|^{\frac{1}{2}}$ and centred about $\mathbf{r} \pm \mathbf{L_k}$.

where $\langle \eta(t)\mathbf{v}(t)\rangle_{\mathbf{k}}$ is the average of the product $\eta\mathbf{v}$ for the group of electrons considered. $\mathbf{L}_T$ thus depends on the rate at which collisions randomize the energy of the electrons, as well as their velocity: inelastic small-angle scattering may randomize η much more rapidly than $\mathbf{v}$, so that $\mathbf{L}_T$ may be much shorter than $\mathbf{L}$.

The form of (15) and (16) is precisely what we should expect on physical grounds. We can rewrite (15) in the form

$$f_{\mathbf{k}} = f_{0\mathbf{k}} - \Delta\mathscr{E}_{\mathbf{k}}(\partial f_{0\mathbf{k}}/\partial\mathscr{E}), \tag{20}$$

where $\Delta\mathscr{E}_{\mathbf{k}} = e\mathbf{E}\cdot\mathbf{L_k}$: the electrons have acquired energy $\Delta\mathscr{E}_{\mathbf{k}}$ in moving distance $\mathbf{L_k}$ through the field $\mathbf{E}$, so that initially, for $t\to-\infty$, they had energy $\mathscr{E}_{\mathbf{k}} - \Delta\mathscr{E}_{\mathbf{k}}$. If at that time they had the unperturbed distribution function f_0 on average, we expect $f_{\mathbf{k}}$ at $t=0$ to be equal to $f_0(\mathscr{E}_{\mathbf{k}} - \Delta\mathscr{E}_{\mathbf{k}})$, and in the linear approximation this at once yields (20). Similarly, we can rewrite (16) in the form

$$f_{\mathbf{k}} = f_{0\mathbf{k}} - k_{\mathrm{B}}\eta_{\mathbf{k}}\Delta T_{\mathbf{k}}(\partial f_{0\mathbf{k}}/\partial\mathscr{E}), \tag{21}$$

where $\Delta T_{\mathbf{k}} = \mathbf{G}\cdot\mathbf{L}_{T\mathbf{k}}$: the electrons have come on average from a region where the temperature is higher by an amount $\Delta T_{\mathbf{k}}$. Using $(\partial f_0/\partial T)_{\mathscr{E}} = -k_{\mathrm{B}}\eta_{\mathbf{k}}(\partial f_0/\partial\mathscr{E})_T$, (21) becomes

$$f_{\mathbf{k}} = f_{0\mathbf{k}} + \Delta T_{\mathbf{k}}(\partial f_{0\mathbf{k}}/\partial T) = f_0(T + \Delta T_{\mathbf{k}}),$$

so that the perturbed distribution function is equal to the unperturbed function at the higher temperature. At this stage, these are no more than plausibility arguments, but they do contain the essential physics behind (15) and (16).

Inserting (15) and (16) into the expressions

$$\mathbf{J}=\frac{e}{4\pi^3}\int d\mathbf{k}\,\mathbf{v}_{\mathbf{k}}g_{\mathbf{k}}, \qquad \mathbf{W}=\frac{k_BT}{4\pi^3}\int d\mathbf{k}\,\eta_{\mathbf{k}}\mathbf{v}_{\mathbf{k}}g_{\mathbf{k}} \tag{22}$$

for the electric current density **J** and heat current density **W**, we at once obtain general expressions for the electrical and thermal conductivity tensors σ_{ij} and κ_{ij}:

$$\sigma_{ij}=-\frac{e^2}{4\pi^3}\int d\mathbf{k}\,\mathbf{v}_{i,\mathbf{k}}\mathbf{L}_{j,\mathbf{k}}\frac{\partial f_{0\mathbf{k}}}{\partial \mathscr{E}} \tag{23}$$

and

$$\kappa_{ij}=-\frac{k_B^2T}{4\pi^3}\int d\mathbf{k}\,\eta_{\mathbf{k}}^2\mathbf{v}_{i,\mathbf{k}}\mathbf{L}_{Tj,\mathbf{k}}\frac{\partial f_{0\mathbf{k}}}{\partial \mathscr{E}}. \tag{24}$$

The d.c. bulk transport properties of a metal can thus be expressed very compactly and elegantly, and without approximation, in terms of the two vectors $\mathbf{L}_{\mathbf{k}}$ and $\mathbf{L}_{T\mathbf{k}}$. As shown by Pippard (1964, 1965), we can in fact equally well express them in terms of the vectors $\mathbf{L}_{\mathbf{k}}^+$, $\mathbf{L}_{T\mathbf{k}}^+$ which describe the mean distance travelled by the electron *after* time $t=0$. If we imagine a field **E** to be applied to the metal at $t=0$ for a very short interval δt, it will change the distribution function from $f_{0\mathbf{k}}$ to $f_{0\mathbf{k}}-\dot{\mathbf{k}}\,\delta t\,(\partial f_{0\mathbf{k}}/\partial \mathbf{k})$ if we neglect the effect of collisions during δt (cf. (1)). The resultant perturbation at $t=0+\delta t$ is thus

$$g_{\mathbf{k}}=-\dot{\mathbf{k}}\,\delta t\,(\partial f_{0\mathbf{k}}/\partial \mathbf{k})=-e\mathbf{E}\cdot\mathbf{v}_{\mathbf{k}}\,\delta t\,(\partial f_{0\mathbf{k}}/\partial \mathscr{E}), \tag{25}$$

and this will thereafter decay back to zero because of collisions. During this decay, the excess charge corresponding to $g_{\mathbf{k}}$ will drift through a mean distance $\mathbf{L}_{\mathbf{k}}^+$ before its motion becomes completely randomized by collisions. If we now think of a steady field **E** as built up of a succession of impulses of this kind, each of which gives rise to a perturbation (25) and a resultant charge displacement, it is easy enough to see that the net result will be a steady current **J** given by

$$\mathbf{J}=-\frac{e^2}{4\pi^3}\int d\mathbf{k}\,(\mathbf{E}\cdot\mathbf{v}_{\mathbf{k}})\mathbf{L}_{\mathbf{k}}^+\frac{\partial f_{0\mathbf{k}}}{\partial \mathscr{E}}, \tag{26}$$

so that

$$\sigma_{ij} = -\frac{e^2}{4\pi^3}\int \mathrm{d}\mathbf{k}\, L^{+}_{i\mathbf{k}} v_{j\mathbf{k}} \frac{\partial f_{0\mathbf{k}}}{\partial \mathscr{E}}. \tag{27}$$

This expression appears to differ significantly from (23), but in fact the two are completely equivalent, because $\mathbf{L}^{+}_{\mathbf{k}}(\mathbf{B}) = \mathbf{L}_{\mathbf{k}}(-\mathbf{B})$, and Onsager thermodynamics tells us that $\sigma_{ij}(\mathbf{B}) = \sigma_{ji}(-\mathbf{B})$. Pippard (1969a) has made elegant use of this 'effective path' approach in a variety of d.c. transport problems.

So far we have assumed without proof that $g_{\mathbf{k}}$ can indeed be written in the form (15) or (16), in terms of a vector mean free path $\mathbf{L}$, and we have given no prescription for actually calculating $\mathbf{L}$. It is easy enough to establish the formal validity of (15) and (16) by using the relaxation time formalism. We recall first that for steady uniform applied fields $\mathbf{E}$ and $\mathbf{G}$ the 'driving' terms in the Boltzmann equation (1) can be written

$$-\mathbf{v}\cdot\frac{\partial f_{\mathbf{k}}}{\partial \mathbf{r}} - \dot{\mathbf{k}}\cdot\frac{\partial f_{\mathbf{k}}}{\partial \mathbf{k}} = -\left[k_{\mathrm{B}}\eta_{\mathbf{k}}\mathbf{v}_{\mathbf{k}}\cdot\mathbf{G} + e\left(\mathbf{E} - \frac{1}{e}\frac{\partial \mathscr{E}_{\mathrm{F}}}{\partial \mathbf{r}}\right)\cdot\mathbf{v}_{\mathbf{k}}\right]\frac{\partial f_{0\mathbf{k}}}{\partial \mathscr{E}}$$

$$-\frac{e}{\hbar c}(\mathbf{v}_{\mathbf{k}}\times\mathbf{B})\cdot\frac{\partial g_{\mathbf{k}}}{\partial \mathbf{k}}, \tag{28}$$

where we have as usual linearized the equation by omitting a term $-(e/\hbar)\mathbf{E}\cdot(\partial g_{\mathbf{k}}/\partial \mathbf{k})$, since this term will vary as E^2, and we are normally interested only in the linear (ohmic) response. For brevity, we can write

$$\mathbf{E} - \frac{1}{e}\frac{\partial \mathscr{E}_{\mathrm{F}}}{\partial \mathbf{r}} = \mathbf{E}^*, \tag{29}$$

where $e\mathbf{E}^*$ is the negative gradient of the electrochemical potential. To avoid the complications of the term in $\mathbf{v}_{\mathbf{k}}\times\mathbf{B}$, we confine ourselves for the moment to the case $\mathbf{B} = 0$. Suppose now that the Boltzmann equation (1), (2) has been solved for an applied field E_x (with $\mathbf{G} = 0$, $\partial\mathscr{E}_{\mathrm{F}}/\partial\mathbf{r} = 0$), and that the resultant values of $g_{\mathbf{k}}$ and $(\partial f_{\mathbf{k}}/\partial t)_{\mathrm{coll}}$ are inserted in (14) to define an effective relaxation time $\tau^{\mathrm{e}}_x(\mathbf{k})$. Then in terms of $\tau^{\mathrm{e}}_x(\mathbf{k})$ we have

$$g_{\mathbf{k}} = -eE_x v_{\mathbf{k}x}\tau^{\mathrm{e}}_x(\mathbf{k})(\partial f_{0\mathbf{k}}/\partial \mathscr{E}).$$

The response to fields E_y, E_z can be written in similar form, and it then follows at once by superposition that in the linear approximation the response to any field $\mathbf{E}$ can be written in the form (15), with $\mathbf{L}_{\mathbf{k}} = (v_{\mathbf{k}x}\tau^{\mathrm{e}}_x(\mathbf{k}),\ v_{\mathbf{k}y}\tau^{\mathrm{e}}_y(\mathbf{k}),\ v_{\mathbf{k}z}\tau^{\mathrm{e}}_z(\mathbf{k}))$. Since the three relaxation times $\tau^{\mathrm{e}}_i(\mathbf{k})$ will not in general be equal, $\mathbf{L}_{\mathbf{k}}$ will not in general be in the same direction as $\mathbf{v_k}$. The response to a field $\mathbf{G}$, with $\mathbf{E}^*=0$, can be shown in the same way to have the form (16).

To obtain an expression for $\mathbf{L_k}$ in terms of the collision integral (9a), we merely have to write the Boltzmann equation in the form

$$e\mathbf{E}\cdot\mathbf{v_k}(\partial f_{0\mathbf{k}}/\partial\mathscr{E})=(\partial f_{\mathbf{k}}/\partial t)_{\text{coll}}=-\int d\mathbf{k}'\,(R_{\mathbf{kk}'}g_{\mathbf{k}}-R_{\mathbf{k}'\mathbf{k}}g_{\mathbf{k}'}) \tag{30}$$

and re-express $g_{\mathbf{k}}$ in terms of $\mathbf{L_k}$, using (15). Since the resulting equation must be true for arbitrary $\mathbf{E}$, we thus find

$$\mathbf{v_k}=\int d\mathbf{k}'\left(R_{\mathbf{kk}'}\mathbf{L_k}-\frac{\partial f_{0\mathbf{k}'}/\partial\mathscr{E}}{\partial f_{0\mathbf{k}}/\partial\mathscr{E}}R_{\mathbf{k}'\mathbf{k}}\mathbf{L}_{\mathbf{k}'}\right).$$

Using (10a), the second term reduces to $-R_{\mathbf{kk}'}\mathbf{L}_{\mathbf{k}'}$, so that we can rewrite this equation in the form

$$\mathbf{L_k}=\mathbf{v_k}\tau(\mathbf{k})+\tau(\mathbf{k})\int d\mathbf{k}'\,R_{\mathbf{kk}'}\mathbf{L}_{\mathbf{k}'}, \tag{31}$$

with $\tau(\mathbf{k})$ given by (12). This integral equation for $\mathbf{L_k}$ can then be solved by iteration (Price 1957; Taylor 1963).

To find an expression for $\mathbf{L}_T$, we replace $e\mathbf{E}\cdot\mathbf{v_k}$ by $k_B\eta_{\mathbf{k}}\mathbf{G}\cdot\mathbf{v_k}$ on the left of (30), and express $g_{\mathbf{k}}$ in terms of $\mathbf{L}_{T\mathbf{k}}$, using (16). In place of (31), we then find

$$\eta_{\mathbf{k}}\mathbf{L}_{T\mathbf{k}}=\eta_{\mathbf{k}}\mathbf{v_k}\tau(\mathbf{k})+\tau(\mathbf{k})\int d\mathbf{k}'\,R_{\mathbf{kk}'}\eta_{\mathbf{k}'}\mathbf{L}_{T\mathbf{k}'}. \tag{32}$$

3.5 Path-integral calculation of L

Equations (31) and (32) hold only for $\mathbf{B}=0$, and it is difficult to generalize them to $\mathbf{B}\neq0$ if we continue to follow the Boltzmann equation approach that we have used so far, because of the rather intractable way that $\mathbf{B}$ appears in (28). Life becomes much simpler if we change now to the path-integral approach, and calculate $\mathbf{L}$ (or $\mathbf{L}^+$) directly by tracing out the history of a group of electrons before (or after) time $t=0$. In a magnetic field, the electrons follow a curved path in both $\mathbf{r}$-space and $\mathbf{k}$-space between collisions, and since $\mathbf{k}$ is changing with time, the scattering rate $1/\tau(\mathbf{k})$ will also change with time. In the linear approximation, $\mathbf{L}$ will be independent of the applied field $\mathbf{E}$, and it suffices to calculate $\mathbf{L}$ in zero electric field, $\mathbf{E}=0$. The net scattering rate (9) then vanishes, of course,

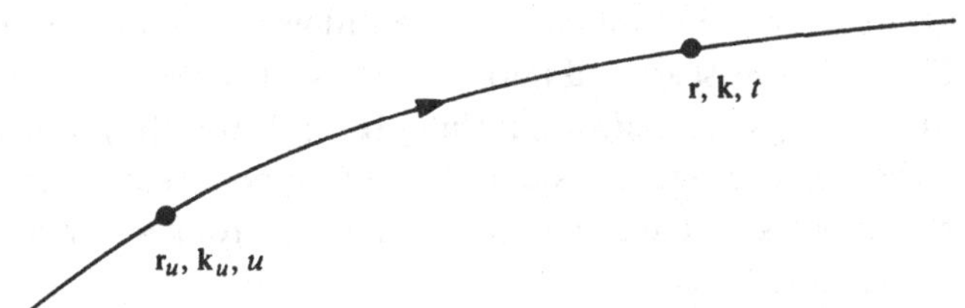

Fig. 3.2. Illustrating the path-integral calculation of $f(\mathbf{r}, \mathbf{k}, t)$.

because $f_\mathbf{k} = f_{0\mathbf{k}}$, $f_{\mathbf{k}'} = f_{0\mathbf{k}'}$, but in the light of the discussion following equation (12) we can regard this as a balance between outward scattering at a rate $f_\mathbf{k}/\tau(\mathbf{k})$ (with $f_\mathbf{k} = f_{0\mathbf{k}}$) and inward scattering at the same rate.

Suppose that an electron which passes through the point $\mathbf{r}$, $\mathbf{k}$ at time t follows a path which, in the absence of scattering, takes it through the point $\mathbf{r}_u$, $\mathbf{k}_u$ at time u. Then we can regard $f(\mathbf{r}, \mathbf{k}, t)$ as made up of all those electrons which have been scattered into this path from other states $\mathbf{k}'$ at times $u < t$, and which have then survived until time t without further scattering (Fig. 3.2). More precisely, the number of electrons in volume element $\mathrm{d}\mathbf{r}\, \mathrm{d}\mathbf{k}$ around $\mathbf{r}$, $\mathbf{k}$ will be $(4\pi^3)^{-1} f(\mathbf{r}, \mathbf{k}, t)\, \mathrm{d}\mathbf{r}\, \mathrm{d}\mathbf{k}$, and these will be the survivors of those scattered at earlier times into a volume element $\mathrm{d}\mathbf{r}_u\, \mathrm{d}\mathbf{k}_u$ about $\mathbf{r}_u$, $\mathbf{k}_u$. But (as in the derivation of the Boltzmann equation) we can put $\mathrm{d}\mathbf{r}_u\, \mathrm{d}\mathbf{k}_u = \mathrm{d}\mathbf{r}\, \mathrm{d}\mathbf{k}$, by Liouville's theorem, so that we can omit the constant factor $(4\pi^3)^{-1}\, \mathrm{d}\mathbf{r}\, \mathrm{d}\mathbf{k}$, and speak loosely of $f(\mathbf{r}, \mathbf{k}, t)$ as the number of electrons at $\mathbf{r}$, $\mathbf{k}$, t.

Because of the variation of $\mathbf{k}_u$ with time, $\tau(\mathbf{k}_u)$ will also vary with time: we have

$$1/\tau(\mathbf{k}_u) = \int \mathrm{d}\mathbf{k}'\, R(\mathbf{k}_u, \mathbf{k}').$$

In equilibrium, then (i.e. for $\mathbf{E} = 0$), the number of electrons scattered into the path in time $\mathrm{d}u$ will be $f_0(\mathbf{k}_u)\, \mathrm{d}u/\tau(\mathbf{k}_u)$, and their chance of survival until time t will be

$$\exp - \int_u^t \mathrm{d}s/\tau(\mathbf{k}_s).$$

But since $\mathbf{E} = 0$, we have $\mathscr{E}(\mathbf{k}_u) = \mathscr{E}(\mathbf{k})$, and hence $f_0(\mathbf{k}_u) = f_0(\mathbf{k})$, so that for the total number of electrons at $\mathbf{r}$, $\mathbf{k}$, t we find

$$f(\mathbf{r}, \mathbf{k}, t) = f_0(\mathbf{k}) \int_{-\infty}^t \mathrm{d}u\, [\tau(\mathbf{k}_u)]^{-1} \exp - \int_u^t \mathrm{d}s/\tau(\mathbf{k}_s), \qquad (33)$$

$=f_0(\mathbf{k})$, as expected. Of more interest, we can use the same approach to find $\mathbf{L}_\mathbf{k}$. The electrons scattered into $\mathbf{k}_u$ from $\mathbf{k}'$ at time u will thereafter travel a distance $\int_u^t \mathbf{v}_s\,ds$ before reaching $(\mathbf{r}, \mathbf{k}, t)$, but they will *already* have travelled, on average, a distance $\mathbf{L}_{\mathbf{k}'}$ before being scattered into $\mathbf{k}_u$. Since the number scattered in time du into $\mathbf{k}_u$ from $\mathbf{k}'$ (or a volume element $d\mathbf{k}'$ around $\mathbf{k}'$) is

$$f_0(\mathbf{k}_u)\,du\,R(\mathbf{k}_u, \mathbf{k}')\,d\mathbf{k}'$$

(from the second of the four terms in (9)), the mean distance previously travelled by electrons arriving at $\mathbf{r}$, $\mathbf{k}$ at $t=0$ will be

$$\mathbf{L}_\mathbf{k} = \int_{-\infty}^0 du\left[\exp - \int_u^0 ds/\tau(\mathbf{k}_s)\right]\int d\mathbf{k}'\,R(\mathbf{k}_u, \mathbf{k}')\left[\int_u^0 \mathbf{v}_s\,ds + \mathbf{L}_{\mathbf{k}'}\right]. \quad (34)$$

In this rather cumbersome-looking expression, the exponential term gives the probability of survival, the term in $R(\mathbf{k}_u, \mathbf{k}')$ gives the rate of scattering in from state $\mathbf{k}'$, and the last term gives the total distance travelled by those scattered in from $\mathbf{k}'$ at time u.

We can simplify (34) a little by noting that

$$\int d\mathbf{k}'\,R(\mathbf{k}_u, \mathbf{k}')\int_u^0 \mathbf{v}_s\,ds = \int_u^0 \mathbf{v}_s\,ds/\tau(\mathbf{k}_u),$$

and that the resulting term in (34) can be integrated by parts: we thus find

$$\mathbf{L}_\mathbf{k} = \int_{-\infty}^0 du\left[\exp - \int_u^0 ds/\tau(\mathbf{k}_s)\right]\left[\mathbf{v}(u) + \int d\mathbf{k}'\,R(\mathbf{k}_u, \mathbf{k}')\mathbf{L}_{\mathbf{k}'}\right] \quad (35)$$

as the integral equation for $\mathbf{L}_\mathbf{k}$ for $\mathbf{B}\neq 0$. We note that when $\mathbf{B}=0$, so that $\tau(\mathbf{k}_u)$, $\mathbf{v}(u)$ and $R(\mathbf{k}_u, \mathbf{k}')$ become time-independent and equal to their values at $\mathbf{k}_u=\mathbf{k}$, (35) reduces at once to (31). The corresponding generalization of (32) is obvious: we find

$$\eta_\mathbf{k}\mathbf{L}_{T\mathbf{k}} = \int_{-\infty}^0 du\left[\exp\left(-\int_u^0 ds/\tau(\mathbf{k}_s)\right)\right]\left[\eta_\mathbf{k}\mathbf{v}(u) + \int d\mathbf{k}'\,R(\mathbf{k}_u, \mathbf{k}')\eta_{\mathbf{k}'}\mathbf{L}_{T\mathbf{k}'}\right]. \quad (36)$$

It is also obvious that in order to calculate $\mathbf{L}^+$, we merely have to follow the evolution of the system in time *after* $t=0$, rather than *up to* $t=0$; in other words, we replace $\int_{-\infty}^0 du$ and $\int_u^0 ds$ by $\int_0^\infty du$ and $\int_0^u ds$. If at the same time we reverse the sign of $\mathbf{B}$, we shall reverse the direction in which the electrons traverse their paths through $\mathbf{k}$-space, so that $\mathbf{k}(u, \mathbf{B})$ becomes $\mathbf{k}(-u, -\mathbf{B})$, $\mathbf{v}(u, \mathbf{B})$ becomes $\mathbf{v}(-u, -\mathbf{B})$, etc., in an obvious notation. The net effect of both changes is thus merely to relabel the time variable throughout by changing its sign, so that we do indeed find $\mathbf{L_k}(\mathbf{B}) = \mathbf{L_k^+}(-\mathbf{B})$, as required to reconcile (23) and (27) with the Onsager relations.

3.6 Thermoelectric effects

It is worth looking briefly at the thermoelectric effects, because in doing so we uncover an interesting and unexpected relationship between $\mathbf{L}$ and $\mathbf{L}_T$.

When an electric field and a temperature gradient are both present, the total perturbation $g_\mathbf{k}$ (in the linear approximation) is simply the sum of those produced by the two fields separately: adding (15) and (16), we find

$$g_\mathbf{k} = -(\partial f_{0\mathbf{k}}/\partial \mathscr{E})(e\mathbf{E}^* \cdot \mathbf{L_k} + k_\mathrm{B}\eta_\mathbf{k}\mathbf{G} \cdot \mathbf{L}_{T\mathbf{k}}), \tag{37}$$

where the replacement of $\mathbf{E}$ by $\mathbf{E}^*$ allows for the diffusion term $\partial\mathscr{E}_\mathrm{F}/\partial\mathbf{r}$ (cf. (29)). Inserting (37) in (22), we find

$$\begin{aligned} J_i &= \sigma_{ij}E_j^* + B_{ij}G_j, \\ W_i &= A_{ij}E_j^* + \kappa_{ij}G_j, \end{aligned} \tag{38}$$

where σ_{ij} and κ_{ij} are given by (23), (24) as before, and

$$A_{ij} = -\frac{ek_\mathrm{B}T}{4\pi^3}\int \mathrm{d}\mathbf{k}\left(\eta v_i L_j \frac{\partial f_0}{\partial \mathscr{E}}\right)_\mathbf{k}, \qquad B_{ij} = -\frac{ek_\mathrm{B}}{4\pi^3}\int \mathrm{d}\mathbf{k}\left(\eta v_i L_{Tj} \frac{\partial f_0}{\partial \mathscr{E}}\right)_\mathbf{k}. \tag{39}$$

The Peltier coefficient Π and the thermoelectric power $\mathcal{E}$ are defined by $W_i = \Pi_{ij}J_j$, with $\mathbf{G}=0$, and $E_i^* = -\mathcal{E}_{ij}G_j$, with $\mathbf{J}=0$, and from (38) it follows that $\Pi_{ij} = A_{ik}\rho_{kj}$ and $\mathcal{E}_{ij} = \rho_{ik}B_{kj}$, where we sum over repeated indices, and ρ_{ij} is the resistivity tensor, so that $\rho_{ij}\sigma_{jk} = \delta_{ik}$. Now the Onsager relations tell us that $\Pi_{ij}(\mathbf{B}) = T\mathcal{E}_{ji}(-\mathbf{B})$, and since $\rho_{ij}(\mathbf{B}) = \rho_{ji}(-\mathbf{B})$, it follows that A_{ij} and B_{ij} must satisfy $A_{ij}(\mathbf{B}) = TB_{ji}(-\mathbf{B})$. Using $\mathbf{L}_T^+(\mathbf{B}) = \mathbf{L}_T(-\mathbf{B})$, this means that the following identity must hold:

$$\int \mathrm{d}\mathbf{k}\left(\eta v_i L_j \frac{\partial f_0}{\partial \mathscr{E}}\right)_\mathbf{k} = \int \mathrm{d}\mathbf{k}\left(\eta v_j L_{Ti}^+ \frac{\partial f_0}{\partial \mathscr{E}}\right)_\mathbf{k}. \tag{40}$$

We encounter exactly the same result, without invoking the Onsager relations explicitly, if we compare (39) with the expressions for A_{ij} and B_{ij} obtained using the 'effective path' method, starting from equation (25) (generalized to include $\mathbf{G}$ as well as $\mathbf{E}^*$). This is not surprising: the equivalence of the path-integral method and the effective path method depends on microscopic reversibility, as do the Onsager relations. (And the equivalence breaks down, as do the Onsager relations, in situations where microscopic reversibility does not apply, in the sense that $\mathbf{L}^+(\mathbf{B}) \neq \mathbf{L}(-\mathbf{B})$: for example, within a distance L of a boundary surface.)

The result (40) is on the face of it a little hard to believe, because, in the presence of small-angle inelastic scattering, for example, we may have $\mathbf{L}_T \ll \mathbf{L}$. But because of the factor $\eta_\mathbf{k}$, which changes sign at $\mathscr{E}_\mathbf{k} = \mathscr{E}_F$, we are dealing here (on each side of the equation) with a delicate balance between the contributions from $\mathscr{E}_\mathbf{k} > \mathscr{E}_F$ and $\mathscr{E}_\mathbf{k} < \mathscr{E}_F$, and intuition is not a safe guide. In fact Price (1957) showed long ago that (40) could be verified explicitly, at least for $\mathbf{B} = 0$. For $\mathbf{B} = 0$, we can replace $\mathbf{L}^+_{Tj}$ by $\mathbf{L}_{Tj}$, and use (31) and (32) to express $\mathbf{v}$ in terms of $\mathbf{L}_T$ on the left, and in terms of $\mathbf{L}$ on the right. Using (10a), the identity (40) then follows.

It is not quite so easy to extend the proof to $\mathbf{B} \neq 0$, because (35) and (36) do not lend themselves to the same kind of manipulation as (31), (32). Instead, we adopt a different approach. Let $p_\mathbf{B}(\mathbf{k}, \mathbf{k}', t)$ be the probability that an electron which is in state $\mathbf{k}$ at $t = 0$ was in state $\mathbf{k}'$ at the (earlier) time t. Then we can write

$$\mathbf{L}_\mathbf{k}(\mathbf{B}) = \int_{-\infty}^{0} \mathrm{d}t \int \mathrm{d}\mathbf{k}' \, \mathbf{v}_{\mathbf{k}'} p_\mathbf{B}(\mathbf{k}, \mathbf{k}', t) \tag{41}$$

and

$$\eta_\mathbf{k} \mathbf{L}^+_{T\mathbf{k}}(\mathbf{B}) = \eta_\mathbf{k} \mathbf{L}_{T\mathbf{k}}(-\mathbf{B}) = \int_{-\infty}^{0} \mathrm{d}t \int \mathrm{d}\mathbf{k}' \, (\eta \mathbf{v})_{\mathbf{k}'} p_{-\mathbf{B}}(\mathbf{k}, \mathbf{k}', t). \tag{42}$$

If we insert these expressions on the two sides of (40), the identity of the two sides follows if

$$p_\mathbf{B}(\mathbf{k}, \mathbf{k}', t)(\partial f_{0\mathbf{k}}/\partial \mathscr{E}) = p_{-\mathbf{B}}(\mathbf{k}', \mathbf{k}, t)(\partial f_{0\mathbf{k}'}/\partial \mathscr{E}). \tag{43}$$

Bearing in mind that the direct scattering rates from $\mathbf{k}$ to $\mathbf{k}'$ and from $\mathbf{k}'$ to $\mathbf{k}$ are related by (10a):

$$R_{\mathbf{k}\mathbf{k}'}(\partial f_{0\mathbf{k}}/\partial \mathscr{E}) = R_{\mathbf{k}'\mathbf{k}}(\partial f_{0\mathbf{k}'}/\partial \mathscr{E}), \tag{10a}$$

and that a reversal of $\mathbf{B}$ reverses the direction of motion of the electrons through $\mathbf{k}$-space, the relation (43) is entirely plausible. Indeed, it is not too difficult to demonstrate it explicitly, by constructing two alternative integral equations for $p_\mathbf{B}(\mathbf{k}, \mathbf{k}', t)$, one involving integration along the path which passes through $\mathbf{k}$ at $t = 0$, and the other along the path which passes through $\mathbf{k}'$ at time t. The derivation is straightforward enough, and we simply quote the results here:

$$p_{\mathbf{B}}(\mathbf{k}, \mathbf{k}', t) = \left| \int_t^0 du\, e^{-|u/\tau|} \int d\mathbf{k}''\, R(\mathbf{k}_u, \mathbf{k}'') p_{\mathbf{B}}(\mathbf{k}'', \mathbf{k}', t-u) \right| + e^{-|t/\tau|} \delta(\mathbf{k}_t, \mathbf{k}') \tag{44}$$

and

$$p_{\mathbf{B}}(\mathbf{k}, \mathbf{k}', t) = \left| \int_t^0 du\, e^{-|u/\tau'|} \int d\mathbf{k}''\, R(\mathbf{k}'', \mathbf{k}'_{t-u}) p_{\mathbf{B}}(\mathbf{k}, \mathbf{k}'', t-u) \right| + e^{-|t/\tau|} \delta(\mathbf{k}_t, \mathbf{k}'), \tag{45}$$

where for brevity we have written $e^{-|u/\tau|}$ and $e^{-|u/\tau'|}$ for

$$\exp\left[-\left|\int_u^0 du/\tau(\mathbf{k}_u)\right|\right] \quad \text{and} \quad \exp\left[-\left|\int_t^{t-u} du/\tau(\mathbf{k}'_{t-u})\right|\right],$$

and the delta-function terms represent electrons which have travelled from $\mathbf{k}'$ to $\mathbf{k}$ without scattering, if $\mathbf{k}'$ and $\mathbf{k}$ happen to lie on the same path through $\mathbf{k}$-space. The routes from $\mathbf{k}'$ to $\mathbf{k}$ followed in (44) and (45) are shown schematically in Fig. 3.3: the relevant scattering rates are $R(\mathbf{k}_u, \mathbf{k}'')$ and $R(\mathbf{k}'', \mathbf{k}'_{t-u})$ rather than $R(\mathbf{k}'', \mathbf{k}_u)$ and $R(\mathbf{k}'_{t-u}, \mathbf{k}'')$ for the same reason that $R(\mathbf{k}_u, \mathbf{k}')$ appears in (34). With the modulus signs included in (44), (45), these equations apply for $t > 0$ as well as for $t < 0$.

If we now interchange $\mathbf{k}$ and $\mathbf{k}'$ in (45), and at the same time replace $\mathbf{B}$ by $-\mathbf{B}$, so that the path of an unscattered electron through $\mathbf{k}$ is reversed, we see from Fig. 3.3 that if $\mathbf{k}_{\mathbf{B}}(0) = \mathbf{k}_{-\mathbf{B}}(t)$, we shall have $\mathbf{k}_{\mathbf{B}}(u) = \mathbf{k}_{-\mathbf{B}}(t-u)$, so that from (45) we obtain

$$p_{-\mathbf{B}}(\mathbf{k}', \mathbf{k}, t) = \left| \int_t^0 du\, e^{-|u/\tau|} \int d\mathbf{k}''\, R(\mathbf{k}'', \mathbf{k}_u) p_{-\mathbf{B}}(\mathbf{k}', \mathbf{k}'', t-u) \right| + e^{-|t/\tau|} \delta(\mathbf{k}_t, \mathbf{k}'). \tag{46}$$

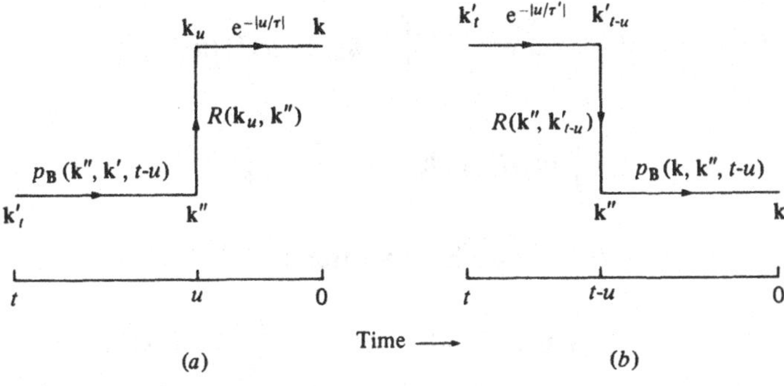

Fig. 3.3. Showing schematically the construction of (*a*) equation (44), (*b*) equation (45).

By comparing (44) and (46), and using (10a), we see that $p_{\mathbf{B}}(\mathbf{k}, \mathbf{k}', t)$ and $p_{-\mathbf{B}}(\mathbf{k}', \mathbf{k}, t)$ are indeed related by (43). We may also note that if (44) is inserted in (41), we regain the expression (35) for $\mathbf{L}_{\mathbf{k}}$, after some manipulation.

3.7 Non-uniform fields

So far we have considered only steady, uniform fields **E** and **G**. It is not difficult to extend the vector mean free path concept to a spatially uniform but time-varying field **E** (or for that matter **G**), and indeed the semi-classical version of the Kubo–Greenwood expression for $\sigma(\omega)$ can be rewritten immediately in terms of an effective path $\mathbf{L}^+$ (see e.g. Lax 1958, equations (4.12) and (4.13)). But for non-uniform fields the vector mean free path ceases to be a very useful concept, and we now turn to the problem of evaluating $f(\mathbf{k}, \mathbf{r}, t)$ in the general case.

Using the path-integral approach of Fig. 3.2 we can write down the formal solution to this problem immediately. As before, we take the first term in the collision integral (9) (involving $R_{\mathbf{kk}'}f_{\mathbf{k}}$) as representing outward scattering from state **k**, and the remaining terms as representing scattering into state **k** from other states **k**′, and we regard $f(\mathbf{k}, \mathbf{r}, t)$ as made up of those electrons which have survived to reach point **k**, **r** at time t after being scattered into earlier points $\mathbf{k}_u$, $\mathbf{r}_u$ on the path at earlier times u. We thus find as our general solution

$$f(\mathbf{k}, \mathbf{r}, t) = \int_{-\infty}^{t} du\, e^{-(t-u)/\bar{\tau}} \int d\mathbf{k}' \, [R(\mathbf{k}_u, \mathbf{k}') f_0(\mathbf{k}_u, \mathbf{r}_u, u) + R(\mathbf{k}', \mathbf{k}_u) g(\mathbf{k}', \mathbf{r}_u, u)] \tag{47}$$

$$= \int_{-\infty}^{t} du\, e^{-(t-u)/\bar{\tau}} \Big[f_0(\mathbf{k}_u, \mathbf{r}_u, u)/\tau(\mathbf{k}_u) + \int d\mathbf{k}' \, R(\mathbf{k}', \mathbf{k}_u) g(\mathbf{k}', \mathbf{r}_u, u) \Big], \tag{48}$$

where $g = f - f_0$; for brevity we have written $e^{-(t-u)/\bar{\tau}}$ for

$$\exp - \int_u^t ds/\tau(\mathbf{k}_s); \quad \text{and} \quad 1/\tau(\mathbf{k}_s) = \int d\mathbf{k}' \, R(\mathbf{k}_s, \mathbf{k}')$$

as usual. Integrating the first term of (48) by parts, we find

$$g(\mathbf{k}, \mathbf{r}, t) = \int_{-\infty}^{t} du\ e^{-(t-u)/\bar{\tau}} \Big[-\partial f_0(\mathbf{k}_u, \mathbf{r}_u, u)/\partial u + \int d\mathbf{k}'\ R(\mathbf{k}', \mathbf{k}_u) g(\mathbf{k}', \mathbf{r}_u, u) \Big], \tag{49}$$

where $\partial/\partial u$ implies differentiation along the path, i.e. taking account of the variation of $\mathbf{k}_u$ and $\mathbf{r}_u$ with u.

Equation (49) was first obtained by Budd (1963) by direct integration of the Boltzmann equation, with the trivial difference that he assumed the slightly less general form (3) for the collision integral. Apart from the assumptions inherent in the semi-classical Boltzmann equation (or path-integral) approach, and the approximations made in deriving (9) or (3) from (2), equation (49) is exact. In deriving (9), we neglected the non-linear terms in (7), and we also neglected the term in $\phi_{\mathbf{q}}$. (It would be a simple matter to restore these terms in (9) and in (49), but less easy to solve the resultant pair of integral equations in $g_{\mathbf{k}}$ and $\phi_{\mathbf{q}}$.) In deriving (3), we assumed $f_{\mathbf{k}} \ll 1$, so that (3) applies only to non-degenerate semiconductors, and in obtaining (49) from (3) we also need to put $Q_{\mathbf{kk}'} = Q_{0\mathbf{kk}'}$, i.e. $\phi_{\mathbf{q}} = 0$, so that we can use (4); but we do not need to neglect non-linear terms. Thus (49) can be used to study non-linear departures from Ohm's law in semiconductors, and indeed the main application of (49) so far has been to the study of these 'hot-electron' effects (Budd 1967; Seeger & Pötzl 1973; Price 1978). The computational problem of solving (49) iteratively for $g_{\mathbf{k}}$ is considerably simplified by using the 'self-scattering' artifice (Rees 1968): the scattering rate $1/\tau(\mathbf{k})$ is replaced by a constant Γ, the same for all electrons, and the excess $\Gamma - 1/\tau(\mathbf{k})$ is treated as 'scattering' from state $\mathbf{k}$ back into state $\mathbf{k}$.

It would take us too far afield to discuss hot-electron effects in detail, but it is worth noting, as a link with the main theme of this volume, the elegant way in which Bauer & Kahlert (1972) have used the reduced amplitude of the Shubnikov–de Haas oscillations in InAs at high currents as a 'thermometer' to measure the temperature of the electron gas.

In the non-linear region, one needs to take into account the effect of the electric field $\mathbf{E}$ itself on the path $\mathbf{k}_u$, $\mathbf{r}_u$ of the unscattered electrons, as well as the effect of $\mathbf{B}$. But in the linear approximation, the direct effect of $\mathbf{E}$ on the path is negligibly small; what matters is the associated effect on the energy $\mathscr{E}(\mathbf{k}_u)$ and hence on f_0 in (49). A temperature gradient will also cause f_0 to vary with $\mathbf{r}$, and so also will a non-uniform charge distribution, through its effect on $\mathscr{E}_F$. Putting $f_0 = f_0(\eta)$, with $\eta = (\mathscr{E} - \mathscr{E}_F)/k_BT$, we thus have

$$-\frac{\partial f_0(\mathbf{k}_u, \mathbf{r}_u, u)}{\partial u} = -\frac{df_0}{d\eta}\frac{\partial \eta}{\partial u} = -\frac{\partial f_0}{\partial \mathscr{E}}\left(\frac{\partial \mathscr{E}}{\partial u} - \frac{\partial \mathscr{E}_F}{\partial u} - \eta k_B \frac{\partial T}{\partial u}\right)$$

$$= -\frac{\partial f_0}{\partial \mathscr{E}}\left[\left(\frac{\partial \mathscr{E}}{\partial \mathbf{r}} - \frac{\partial \mathscr{E}_F}{\partial \mathbf{r}}\right) \cdot \mathbf{v}(u) + \eta(u) k_B \mathbf{G} \cdot \mathbf{v}(u)\right]$$

$$= -\frac{\partial f_0}{\partial \mathscr{E}}[e\mathbf{E}^*(\mathbf{r}_u, u) \cdot \mathbf{v}(u) + \eta(u) k_B \mathbf{G} \cdot \mathbf{v}(u)], \qquad (50)$$

where $\mathbf{E}^* = \mathbf{E} - e^{-1}\, \partial\mathscr{E}_F/\partial \mathbf{r}$, as before, and in the linear approximation we can put $\partial f_0/\partial \mathscr{E} = \partial f_0(\mathbf{k}_u, \mathbf{r}_u)/\partial \mathscr{E} = \partial f_0(\mathbf{k}, \mathbf{r})/\partial \mathscr{E}$, independent of u. We have written $\mathbf{G}$ rather than $\mathbf{G}(\mathbf{r}_u, u)$ because (unlike $\mathbf{E}^*$ or $\mathbf{E}$) $\mathbf{G}$ will seldom if ever vary appreciably over a distance L or a time τ, and it can therefore be treated as effectively constant. Substitution of (50) in (49) then yields the general linearized path-integral solution to transport problems.

The effect of boundaries is readily included in (48): if the electron path intersects the surface of the metal at time t_0 ($<t$), we replace the lower limit of integration by t_0, and add a term to describe the electrons scattered into the path at the boundary. If electrons approaching the boundary in state $\mathbf{k}'$ have a probability $S[\mathbf{k}', \mathbf{k}(t_0)]$ of being scattered there into the emergent state $\mathbf{k}(t_0)$, the additional term will be

$$e^{-(t-t_0)/\bar{\tau}} \int_{\text{in}} d\mathbf{k}'\, S[\mathbf{k}', \mathbf{k}(t_0)] f(\mathbf{k}', \mathbf{r}_0, t_0),$$

where the subscript 'in' confines the integration to states $\mathbf{k}'$ heading *towards* the surface. In thermal equilibrium, when $f(\mathbf{k}', \mathbf{r}_0, t_0) = f_0(\mathbf{k}')$, this term must reduce to $e^{-(t-t_0)/\bar{\tau}} f_0(\mathbf{k}(t_0))$, and this contribution will just cancel the term which arises, when we integrate (48) by parts, from the lower limit t_0. The net result is that to take account of surface scattering, we must replace the lower limit in (49) by t_0 and add a term

$$e^{-(t-t_0)/\bar{\tau}} \int_{\text{in}} d\mathbf{k}'\, S[\mathbf{k}', \mathbf{k}(t_0)] g(\mathbf{k}', \mathbf{r}_0, t_0). \qquad (51)$$

If the surface scattering randomizes the electron distribution completely, so that $S(\mathbf{k}', \mathbf{k})$ is independent of $\mathbf{k}'$, we can argue that particle conservation requires (51) to vanish, and the only effect of surface scattering is then to replace the lower limit in (49) by t_0. This is the approximation of 'diffuse' surface scattering which has frequently been used for simplicity in discussing surface effects in transport problems.

Likewise, if $R(\mathbf{k}', \mathbf{k}_u)$ in (49) is independent of $\mathbf{k}'$, we can argue that the $\int d\mathbf{k}'$ term in that equation must vanish, and (49), (50) reduce to the much simpler form derived by Chambers (1952). This is the approximation of 'catastrophic' scattering, which as we have seen is equivalent to the assumption of a universal relaxation time. It has indeed been suggested (Richards 1974) that the resultant simple equation still remains valid for arbitrary $R(\mathbf{k}', \mathbf{k})$, even when $\mathbf{E} = \mathbf{E}(\mathbf{r}, t)$ and $\mathbf{B} \neq 0$, if the definition of $\tau(\mathbf{k})$ is suitably generalized. But this suggestion must be incorrect: it would imply, for example, that when $\mathbf{B} = 0$, $g_{\mathbf{k}}$ must be zero if $\mathbf{v}_{\mathbf{k}} \cdot \mathbf{E} = 0$, and this is certainly untrue.

As an elementary application of (49), (50), we note that, if a steady uniform electric field is applied to the system (with $\mathbf{G} = 0$), or a steady uniform temperature gradient (with $\mathbf{E} = 0$), and we describe the resultant perturbation $g_{\mathbf{k}}$ in (49) in terms of a vector mean free path $\mathbf{L}_{\mathbf{k}}$ or $\mathbf{L}_{T\mathbf{k}}$, using (15) or (16), we immediately regain the integral equations (35), (36) for $\mathbf{L}_{\mathbf{k}}$ and $\mathbf{L}_{T\mathbf{k}}$. Indeed, we might have presented the material of this chapter more economically by first deriving (49), and then turning to the uniform-field situation afterwards. But the approach we have followed has perhaps been more instructive, and certainly more in line with the historical development of the path-integral method. Because of its complexity, equation (49) has been little used so far, except in the study of the hot-electron problem, but it is nevertheless probably the simplest starting point for a realistic treatment of transport problems generally.

4

The Shoenberg effect (magnetic interaction)

A. B. PIPPARD

4.1 Introduction

In the original version of the theory of the de Haas–van Alphen (dHvA) effect, it was taken for granted that the magnetic field acting on the conduction electrons was the applied field **H** and that each oscillatory contribution to $\tilde{M}$, the magnetic moment per unit volume, could be written in the form $M_0 \sin(2\pi F/H + \psi)$, plus harmonics at multiples of the fundamental frequency F. These harmonics become prominent at lower temperatures where the details of the energy level structure are not so smoothed by thermal excitation as to eliminate all but the fundamental Fourier component. The impulsive field method used by Shoenberg (1962), in his classical attack on the Fermi surface of the noble metals, was peculiarly adapted to reveal harmonic content in the oscillatory magnetization. He found an unexpectedly high amplitude for the second harmonic, and proposed magnetic interaction between the electrons as the explanation of the line-shape distortion. The mechanism is simple enough – the magnetizing field is not **H** but **B**; this is one of the rare examples of a cooperative system which is both interesting and adequately described by a mean-field theory, short-range interactions being negligible. It is thus possible to account in considerable detail for a variety of observations, including the phase-instability and domain formation that make the phenomenon interesting, by use of only elementary physical arguments.

To illustrate the effect let us adopt a simplified model in which the oscillations, in the absence of magnetic interaction, are assumed periodic in H rather than in $1/H$:

$$\tilde{M} = M_0 \sin(H/H_0). \tag{1}$$

This is perfectly adequate in most cases, where H_0 hardly varies over one cycle of oscillation; the phase constant ψ is dropped as irrelevant to the discussion. Shoenberg's proposal is to replace H in (1) by B, so that

$$\tilde{M} = M_0 \sin (B/H_0) = M_0 \sin [(H + 4\pi\tilde{M})/H_0]. \qquad (2)$$

For a long, thin sample, set parallel to the applied field, H is the same inside and out; the solution of (2) then shows how $\tilde{M}$ varies with applied field. A graphical construction makes the essence of the effect clear. In Fig. 4.1 the broken curves, sinusoids in accordance with (2), represent $\tilde{M}(B)$ for three choices of M_0. To convert them into curves of $\tilde{M}(H)$ all that is needed is to shift each point to the left by an amount $4\pi\tilde{M}$, thus changing B into H while keeping $\tilde{M}$ constant. In this way the full curves are drawn, and it is to these, with their distorted form, that Shoenberg draws attention as the origin of the harmonic signals. In the second example, $4\pi M_0/H_0$ has been chosen equal to unity, and in the third, where $\tilde{M}$ has become a multivalued function of H, $4\pi M_0/H_0 > 1$. In this multivalued behaviour lies the origin of phase-instability, but before discussing this we must be satisfied of the validity of Shoenberg's proposal.

Holstein, Norton & Pincus (1973) have given a searching microscopic analysis, which perhaps is best regarded as a labour of love, since the physical processes responsible for the dHvA effect are clear enough to make so rigorous an investigation unnecessary. It has long been recognized (e.g. Van Vleck 1932) that the vanishing of magnetic moment in a classical electron assembly is the result of cancellation between two groups of electrons; on the one hand the many interior electrons which can complete their diamagnetic orbits, and on the other the relatively few

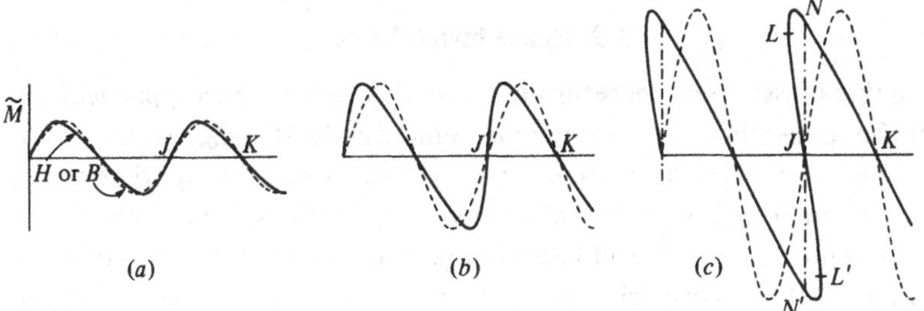

Fig. 4.1. $\tilde{M}(B)$ (broken curves) and $\tilde{M}(H)$ (full curves) for three values of a $(= 4\pi M_0/H_0)$: (a) $a = \frac{1}{2}$, (b) $a = 1$, (c) $a = 2$.

electrons that generate large paramagnetic moments by bouncing round the periphery of the sample. When the assembly is quantized the latter are virtually unaffected, their energy levels being much closer together than $k_B T$, but the former fall into more widely separated levels and the distribution of orbit sizes is slightly modified from the classical result. Thus the exact cancellation fails, and the oscillatory moment is precisely accounted for in this way (Pippard 1969b). Now the quantization rule for the closed orbits can be expressed (Onsager 1952) as a requirement that the enclosed flux shall be a multiple (with a fractional correction) of the flux quantum $2\pi\hbar/e$, and it is through this that **B** makes its appearance instead of **H**. For whatever field fluctuations there may be on the atomic scale, the mean flux density is certainly **B**, and the orbit diameters are such (typically 10^3 atomic spacings) that these fluctuations play an insignificant role. Another way of saying the same thing is to recall that the momentum that enters into Schrödinger's equation is $m\mathbf{v}+e\mathbf{A}$; it is not the local magnetic field that matters but the local vector potential, whose integral round an orbit equals the enclosed flux. Thus the fluctuations in local field experienced by an electron may be large and yet unimportant since, occurring as they do on a small scale, they contribute very little to fluctuations of **A**. There is every reason for confidence, then, in the assertion that the quantization of complete orbits is determined by **B**, which must therefore replace **H** in any expression for the magnetization due to interior electrons. It is unnecessary to ask whether the same holds for the peripheral electrons since, whatever the effective field for them, they contribute no oscillatory terms. It is conceivable, though unlikely, that the magnetic interaction has an appreciable influence on the mean cancellation, and therefore on the Landau diamagnetism, but that is not our concern.

4.2 Phase instabilities

With this reassurance we return to (2) and its consequences, particularly what happens when the parameter a, defined as $4\pi M_0/H_0$, exceeds unity. There is no great problem about achieving this condition experimentally. If an extremal area of the Fermi surface is $\mathscr{A}$, and the second derivative at the extremum, with respect to variation of **k** along the field direction, is $\mathscr{A}''$, the Lifshitz–Kosevich (1955) theory gives the values, at 0 K and in the absence of collision broadening,

$$M_0=\frac{e\hbar\mathscr{A}}{4\pi^4 m_c}\left(\frac{2\pi eB}{\hbar}\right)^{\frac{1}{2}}\Big/\mathscr{A}''^{\frac{1}{2}}, \qquad (3)$$

and

$$H_0 = eB^2/\hbar\mathscr{A}, \tag{4}$$

where m_c is the cyclotron mass of the extremal orbit. Hence

$$a = \frac{\hbar^2\mathscr{A}^2}{\pi^3 m_c B^2}\left(\frac{2\pi eB}{\hbar}\right)^{\frac{1}{2}} \Big/ \mathscr{A}''^{\frac{1}{2}}. \tag{5}$$

For a spherical Fermi surface of the same volume as that of copper, a takes the value 16 in a field of 100 kG, and although this falls as the temperature is raised, it is not until about 3 K that it reaches unity. Moreover there are metals, especially beryllium, more favourable than this for observing the effects; it is therefore highly pertinent to enquire into the expected behaviour when $a > 1$.

When $\tilde{M}$ is multivalued not all the solutions can be expected to represent stable states of homogeneous magnetization. Let us imagine the response to a weak perturbation in the form of a field, parallel to the main field (directed along the z-axis) but varying sinusoidally as $b \sin qx$. If the main field is chosen to have such a value that $B/H_0 = n$, then (2) shows that initially the sample is unmagnetized, but the perturbing field produces a sinusoidally varying magnetization $\mathscr{M} \sin qx$, say. This in its turn modifies the field, so that the resultant perturbing field is given by $b_r \sin qx$, where

$$b_r = b + 4\pi\mathscr{M} = b + 4\pi\chi_q b_r,$$

and χ_q is the susceptibility, $\mathscr{M}/b_r$, appropriate to a perturbation of wave-number q. Hence

$$b_r = b/(1 - a_q), \tag{6}$$

a_q being written for $4\pi\chi_q$. When q is small and the perturbation has a long wavelength, χ_q tends to the value of $\mathrm{d}\tilde{M}/\mathrm{d}B$ for a uniform field, M_0/H_0 at this particular phase of the oscillation, and $a_q \approx a$. But χ_q falls off, with larger q, since perturbations on a scale smaller than the orbit diameter have little effect on the electronic motion. It is clear, then, that once a exceeds unity there will be some value of q for which $a_q = 1$ and the system can acquire spontaneous magnetization as the result of any chance fluctuation, however small. There is no reason to suppose that the periodicity determined by this special value of q has any particular significance in determining the character of the pattern of magnetization; fluctuations of smaller q are even more unstable. All the argument demonstrates is the instability of the system at the point shown as J in Fig. 4.1(c). It is readily extended to the whole stretch of the $\tilde{M}(H)$ curve between L and L', the points where the slope is infinite; when three values of $\tilde{M}$ are possible at a given H, the configuration corresponding to the central one is always unstable; in this the behaviour resembles that of a ferromagnet. And in like manner, of the remaining two configurations one may be preferred to the other, but both are intrinsically stable and the mechanism by which the more stable achieves dominance requires consideration of possible domain structures.

We shall return to this presently, but as a necessary preliminary the magnetic Gibbs function per unit volume must be calculated. This is the thermodynamic function $G = U - TS - H\tilde{M}$, which takes the same value for two phases in equilibrium in a solid (where volume changes are negligible), and a minimal value in the stable configuration. For changes at constant temperature $(\partial G/\partial H)_T = -\tilde{M}$, and therefore the behaviour of $G(H)$ is found by integrating the full curves of Fig. 4.1, to give the result shown in Fig. 4.2. The three cases may be imagined as representing the behaviour of the same sample under identical conditions except for being purer on the right than on the left, so that the dHvA oscillations are stronger. If the magnetic interaction could be imagined switched off, G for each would oscillate sinusoidally about the same mean:

$$\text{if} \quad \tilde{M} = M_0 \sin(H/H_0),$$

$$\text{then} \quad G = \bar{G} + H_0 M_0 \cos(H/H_0).$$

In particular, at points such as K, where $H/H_0 = (2n+1)\pi$, G takes its minimum value of $\bar{G} - H_0 M_0$. The reason for drawing attention to the point K in particular is that since $\tilde{M}$ vanishes here, and there is no

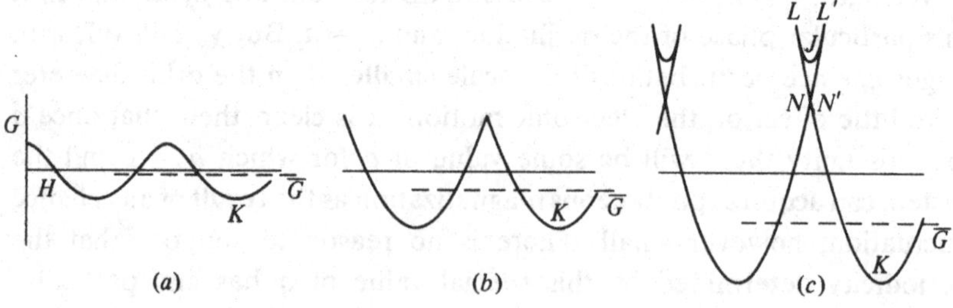

Fig. 4.2. Magnetic Gibbs function per unit volume, G, corresponding to the three curves in Fig. 4.1. The value shown for $\bar{G}$ in (c) assumes that only the curve below NN' is realized.

stability problem as at J, no change occurs on switching on the interaction; in this way the three curves for G in the presence of interaction, derived by integration along the curves for $\tilde{M}(H)$, are located vertically with respect to each other. This will be found significant when we come to discuss inhomogeneous magnetization.

In Fig. 4.2(c) the problem raised by multiple solutions for $\tilde{M}(H)$ when $a > 1$ is seen to be reflected in the behaviour of G, but the argument given above shows that the local minimum of G at J is essentially unstable, so that we have to consider only the lower branches of the curve up to the cusps at L and L'. The thermodynamically stable solution lies below N, and if this is realizable, the magnetization suffers discontinuous reversals every cycle at N. The resulting sawtooth form of $\tilde{M}(H)$ is then reversible and evenly disposed with respect to positive and negative values. To achieve this behaviour, however, requires nucleation of a region of reversed magnetization as soon as H reaches the value represented by J, but if, as is usual in first-order phase transitions, there is a surface energy at the boundary there may be a delay analogous to superheating or supercooling. It is even conceivable that as H is increased the magnetization will follow the full curve right to the point L', but it can go no further, and a homogeneous transition, requiring no nucleation, must then occur spontaneously. If this happens, the sample spends more than half of each cycle with negative magnetization. Conversely, in a falling field the positive magnetization curve is followed as far as L; the behaviour is irreversible, with a hysteresis loop between the values of H corresponding to L and L'.

The experiments of Condon (1966a), however, and his analysis of them, indicate strongly that nucleation of the favoured phase takes place without much difficulty. Condon worked with beryllium, in which magnetic interaction is very strong and, what is significant for this discussion, with disc-shaped samples set normal to the magnetic field. It might be thought that, in the extreme case of a thin disc, the demagnetizing effect would inhibit magnetic interaction; for when the applied field is H_{ext} the internal field, H_{int}, in a sample whose demagnetizing coefficient is $\mathscr{D}$, is $H_{ext} - 4\pi\mathscr{D}\tilde{M}$, and

$$B = H_{int} + 4\pi\tilde{M} = H_{ext} + 4\pi(1-\mathscr{D})\tilde{M}. \quad (7)$$

In particular a thin disc, for which $\mathscr{D}$ is unity, has $B = H_{ext}$ and if (2) is obeyed we are tempted to expect $\tilde{M}$ to be an undistorted sinusoid, however large a may be. The fallacy of this argument becomes clear when one realizes that if the magnetization traces out the complete sinusoid the material must periodically find itself in the essentially unstable regime LJL' in Fig. 4.1(c). Spontaneous nucleation of regions of reversed magnetization must then occur, though the demagnetizing effect then causes the value of H_{int} to change, and before complete reversal has taken place the change is enough to take the material out of the unstable regime. The sample thus forms domains of alternate positive and negative magnetization corresponding, if the domain boundaries are plane, to points N and N'; for it is only here that the two values of G are equal, a condition for phase equilibrium with plane boundaries (under which conditions surface energy plays no part). The equilibrium magnetization curve for any value of $\mathscr{D}$ can be constructed as in Fig. 4.3, by noting that so long as there are domains present, H_{int} takes a constant value while the mean value of $\tilde{M}$ lies anywhere between N and N' in Fig. 4.1(c); for constant H_{int}, $d\tilde{M}/dH_{ext} = 1/4\pi\mathscr{D}$, and this gives the line $\mathscr{L}_1$. Otherwise, along $\mathscr{L}_2$, $\tilde{M}$ is homogeneous and its variation follows from (2) and (7):

$$\tilde{M} = M_0 \sin [(H_{ext} + 4\pi(1-\mathscr{D})\tilde{M})/H_0],$$

as if the value of a were $4\pi(1-\mathscr{D})M_0/H_0$. This behaviour is reminiscent of the formation of the intermediate state in type 1 superconductors, of which a good account is to be found in Shoenberg's (1952b) book. Here also we might expect surface energy to create nucleation problems, but in

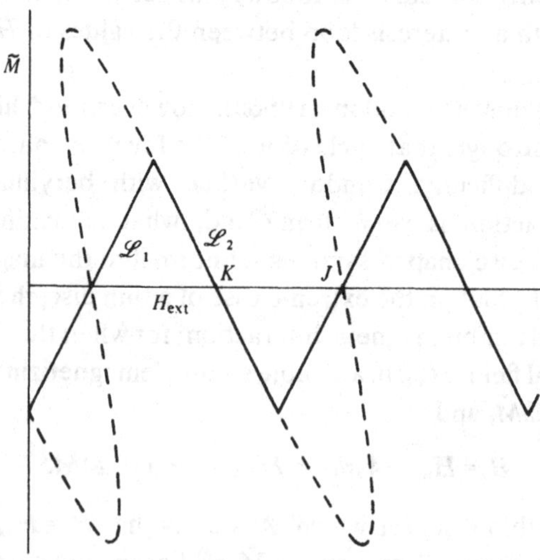

Fig. 4.3. $\tilde{M}(H_{ext})$ when domains are formed along the lines $\mathscr{L}_1$. In this example a is about 5 and the demagnetizing coefficient $\mathscr{D}$ about 0.6.

practice it is rare to encounter difficulty in creating the normal domains which break up the homogeneity of the superconducting phase. Since it is fairly well understood how these domains are nucleated at the surface of the sample, we may look to an analogous process for creating Condon's domains. But let us first briefly discuss their size and form, though it will soon become clear that the discussion must remain incomplete.

4.3 Domain structures

As in superconductors and ferromagnets, we expect the domains to run rather closely parallel to the applied field, since normal components of $\tilde{\mathbf{M}}$ at a domain wall are energetically unfavourable. This does not exclude complex structure where the domains run to the surface, but over the greater part of a not-too-small sample parallel laminae are the norm. Were it not for the positive surface energy of the domain wall the laminae would be very thin, but clearly surface energy works against too fine a subdivision. No detailed calculations of the surface energy have been published, but the order of magnitude may be estimated by considering the electron orbits which span a domain wall and experience a non-uniform field. If there were a sharp reversal of $\tilde{\mathbf{M}}$, and hence a step in $\mathbf{B}$, a classical free-electron orbit would not be a closed circle but an alternating sequence of slightly different circular arcs. The behaviour would approximate to a closed orbit drifting slowly along the domain wall. All orbits spanning the wall would contribute to a drift current along the wall, but this current, being distributed over a thickness of roughly two orbit diameters, would not be compatible with the assumption of an abrupt step in $\mathbf{B}$, which demands a thin current sheet. One must expect therefore the reversal of $\tilde{\mathbf{M}}$ to be distributed across a layer at the domain wall at least one or two orbit diameters thick, so that the current density and the gradient of $\mathbf{B}$ shall be compatible. The electrons in this layer will have their orbits quantized, but with an energy level structure that does not give as low a value of G as is achieved in the body of the domains. For well away from the domain walls $\mathbf{B}$ takes such values as place the system at N or N' in Fig. 4.2(c); in crossing the wall the mean value seen by an electron in orbit changes progressively from one value to the other, and we may expect the local density of G to run, at least approximately, through the corresponding range of values represented by the cusped region $NLJL'N'$. It is from this that the excess boundary energy arises, and the argument leads one to expect the magnitude of the energy per unit area to be of the order of H_0M_0 (the amplitude of G-variations) times the orbit radius r_0. Without a detailed analysis of orbit quantization in a field varying on the scale of the orbit radius, it is not possible to calculate the numerical factor involved. Privorotskii's (1967) analysis for a more slowly varying field confirms the physical picture, but hardly supplies a reliable number. In a brief abstract Condon (1966b) refers to a more relevant calculation, but gives no details except for remarking that the wall thickness is about two orbit diameters and the surface energy is positive. There is no doubt that the system can be analysed in detail if numbers are needed, but for most purposes the estimated order of magnitude is enough.

Consider now the idealized domain structure shown in Fig. 4.4, in which $X \ll Z \ll Y$. The domain wall area per unit volume of the sample is $2/X$, so that the surface energy per unit volume is $\alpha r_0 H_0 M_0/X$, where α is a numerical constant of order unity. Over most of the sample the field at the domain wall can take the required value, but at the surface there will be regions around each wall where the field is distorted, and we may expect an extra energy proportional to the volume of these regions. Provided there is no branching structure whose pattern varies with domain size, the regions of field distortion scale with X, both in width and depth, and the proportion of the volume they occupy is of order X/Z; their excess energy then takes the form $\beta H_0 M_0 X/Z$, where β is another numerical constant of order unity. Both terms together contribute ΔG to the magnetic Gibbs function:

$$\Delta G = \alpha r_0 H_0 M_0/X + \beta H_0 M_0 X/Z,$$

with a minimum value of $2(\alpha\beta r_0/Z)^{\frac{1}{2}} H_0 M_0$ when $X = (\alpha r_0 Z/\beta)^{\frac{1}{2}}$. The scale of the domain pattern, X, is roughly the geometric mean between the sample thickness and the orbit radius.

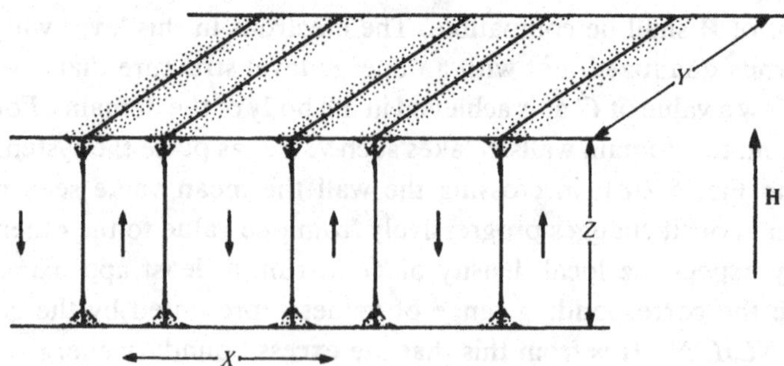

Fig. 4.4. Hypothetical domain structure showing, as stippled areas, the regions where field distortion increases the magnetic energy.

In one of the few experiments to demonstrate the existence of the domains directly, Condon & Walstedt (1968) used a silver sample 0.8 mm thick in a field of 90 kG. With r_0 about 10^{-3} mm in this case, the domain width might be expected to be about 3×10^{-2} mm, at least an order of magnitude greater than the wall thickness. They carried out a standard NMR measurement and found two close but well-resolved frequencies, consistent with most of the sample being composed of domains in which B differed by 11.5 G. The observation of two lines is in itself good direct evidence for the domain model, quantitatively confirmed by relating the observed value of ΔB to the dHvA periodicity of 16.7 G. This calls for a little analysis of the magnetization curve, Fig. 4.1(c).

According to (2), at N and N', where $H/H_0=2n\pi$,

$$\tilde{M}/M_0=\sin\,(a\tilde{M}/M_0). \tag{8}$$

If $\pm M'$ are the non-zero solutions of (8), the difference in B between N and N' is $8\pi M'$. Hence ΔB is related to the oscillation period, $2\pi H_0$, by the expression

$$\Delta B/2\pi H_0=aM'/\pi M_0. \tag{9}$$

When a is very large, (8) shows that $aM'/M_0\approx\pi$, so that $\Delta B\approx 2\pi H_0$. As the applied field is changed, B remains almost constant until it suddenly switches to a new value, once every period. With smaller values of a, however, the jump is less and some of the change in B takes place in between jumps. The evidence from the experiment is that ΔB is 0.69 times the period so that, from (9), $aM'/M_0=0.69\pi$ and, from (8), $\tilde{M}/M_0=\sin 0.69\pi=0.83$. Hence $a=0.69\pi/0.83=2.6$. According to Condon and Walstedt this value accords well with other determinations of a in silver of comparable purity. Although this provides very convincing evidence for the general picture of the domain-forming process, it does not validate the estimate of domain size and its dependence on r_0 and sample thickness. Other models, involving domain branching, can be devised in parallel with well-investigated models of the intermediate state of superconductors, and different formulae for the domain size thereby obtained. All agree, however, on the one point that the experiments verify, that the domains are much wider than the domain walls in any sample of the size normally studied.

We may now return to the point made earlier, that to appreciate how the domain structure may be nucleated we can glean a hint from superconductors. In both cases the surface energy is normally so high that thermal fluctuations cannot be expected to cause spontaneous nucleation in the heart of a homogeneous, domain-free sample. For a nucleus to grow the thermodynamic pressure acting outwards, which is the difference between G outside and inside, must exceed the inward pressure due to surface tension of the curved interface, and the radius of a viable nucleus is likely to be several times r_0; such a nucleus requires an energy fluctuation many times $k_B T$. Conditions are, however, much less severe at the surface of the sample, especially at the equator (Q in Fig. 4.5). In an increasing field the new phase to be nucleated has $\tilde{\mathbf{M}}$ parallel to $\mathbf{H}$, and is most favoured at Q where the reverse magnetization has enhanced the equatorial field. The phase boundary indicated by a broken line is nucleated without difficulty, since within a distance of something like r_0 from the surface the orbits are disrupted by collisions with the surface and the local value of a is small; there is in fact no possibility of a phase boundary until a exceeds unity, perhaps $2r_0$ inside the surface, and then the surface energy rises from zero. But it is not enough to see how the boundary may be formed about $2r_0$ below the surface; it must also be able to move into the material, and here an obstacle appears. Inward movement of the boundary is opposed by a pressure equal to the gradient of surface energy, and if a rises, as we expect, to its full value in a few times r_0 this pressure is of the same order as with a curved surface with radius a few times r_0. We appear to be no further forward in finding a nucleation mechanism. It must be remembered, however, that the surface round Q is probably rough on a scale distinctly greater than r_0. The electrons within the asperities suffer many surface collisions, and a is likely to remain small until the interface has cleared most of the roughness. Even then, since different regions are pitted to different depths, the mean surface energy is liable to rise to its full value only over quite a considerable distance, measured in terms of r_0. This is enough to allow the inward motion of the interface with only a very modest thermodynamic pressure driving it. It is worth recalling that in superconductors persistence of the superconducting phase in fields greater than critical is only rarely observed, showing that, here too, surface nucleation is relatively easy, but

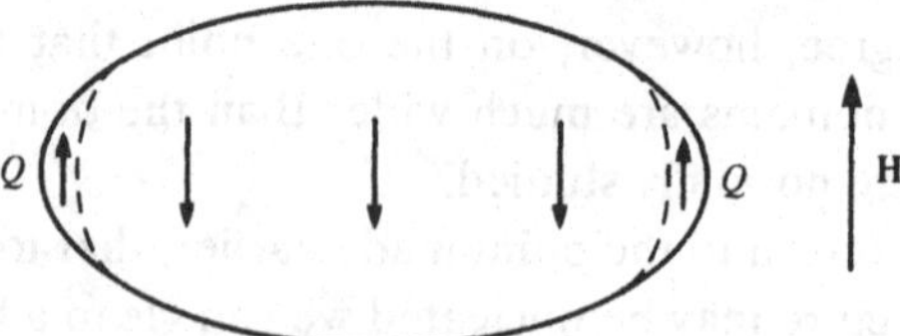

Fig. 4.5. Ellipsoidal sample, showing the equatorial zone where new domains are nucleated.

when it is observed it is with smooth samples obtained, for example, by electropolishing. Once the first phase boundary has moved off into the interior, the field at Q falls and a new phase of the original magnetization may form. As this process continues, a succession of alternately magnetized domains travels into the sample, equilibrium being reached only when the whole is filled with the appropriate proportion of each.

The rate at which this process occurs is limited in principle by time delays in nucleation and by eddy current braking of the domain-wall migration. If the model presented here is correct, the former should be negligible since domain walls have no difficulty in appearing when required just below the surface. Eddy current braking becomes significant when the time-scale of the experiment is so short that any change in applied field cannot spread evenly throughout the sample. This is a typical skin-effect problem and no further discussion is needed to recognize that the only experiments in which difficulties are likely to arise are those using field-modulation or impulsive fields. In such experiments it is desirable to understand what field-inhomogeneities are produced by eddy currents, and how they distort the measurements, and it is not difficult to solve the diffusion equation that governs the problem if the process can be treated as a classical skin-effect problem in a material of uniform conductivity. It is another matter when a steady increase of field results in the nucleation and propagation inwards of domain walls. A few simple models have been analysed (Pippard 1963), enough to show that the pattern of eddy currents is modified by magnetic interaction, so that care must be taken in interpreting experiments where this effect may exist. This cautionary remark does not apply solely to dHvA studies but may also include measurements of helicon damping which may well be affected by domain wall oscillations in phase with the helicon. We shall not, however, pursue this topic further, simply noting the potential of magnetic interaction for generating interesting phenomena.

4.4 Experimental consequences

So far, apart from describing the experimental demonstration of domains by NMR, our discussion has proceeded as if fully confident of the theoretical formulation. It is appropriate at this point to ask the extent to which careful measurement bears out this belief. The answer is, very well. Shoenberg's original observation of unexpectedly high harmonic content was soon followed by studies of the line-shape of dHvA oscillations. Examples are shown and discussed by Shoenberg & Vuillemin (1966), and although they point to minor discrepancies and catalogue possible causes (sample inhomogeneity, recording infidelities) it is clear from many subsequent papers that agreement has been thought good enough to justify technical developments that depend crucially on fully understanding the effect. This is illustrated by the use of dHvA oscillations for measuring the g-value for the spins of conduction electrons and is discussed more fully in Chapter 10. The effect of electron spin is to split each Landau level into two, and thus to modify the amplitude of the oscillations. For if the density of states as a function of electron energy is Fourier analysed, the fundamental determines the amplitude of the fundamental magnetic oscillation. At not too low a temperature this is the only component to survive smoothing by the Fermi tail. Clearly spin-splitting modifies the amplitude; in an extreme case when the split levels are separated by half the original Landau level spacing the fundamental vanishes, for the new level system is evenly spaced at half the original separation. At the same time the harmonics in the Fourier analysis of energy density, which generate harmonics of the magnetic oscillation (and invalidate the simple form of $\tilde{M}$ assumed in (1) and the subsequent discussion), are also changed by the spin-splitting. In principle, therefore, by working at a low enough temperature for the first two or three Fourier components to produce a measurable distortion of the line-shape, it is possible to determine the spin-splitting and hence the g-factor from the ratio of amplitudes of harmonics and fundamental. This technique was successfully used by Randles (1972), among others, but he was well aware that magnetic interaction would also modify the line-shape, and confined his work to conditions where it was inappreciable. Not long after, however, Knecht (1975) measured the g-factor in potassium under conditions where magnetic interaction, though not strong, was actually more significant than spin in determining the harmonic content. He was nevertheless able to separate the two effects, since the phase difference between harmonics is not the same in each. A maximal cross-sectional area of the Fermi surface gives rise, at 0 K and when spin is neglected, to the magnetization waveform shown in Fig. 4.6; if this is analysed into a Fourier series of the form $\sum_r A_r \sin(rB/H_0 + \psi_r)$, ψ_r takes the value $r\pi/4$. Spin-splitting changes the magnitudes of the A_n, even reversing the sign of some, but does not alter the phases, ψ_r. There is thus a $\pi/4$ phase difference between second harmonic and fundamental, in contrast to the effect of magnetic interaction which generates a waveform analysable into sine terms only, with no phase differences. This is discussed in detail in Chapter 10 and it suffices to note here that, by measuring the actual

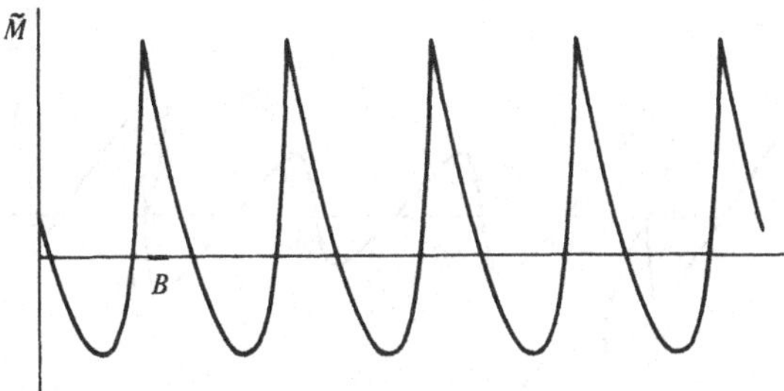

Fig. 4.6. Theoretical form of $\tilde{M}(B)$ at 0 K.

phase shift of the second harmonic, the amplitudes to be assigned to spin-splitting and to magnetic interaction are determined. It is the latter that concerns us here, for the value of a found in this way from the harmonic content of the oscillation may be compared with other estimates. Knecht calibrated his apparatus to give a direct measure of the amplitude of the fundamental, and in addition knew the parameters of the Fermi surface so well as to be able to calculate the amplitude by use of (3). The excellence of the agreement between the three determinations is attested not just by the plausible value for the spin g-factor obtained after the major contribution to the harmonic content has been subtracted, but by the absence of any hint by Knecht or others similarly engaged that any doubt could attach to the theory of magnetic interaction. This is not to imply that the matter is as straightforward in practice as may have been suggested by this abbreviated account, but complications due to scattering and sample inhomogeneity are not intrinsic to the interaction and can be estimated in separate experiments. After allowance has been made for the depression in the value of a which these effects produce, the Lifshitz–Kosevich formula (3) appears to predict the observed amplitude correctly; moreover this is the amplitude needed to give the observed distortion of line-shape by magnetic interaction.

This test relates to rather weak magnetic interaction, with $a < 1$, and for stronger effects we must look elsewhere for confirmation. Perhaps the best evidence for discontinuities in the response is to be found in the non-linear mixing effects that they produce. Let us consider a fairly common situation when two dHvA frequencies, simultaneously present, beat to give an amplitude-modulated oscillation of $\tilde{M}$. Clearly this can be thought of in terms of relatively slow oscillations of a about a mean value. The resulting magnetization curve, in a sample with a small demagnetizing coefficient, might then have the form shown in Fig. 4.7, with domain

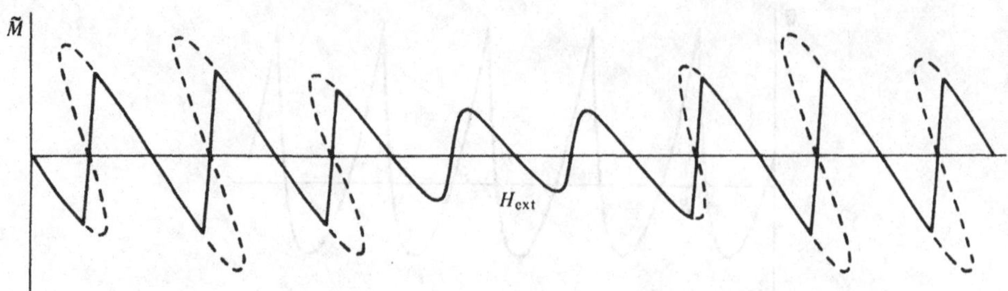

Fig. 4.7. $\tilde{M}(H_{ext})$ when a varies through the beat period between values larger and smaller than unity, and domains are formed as required.

formation when a is large, but not when it is small. The mean level hardly changes, and consequently there is little mixing of the two frequencies to form a difference tone. In view of this it is at first sight surprising to find reported a very strong component at the difference frequency in a sample of silver. There is, however, an important difference between the components of magnetic moment, $M_{\parallel}$ and $M_{\perp}$, parallel and perpendicular to **H**, in the way they are affected by magnetic interaction. We shall return to this presently, to show that in principle the observation of Joseph and Thorsen, which were made with a torsion balance and therefore concern $M_{\perp}$, could be due to magnetic interaction with a rather high value of a. A more likely explanation, however, that this is an instrumental effect, is due to Vanderkooy & Datars (1967), who point out that measurements of magnetic moment with a torsion balance can be confused by a gross mechanical instability quite unrelated to magnetic interaction. The torque on a dHvA sample suspended in a uniform magnetic field H is $HM_{\perp}$, and $M_{\perp}$ is liable to vary rapidly with orientation, because of variations of the extremal cross-sections. Even with a stiff suspension the small rotations consequent upon the oscillations of $M_{\perp}(H)$ may be significant. Let us suppose that the intrinsic magnetization of the sample behaves like (1), $M_{\perp} = M_0 \sin qH$, with q $(=1/H_0)$ a linear function of angular displacement θ over small variations of θ. Then when the torsion constant of the suspension is C, the angle θ through which the sample of volume V is turned is determined by the equation

$$C\theta = VM_0H \sin qH,$$

where $q = q(\theta) = q_0 + q'\theta$, if θ is small. Hence

$$C\theta = VM_0H \sin (q_0H + q'H\theta). \qquad (10)$$

This has the same form as (2), the part of a now being played by $q'M_0H^2V/C$. Writing VM_0H/C as θ_0, the amplitude of angular oscillation in the absence of this simulated magnetic interaction, and $1/q'H$ as θ', the angle through which the crystal must be turned to change the oscillation phase by one radian, we have that $a = \theta_0/\theta'$. In the experiments of Joseph & Thorsen (1965) a typical run had θ' equal to 0.004°, 45 cycles being observed while the sample was turned through 1.8°. Clearly the torsion suspension has to be stiff for the oscillation movement to be less than θ', when M_0HV is probably of the order of 100 dyne cm, but unfortunately no information is provided. Vanderkooy and Datars, however, from their experience of torquemeters, are persuaded that this is the explanation, rather than magnetic interaction. It may be added that, to judge from other observations of magnetic interaction in silver, it is unlikely that the true value of a in Joseph and Thorsen's experiment was as large as unity, let alone the considerable magnitude that is implied by the very strong difference frequency observed.

If we accept the instrumental origin of the effect, it is still necessary to explain the difference between a curve like Fig. 4.7, where no strong difference frequency is generated, and what results from torsional instability. The explanation lies in the possibility in the former case of making a continuous branch transition from N' to N in Fig. 4.2(c) by nucleation and growth of a domain of reversed magnetization. There is no analogue to this in torsional instability and one now expects the canted sinusoid in Fig. 4.1(c) to be traversed, in rising field, right up to the point of essential instability, L'. This immediately introduces an up–down asymmetry and generates a difference frequency from a modulated signal, as illustrated in Fig. 4.8 for a case where θ_0/θ' rises to the rather high value of 12. In a rising field one expects the lower curve to be followed; in a falling field, the upper. In both cases the difference frequency has a higher amplitude than what is left of the original

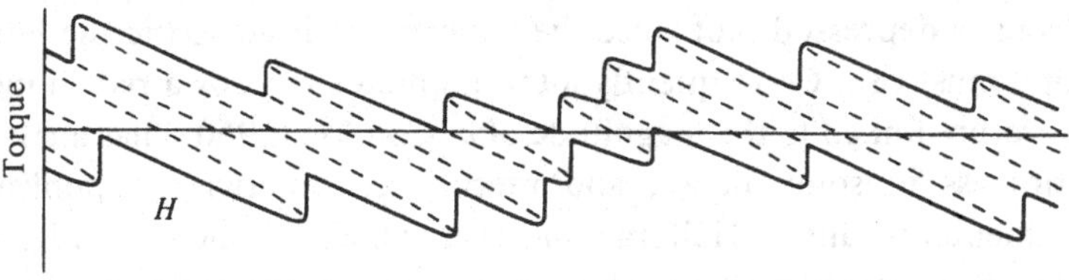

Fig. 4.8. Theoretical behaviour of a torque balance when θ/θ' is large. The broken curve is a canted sinusoid and the full curves show which parts are realized in rising field (lower curve) and falling field (upper curve).

modulated signal. Joseph and Thorsen do not report whether they made observations in both rising and falling fields, or whether any hysteresis was found. The diagram exaggerates a number of features that might be looked for; with a longer beat period, as with silver, the hysteresis would show up as a phase shift (nearly π) in the difference frequency observed in rising and falling fields, and frequency modulation of what is left of the original oscillations should also be observed, though not so strongly as in the diagram. It anything of this nature were found it would confirm that the component arises either through lack of perfect rigidity, or because domains do not form; but the latter explanation seems improbable in view of Condon and Walstedt's observation.

When mixing occurs with firmly mounted samples it is likely that magnetic interaction is the cause, and there are indeed oscillatory effects that are much more favourable than magnetization for revealing it. The magnetic Gibbs function G, for example, has strong up–down asymmetry, and it is obvious from Fig. 4.2 that as a increases the mean value of G along the stable portions of the curves, below N, falls steadily. If a is slowly modulated, any measurement that picks up G or some similar quantity will contain a real component at the beat frequency. Of course, this means that the parallel component of magnetic moment also contains such a component, but since $\tilde{M} = -\partial G/\partial H$, it is weaker in proportion to the ratio of beat frequency to carrier frequency. And measurements that detect $\partial\tilde{M}/\partial H$, such as the popular modulation method introduced by Shoenberg & Stiles (1964), have yet another such factor to hide the beat. On the other hand, the difference frequency is favoured in the more rarely employed observations of the magnetothermal effect, temperature oscillations in an isolated sample as the field is steadily varied. This is because the entropy of the electron assembly oscillates in phase with G. For $(\partial G/\partial T)_H = -S$ and the oscillations of G have a temperature-dependent amplitude but temperature-independent phase. If the positive peaks of G are eliminated by magnetization-reversals, so also are the positive peaks of S. Where a is large, then, the mean value of S for the electrons is depressed, but since the experiment is adiabatic the lattice entropy must rise. Consequently a thermometer records a rise in mean temperature linked to the magnitude of a. Condon (1966a) has analysed the process in some detail, and reproduces otherwise unpublished experimental results of Halloran and Hsu, which are shown in Fig. 4.9. The two frequencies present in beryllium give a strong amplitude modulation, and as the temperature is lowered the magnetothermal oscillations grow but show no sign of magnetic interaction until below 2.9 K. If there

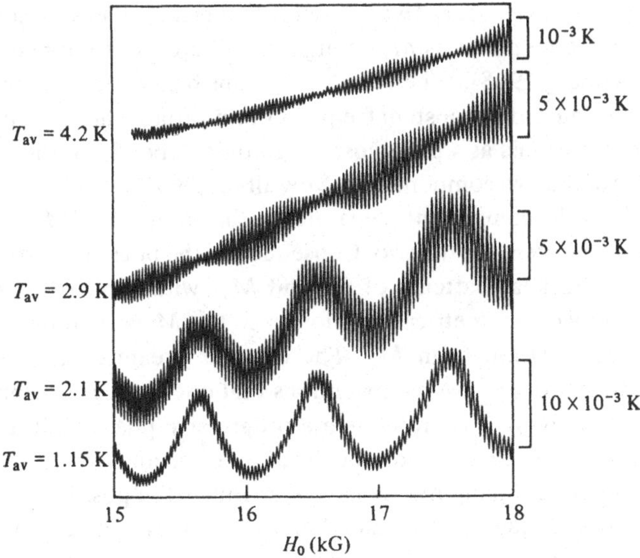

Fig. 4.9. Magnetothermal oscillations in beryllium (M. H. Halloran and F. S. L. Hsu, quoted by Condon (1966a)) at four different temperatures as indicated on the left-hand side.

were no interaction at 2.1 K the envelope would be large but still symmetrical; elimination of part of each cycle of the carrier, however, once a exceeds unity removes most of the colder phases of the oscillation, and accounts for the curiously shaped envelope observed at this temperature. At 1.15 K, $a \gg 1$ over most of the beat cycle, and the beat frequency now dominates what is left of the carrier.

This experiment illustrates admirably how magnetic interaction, if strong enough, can excise most of the cycle of oscillation, and give a strong bias to any property that oscillates in phase with G. This does not apply to dHvA experiments that measure $\partial M/\partial H$ since, although the phase is right for excision to generate a difference frequency, excision does not occur in the sense used here; for while the switch from N' to N in Fig. 4.2(c) leaves S unchanged (if it occurs reversibly) and passes unnoticed in the magnetothermal experiments, the accompanying magnetization reversal induces a strong signal in a pick-up coil, such as to annul the bias that would otherwise show up in $\partial M/\partial H$.

This argument does not apply, of course, to direct measurements of $M_{\parallel}$ and $M_{\perp}$, by mechanical means. The former is not only very hard to measure directly but, as we have seen, is so phased as to possess no rectifying bias to generate a strong difference frequency. On the other hand $M_\perp$ is readily measured with a torsion balance, and in addition need not have the same phase as $M_\parallel$, though the difference is not necessarily large (Shoenberg 1968). When two oscillations beat together, the resultant is not a sinusoid of constant frequency and variable height, but there are phase distortions as well, whose magnitude depends on the relative amplitudes of the two components.† Now although $M_\parallel$ and $M_\perp$ oscillate in phase for each component separately, the ratio $M_\perp/M_\parallel$ may be significantly different for the two. Consequently the phase distortions are different in the beat patterns of $M_\parallel$ and $M_\perp$; with a phase difference between the two, symmetrical excision from the $M_\parallel$ oscillations implies asymmetrical excision from $M_\perp$. The effect is greatest when $M_\perp/M_\parallel$ differs so much for the two contributions that one frequency dominates $M_\parallel$ and the other $M_\perp$. There is then a progressive phase shift, running through 2π for each cycle of the beat pattern, and this provides a reasonably strong mixing mechanism; sometimes the positive phases of the oscillatory torque are eliminated and sometimes, half a beat period later, the negative, so that the mean torque oscillates at the beat frequency. However, as already discussed, it does not seem a likely explanation of what has actually been observed in silver. A really striking example of something very similar occurs in the resistance oscillations of beryllium which we shall turn to presently.

Magnetostrictive oscillations (Chandrasekhar, Fawcett, Sparlin & White 1966) would not be expected to reveal magnetic interaction in any dramatic fashion, since the length of the sample oscillates in phase with the magnetization, without bias; and no difference frequency has been reported, though perhaps the search has not been assiduous. Any property, however, linked to the density of states is a candidate for showing frequency-mixing in the same way as magnetothermal oscillations, since the density of states and entropy are directly related. Acoustic attenuation is such a property, with the particular manifestation of giant quantum oscillations as potentially the most striking case. When the frequency is high, the electronic free path long, and **B** parallel to the propagation vector of the wave, the only electrons that can interact and damp the wave are those in orbits so near the extremal orbit that their mean velocity along **B** closely matches the wave velocity. As B is changed, these effective orbits periodically satisfy the quantization condition and momentarily acquire a high density of states; correspondingly the

† The details of this argument are most readily appreciated by drawing vector triangles, to represent the two components and their resultant; the angle between the two components advances progressively through the beat cycle, and the phase of the resultant is a compromise between their phases.

attenuation of the wave rises to a sharp peak. Good examples of the effect are to be found in the review article by Shapira (1968). Now if the effective orbit is so close to the extremal orbit (as it would be, for example, in copper when $B = 100$ kG and is parallel to a cube axis) that the two are virtually identical for the purpose of quantization, magnetic interaction has a profound effect, excising just those phases of oscillation in which the giant peaks reside. One could therefore hope to find, under the right conditions, oscillation of attenuation whose amplitude increased on lowering of the temperature as a rose towards unity, but instead of growing into giant oscillations began to decrease and even ultimately disappeared. This process does not appear to have been observed.

On the other hand magnetoresistance, which also depends on density of states, provides a fine example of frequency-mixing, once more in beryllium (Reed & Condon 1970). But this is a very special case, and a worthy climax to our catalogue. First let us note that conditions for strong magnetic interaction and strong magnetoresistance are to some extent incompatible. Resistance oscillations which depend on variations of electron scattering, due to variations in density of states for scattering into, are strong only when the quantum number of the oscillation is low, i.e. for small Fermi surfaces and strong fields (as in bismuth, graphite and some semiconductors). But these are the conditions for long period oscillations and weak magnetization, so that a is likely to be small. To see substantial resistance oscillations with larger Fermi surfaces magnetic breakdown is virtually essential. The oscillations in density of states for the orbits round one arm of the Fermi surface modulate the orbit-switching probabilities for other electrons and give rise to effects very similar to scattering, but periodic in $1/B$. To observe the effect with only one type of extremal orbit playing a part would involve some compromise between having enough breakdown to give a good magnetoresistance, but not so much as to reduce the amplitude of magnetization oscillations so that a became rather small. In beryllium, however, there are two types of extremal orbit on the 'cigar', of which one (non-central) dominates the magnetization oscillations and the other (central) undergoes magnetic breakdown. The beating of their frequencies produces powerful effects such as are shown in Fig. 4.10. A qualitative explanation can be given quite simply in terms of the more detailed curves in Fig. 4.11. At 4.2 K the magnetization oscillations are too weak to give significant magnetic interaction, and the resistance oscillations, being determined by the central orbit only, proceed hardly affected by the modulations of the former. At 1.35 K, however, $a \gg 1$, as we have already seen

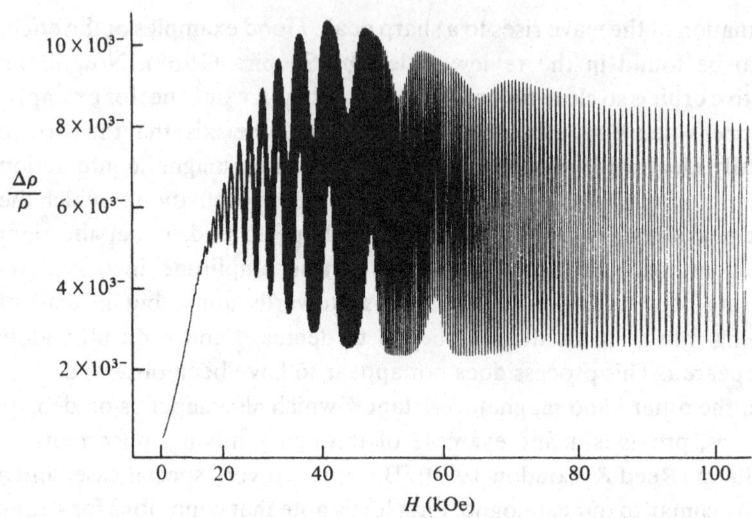

Fig. 4.10. Magnetoresistance of beryllium ($T = 1.39$ K) (Reed & Condon 1970).

from the magnetothermal behaviour. The magnetization is dominated by the non-central orbits, and theirs is the carrier frequency which is modulated by the central orbit. The fact that a is large and varying does not show up as a difference frequency in $\mathrm{d}M/\mathrm{d}H$; this is just what we expect. Nevertheless a substantial portion of each cycle of magnetization is missing, and because the frequency of magnetization oscillations does not match the frequency due to the central orbit, the latter loses a slightly different portion of each successive cycle. Since it is the central orbit that produces resistance oscillations, sometimes the low and sometimes the high resistance phases are eliminated. This is the same mechanism as we drew attention to as a possibility in torsion measurements, but here it is more than a mere possibility – the experimental realization is obvious as a wavy background on which what is left of the high-frequency oscillation is superimposed. Moreover, at the peaks of the beat pattern for the magnetization a takes its highest value and the excision is strongest; what is left of the fast resistance oscillation is correspondingly weaker, and this also shows clearly. Reed and Condon have computed model simulations of the system, and their curves give very satisfactory support to this explanation.

The general conclusion must be that magnetic interaction is a well-understood phenomenon. One cannot predict there will be no surprises in the future, but it seems safe to accept the basic formulation by Shoenberg as correct, needing no modification in principle but only a reasonable measure of care in applying it to any specific case.

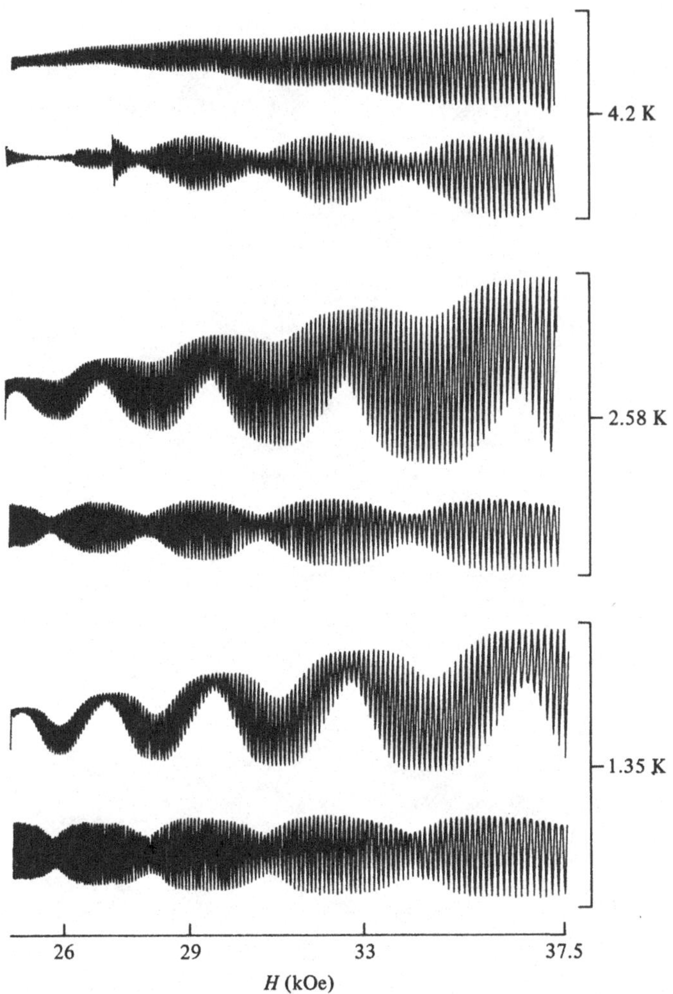

Fig. 4.11. Details of curves like Fig. 4.10 at three temperatures, with corresponding records of dM/dH (Reed & Condon 1970). In each pair, the upper curve is resistivity, the lower dM/dH.

The emphasis in this account has been on the phenomenon as a physically interesting thing in itself, with little attention paid to detailed calculations of amplitudes, harmonic content and the disturbing effects of sample inhomogeneity. These are important to the experimenter who employs or is bedevilled by magnetic interaction, and for him the presentations by Shoenberg (1968, 1976), Gold (1968) and Crabtree, Windmiller & Ketterson (1975) contain much of value.

Part II The electronic structure

5

The electronic structure of transition metals

A. R. MACKINTOSH AND O. K. ANDERSEN

If one had a great calculating machine, one might apply it to the problem of solving the Schrödinger equation for each metal and obtain thereby the interesting physical quantities, such as the cohesive energy, the lattice constant, and similar parameters. It is not clear, however, that a great deal would be gained by this. Presumably the results would agree with the experimentally determined quantities and nothing vastly new would be learned from the calculation. It would be preferable instead to have a vivid picture of the behaviour of the wave functions, a simple description of the essence of the factors which determine cohesion and an understanding of the origins of variation in properties from metal to metal. The present article deals with an attempt to achieve at least part of this objective. The task before us is not a purely scientific one; it is partly pedagogic.

E. P. Wigner & F. Seitz (1955)

5.1 Introduction

During the last twenty years, sophisticated experimental techniques have been developed for the preparation and purification of transition metal single crystals and for measuring their physical properties. The information so obtained has been complemented by energy-band calculations to produce a remarkable extension in our understanding of their electronic structure. The work of David Shoenberg, and of his students and colleagues, has been of crucial significance for this development.

The purpose of this chapter is to review our present knowledge of the electronic band structure of the transition metals. The characteristic properties of these metals are a consequence of the d-states in the conduction bands, and we shall use a broad interpretation of the term transition metal to include all metals with occupied d-like conduction band states, thereby including the noble metals, the alkaline earth metals and the lanthanides in our discussion. The most detailed experimental information which has been obtained concerns the electrons at the Fermi surface, but we shall also consider some of the physical properties which are directly determined by the electron energy bands over a range of energies.

The foundations for our understanding of the electronic structure of transition metals were laid before the Second World War, principally by Mott, Stoner and Slater. In particular, Slater (1937) developed the augmented plane-wave (APW) method which has been used for the great majority of band-structure calculations on d-band metals. Chodorow (1939) applied this method to Cu, using a semi-empirical potential, and determined energy eigenvalues at certain symmetry points. From these results, it is possible to predict that the Fermi surface contacts the fcc Brillouin zone boundary (see Fig. 5.1) near *L*. However, it was not until the development of electronic computers allowed the calculations of the full band structure by Burdick (1961), using the APW method, and Segall (1961), who used the KKR method of Korringa (1947), and Kohn & Rostocker (1954), that the remarkable success of Chodorow's potential, which gives Fermi surface radii within a few per cent of the most accurate experimental values, became apparent. It is interesting that the pioneering measurements of Justi & Scheffers (1938) on the magnetoresistance of pure single crystals of Au clearly show, with the benefit of hindsight, that its Fermi surface must be open. However, it required a development of the theory of galvanomagnetic effects by Lifshitz, Azbel' & Kaganov (1956a) before the significance of their results could be appreciated.

After the war, although a steady improvement occurred in the qualitative understanding of the significance of the d-bands, quantitative progress was slow until the late 1950s, when the experiments on Cu by Pippard (1957b), the first complete Fermi surface determination for any metal, heralded a period of very rapid progress. By modern standards, the anomalous skin effect is a very blunt instrument for measuring Fermi surfaces, and it is a tribute to Pippard's skill that, although the shape

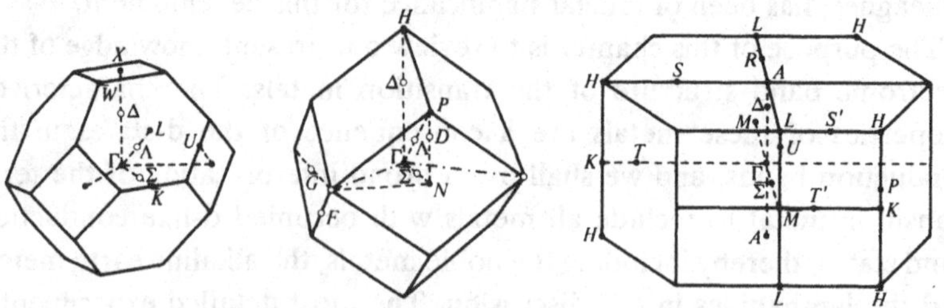

Fig. 5.1. The Brillouin zones for the fcc, bcc and hcp structures.

which he derived differs somewhat from that which is now accepted as correct, he was able to deduce the essential feature of contact with the zone boundary, illustrated in Fig. 5.2, which gives rise to so many interesting physical effects, especially in a magnetic field. This feature was shown to be common to all the noble metals by the dHvA experiments of Shoenberg (1960b, 1962), the galvanomagnetic measurements of Alekseevskii & Gaidukov (1959) and a variety of other experimental studies. Shoenberg's pioneering determination of the precise Fermi surface dimensions of the noble metals by the dHvA effect provided the impetus for the development which has resulted in our present detailed knowledge of the Fermi surfaces of almost all transition metals.

Fig. 5.2. The Fermi surface of Cu in the periodic zone scheme, from a model by David Shoenberg.

The Cooperstown Fermi Surface Conference (Harrison & Webb 1960) set the stage for the impressive progress of the 1960s. The studies on the noble metals, which were extensively reviewed at the Conference, showed what was possible, and it was apparent that the time was ripe for rapid advances in both theory and experiment. Shoenberg (1957) had observed the dHvA effect in Mo and W and had thereby demonstrated that the method could be used for Fermi surfaces with strong d-character. A number of groups were stimulated to begin investigations of various transition metals, of which single crystals of rapidly improving quality were becoming available. The first progress was made by means of studies of galvanomagnetic effects, largely through the work of Fawcett (1961). He established the Fermi surface topology of a number of transition metals, using his own results and those of Alekseevskii and Gaidukov, and showed clearly that the nearly-free electron model, which was being used successfully in the interpretation of experiments on simple metals, could not be applied to d-band metals. Furthermore Fawcett & Reed (1962) demonstrated that the galvanomagnetic properties of Ni could be understood by considering separately the Fermi surfaces corresponding to the two spin states, thus providing the first explicit experimental evidence for the Stoner model. In addition, they showed that one sheet of the Fermi surface is qualitatively similar to that of the noble metals. Their determination of the size of the contact region with the zone boundary, or 'neck', was confirmed by the dHvA measurements of Joseph & Thorsen (1963). These results and the observation of the dHvA effect in Fe by Anderson & Gold (1963) were the first steps on the road which has led, largely through the work of Gold, Stark and their respective collaborators, to the picture of the electronic structure of the ferromagnetic metals which is described in the next chapter.

Fe was in fact the first transition metal for which a complete band structure was obtained, when Wood (1962 and references therein) performed calculations for both fcc and bcc structures, in the paramagnetic phases. At that time, little experimental information on Fe was available, but Lomer (1962), with remarkable insight, used the energy bands for the bcc phase and the rigid-band model to predict the Fermi surfaces for the Cr-group metals. It was not long before the efforts of a number of groups revealed that the Lomer model is qualitatively correct for Mo and W, whose Fermi surfaces were known in considerable detail by the middle of the decade. A particularly interesting detail was the observation by Walsh & Grimes (1964) that the electron and hole surfaces in W do not touch in the ΓH direction of the zone, due to spin–orbit coupling (see Fig. 5.24). This was the first convincing demonstration of the importance of spin–orbit coupling in a metal, and it established that relativistic effects must be taken into account in calculating the electronic structure of transition metals.

This problem was solved by Loucks (1965a), whose relativistic APW (RAPW) method, together with the crystal potential construction devised by Mattheiss (1964a), has provided the basis for most of the accurate Fermi surface calculations on transition metals performed since that time. Loucks first applied his method to W, thus extending his earlier non-relativistic calculations on the Cr-group (Loucks 1965b), while Mattheiss (1966) extended it to the hcp structure and provided a rather complete interpretation of the dHvA measurements on Re of Thorsen, Joseph & Valby (1966). Vuillemin & Priestley (1965) measured the dHvA effect in Pd, the first detailed study of an fcc metal with an incomplete d-band, and showed that their results and the galvanomagnetic measurements of Alekseevskii, Karstens & Mozhaev (1964) could be qualitatively explained in terms of the Cu band structure, with an appropriate shift of the Fermi level. Andersen & Mackintosh (1968) made quantitative calculations of the band structures of the fcc transition metals, and their predictions of the Fermi surface dimensions have proved to be in generally excellent agreement with the very precise experimental results of Ketterson and his colleagues. Keeton & Loucks (1968) extended the earlier non-relativistic calculations on the rare earth metals by Dimmock & Freeman (1964), who first demonstrated the importance of the d-bands in their electronic structure. Although the positron annihilation measurements of Williams & Mackintosh (1968) were in qualitative agreement with these calculations, it was not until the recent work of Young, Jordan & Jones (1973) on Gd that the Fermi surface of a rare earth metal was studied in detail.

Although the potentialities of electromagnetic radiation for exploring the band structure away from the Fermi level had long been appreciated, they were not extensively exploited until the work of Spicer & Berglund (1964) on Cu revealed the power of ultraviolet photoemission for this purpose. The technique has been developed and applied by a number of groups and, particularly if higher energy-derivatives of the photocurrent are measured, can give precise information about the band structure within about one Rydberg of the Fermi level. The same may be said of the optical absorption and its temperature and strain dependence. In both cases, however, a full interpretation of the results requires extensive calculations based on an assumed band structure, preferably including optical matrix elements, since an analysis taking into account critical points in symmetry directions alone may be misleading. The range over which the band structure may be studied experimentally has recently been extended by measurements of the fine structure near X-ray absorption edges, especially those of Kostroun, Fairchild, Kukkonen & Wilkins (1976). As shown by Muller, Jepsen, Andersen & Wilkins (1978), these may give the positions and widths of the broadened but still distinguishable energy bands up to several Rydbergs above the Fermi level.

As the basic reliability of band-structure calculations for transition metals has become established, attempts have been made to simplify their formalism, in order to reduce computational effort and allow the physical content of the results to become more transparent. The tight-binding formalism of Slater & Koster (1954) for the d-bands was combined with a pseudopotential treatment of the sp-bands by Hodges & Ehrenreich (1965) and Mueller & Phillips (1967), resulting in an interpolation scheme with a number of adjustable parameters, suitable for rapid calculations. The justification of such schemes from first principles was considered by, among others, Heine (1967) and Hubbard (1967), who exploited the idea that the d-bands arise from a resonance in the d phase shift associated with the crystal potential within a unit cell (Ziman 1965). At the beginning of the present decade, Andersen (1973) introduced the ideas of canonical bands and potential parameters which we shall use in this review.

Over the same period, there has been a continuous effort to improve the reliability of the crystal potentials used in band-structure calculations. Slater (1951) first proposed that the exchange and correlation might be taken into account by means of a potential depending only on the local charge density $\rho(\mathbf{r})$, and his $\rho^{\frac{1}{3}}$ potential has been extensively used, frequently multiplied by a parameter to yield the so-called $X\alpha$-potential. A systematic attempt to justify such a one-electron picture was made by Kohn and his colleagues and, using their approach, Hedin & Lundqvist (1971) constructed a local exchange-correlation potential for self-consistent calculations. This potential, combined with a suitably efficient procedure for calculating self-consistent band structures, has been remarkably successful in accounting for bulk ground-state phenomena, such as magnetic and cohesive properties, as in the work of Moruzzi, Williams & Janak (1977), who employed the KKR method, and of Poulsen, Kollár & Andersen (1976), in which canonical band theory was used.

In the following pages, we will attempt to describe in a synoptic way the principal results of this development. We will begin by summarizing the most common methods for constructing the periodic crystal potential and solving the Schrödinger equation for an electron moving in it. The reader is also referred to Chapter 2 (Section 2.2) where the many-body underpinning for band-structure theory is discussed. We will explain how the contributions to the band structure which depend on the lattice structure and potential may be separated by using the concepts of canonical bands and potential parameters. The importance of the relativistic effects will be discussed and values of the potential parameters for the transition metals presented, emphasizing the trends in the periodic table. In the following section we will proceed to describe how the lattice and electronic structures are related and how the energy bands may be used to calculate ground-state properties such as cohesive energies, lattice parameters and magnetism. We will discuss the significance of the d-band for these properties and again show the trends in the periodic table. We will then give a concise description of our present knowledge of the band structures and Fermi surfaces of the individual metals, including both calculations of the electronic structure and experimental evidence. In conclusion we will attempt to summarize the present state of our understanding of the electronic structure of transition metals and suggest the broad lines along which the subject may be expected to develop in the future.

5.2 Crystal potentials and energy-band theory

5.2.1 One-electron potentials

At present, the most satisfactory way of arriving at a one-electron description of ground-state properties, including the Fermi surface, is the density-functional approach suggested by Hohenberg, Kohn and Sham. These authors considered a system of N electrons moving in some external potential, v_{ext}. The Hamiltonian (in atomic Rydberg units),

$$\mathcal{H} = T + U + V$$
$$= \sum_i (-\nabla_i^2) + \sum_{i\neq j}\sum r_{ij}^{-1} + \sum_i v_{ext}(\mathbf{r}_i), \quad (1)$$

comprises the kinetic energy, the electron–electron interaction, and the interaction with the external potential, which includes the electrostatic interaction with the fixed nuclei. Hohenberg & Kohn (1964) first showed that, for specified operators T and U, the ground state $\Phi(\mathbf{r}_1\sigma_1, \mathbf{r}_2\sigma_2, \ldots)$ is a unique functional of the electron density only,

$$n(\mathbf{r}) = N \sum_{\sigma_1} \sum_{\sigma_2} \ldots \int d^3 r_2 \ldots |\Phi(\mathbf{r}_1\sigma_1, \mathbf{r}_2\sigma_2, \ldots)|^2. \tag{2}$$

Their proof hinges on the result that the density uniquely specifies the external potential, and hence the Hamiltonian and the ground state. Since Φ is a functional of n, so evidently is $\langle\Phi|T+U|\Phi\rangle$, which is a universal functional, valid for any external potential and for any number of electrons. They showed secondly that the energy functional,

$$\langle\Phi|\mathscr{H}|\Phi\rangle = \langle\Phi|T+U|\Phi\rangle + \int d^3 r \, v_{\text{ext}}(\mathbf{r}) n(\mathbf{r}), \tag{3}$$

attains its minimum, the ground-state energy, for the proper ground-state density. If we knew the universal functional, we could thus rather easily use the variational principle to determine the ground-state energy and density for any system specified by an external potential. However, the functional is not known and the major part of the complexity of the many-electron problem is associated with its determination.

Guided by the success of the one-electron picture in describing the properties of atoms, molecules, and solids, Kohn & Sham (1965) showed that the variation of the energy functional (3) with respect to density could lead to a set of one-electron Schrödinger equations to be solved self-consistently. Together with the real system of electrons, they considered a system of non-interacting electrons with the same density, $n(\mathbf{r})$, i.e. they considered a system with the Hamiltonian

$$\mathscr{H}' = \sum_i (-\nabla_i^2) + \sum_i v(\mathbf{r}_i) \tag{4}$$

in which the external potential, v, is determined in such a way that the density in the ground state of the non-interacting system is the same as the density in the ground state of the real system. The fact that the external potential is a unique functional of the density shows that the Hamiltonian (4), and hence the non-interacting system, is uniquely defined by $n(\mathbf{r})$. Now, for a fixed potential, the ground state of the non-interacting system is simply the Slater determinant obtained by occupying the lowest-lying one-electron states, defined by the Schrödinger equation

$$[-\nabla^2 + v(\mathbf{r})]\psi(\mathbf{k}, \mathbf{r}) = \mathscr{E}_{\mathbf{k}}\psi(\mathbf{k}, \mathbf{r}), \tag{5}$$

and the density is therefore given by

$$n(\mathbf{r}) = \sum_{\mathbf{k}} \theta(\mathscr{E}_F - \mathscr{E}_{\mathbf{k}})|\psi(\mathbf{k}, \mathbf{r})|^2. \tag{6}$$

It now remains to determine the potential v in such a way that (6) is the density in the ground state of the real system. For this purpose Kohn & Sham (1965) wrote the energy functional (3) of the real system as follows

$$\begin{aligned}\langle\Phi|\mathcal{H}|\Phi\rangle = \langle\Phi'|T|\Phi'\rangle \\ +\iint d^3r_1\, d^3r_2\, r_{12}^{-1} n(\mathbf{r}_1) n(\mathbf{r}_2) \\ + E_{xc}\{n(\mathbf{r})\} + \int d^3r\, v_{ext}(\mathbf{r}) n(\mathbf{r}).\end{aligned} \tag{7}$$

Here, the first term

$$\langle\Phi'|T|\Phi'\rangle = \sum_{\mathbf{k}} \theta(\mathscr{E}_F - \mathscr{E}_{\mathbf{k}}) \int d^3r\, \psi(\mathbf{k}, \mathbf{r})^*(-\nabla^2)\psi(\mathbf{k}, \mathbf{r}) \tag{8}$$

is the kinetic energy of the non-interacting system, the second term is the Coulomb energy of the electronic charge cloud, and the third, so-called exchange-correlation term is the difference between the true kinetic energy and that of the non-interacting system, plus the difference between the true interaction energy of the electrons and that of the charge cloud. This term, which presumably is rather small, thus contains all our ignorance of this problem. If we now minimize (7) and the corresponding functional for the non-interacting system with respect to density, and require that they attain their respective minima for the same density, we find that the non-interacting electrons must move in the potential

$$\begin{aligned}v(\mathbf{r}_1) &= \partial[\langle\Phi|\mathcal{H}|\Phi\rangle - \langle\Phi'|T|\Phi'\rangle]/\partial n(\mathbf{r}) \\ &= \int d^3r_2\, 2r_{12}^{-1} n(\mathbf{r}_2) + v_{ext}(\mathbf{r}_1) + v_{xc}\end{aligned} \tag{9}$$

given by the electrostatic potential from the electronic charge cloud and the nuclei, plus the (unknown) operator

$$v_{xc} = \partial E_{xc}\{n(\mathbf{r})\}/\partial n(\mathbf{r}). \tag{10}$$

It should be emphasized that the formally exact, self-consistent one-electron scheme, as defined by (5), (6), and (9), (10), is designed to yield ground-state properties such as the density, the total energy, etc., and that the one-electron energies and wave-functions obtained from the Schrödinger equation (5) have no direct meaning. Nevertheless, they are useful concepts for understanding cohesive and magnetic properties of solids. Energies and lifetimes of quasi-particle excitations should, in principle, be obtained from an equation containing the Dyson self-energy operator $\Sigma(\mathbf{r}, \mathbf{r}'; \mathscr{E}_\mathbf{k})$ instead of v_{xc} but, at the Fermi level, Σ is identical with v_{xc} and the Fermi surface obtained in a ground-state calculation is therefore the correct one. Being a ground-state property, the Dyson operator may be obtained from the self-consistently determined electron density.

In order to convert the density-functional scheme into a practical method, the exchange-correlation energy functional must be specified. The presently used, *local* approximation consists of writing

$$E_{xc}\{n(\mathbf{r})\} \approx \int d^3r\, \varepsilon_{xc}(n(\mathbf{r}))n(\mathbf{r}), \tag{11}$$

where $\varepsilon_{xc}(n)$ is the best possible estimate of the exchange plus correlation energy per electron for a homogeneous electron gas of density n. This local density (LD) approximation is exact in the limit of slowly varying densities and, in contrast to the Thomas–Fermi approximation, it is also exact in the limit of high densities, because it correctly includes the kinetic energy of the non-interacting system. The only source of error therefore lies in those regions where the density is low and rapidly varying. For the exchange-correlation potential we obtain (cf. equation (7a) of Chapter 2)

$$v_{xc} \approx d[n\varepsilon_{xc}(n)]/dn \equiv \mu_{xc}(n(\mathbf{r})), \tag{12}$$

which is a local potential, namely the exchange-correlation part of the chemical potential in a homogeneous electron gas. The exchange part of the potential is

$$\mu_x(n) = \tfrac{4}{3}\varepsilon_x(n) = -2(3\pi^{-1}n)^{\frac{1}{3}} \tag{13}$$

with $n \equiv n(\mathbf{r})$, and is often named the Gaspar (1954), the Kohn–Sham, or the $\alpha = \frac{2}{3}$ potential. Hedin & Lundqvist (1971) have given useful estimates of ε_{xc}, μ_{xc}, and Σ. Their exchange-correlation potential may be expressed as

$$v_{xc} \approx \mu_{xc}(n) = \tfrac{3}{2}\alpha(n)\mu_x(n), \tag{14}$$

where α increases from $\frac{2}{3}$ in the high-density limit, where correlations can be neglected, to about 0.85 for $(\frac{4}{3}\pi n)^{-\frac{1}{3}} \equiv r_s = 4$, i.e. for a density typical of the interatomic region in a metal.

Although, in principle, the density functional scheme can yield a ground state with net spin, in practice this requires an approximation far more sophisticated than (11). It is, however, a rather simple matter to generalize the formalism to a spin-density scheme in which the element,

$n(\mathbf{r}\sigma, \mathbf{r}\sigma')$, of the density matrix is the independent variable, and in which the external potential has the form $v_{ext}(\mathbf{r}\sigma, \mathbf{r}\sigma')$. This generalization furthermore allows inclusion of the spin–orbit coupling and a magnetic field in the external potential, and hence the calculation of spin-susceptibilities. The one-electron equation now takes the form

$$[-\nabla^2 + v(\mathbf{r})]\psi(\mathbf{k}, \mathbf{r}\sigma) + \sum_{\sigma'} [v'_{ext}(\mathbf{r}\sigma, \mathbf{r}\sigma') + v'_{xc}(\mathbf{r}\sigma, \mathbf{r}\sigma')]\psi(\mathbf{k}, \mathbf{r}\sigma') = \mathscr{E}_{\mathbf{k}}\psi(\mathbf{k}, \mathbf{r}\sigma), \quad (15)$$

where v'_{ext} and v'_{xc} are the spin-dependent parts of the external and exchange-correlation potentials. In the local spin density (LSD) approximation, the energy density entering (11) is replaced by $\varepsilon_{xc}(n\uparrow, n\downarrow)$, which is the exchange-correlation energy per electron for a homogeneous electron gas with density $n \equiv n\uparrow + n\downarrow$ and spin density $m = n\uparrow - n\downarrow$, as created by an external, homogeneous magnetic field. The corresponding exchange-correlation potential is diagonal in the spin variable and it has been estimated by von Barth & Hedin (1972) and by Gunnarsson & Lundqvist (1976). To first order in the spin polarization

$$v'_{xc}(\mathbf{r}\sigma) = \text{sign}(\sigma)\,\tfrac{1}{3}\delta(n)\mu_x(n)m/n, \quad (16)$$

where $n \equiv n(\mathbf{r})$ and $m \equiv m(\mathbf{r})$, and the effect of correlation is to reduce δ from 1, in the high-density limit, to about 0.55 for $r_s = 4$.

Similar to the LD and LSD schemes is the Xα-method of Slater (1974), which is based on the so-called hyper-Hartree–Fock method. The Xα-potential is given by (14) but α is a constant depending only on the atomic number and determined by the requirement that, for the isolated atom, the total energy obtained by the Xα-method agrees with that obtained by the Hartree–Fock method. The values of α thus obtained range from 0.78 in Li to about 0.70 for atoms heavier than Cu. In the spin-polarized version of the Xα-scheme, $\delta = \frac{3}{2}\alpha$ exceeds 1, and the tendency towards magnetism is therefore overestimated.

The potential corresponding to $\alpha = 1$ is the local exchange potential originally suggested by Slater (1951) and it was derived as the average of the exchange operator over all occupied states for a homogeneous electron gas. This potential has been used in self-consistent relativistic (Liberman, Waber & Cromer 1965) and non-relativistic (Herman & Skillman 1963) calculations for atoms and in the great majority of existing band-structure calculations. Due to the large amount of computing involved when solving (5) for a number of Bloch vectors and bands, such calculations are rarely carried to self-consistency, but the crystal

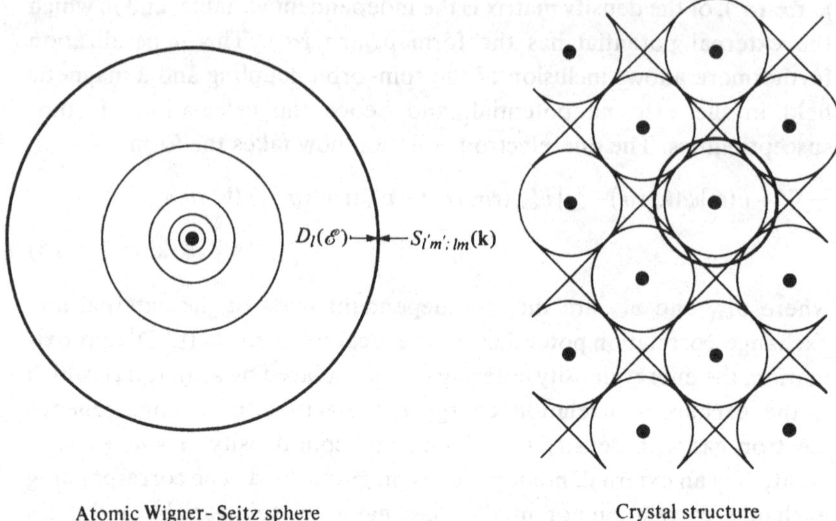

Fig. 5.3. The atomic sphere approximation: by considering the atomic Wigner–Seitz spheres (thick line), rather than the muffin-tin spheres (thin lines), the band-structure problem may be approximated by a boundary-value problem in which all information about the size of the atomic sphere and the potential in it is carried by the logarithmic derivative functions, $D_l(\mathscr{E})$, while the information about the crystal structure is carried by the structure constants, $S_{l'm';lm}(\mathbf{k})$, which are independent of energy and atomic volume.

potential is constructed by using in (9) a superposition in the crystal of self-consistent atomic densities. Moreover, for ease of solving the band-structure problem, the density and potential in the crystal are spatially averaged to comply with the so-called muffin-tin (MT) geometry, which is spherically symmetric within non-overlapping spheres surrounding the atoms and constant in the interstitial regions, as shown in Fig. 5.3. This *ad hoc* construction of a non-self-consistent, $\alpha = 1$ potential was first suggested by Mattheiss (1964a) and it has been surprisingly successful. In the following we shall refer to such a crystal potential as the standard potential, when the atomic densities are relativistic and correspond to the $d^{n-1}s^1$ configuration in the 3d- and 4d-series, and to the configuration $d^{n-2}s^2$ in the 5d-series.

5.2.2 The atomic sphere approximation

We shall now derive a picture according to which the energy bands of a closely packed crystal such as a transition metal may be constructed from 'canonical bands' and 'potential parameters' (Andersen 1973; Andersen

& Jepsen 1977). The canonical bands depend only on the crystal structure, and contain all the complicated **k**-space dependences. The dependence on the atomic species is contained in the potential parameters, and these convert the canonical bands into energy bands by specifying their positions and widths on the energy scale. This picture is achieved through the atomic sphere approximation (ASA) to the KKR and muffin-tin orbital (MTO) (Andersen 1975) methods.

For a monatomic, closely-packed crystal it is reasonable to assume that the atomic Wigner–Seitz cells are spheres and that the electron density is spherically symmetric inside each atomic sphere. Since the spheres are neutral, there will be no electrostatic interaction between them and, including now the inter-nuclear repulsion, we shall thus approximate the functional (7) for the total energy per atom by

$$\langle\Phi|\mathcal{H}|\Phi\rangle_{\mathrm{ASA}} = \langle\Phi'|T|\Phi'\rangle_{\mathrm{ASA}} + \int_s d^3r_1 \left[\int_s d^3r_2\, r_{12}^{-1} n(r_2) - 2r_1^{-1}Z + \varepsilon_{\mathrm{xc}}(n(r_1))\right] n(r_1). \tag{17}$$

Here the integrals extend over the atomic sphere of radius $s = (3V/4\pi)^{\frac{1}{3}}$. This functional of the spherically averaged density leads to a band-structure problem (5) for a potential (9) which is spherically symmetric inside each atomic sphere.

In order to explain how the band-structure problem is solved in the ASA, we shall first consider the classical Wigner–Seitz cellular method:

Inside any atomic cell the solutions of Schrödinger's differential equation at any chosen energy, $\mathscr{E}$, may be taken as the partial waves

$$\phi_{lm}(\mathscr{E}, \mathbf{r}) \equiv \phi_l(\mathscr{E}, r)\mathrm{i}^l Y_{lm}(\hat{r}), \tag{18}$$

where the radial function ϕ_l is a solution of the appropriate radial Schrödinger equation. We may now place a linear combination of these functions, multiplied by the appropriate Bloch factor, in each atomic cell throughout the crystal, i.e.

$$\psi(\mathbf{k}, \mathscr{E}, \mathbf{r}) = \sum_{lm} B_{lm}(\mathbf{k}) \sum_{\mathbf{R}} \mathrm{e}^{\mathrm{i}\mathbf{k}\cdot\mathbf{R}} \theta(\mathbf{r}-\mathbf{R}) \phi_{lm}(\mathscr{E}, \mathbf{r}-\mathbf{R}). \tag{19}$$

Here, $\mathbf{R}$ are the lattice translations and $\theta(\mathbf{r})$ is unity in the cell at the origin and zero outside. For simplicity, we consider a structure with only one atom per primitive cell. If, for a given $\mathscr{E}$ and wave-vector $\mathbf{k}$, we can find coefficients B, such that $\psi(\mathbf{k}, \mathscr{E}, \mathbf{r})$ is continuous and differentiable as we pass from one cell to the next, we have constructed a solution of

Schrödinger's equation for the entire solid, and $\mathscr{E}$ is an eigenvalue at wave-vector **k**. This matching condition must depend on **k** and the crystal structure but not on the lattice constant, and, in the approximation where the cell is substituted by the equivalent sphere, it is a condition on the radial logarithmic derivatives,

$$D_l(\mathscr{E}) \equiv s\phi_l'(\mathscr{E}, s)/\phi_l(\mathscr{E}, s), \tag{20}$$

depending on the atomic-sphere potential, the energy, and the magnitude of the local angular momentum.

It was found to be very difficult to express the matching condition in a practical way for a general **k**-point. The so-called spherical approximation of Wigner and Seitz using an isotropic **k**-space was very crude, and the ASA, or any similar approximation, was not proposed until forty years later. By that time the matching problem had been reformulated as the problem of multiple scattering between non-overlapping MT spheres, and much experience had been gained by application of the derived KKR (and the related APW) technique to nearly all elemental metals, using large amounts of time on modern computers.

In the ASA we join a tail continuously and differentiably onto the partial wave at the atomic sphere, so that

$$\frac{\phi_l(\mathscr{E}, r)}{\phi_l(\mathscr{E}, s)} = \frac{D_l(\mathscr{E})+l+1}{2l+1}\left(\frac{r}{s}\right)^l + \frac{l-D_l(\mathscr{E})}{2l+1}\left(\frac{r}{s}\right)^{-l-1} \tag{21}$$

for $r \geq s$. This tail has been chosen to have zero kinetic energy, i.e. it is a harmonic function, a solution of Laplace's equation, and the consequences of this choice will later be evident. From the partial wave we may subtract $(D+l+1)(r/s)^l/(2l+1)$, and thus obtain the following function, which is continuous, differentiable and regular in all space:

$$\chi_{lm}(\mathscr{E}, \mathbf{r}) = \mathrm{i}^l Y_{lm}(\hat{r})\chi_l(\mathscr{E}, r), \tag{22}$$

with the radial part

$$\chi_l(\mathscr{E}, r) = \phi_l(\mathscr{E}, s)\begin{cases} \dfrac{\phi_l(\mathscr{E}, r)}{\phi_l(\mathscr{E}, s)} - \dfrac{D_l(\mathscr{E})+l+1}{2l+1}\left(\dfrac{r}{s}\right)^l, & r \leq s, \\[2ex] \dfrac{l-D_l(\mathscr{E})}{2l+1}\left(\dfrac{r}{s}\right)^{-l-1}, & r \geq s. \end{cases}$$

This 2^l-pole field is essentially what we call a muffin-tin orbital (MTO), although in the present context it would be preferable to use the name atomic-sphere orbital. The tail of an MTO centred at **R** may be expanded in angular momenta about the origin, and we may write

$$\sum_{\mathbf{R}\neq 0} e^{i\mathbf{k}\cdot\mathbf{R}}\chi_{lm}(\mathscr{E}, \mathbf{r}-\mathbf{R}) = -\sum_{l'm'}\left(\frac{r}{s}\right)^{l'}\frac{i^{l'}Y_{l'm'}(\hat{r})}{2(2l'+1)} \times S_{l'm';lm}(\mathbf{k})\chi_l(\mathscr{E}, s), \tag{23}$$

which converges inside the sphere passing through the nearest-neighbour positions. The expansion coefficients

$$S_{l'm';lm}(\mathbf{k}) = \sum_{\mathbf{R}\neq 0} e^{i\mathbf{k}\cdot\mathbf{R}} S_{l'm';lm}(\mathbf{R}) \tag{24}$$

are the Hermitian, canonical structure constants. Here

$$S_{l'm';lm}(\mathbf{R}) = g_{l'm';lm}(4\pi)^{\frac{1}{2}}(-i)^{\lambda}Y^{*}_{\lambda\mu}(\hat{R})(R/s)^{-\lambda-1},$$

$$g_{l'm';lm} \equiv (-1)^{m+1}2\left[\frac{(2l'+1)(2l+1)}{2\lambda+1}\times\frac{(\lambda+\mu)!(\lambda-\mu)!}{(l'+m')!(l'-m')!(l+m)!(l-m)!}\right]^{\frac{1}{2}}$$

and

$$\lambda \equiv l+l', \quad \text{and} \quad \mu \equiv m'-m.$$

Therefore, the condition that the linear combination

$$\psi(\mathbf{k}, \mathscr{E}, \mathbf{r}) \equiv \sum_{lm} B_{lm}(\mathbf{k})\sum_{\mathbf{R}} e^{i\mathbf{k}\cdot\mathbf{R}}\chi_{lm}(\mathscr{E}, \mathbf{r}-\mathbf{R}) \tag{25}$$

of MTOs is a solution of Schrödinger's equation for the crystal, is that (19) and (25) are identical. This means that, inside the sphere at the origin (and hence inside any other atomic sphere), the sum of the tails coming from all other sites must interfere destructively with the terms proportional to $i^l Y_{lm}(\hat{r})r^l$ from the MTOs at the origin. This condition leads to the set of linear, homogeneous equations

$$\sum_{lm}[S_{l'm';lm}(\mathbf{k}) - P_{l'}(\mathscr{E})\delta_{l'l}\delta_{m'm}]\chi_l(\mathscr{E}, s)B_{lm}(\mathbf{k}) = 0 \tag{26}$$

for all $l'm'$ and with the potential functions defined as

$$P_l(\mathscr{E}) \equiv 2(2l+1)\frac{D_l(\mathscr{E})+l+1}{D_l(\mathscr{E})-l}. \tag{27}$$

A non-trivial solution may be obtained only if the values of $\mathbf{k}$ and $\mathscr{E}$ are such that the determinant of the matrix in the square bracket of (26) vanishes. This is the matching condition in the ASA between the structure constants (24), which depend on $\mathbf{k}$, and the logarithmic derivatives (20), which depend on $\mathscr{E}$.

Assuming that the partial waves are normalized to unity in the atomic sphere, i.e. that

$$\int_0^s \phi_l^2(\mathscr{E}, r) r^2 \, dr = 1, \tag{28}$$

the wave-function (19) or (25) will be normalized in the same way, provided that the solutions of (26) are normalized according to

$$\sum_{lm} |B_{lm;j}(\mathbf{k})|^2 = 1. \tag{29}$$

The logarithmic derivatives (20) and the potential functions (27) are respectively decreasing and increasing functions of energy and it may be shown that their energy derivatives, $\dot{D}$ and $\dot{P}$, are related to the amplitudes at the sphere boundary of respectively the partial waves and the MTOs through

$$\dot{D}_l(\mathscr{E}) = -[s\phi_l^2(\mathscr{E}, s)]^{-1} \tag{30}$$

and

$$\dot{P}_l(\mathscr{E}) = [\tfrac{1}{2}s\chi_l^2(\mathscr{E}, s)]^{-1}. \tag{31}$$

The simple picture offered by the ASA for a monatomic, closely-packed solid is shown in Fig. 5.3. The band-structure problem is reduced to that of finding the eigenvalues and eigenfunctions for a single atomic sphere, with its spherically symmetric potential, subject to a **k**-dependent and angular-momentum destroying boundary condition imposed by the surroundings. The information about the atomic sphere is carried by the logarithmic derivatives, while the information about the crystal structure is carried by the structure matrix. A consequence of choosing the MTO tails as harmonic functions is that the structure matrix, considered as a function of **k***s*, is independent of the lattice constant; that is the reason why in Fig. 5.3 the atomic sphere and the surrounding structure are drawn on different scales.

At a solution of (26), tail-cancellation has taken place inside each sphere and this is the reason why we could make a convenient choice for the tails. The choice is, however, not completely arbitrary; for the ASA to be useful, the determinantal equation yielding the energies must be substantially converged for $(l', l) \leq 3$. At that point tail cancellation takes place only at the interior of the spheres, so that the tails do in fact play a role in making the wave-function (25) continuous and differentiable as we pass from one sphere to the next. Therefore, in order that the l-convergence of the energies be fast, the kinetic energy of the tails should not differ by more than a Rydberg or so from the kinetic energy, $\mathscr{E}-v$, of the electron in the outer region of the spheres. Since we are generally interested in energies for which the electron is barely able to move from one atom to another, it is natural to choose tails with vanishing kinetic energy, and this implies that the band shapes are independent of the lattice constant. For a quantitative discussion of the errors involved in the ASA and for a correction technique we refer to Andersen (1975).

Before returning to the implications of Fig. 5.3 we shall make a brief digression and point out the relation to the KKR method.

5.2.3 The KKR method

After the first attempts with the cellular method of Wigner and Seitz and the spherical approximation, Slater (1937) suggested using touching MT-spheres instead of the atomic spheres, letting the potential take the constant value v_{MTZ} in between. In that interstitial region, which covers about 30 per cent of the volume in closely packed structures, there is a constant kinetic energy

$$\kappa^2 = \mathscr{E} - v_{\mathrm{MTZ}}. \tag{32}$$

For the MT-problem we can define MTOs essentially as before, the only differences being that the tails are attached at the MT rather than at the atomic spheres, and that they are wave-equation solutions with the kinetic energy κ^2. The condition of tail-cancellation becomes the KKR equations, which have the same form as (26). When the KKR condition is satisfied, the tails play the role of forming the proper solutions in the region between the MT-spheres, but the price for not violating the geometrical constraints is high: The KKR structure constants depend on the lattice constant and on the energy (through (32)), and they have poles whenever the magnitude of **k** plus any reciprocal lattice vector equals κ. Also the KKR potential functions,

$$\kappa \cot \eta_l = \kappa \frac{n_l}{j_l} \frac{D_l(\mathscr{E}, s_{\mathrm{MT}}) - \kappa s_{\mathrm{MT}} n_l'/n_l}{D_l(\mathscr{E}, s_{\mathrm{MT}}) - \kappa s_{\mathrm{MT}} j_l'/j_l}, \tag{33}$$

where η_l is the phase shift and where the argument of the spherical Bessel (j_l) and Neumann (n_l) functions is κs_{MT}, depend strongly on κ; and yet, when tail cancellation occurs, there must be a substantial cancellation between these two κ-dependences, provided that the extent of the interstitial region is much smaller than the wavelength $2\pi/\kappa$.

If the interstitial region is taken approximately into account by substituting the MT-radius, s_{MT}, by the atomic-sphere radius, the cancellation between the κ-dependences becomes nearly complete. This is the basis for the ASA.

Although the KKR method can provide one-electron solutions of arbitrary accuracy for a MT-potential, it should be kept in mind that the MT-form itself represents an approximation. Moreover, an approximate density functional (like (17)) which upon minimization gives rise to an MT-potential has not yet been devised. We note that the KKR method provides a formulation within which the electronic structures of alloys may be discussed (see Chapter 8).

5.2.4 Canonical bands

The ASA is convenient because it completely separates the potential- and structure-dependences of the one-electron energies and wave-functions. The ASA equations (26) may be regarded as generalized eigenvalue equations in which the eigenvalue has been replaced by the l-dependent, increasing functions of energy, $P_l(\mathscr{E})$, and canonical band theory exploits the fact that the ASA equations may be solved without specifying the potential functions, but regarding them, rather than the energy, as the independent variables. In this section we shall discuss the boundary condition at the atomic sphere, as expressed by the canonical structure constants defined in (24). In the following section, we shall consider the potential functions.

Let us first assume that the magnitude of the local angular momentum is conserved, i.e. we neglect the structure constants with $l' \neq l$. For each value of l we may then diagonalize the corresponding sub-block $S_{lm';lm}(\mathbf{k})$, obtaining the $2l+1$ *unhybridized* or *pure, canonical sub-bands*, $S_{li}(\mathbf{k})$, and the eigenvectors, $U_{lm;li}(\mathbf{k})$. This unitary transformation will also diagonalize the ASA equations (26) because the potential functions are derived from a spherically symmetric potential and hence do not depend on m, but only on l. The pure nl-energy band, $\mathscr{E}_{nli}(\mathbf{k})$, is therefore the nth solution of

$$S_{li}(\mathbf{k}) = P_l(\mathscr{E}), \tag{34}$$

which is merely a monotonic scaling, specified by the nth branch of the potential function, of the energy axis.

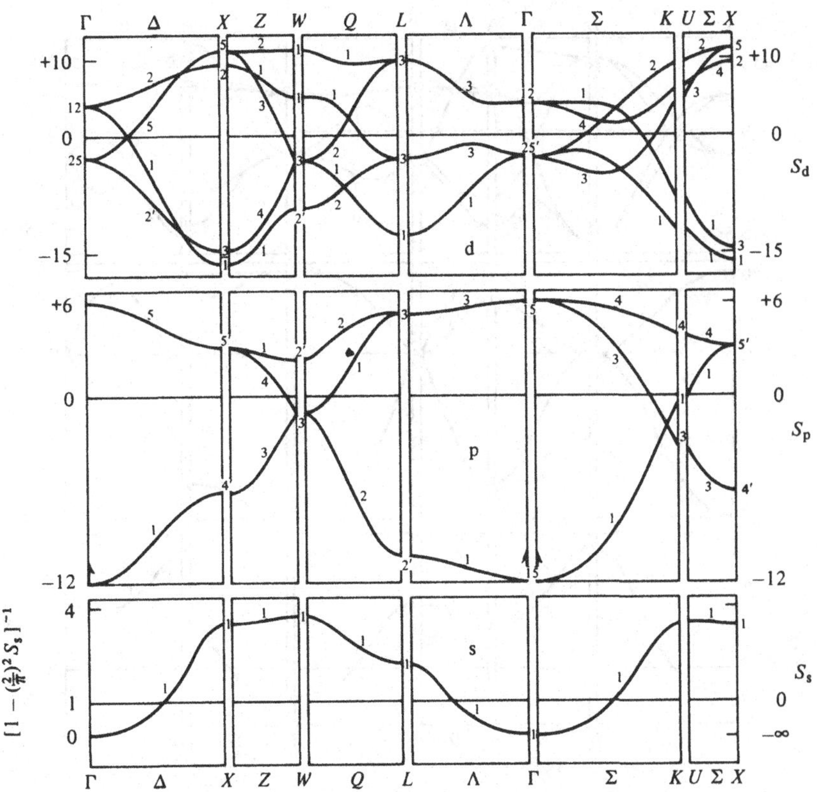

Fig. 5.4. The unhybridized canonical bands for the fcc structure.

The pure canonical s-, p-, and d-bands of the fcc, bcc, and hcp structures are shown in Figs. 5.4–5.6 along the lines of high symmetry in the respective Brillouin zones and in Fig. 5.7 we show the densities of states for the pure, canonical d-bands. These figures may be used to construct an estimate of the energy-band structure specified by a set of potential functions or, more conveniently, from the corresponding set of potential parameters given in the following section. From the **k**-independent scalings (34) we first obtain the pure energy bands, whereafter those carrying the same symmetry label are allowed to hybridize, as illustrated in Fig. 5.8. The pure canonical bands may alternatively be used to interpret band structures existing in the literature in terms of s-, p-, and d-bands and, by fitting to the existing bands at **k**-points of high symmetry, the potential parameters may be deduced.

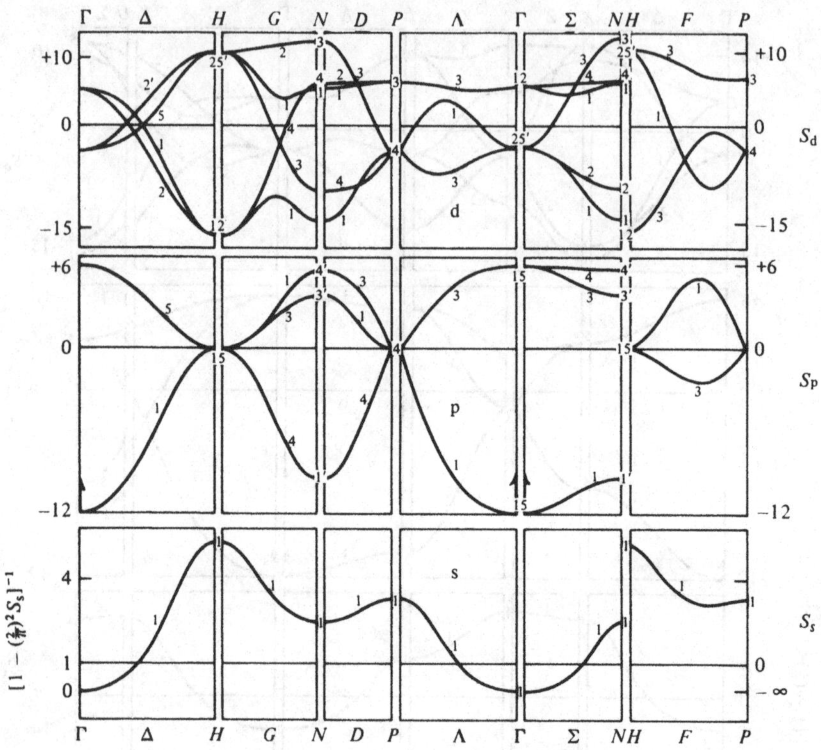

Fig. 5.5. The unhybridized canonical bands for the bcc structure.

The following general properties may be derived from (24):

The centre of gravity of a pure canonical band is zero. Except for canonical p-bands at $\mathbf{k}=0$ and for canonical s-bands, this holds at any $\mathbf{k}$-point and it always holds for the average over the Brillouin zone.

Due to the infinite range of the s-MTO, a pure canonical s-band diverges at the centre of the Brillouin zone. Specifically

$$S_s(\mathbf{k}) \to -6(ks)^{-2} + \text{const.} \tag{35}$$

and in Figs. 5.4–5.6 we have therefore used the free-electron-like scale: $[1-(2/\pi)^2 S_s]^{-1}$. Using (34) and (35), together with (27) and (30), we recover the well-known result that the bottom, V_s, of a pure s-band corresponds to the boundary condition $D_s(V_s)=0$ and that the bottom of this band is parabolic with the relative band mass $\tau_s = 3[s^3 \phi_s^2(V_s, s)]^{-1}$.

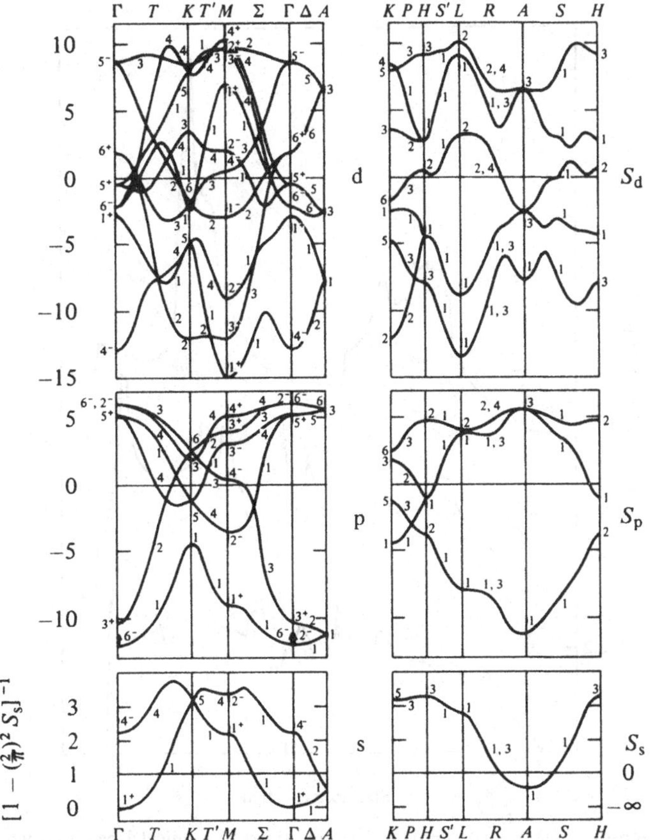

Fig. 5.6. The unhybridized canonical bands for the hcp structure ($c/a = (\frac{8}{3})^{\frac{1}{2}}$).

The width of a pure canonical band may conveniently be estimated from the following expression for the second moment

$$S_l^2 \equiv (2l+1)^{-1} \sum_{i=1}^{2l+1} (2\pi)^{-3} V \int \mathrm{d}^3k\, S_{li}^2(\mathbf{k})$$

$$= 2^{l+2}(2l+1) \times \frac{(2l+1)(2l+3)\ldots(4l-1)}{1\times 2\times \ldots \times l} \sum_{\mathbf{R}\neq 0} \left(\frac{s}{R}\right)^{2(2l+1)}, \quad (36)$$

which depends only on the number of atoms in the various shells. The second moment yields the measure $(12S_l^2)^{\frac{1}{2}}$ of the canonical bandwidth, and for the bcc, fcc, and hcp ($c/a = (\frac{8}{3})^{\frac{1}{2}}$) structures we obtain, respectively, $(12S_\mathrm{p}^2)^{\frac{1}{2}} = 18.8$, 18.7, and 18.6, and $(12S_\mathrm{d}^2)^{\frac{1}{2}} = 23.8$, 23.5, and 23.5. By comparison with the results of Figs. 5.4–5.6, we realize that the estimates of the bandwidths given above are quite accurate. Moreover,

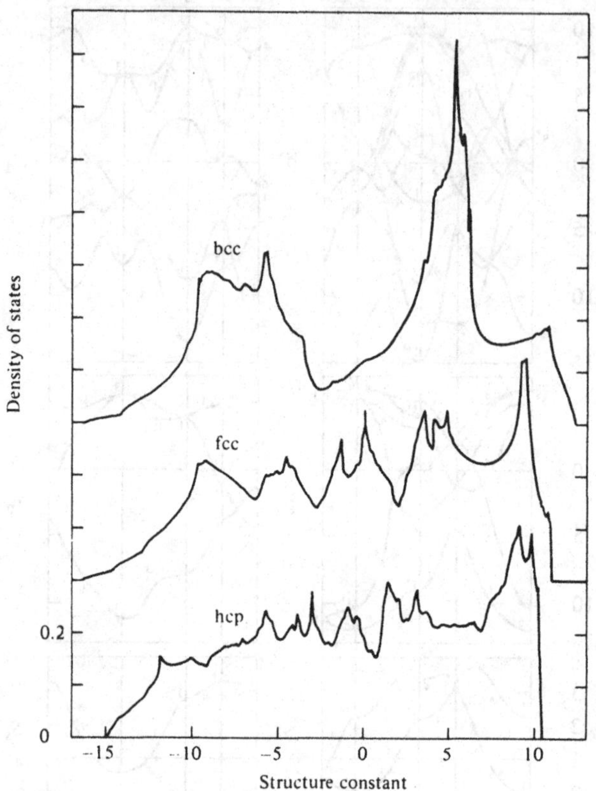

Fig. 5.7. Densities of states (per spin) for the unhybridized canonical d-bands shown in Figs. 5.4–5.6.

they confirm the intuitive rule, used in the renormalized atom approximation of Hodges, Ehrenreich & Watson (1972), that a band of l-character extends from the boundary condition: $\phi'_l(s) = 0$, corresponding to maximal bonding between nearest neighbours, to the condition: $\phi_l(s) = 0$, corresponding to maximal anti-bonding. According to (20) and (27) this so-called Wigner–Seitz rule yields the ranges $(-\infty \to 2)$, $(-12 \to 6)$, and $(-15 \to 10)$ for pure canonical s-, p-, and d-bands, regardless of crystal structure.

The hybridization is taken into account by including the structure constants with $l \neq l'$ in the ASA equations (26) and, as an example, we show in the first two panels of Fig. 5.8 the effect of sp–d hybridization on the band structure of hcp Os. Since the effects of hybridization depend on the relative positions of the hybridizing bands, there seems to be no

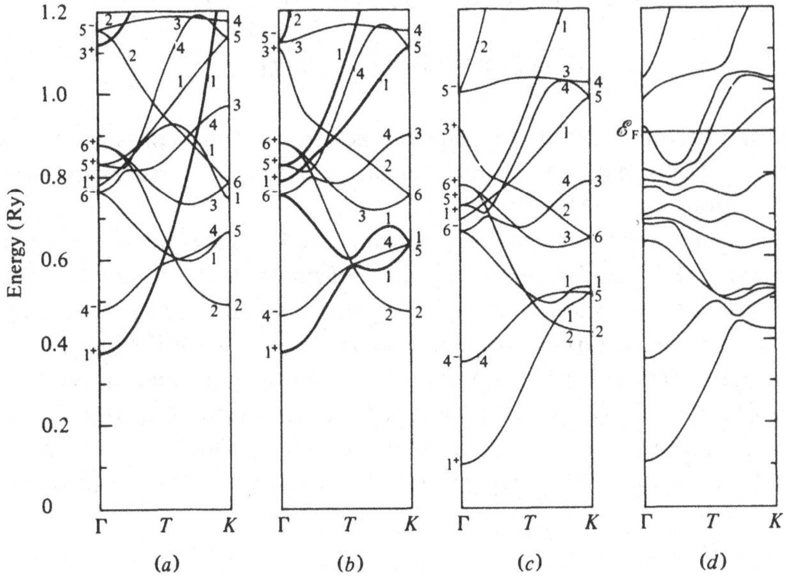

Fig. 5.8. The development of the Os band structure. (*a*) The non-relativistic bands, neglecting hybridization between the sp-bands, indicated by thick lines, and the d-bands. (*b*) The hybridized non-relativistic bands. The thick lines denote strongly hybridized bands. (*c*) The relativistic band structure neglecting spin–orbit coupling. (*d*) The fully relativistic energy bands.

potential-independent, i.e. canonical, way of including them. Nevertheless, the energy- and potential-dependence enters (26) exclusively through the few potential functions $\{P_s, P_p, P_d, \ldots\} \equiv \mathbf{P}$ and, if we regard this *vector*, or any uniquely related vector such as $\mathbf{D}$, as the independent variable, a canonical theory which includes hybridization may in fact be developed (Andersen, Klose & Nohl 1978). An important result is that the l-projected densities of states (per spin)

$$N_l(\mathscr{E}) \equiv \sum_j (2\pi)^{-3} V \int_{BZ} d^3k \sum_m |B_{lm;j}(\mathbf{k})|^2 \delta(\mathscr{E} - \mathscr{E}_{\mathbf{k},j}) \tag{37}$$

may be expressed simply as

$$2N_l(\mathscr{E}) = \dot{P}_l(\mathscr{E})[\partial n(\mathbf{P})/\partial P_l]_{\mathbf{P}(\mathscr{E})} \tag{38}$$

where the structural information is carried by a *canonical number-of-states function, n*(**P**). According to (19) and (37) the density of conduction electrons, spherically averaged in the atomic sphere, is

$$n(r) = (4\pi)^{-1} 2 \sum_l \int^{\mathscr{E}_F} \phi_l^2(\mathscr{E}, r) N_l(\mathscr{E})\, d\mathscr{E}, \tag{39}$$

and the canonical number-of-states function therefore contains all the structural information needed to perform self-consistent band-structure calculations in the ASA.

In Fig. 5.9 we show the s-, p-, and d-projected densities of states for bcc Nb and fcc Pd. The d-projected densities of states may be compared with those of the unhybridized, canonical d-bands which were shown in Fig. 5.7 and which correspond to the potential path: $P_{l'} = \infty$ for $l' \neq 2$. We note that the effect of hybridization on the d-band is very small in the late transition metal Pd, where the pure d-band lies entirely below the pure p-band, and fairly small in Nb where the d-band lies higher. Conversely, in Pd the p-projected density of states arises mainly from the partial-wave

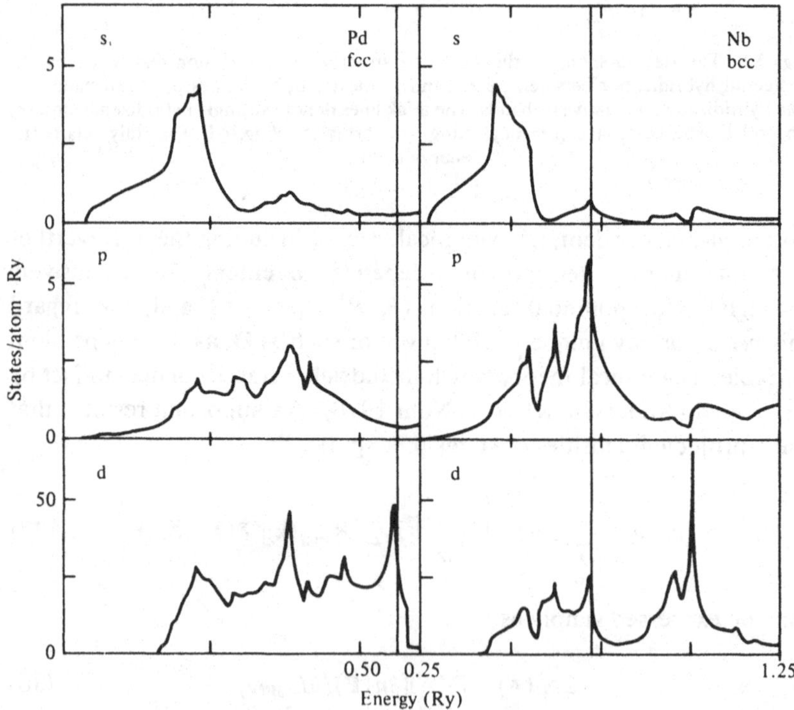

Fig. 5.9. s-, p- and d-projected electron state densities for fcc Pd and bcc Nb, calculated by J. Kollar with the standard potential. The zero of energy is the value of the one-electron potential at the atomic sphere.

expansion of the s- and d-MTOs centred at neighbouring sites, while in Nb this is only approximately true (weak hybridization). For the f- and higher partial waves, the projected densities of states arise exclusively from the tails of s- and d-MTOs (very weak hybridization). The pure d-band cuts through the pure s-band in all transition metals, and the s-projected density of states therefore always exhibits a deep hybridization gap in the d-band region (strong hybridization). The effect of s–d hybridization is smaller on the d- than on the s-projected density of states because there are five times more d-bands than s-bands.

The details of transition metal band structures are often discussed using the formalism of the two-centre approximation of Slater & Koster (1954) to the LCAO method. The ASA may conveniently be written in the two-centre form because, over an energy range so narrow that the potential functions are linear, i.e. that

$$P_l(\mathscr{E}) = P_l(\mathscr{E}_\nu) + (\mathscr{E} - \mathscr{E}_\nu)\dot{P}_l(\mathscr{E}_\nu)$$
$$\equiv [\mathscr{E} - \hat{C}_l(\mathscr{E}_\nu)][\tfrac{1}{2}s\chi_l^2(\mathscr{E}_\nu, s)]^{-1} \qquad (40)$$

the ASA equations become

$$\sum_{lm} [H_{l'm';lm}(\mathbf{k}) - \mathscr{E}\delta_{l'l}\delta_{m'm}]B_{lm}(\mathbf{k}) = 0 \qquad (41)$$

with the effective Hamiltonian

$$H_{l'm';lm}(\mathbf{k}) = \hat{C}_l\delta_{l'l}\delta_{m'm} + \tfrac{1}{2}s\chi_{l'}\chi_l S_{l'm';lm}(\mathbf{k}). \qquad (42)$$

Hence $\hat{C}_l(\mathscr{E}_\nu)$ is the effective centre of gravity of the unhybridized l-band and

$$[\tfrac{1}{2}s\chi_{l'}^2(\mathscr{E}_\nu, s)]^{\frac{1}{2}} S_{l'm';lm}(\mathbf{R})[\tfrac{1}{2}s\chi_l^2(\mathscr{E}_\nu, s)]^{\frac{1}{2}}$$

are the effective transfer integrals, which factorize into potential parameters and structure constants. The structure-constant factors, $S_{l'm;lm}(R\hat{\mathbf{z}})$, of the effective two-centre integrals are simply

$$\left.\begin{array}{l} \mathrm{dd}(\sigma, \pi, \delta) = 10(s/R)^5(-6, 4, -1), \\ \mathrm{pd}(\sigma, \pi) = 6\sqrt{5}(s/R)^4(-\sqrt{3}, 1), \\ \mathrm{pp}(\sigma, \pi) = 6(s/R)^3(2, -1), \qquad \mathrm{sd}\sigma = -2\sqrt{5}(s/R)^3, \\ \mathrm{sp}\sigma = 2\sqrt{3}(s/R)^2, \quad \text{and} \quad \mathrm{ss}\sigma = -2(s/R). \end{array}\right\} \qquad (43)$$

The canonical structure constants may thus be regarded as potential-independent transfer integrals and, in this context, the assumption (40) concerning the potential functions is irrelevant. It should be kept in mind, however, that whereas the dd-interaction has fairly short range, this is not so when $l+l' \lesssim 4$ and, in such cases, the lattice summations (24) must be performed by the Ewald technique.

5.2.5 Potential parameters

Returning to the eigenvalue problem illustrated in Fig. 5.3, we now focus on the atomic sphere. For each value of l we wish to parametrize the energy dependence of the radial wave-function, $\phi_l(\mathscr{E}, r)$, and its logarithmic derivative (20) or potential function (27) in a range corresponding to a particular value, ν_l, of the principal quantum number. In this range the radial wave-function is quite accurately described by the first few terms of its Taylor series

$$\phi(\mathscr{E}, r) = \phi_\nu(r) + (\mathscr{E} - \mathscr{E}_\nu)\dot{\phi}_\nu(r) + o(\mathscr{E} - \mathscr{E}_\nu) \tag{44}$$

provided that $\mathscr{E}$ is an energy in the middle of the range and that $\phi_\nu(r) \equiv \phi(\mathscr{E}_\nu, r)$, etc. Here and in the following we often drop the subscript l, and use $o(\ )$ to denote higher order terms. The 4d-energy derivative functions for Y are shown in Fig. 5.10, from where it appears that the term proportional to $\ddot{\phi}_\nu(r)$ may be neglected over an energy range comparable to the 4d-bandwidth. The radial wave-function is therefore spanned by ϕ_ν and $\dot{\phi}_\nu$ which, due to the normalization (28), are orthogonal in the sphere. The electron density (39) may now be expressed as

$$n(r) = (4\pi)^{-1} \sum_l n_l[\phi_{\nu l}^2(r) + 2(\bar{\mathscr{E}}_l - \mathscr{E}_{\nu l})\phi_{\nu l}(r)\dot{\phi}_{\nu l}(r) + \ldots], \tag{45}$$

where

$$n_l \equiv 2 \int^{\mathscr{E}_F} N_l(\mathscr{E})\, d\mathscr{E} \tag{46}$$

is the number of l-electrons and

$$\bar{\mathscr{E}}_l \equiv n_l^{-1} 2 \int^{\mathscr{E}_F} \mathscr{E} N_l(\mathscr{E})\, d\mathscr{E} \tag{47}$$

is the energy averaged over the occupied part of the l-band. The second term in (45) represents the radial displacement of charge, e.g. in bond formation, due to the broadening of the band about $\mathscr{E}_{\nu l}$.

Since the logarithmic derivatives form the link between the atomic-sphere potential and the crystal structure, it is again convenient to

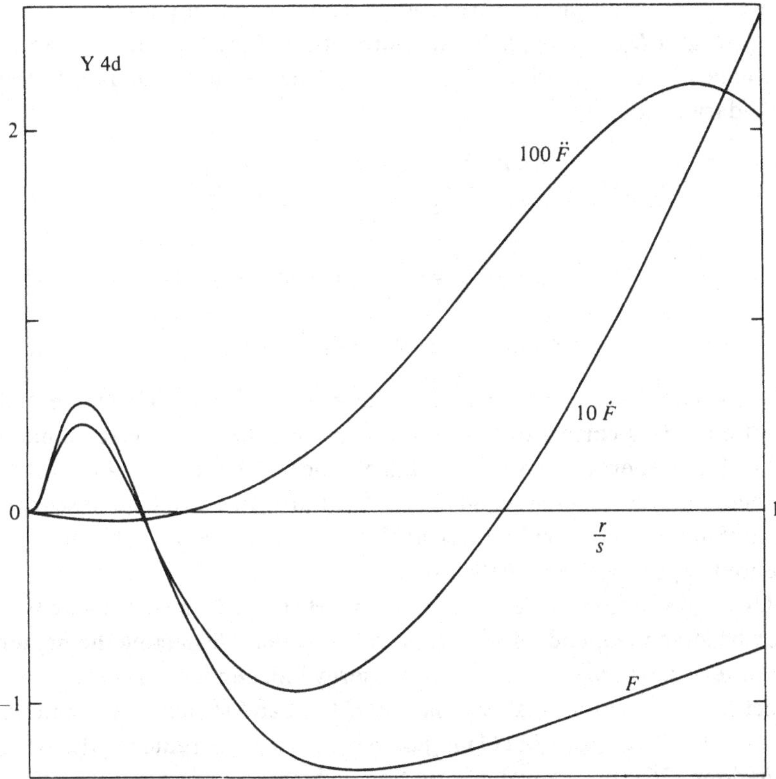

Fig. 5.10. Energy derivatives of the radial 4d-wave-function of Y, corresponding to $D_\nu = -l-1 = -3$, and where $F \equiv s^{\frac{1}{2}} r\phi_\nu$, $\dot{F} \equiv s^{-2}\,\partial F/\partial \mathscr{E}$, $\ddot{F} \equiv s^{-4}\,\partial^2 F/\partial \mathscr{E}^2$, and the 4d band mass $\mu = 25/Ws^2 = 2/F^2(s)$.

consider these, rather than the energy, as the independent variables. The radial trial function, which satisfies the boundary condition specified by D, is

$$\Phi(D, r) = \phi_\nu(r) + \omega(D)\dot{\phi}_\nu(r) \tag{48}$$

with

$$\omega(D) = -\frac{\phi_\nu(s)}{\dot{\phi}_\nu(s)}\frac{D - D_\nu}{D - D_{\dot{\nu}}}. \tag{49}$$

Here, $D_\nu \equiv s\phi'_\nu(s)/\phi_\nu(s) = D(\mathscr{E}_\nu)$, $D_{\dot{\nu}} \equiv s\dot{\phi}'_\nu(s)/\dot{\phi}_\nu(s)$, and it may be shown that

$$D_{\dot{\nu}} - D_\nu = -[s\phi_\nu(s)\dot{\phi}_\nu(s)]^{-1}, \tag{50}$$

so that (49) only contains three independent potential parameters, e.g. D_ν, $s\phi_\nu^2$, and $D_{\dot\nu}$. The explicit parametrization of the function inverse to the logarithmic derivative, $D(\mathscr{E})$, i.e. the energy corresponding to the boundary condition D, is

$$\mathscr{E}(D) = \frac{\langle\Phi(D)|-\nabla^2 + v|\Phi(D)\rangle}{\langle\Phi^2(D)\rangle} + o((\mathscr{E}-\mathscr{E}_\nu)^3)$$

$$= \mathscr{E}_\nu + \frac{\omega(D)}{1+\langle\dot\phi_\nu^2\rangle\omega^2(D)} + o((\mathscr{E}-\mathscr{E}_\nu)^3) \tag{51}$$

$$= \mathscr{E}_\nu + \omega(D) + o((\mathscr{E}-\mathscr{E}_\nu)^2), \tag{52}$$

where $\langle\ \rangle$ denotes an integral in the sphere. Since, for $D = D(\mathscr{E})$, the trial function (48) is correct to first order in $\mathscr{E}-\mathscr{E}_\nu$, the variational estimate (51) of the energy is correct to third order and the estimate (52) is consequently correct to second order. The third-order estimate is seen to be confined to a 'window' centred at $\mathscr{E}_\nu$ and of width $\langle\dot\phi_\nu^2\rangle^{-\frac{1}{2}}$, which is then the fourth potential parameter.

Using (48) it is possible to redefine the MTOs (22) in such a way that they become independent of energy to first order. Neglecting the higher order terms, one may then use these energy independent MTOs as basis functions for a variational calculation of the band structure accurate to third order. This *linear* MTO method has the virtue of reducing the ASA equations (26) to eigenvalue equations which, in contrast to the two-centre form (41), are valid in a wide energy range but contain three-centre integrals and a non-diagonal overlap matrix. Furthermore, with this technique it is possible to treat non-spherical terms in the potential.

For the purposes of displaying the scaling (34) of the canonical bands in a transparent way, and of comparing parameters for different transition-metal potentials, we shall use the conventions $D_\nu = -l-1 = -3$, for $l=2$, and $D_\nu = l$, for $l \neq 2$, whereby the d-band is expanded about its *centre*

$$\mathscr{E}_\nu = \mathscr{E}(-l-1) \equiv C \tag{53}$$

and the s-band is expanded from its bottom,

$$\mathscr{E}_\nu = \mathscr{E}(l) \equiv V. \tag{54}$$

The energies V_l are named the *square-well pseudopotentials* because, for free electrons, they would all coincide with the value of the uniform potential. The bandwidth parameters are

$$s\phi^2(C, s) \equiv 2(\mu s^2)^{-1}, \tag{55}$$

for $l=2$, and

$$s\phi^2(V,s)\equiv(2l+3)(\tau s^2)^{-1}, \tag{56}$$

for $l\neq 2$. The dimensionless parameters μ and τ are named the *intrinsic band masses* because, for free electrons, they would all equal unity. In Figs. 5.11 and 5.12, we show values of the principal potential parameters C, V, μ, and τ, and we shall later return to a discussion of them. Using (49), (50), (52) and (27) we realize that the function inverse to the potential function, $P_l(\mathscr{E})$, may be expressed as

$$\mathscr{E}(P)=C+\frac{1}{\mu s^2}\frac{P}{1-\gamma P}+o((\mathscr{E}-C)^2) \tag{57}$$

$$=V+\frac{1}{\tau s^2}\frac{2(2l+1)^2(2l+3)\gamma}{1-\gamma P}+o((\mathscr{E}-V)^2), \tag{58}$$

where

$$\gamma\equiv\left[2(2l+1)\frac{D_{\dot\nu}+l+1}{D_{\dot\nu}-l}\right]^{-1} \tag{59}$$

is a potential parameter expressing the non-linearity of the potential function and hence the 'distortion' of the canonical bands (34).

The distortion of transition-metal d-bands is fairly small and, if we neglect it, we realize, using (57) and the Wigner–Seitz rule or (36), that the width W_{d} is related to the band mass and the squared amplitudes at the sphere boundary by

$$W_{\mathrm{d}}/25=(\mu_{\mathrm{d}}s^2)^{-1}=\tfrac{1}{2}s\phi_{\mathrm{d}}^2(C_{\mathrm{d}},s)=\tfrac{1}{2}s\chi_{\mathrm{d}}^2(\mathscr{E},s). \tag{60}$$

Using (58) together with (34) and (35) we verify that, regardless of the value of γ_{s}, the bottom of the pure s-band falls at V_{s} and the mass is τ_{s}. Including the hybridization with the p-band, it may be shown that the position of the bottom of the s-band is unchanged and that the mass becomes

$$\tau_{\mathrm{sp}}=\tau_{\mathrm{s}}/D_{\mathrm{p}}(V_{\mathrm{s}})\approx\tau_{\mathrm{s}}[1-\tau_{\mathrm{p}}s^2(V_{\mathrm{p}}-V_{\mathrm{s}})/5]. \tag{61}$$

The squared s- and p-MTO amplitudes entering the two-centre formulation (42) are

$$\tfrac{1}{2}s\chi_l^2(\mathscr{E},s)=[2(2l+1)^2(2l+3)]^{-1}\tau_l s^2(\mathscr{E}-V_l)^2 \tag{62}$$

and the sd-hybridization is therefore proportional to

$$(\tau_{\mathrm{s}}/\mu_{\mathrm{d}})^{\frac{1}{2}}(\mathscr{E}-V_{\mathrm{s}})\approx(\tau_{\mathrm{s}}\mu_{\mathrm{d}})^{-\frac{1}{2}}k^2.$$

The values of the 'small' potential parameters $D_{\dot\nu}$ and $\langle\dot\phi_\nu^2\rangle^{-\frac{1}{2}}$ are rather insensitive to the choice of D_ν and their variation among different transition metals is fairly small. Going from the alkaline earth to the noble metals, $D_{\dot\nu \mathrm{d}}$ decreases from 5 to 2, the latter value characterizing a resonant d-band. For such a band $\langle\dot\phi_{\nu\mathrm{d}}^2\rangle^{-\frac{1}{2}}/W_\mathrm{d}=(\mu_\mathrm{d}/2)^{\frac{1}{2}}$, and for real transition-metal d-bands the energy window is about twice as wide as the band. For all transition metals, $D_{\dot\nu\mathrm{s}}=6\pm1$ and $D_{\dot\nu\mathrm{p}}=9\pm2$, which are close to the values $3l+5=5$ and 8 characterizing free electrons. The s- and p- energy windows are much wider than the d-window and may therefore be neglected. As a consequence, only the values of the 'large' parameters C, V, μ, and τ vary significantly among different metals and we may therefore choose typical values, such as the free electron and resonant values, for $D_{\dot\nu}^*$ to define yet a third canonical scale, Ω, related to the D- and P-scales by: $\Omega\equiv\mu s^2\omega^*(D)=P$ for $l=2$, and by: $\Omega\equiv\tau s^2\omega^*(D)$ for $l\neq2$. This Ω-scale then has the property that the 'potential functions',

$$\mathscr{E}(\Omega)=C+(\mu s^2)^{-1}\Omega+o(\Omega), \tag{63}$$

for $l=2$, and

$$\mathscr{E}(\Omega)=V+(\tau s^2)^{-1}\Omega+o(\Omega), \tag{64}$$

for $l\neq2$, are nearly linear, provided that $D_{\dot\nu}$ is nearly equal to $D_{\dot\nu}^*$, and that $(\mathscr{E}-\mathscr{E}_\nu)^2\langle\dot\phi_\nu^2\rangle$ is much smaller than one. As a consequence, the average energy (47) may be expressed to first order in terms of the moment

$$n_l\bar\Omega_l\equiv\oint^{n_l}\Omega_l[\partial n(\boldsymbol{\Omega})/\partial\Omega_l]\,\mathrm{d}\Omega_l, \tag{65}$$

where $n(\boldsymbol{\Omega})$ is the canonical number-of-states function, as

$$\bar{\mathscr{E}}_\mathrm{d}=C_\mathrm{d}+\frac{\bar\Omega_\mathrm{d}}{\mu_\mathrm{d}s^2},\quad\text{and}\quad\bar{\mathscr{E}}_l=V_l+\frac{\bar\Omega_l}{\tau_l s^2} \tag{66}$$

when $l\neq2$. Furthermore, it may be shown that the change in the sum of the one-electron energies caused by a change of the potential parameters, which conserves the total number of electrons, is simply

$$\begin{aligned}\delta\int^{\mathscr{E}_\mathrm{F}}\mathscr{E}N(\mathscr{E})\,\mathrm{d}(\mathscr{E})\\&=\sum_{l\neq2}n_l[\delta V_l+\bar\Omega_l\delta(\tau_l s^2)^{-1}]+n_\mathrm{d}[\delta C_\mathrm{d}+\bar\Omega_\mathrm{d}\delta(\mu_\mathrm{d}s^2)^{-1}]\\&=\sum_{l\neq2}n_l[\delta V_l-(\bar{\mathscr{E}}_l-V_l)\delta\ln\tau_l s^2]+n_\mathrm{d}[\delta C_\mathrm{d}+(\bar{\mathscr{E}}_\mathrm{d}-C_\mathrm{d})\delta\ln W_\mathrm{d}]\end{aligned} \tag{67}$$

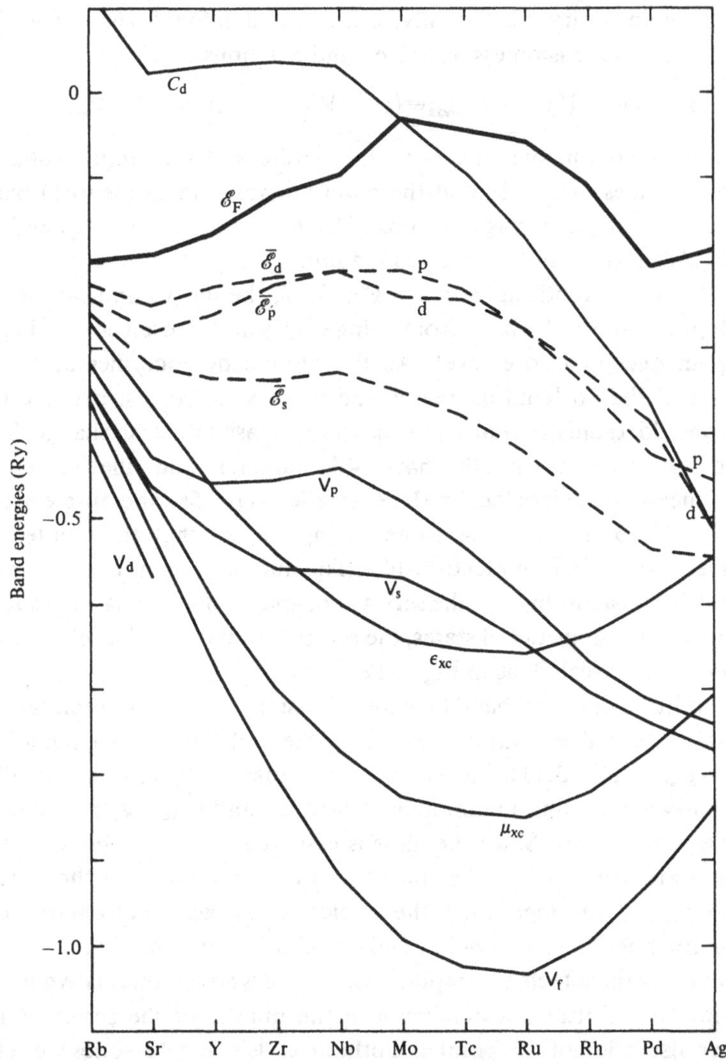

Fig. 5.11. Band energies in the 4d-series obtained by self-consistent LD–ASA calculations (Y. Glötzel, D. Glötzel & O. K. Andersen, unpublished data). V_s is the bottom of the s-band, C_d the centre of the d-band, $\mathscr{E}_F$ the Fermi energy, and μ_{xc} is the value of the one-electron potential at the atomic sphere, approximately equal to the muffin-tin zero.

to first order in $\bar{\mathscr{E}}_l - \mathscr{E}_{\nu l}$. This expression will prove useful when calculating the electronic pressure in Section 3.

The band energies obtained from self-consistent calculations (Y. Glötzel, D. Glötzel & O. K. Andersen, unpublished data) employing the LD and the ASAs are shown in Fig. 5.11 for the 4d-series. Since the atomic volumes vary substantially, as illustrated in Fig. 5.16, we show in Fig. 5.12 the dimensionless, relative band positions

$$\Delta_{ps} \equiv (V_p - V_s)s^2, \qquad \Delta_{ds} \equiv (C_d - V_s)s^2, \qquad \Delta_{fs} \equiv (V_f - V_s)s^2 \quad (68)$$

which are scaled in such a way that, regardless of the atomic volume, identical values of the Δs and the band masses lead to identical band structures when plotted as functions of $\mathbf{k}s$. For free electrons, Δ_{ps} and Δ_{fs} vanish, Δ_{ds} is 20.2, and the masses are unity.

Of the many trends apparent in Fig. 5.12, the most significant is the development of the 4d-band from being empty and free-electron-like, to being an occupied core level. As the atomic number increases, the position of the 4d-band decreases and its mass increases, because the electron added only partially screens the increased nuclear change. The added electrons essentially have 4d-character and the screening experienced by the less localized, i.e. smaller mass, 5s- and 5p-electrons (see Fig. 5.13) is therefore rather complete, so that their potential parameters remain free-electron-like. The number of non-d-electrons is thus fairly constant through the series and, apart from fluctuations due to structure in the densities of states, the position of the Fermi level is nearly constant, when scaled, as in Fig. 5.12.

The bottom of the 5s-band falls well above the 4f-pseudopotential and well above the value of the one-electron potential at the sphere boundary ($v(s) = \mu_{xc}$ in Fig. 5.11). Moreover, the 5s-mass is significantly smaller than unity and attains a minimum, as does Δ_{fs} and $(\mu_{xc} - V_s)s^2$, near the middle of the series. Since the mass is essentially the ratio between the normalization integral in the sphere and the probability at the sphere boundary (56), this means that the 5s-electron is excluded from the core region (orthogonalization hole), and the above mentioned facts may be explained as the effects of a repulsive ion-core whose size, relative to the atomic sphere, attains a maximum in the middle of the series. For a further discussion of this point and other trends in the 4d-series we refer to Pettifor (1977).

In Fig. 5.11, the zero of energy is the electrostatic potential between the atoms in an infinite crystal. This value is well-defined in the ASA, where the charge density is spherical and the spheres neutral. At the sphere boundary the one-electron potential (9) therefore equals the exchange-correlation potential which, in the LD approximation (14), is $\mu_{xc}(n(s)) \equiv \mu_{xc}$. In the Wigner–Seitz approximation the exchange-correlation hole is centred at the atom in question, such that $v_{xc}(s) = -2/s$, and this accounts fairly well for the trend observed for μ_{xc}. In the following section we shall make use of $\varepsilon_{xc}(n(s)) \equiv \varepsilon_{xc}$ which, according to (13), is approximately $\frac{3}{4}\mu_{xc}$. The parabolic trend followed by the Fermi energy results from the combined effects of filling up the d-band and lowering its centre of gravity. The Fermi energy is the negative of the internal work function, and the difference between this and the external work function, which experimentally increases from about 0.23 Ry in Y to about 0.40 Ry in Pd, is due to the surface dipole. The empirical ionicity scale of Miedema (1976), used to categorize heats of formation of binary transition metal alloys, closely resembles the experimental, external work functions although, in principle, it should be related to the internal work functions. The theoretical Fermi energies are, however, extremely sensitive functions of atomic volume and the apparent discrepancy points to the problem of defining atomic volumes in an alloy (Hodges 1977).

The potential parameters indicated by open circles in Fig. 5.12 result from the standard, non-self-consistent potential construction described at the end of Section 5.2.1. For the transition metals proper the band structures derived from the LD and the standard potentials differ by less than 20 mRy, but for the simple-, alkaline earth-, and noble metals the differences are larger and here the LD potential generally yields the better Fermi surfaces, as we shall see in Section 5.4.

Crude estimates of the *central* potential parameters $C_d - C_s$, μ_s, and μ_d may be obtained directly from 5s- and 4d-*atomic* energies and wavefunctions by renormalizing the latter to the atomic sphere and by using first-order perturbation theory. This procedure works because the standard atomic-sphere potential and the atomic potential used to construct it are rather similar, and the shapes of the 5s-, and 4d-partial waves shown in Fig. 5.13 resemble those of the corresponding atomic orbitals. This means, for instance, that the radii, s_{5s} and s_{4d}, determined by the condition that the logarithmic derivative of the *atomic* orbital is $-l-1$, both differ less than about 15 per cent from the equilibrium radius in the metal, and these, purely atomic, radii thus crudely display the trend shown in Fig. 5.16.

The relativistic corrections to the band structures of transition metals have their origin in the regions close to the nuclei, where the velocity of the electrons is high, and they may therefore be separated into mass-velocity plus Darwin shifts, and the spin–orbit coupling. For a given metal the relativistic effects decrease with increasing angular momentum and principal quantum number and, for a given *nl*-shell, they increase with atomic number, approximately as Z^2. Due to the normalization (28) of the partial wave to the atomic sphere, the magnitude of the relativistic

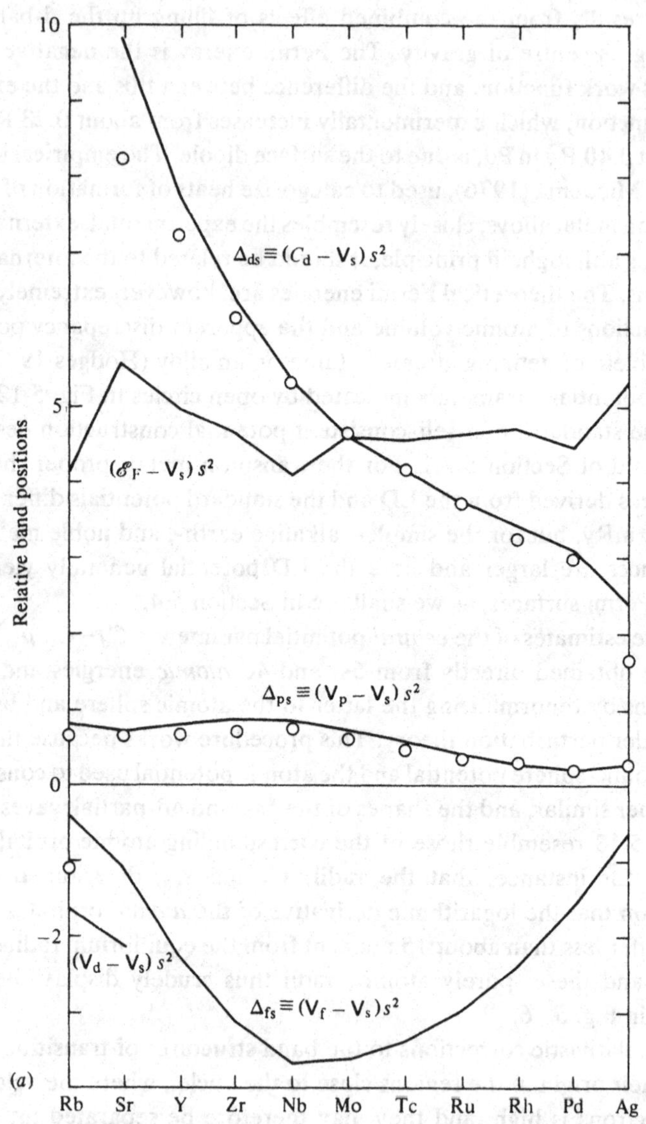

Fig. 5.12. Band positions relative to the bottom of the s-band, in dimensionless units, and intrinsic band masses, on a reciprocal scale, in the 4d-series. The full lines represent the results of self-consistent LD–ASA calculations, while the open circles were obtained from a standard potential construction (Andersen & Jepsen 1977).

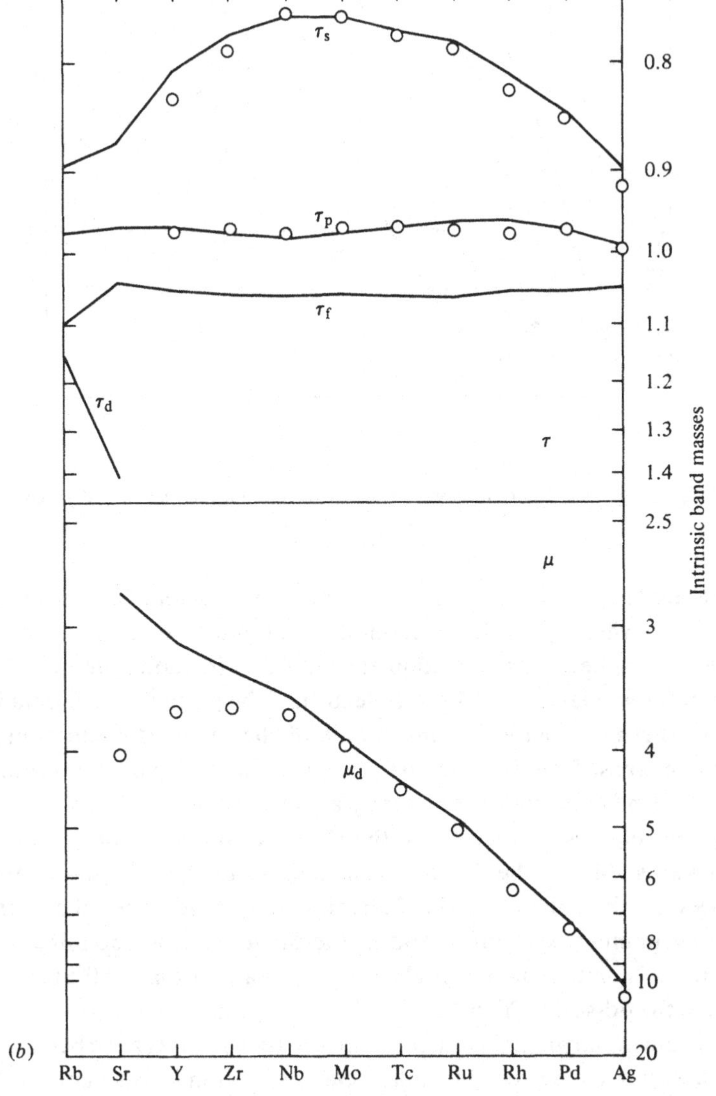

τ_s
τ_p
τ_f
τ_d
τ
μ
μ_d
0.8
0.9
1.0
1.1
1.2
1.3
1.4
2.5
3
4
5
6
8
10
20
Intrinsic band masses
(b)
Rb
Sr
Y
Zr
Nb
Mo
Tc
Ru
Rh
Pd
Ag

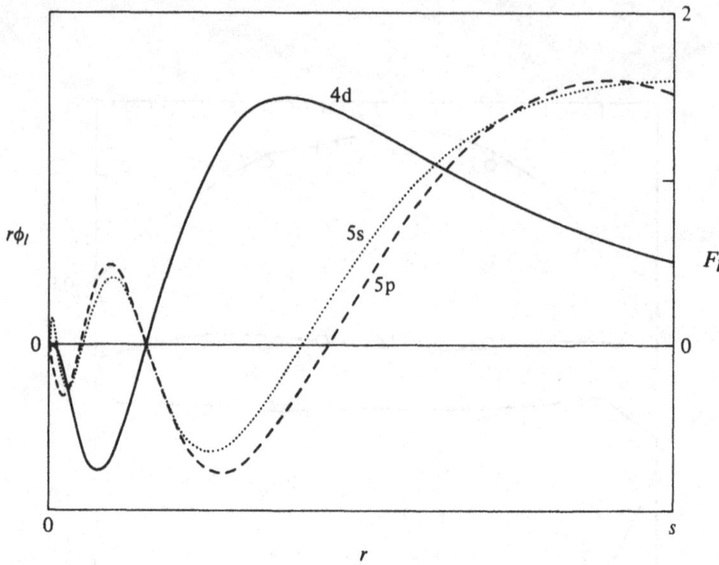

Fig. 5.13. s-, p- and d-partial waves of Pd, corresponding to the boundary condition $D_\nu = -l-1$.

effects are larger in the solid than in the atom by a factor of approximately $(1-q_{nl})^{-1}$, where q_{nl} is the fraction of the atomic nl-electron which lies outside the sphere. For transition elements this fraction is about 0.5 for s-electrons and less than 0.1 for d-electrons. The partial wave normalization furthermore causes the magnitude of the relativistic effects in the solid to increase from the bottom to the top of the band and, for transition metal d-bands, this increase is nearly a factor of two.

Of the relativistic corrections, the shifts are the most important; the downwards shift of the 4s-band with respect to the 3d-band is about 15 mRy in the middle of the 3d-series, in the 4d-series the corresponding number is 75 mRy, and in the 5d-series it is 250 mRy. As a result, the number of non-d-electrons increases from 1.50 ± 0.10 per atom in the 4d-series (Y to Ag) to 1.75 ± 0.15 in the 5d-series (Lu to Au). In the figures shown in this chapter, the relativistic shifts have been taken into account by performing fully relativistic atomic calculations, but leaving out the spin–orbit coupling effects on the band states, by deriving the potential parameters from a modified radial Dirac equation. As shown in Fig. 5.14, the spin–orbit coupling parameters are typically one order of magnitude smaller than the relativistic shifts. Moreover, for crystals with inversion symmetry, the spin–orbit coupling cannot split the

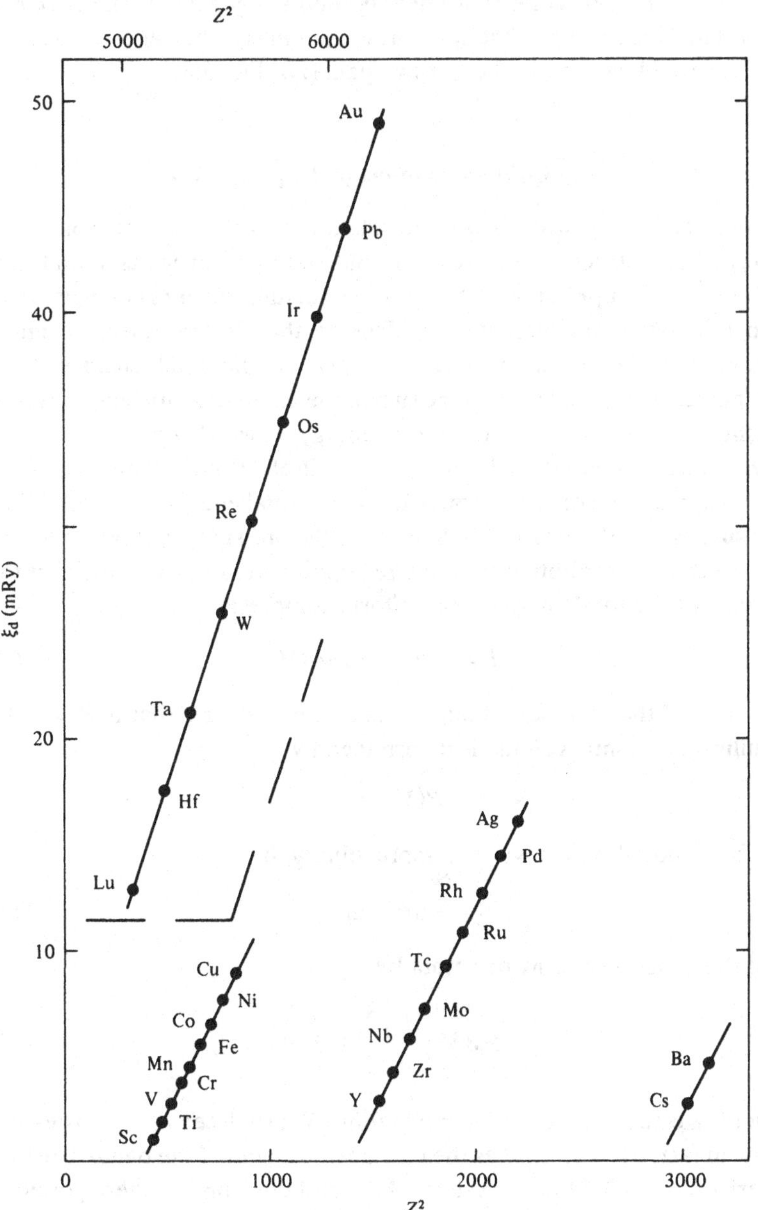

Fig. 5.14. Spin–orbit coupling parameter, $\xi_d(C_d)$, for the centre of the d-band, as a function of the square of the atomic number.

two spin bands, and it therefore gives rise to splittings of first order in ξ only in small regions of **k**-space near points of degeneracy. The effects of the relativistic shifts and the spin–orbit coupling on the band structure of hcp Os are illustrated in the last two panels of Fig. 5.8.

5.3 Cohesive and magnetic properties

In this section we will discuss the relation between the self-consistent energy-band structure and such ground-state properties as the atomic volume, the compressibility, the crystal structure, the magnetic moment, and the static spin-susceptibility. Due to the electron–electron interaction, there is no simple relation between the total electron- and inter-nuclear energy (17) and the sum of the band-structure energies, and furthermore the changes of the total energy, in which we are interested, are minute compared to the total energy itself. Rather than basing our discussion of the cohesive properties on the total energy, we shall follow the suggestion of Pettifor (1978) and use the pressure, P, which, neglecting the small contribution from the zero-point motion of the nuclei, is the *change* of the total energy with uniform compression, i.e.

$$P \equiv -\mathrm{d}\langle\Phi|\mathcal{H}|\Phi\rangle/\mathrm{d}V. \tag{69}$$

In terms of the equation of state at zero temperature, that is $P(V)$, the equilibrium atomic volume is determined by

$$P(V_0)=0, \tag{70}$$

the bulk modulus, or inverse compressibility, by

$$B=-\mathrm{d}P/\mathrm{d}\ln V|_{V_0}, \tag{71}$$

and the cohesive energy per atom by

$$-\int_{V_0}^{\infty} P\,\mathrm{d}V = -\int_{s_0}^{\infty} 3PV\,\mathrm{d}\,(\ln s), \tag{72}$$

where s_0 is the equilibrium atomic radius. We shall see that the pressure may, in fact, be expressed as the change of the sum of the band-structure energies, provided that we use the ASA and the atomic-sphere potential is *frozen*, i.e.

$$P=-\delta\left[\int^{\mathscr{E}_F} \mathscr{E}N(\mathscr{E})\,\mathrm{d}\mathscr{E}\right]\Big/\delta V. \tag{73}$$

5.3.1 Force and pressure relations

Before considering a uniform expansion and the pressure, let us consider the simpler and more general situation in which we divide space into two separable regions, a and b, and wish to calculate the change of the total energy caused by an infinitesimal displacement, $\mathbf{dc}$, of the corresponding two groups of nuclei. In the expression for the energy functional (7) within the LD approximation (11), we divide all integrals into parts a and b, and, for each part, we use *local coordinates*, $\mathbf{r}_a$ and $\mathbf{r}_b$. The kinetic energy term may be expressed as

$$\langle\Phi'|T|\Phi'\rangle = \int^{\mathscr{E}_F} \mathscr{E}N(\mathscr{E})\,\mathrm{d}\mathscr{E} - \int_a \mathrm{d}^3 r_a\, v_a(\mathbf{r}_a) n_a(\mathbf{r}_a) - \int_b \mathrm{d}^3 r_b\, v_b(\mathbf{r}_b) n_b(\mathbf{r}_b) \tag{74}$$

using (8), (5) and (6). Here, n_a, n_b, v_a and v_b are the self-consistent densities and potentials in regions a and b respectively. We now separate the regions by the translation, $\mathbf{dc}$, inserting the infinitesimal region, c, between them, and seek the first-order change of the total energy. An allowable trial density may be obtained from (6) by keeping the number of electrons constant, and by using the solutions of a one-electron problem (5) for a potential which is v_a in region a and v_b in region b, and is arbitrary in region c. If we insert this density, $\tilde{n}$, into the expression for the perturbed energy functional, we find the following change of the kinetic energy

$$\delta\langle\Phi'|T|\Phi'\rangle = \delta \int^{\mathscr{E}_F} \mathscr{E}N(\mathscr{E})\,\mathrm{d}\mathscr{E} - \int_c \mathrm{d}^3 r_c\, v_c(\mathbf{r}_c) n_c(\mathbf{r}_c)$$
$$- \int_a \mathrm{d}^3 r_a\, v_a(\mathbf{r}_a)\delta n_a(\mathbf{r}_a) - \int_b \mathrm{d}^3 r_b\, v_b(\mathbf{r}_b)\delta n_b(\mathbf{r}_b), \tag{75}$$

to first order in $\delta n_a \equiv \tilde{n}_a - n_a$ and $\delta n_b \equiv \tilde{n}_b - n_b$. The change of the remaining terms in (7), including the inter-nuclear repulsion, may be separated into terms arising from: (i) the density changes δn_a and δn_b, (ii) integrals over region c, and (iii) the change in the position of the nuclei. The terms (i) simply cancel the two last terms in (75), as may be seen from the arguments leading to (9). The terms (ii) are

$$\int \mathrm{d}^3 r_c\, [v_{\mathrm{Coul}}(\mathbf{r}_c) + \varepsilon_{\mathrm{xc}}(n(\mathbf{r}_c))]n(\mathbf{r}_c)$$

according to (7) and (11), and where v_{Coul} is the electrostatic potential from the electrons and the nuclei. The term (iii) is simply the electrostatic force between the nuclei in region a, screened by the part of the electronic charge cloud which lies in region *a*, and the nuclei in region *b*, screened by the electronic charge cloud in region *b*.

Within the LD approximation, we thus have the rigorous result that the cleavage force equals the change in the sum of the *one*-electron energies upon virtual displacement of *frozen* potentials, a surface term, and the electrostatic force between the nuclei, *screened* by the electronic charge clouds in the two regions, i.e.

$$\begin{aligned} \mathrm{d}\langle\Phi|\mathcal{H}|\Phi\rangle = \delta\int^{\mathscr{E}_F} \mathscr{E}N(\mathscr{E})\,\mathrm{d}\mathscr{E} \\ + \int_c \mathrm{d}^3 r_c\,[v_{\mathrm{Coul}}(\mathbf{r}_c) + \varepsilon_{\mathrm{xc}}(n(\mathbf{r}_c)) - v_c(\mathbf{r}_c)]n(\mathbf{r}_c) \\ + \text{electrostatic force}\cdot(-\mathbf{dc}). \end{aligned} \tag{76}$$

The sum of the two first terms is independent of v_c, and we may choose v_c equal to $v_{\mathrm{Coul}} + \varepsilon_{\mathrm{xc}}$ (whereby $\tilde{v}$ is discontinuous at the surface by the amount $\varepsilon_{\mathrm{xc}} - \mu_{\mathrm{xc}}$) such that the surface term vanishes. Moreover, the total force only depends on the virtual displacement of the nuclei and not on where, in the region between the two groups of nuclei, we choose to make the cut; if it is made through regions of low electron density, we ensure that only the energies of valence and conduction electrons contribute to (76). In this apparent one-electron expression, double counting of the electron–electron interaction is avoided by using the frozen, rather than the self-consistently readjusted potentials.

The 'force relation' (76) yields a useful expression for the pressure of an infinite crystal if we perform the cuts along the boundaries of the atomic Wigner–Seitz cells and then let them all expand. For a monatomic solid, and in the ASA, the electrostatic term vanishes and, with the above mentioned choice of v_c, we recover the result (73). Moreover, the differentiation with respect to atomic volume may be performed analytically because, in the ASA, it only applies to the potential functions, since the boundary conditions are independent of atomic volume. In terms of the canonical number-of-states function

$$\begin{aligned} 3PV &= -\frac{\delta\int^{\mathscr{E}_F}\mathscr{E}N(\mathscr{E})\,\mathrm{d}\mathscr{E}}{\delta\ln s} = \int^{\mathscr{E}_F}\mathrm{d}\mathscr{E}\,\frac{\delta n(\mathbf{D}(\mathscr{E}))}{\delta\ln s} \\ &= \sum_l\int^{\mathscr{E}_F}\mathrm{d}\mathscr{E}\,\frac{\partial n(\mathbf{D})}{\partial D_l}\,\frac{\delta D_l(\mathscr{E})}{\delta\ln s} \\ &= \sum_l\int^{\mathscr{E}_F}\mathrm{d}\mathscr{E}\,N_l(\mathscr{E})s\phi_l^2(\mathscr{E},s)\frac{-\delta D_l(\mathscr{E})}{\delta\ln s} \\ &\equiv \sum_l 3P_lV, \end{aligned} \tag{77}$$

where, from the radial Schrödinger equation,

$$-\delta D_l(\mathscr{E})/\delta \ln s = [D_l(\mathscr{E})+l+1][D_l(\mathscr{E})-l]+[\mathscr{E}-\varepsilon_{xc}]s^2, \quad (78)$$

and $\varepsilon_{xc} \equiv \varepsilon_{xc}(n(s))$. This is the pressure relation of Nieminen & Hodges (1976), and Pettifor (1976), who derived it under more restrictive conditions and in a different way. For conceptual simplicity we shall use the first-order expression (67) for $-P\,\mathrm{d}V = -3PV\,\mathrm{d}\ln s$, together with the results

$$\frac{\delta V}{\delta \ln s} = -(2l+3)\frac{V-\varepsilon_{xc}}{\tau}, \qquad \frac{\delta C}{\delta \ln s} = -2\frac{C-\varepsilon_{xc}}{\mu}$$

$$\frac{\delta \ln \tau s^2}{\delta \ln s} = -(2l+1)+\frac{2l+3}{\tau}-\frac{2(V-\varepsilon_{xc})s^2}{D_{\dot{\nu}}-l} \quad (79)$$

$$\frac{\delta \ln \mu s^2}{\delta \ln s} = -\frac{\delta \ln W}{\delta \ln s} = (2l+1)+\frac{2}{\mu}-\frac{2(C-\varepsilon_{xc})s^2}{D_{\dot{\nu}}+l+1},$$

for the dependence of the band positions and masses on atomic volume obtained from (78), (30), (55) and (56) for the frozen atomic-sphere potential. The equations (67) and (79) thus provide the relation between the band structure and the pressure.

5.3.2 Cohesive properties

The partial pressures for the 4d-series, obtained from the self-consistent LD–ASA calculations of Y. Glötzel, D. Glötzel & O. K. Andersen (unpublished data), are shown in Fig. 5.15 as functions of $(s-s_o)/s_o$, where s_o is the experimentally observed equilibrium radius. The radius at which the theoretical total pressure vanishes is denoted by 'c' and is seen to be generally a few per cent too small. Using the band parameters, shown in Figs. 5.11 and 5.12 for the experimental radius, together with (79) we shall now discuss the balance between the partial contributions to the pressure, as well as the relative importance of the band-position and band-broadening terms exhibited by (67).

For the homogeneous electron gas, $V = \mu_{xc} < \varepsilon_{xc}$ and $\tau = 1$, for all values of l and s. The δV-term in (67) therefore provides an attractive, i.e. negative, pressure which is the correction, due to exchange and correlation, to the kinetic, free-electron term proportional to $\bar{\mathscr{E}} - V \propto k_F^2 \propto s^{-2}$. The equilibrium occurs at $s_o = 4.1$ a.u. for $n = 1$. Crudely speaking, this is the situation found in the simple metals. As we proceed into the 4d

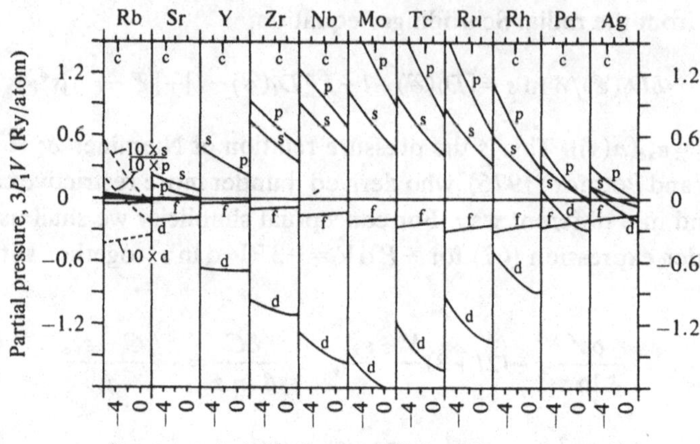

Fig. 5.15. The product of the partial pressure and atomic volume as a function of the relative deviation from the observed atomic radius in the 4d-series, obtained from self-consistent LD–ASA calculations. The calculated total pressure vanishes at the radii denoted by 'c'.

transition series, where the number of s- and p-electrons is about 1.5, the atomic volumes decrease due to increased d-electron binding, as will be explained below. The relative core size thereby increases with the result that the s-, and p-pseudo-potentials, V, increase and reach maxima well above ε_{xc} near the middle of the series, and that τ_s decreases to a minimum of about 0.76. The s- and p-pressures contributed by the δV-term therefore change sign and reach positive maxima near the middle of the series. Also the positive, kinetic s- and p-contributions to $3PV$, arising from the $(\bar{\mathscr{E}} - V)$ term, increase with decreasing s. The s^{-2} behaviour found for free electrons is, however, somewhat modified, due to the presence of the hybridization gap created by the d-bands and shown in Fig. 5.9. Turning now to the d-electrons, we realize that C_d lies above ε_{xc} and therefore increases with pressure. This means that the δC-term in (67) is repulsive, as are the s- and p-terms, and it reaches a maximum near the middle of the series due to the combined behaviour of $C_d - \varepsilon_{xc}$, μ_d, and n_d. The attraction is provided by the band-broadening term and, for a rectangular density of d-states, $\bar{\mathscr{E}}_d - C_d \approx -W_d(10 - n_d)/20$ and $\delta \ln W_d / \delta \ln s \approx -5$, as obtained from (79). Therefore, the d-band broadening gives rise to the well-known parabolic behaviour

$$3P_d V \approx -W_d n_d (10 - n_d)/4, \tag{80}$$

which is about 30 per cent reduced by the δC-term. The negative f-pressure seen in Fig. 5.15 arises essentially from the tails of the d-orbitals, as discussed in Section 5.2.4.

In order to obtain the cohesive energy from (72), the band parameters must be determined self-consistently as functions of atomic volume. Although the electron–electron interaction causes the d-band energies to rise rapidly with compression, we have seen that it is the change for a *frozen* potential which enters the pressure expression. This change is at least one order of magnitude smaller than the self-consistent change, and we have argued that it is insignificant compared with the contribution from the broadening of the d-band. The effect of electron–electron interaction is nearly the same for all d-electrons throughout the band and this means that

$$\mathrm{d} \ln W_{\mathrm{d}}/\mathrm{d} \ln s \approx \delta \ln W_{\mathrm{d}}/\delta \ln s. \tag{81}$$

In Pd, for instance, the values in (81) are respectively 5.2 and 4.5. The d-contribution to the cohesive energy is therefore roughly minus the sum of the one-electron energies measured relative to C_{d}. The sum of the s-, and p-pressures becomes negative for $s \gtrsim 4$, as we have seen, and their integral from s_{o} to infinity turns out to be rather small. The major contribution to the cohesive energy in the transition metals thus arises from the broadening of the d-band as argued originally by Wigner & Seitz (1955). LD–KKR calculations of the cohesive energies in 3d- and 4d-metals have been performed by Moruzzi *et al.* (1977), who find good agreement in the 4d-series but less satisfactory agreement in the 3d-series, where a more careful treatment of the atomic energies seems to be needed. Pettifor (1978) has performed non-hybridized LD–ASA calculations for the 4d-elements and we refer to this work, and that of Gelatt, Ehrenreich & Watson (1977) for a careful discussion.

The LD–ASA results for the atomic volumes and bulk moduli of the 4d- and 5d-metals of Y. Glötzel, D. Glötzel & O. K. Andersen (unpublished data) are summarized in Fig. 5.16. The errors in the radii are somewhat larger than found by Moruzzi *et al.* (1977), and this suggests that the major part of the discrepancy is due to the assumption made in the ASA that each nucleus is perfectly screened by the electronic charge cloud in the surrounding cell. The bulk modulus varies substantially with atomic volume and the bulk moduli evaluated at the experimental rather than at the theoretical equilibrium volumes therefore have by far the smaller errors.

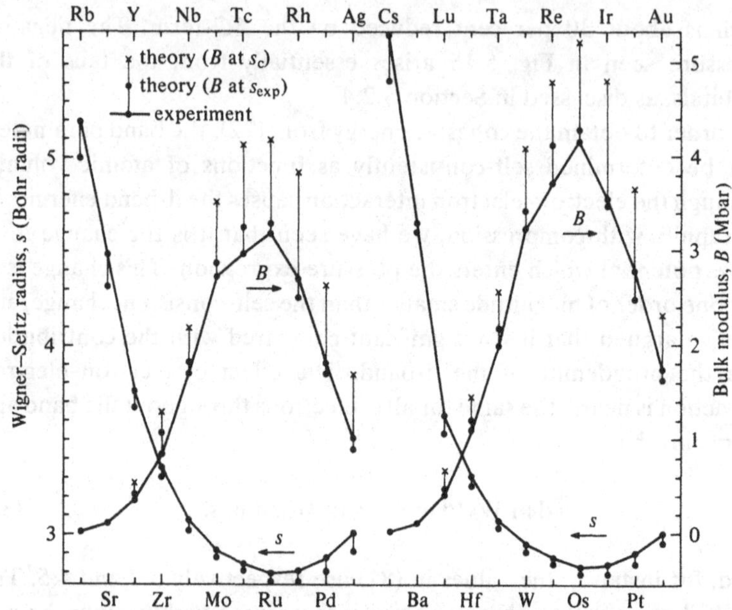

Fig. 5.16. Theoretical LD–ASA atomic radii and bulk moduli for the 4d- and 5d-series, compared with the experimental low-temperature values (Pearson 1964, Gschneidner 1964). The theoretical bulk moduli evaluated at the calculated radii are denoted by crosses while those calculated at the experimental radii are denoted by dots. The theoretical results for the hcp metals were obtained in the fcc structure.

Structural energy differences are smaller than cohesive energies by one or two orders of magnitude and no successful attempt has so far been made to explain the crystal structures of the transition metals (shown in the upper frame in Fig. 5.18) from first principles. If the rearrangement of the atoms needed to go from one crystal structure to another were an infinitesimal one, we could use the force relation (76) and calculate the structural energy difference as the difference between the sum of the one-electron energies, calculated for the same potential in the two different crystal structures. In the ASA, we would furthermore neglect the difference in the electrostatic Madelung terms and, considering the unhybridized d-band only and neglecting the small potential parameters γ_d and $\langle \dot{\phi}^2_{\nu d} \rangle$, the structural energy differences are simply the difference between

$$n_d \bar{S}_d = 2 \int^{S(n_d)} SN(S)\, dS \tag{82}$$

in units of $(\mu_d s^2)^{-1}$, N being the canonical d-densities of states given in Fig. 5.7. The result shown in Fig. 5.17 is similar to one obtained by

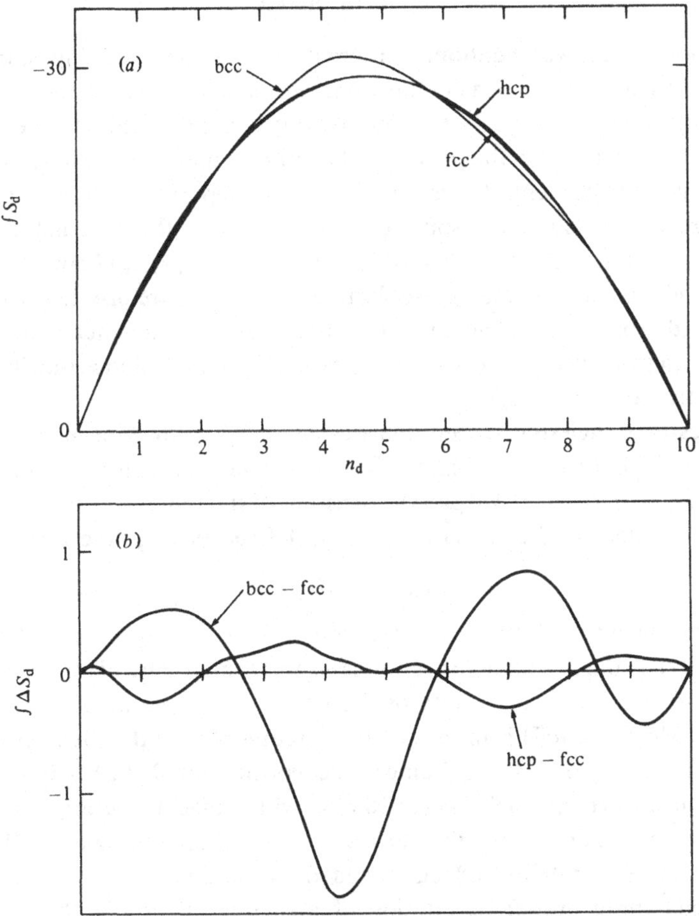

Fig. 5.17. (*a*) First moment of the unhybridized canonical d-electron state density as a function of the d-band occupancy. (*b*) The differences in this quantity between the different structures.

Pettifor (1970) and it accounts qualitatively for the crystal structures of the non-magnetic transition metals, apart from the occurrence of the fcc structure at the end of the series. A careful numerical investigation of the various approximations involved in using (82) reveals that this discrepancy must be ascribed to the use of the force relation, which is not strictly justified. This implies that a very accurate calculation of the total energy in the different phases, going beyond the ASA and MT approximations, is necessary to resolve this problem.

5.3.3 Magnetic effects

In Section 5.2.1 we mentioned a practical way of including spin-dependent effects in the density functional formalism. If we use the local (LSD) approximation, neglect the spin–orbit coupling, and consider a para- or ferro-magnet in the presence of a uniform magnetic field, H, the band-structure problem (15) is decoupled into separate equations for each direction of spin; the spin-dependent parts of the external and exchange-correlation potentials are given by $-\text{sign}\,(\sigma)\mu_B H$ and (16) respectively. By solving this spin-polarized band-structure problem self-consistently, one may calculate zero-temperature spin-magnetizations, m, uniform spin-susceptibilities, $\chi \equiv \mu_B\, \mathrm{d}m/\mathrm{d}H$, and magnetic contributions to the cohesive properties.

The result of such spin-susceptibility calculations for the non-magnetic 4d-metals (Y. Glötzel, D. Glötzel & O. K. Andersen, unpublished data) is shown in Fig. 5.18, where we plot the densities of states at the Fermi level, N, and the effective Stoner parameters, I, defined through the relation

$$\chi = 2\mu_B^2 N(1 - IN)^{-1} \tag{83}$$

for the enhanced Pauli susceptibility. Only in Pd is the enhancement so large that the spin-susceptibility dominates the orbital- and diamagnetic contributions. The susceptibility of 7.1×10^{-4} emu/mole measured by Foner & McNiff (1967) is in reasonable agreement with the theoretical spin-susceptibility of 4.3×10^{-4} emu/mole, considering that $IN = 0.815$, and that an uncertainty in $C_d - V_s$ of 10 mRy will change the Fermi energy by about 1 mRy, and hence the density of states (Fig. 5.9) by about 10 per cent. The correlation effects included in the LSD approximation are crucial for a proper description of magnetic effects; in the $X\alpha$-approximation, Pd would be a strong ferromagnet.

The LSD theory may be approximated by a Stoner formalism. If, for instance, we treat the spin-dependent parts of the potential in (15) by first-order perturbation theory (Gunnarsson 1976; Andersen *et al.* 1977) and use (16) and (13), we find that the exchange splitting is given by

$$\begin{aligned}
\Delta\mathscr{E} &\equiv \mathscr{E}_{\mathbf{k}\downarrow} - \mathscr{E}_{\mathbf{k}\uparrow} \\
&= 2\mu_B H + m\left\langle \psi(\mathbf{k}, \mathbf{r}) \left| \frac{4}{(9\pi)^{\frac{1}{3}}} \frac{\delta(n(\mathbf{r}))}{n^{\frac{2}{3}}(\mathbf{r})} \frac{m(\mathbf{r})}{m} \right| \psi(\mathbf{k}, \mathbf{r}) \right\rangle \\
&\equiv 2\mu_B H + mI.
\end{aligned} \tag{84}$$

Here I, and hence $\Delta\mathscr{E}$, are only strictly independent of $\mathbf{k}$ if we assume that the electron and spin densities $n(\mathbf{r})$ and $m(\mathbf{r})$, are spherically symmetric in the atomic sphere, and that only one partial wave contributes to the expansion (19) of the wave-function. For a **k**-independent exchange splitting the self-consistency condition is, of course, that the integral of the paramagnetic state density (per spin), over the energy range, $\Delta\mathscr{E}$, equals the magnetization, m. In terms of a function $\bar{N}(n, m)$, which is the density of states averaged about the Fermi level corresponding to an occupancy of $n/2$ spins, over a range corresponding to m spins, self-consistency requires that $m/\Delta\mathscr{E} = \bar{N}(n, m)$ and, when combined with (84), this yields the Stoner condition

$$(I + 2\mu_B H/m)\bar{N}(n, m) = 1, \tag{85}$$

which determines the magnetization and leads to (83) when there is no spontaneous magnetization.

The trend shown for I in Fig. 5.18 may be explained from (84) by the fact that I is larger for free-electron-like s- and p-waves than for resonant

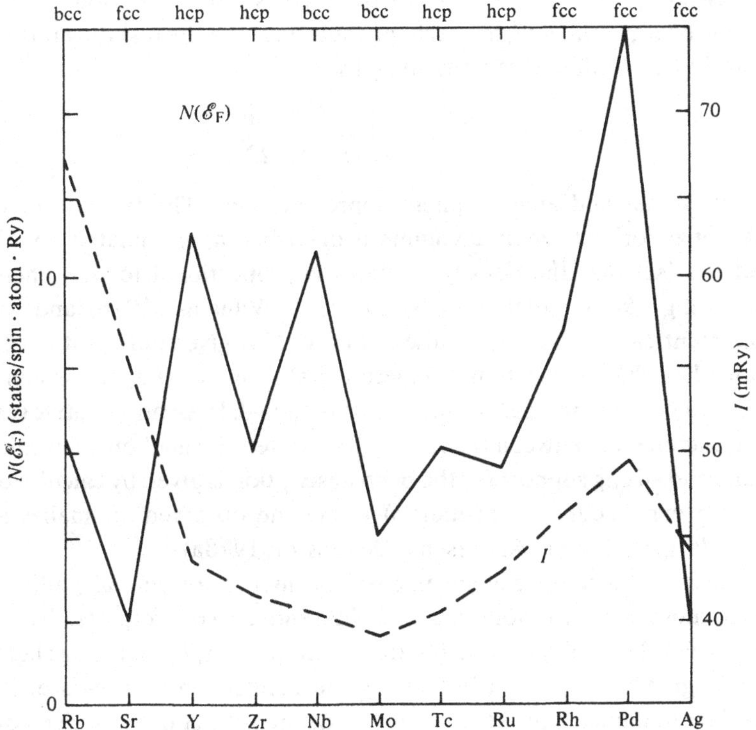

Fig. 5.18. Density of states at the Fermi level and effective Stoner parameter for the 4d-series, obtained from self-consistent LSD–ASA calculations,

d-waves (Fig. 5.13), because the former include a large contribution from the factor $n^{-\frac{2}{3}}(\mathbf{r})$ in the outer region of the cell. The minimum found in the middle of the series is caused by the mixing of the s-, p-, and d-waves. This trend was first pointed out by Janak (1977), who calculated the Is for all metals up to and including In using, instead of (84), the more elegant formalism of Vosko & Perdew (1975). The results of Janak are in close agreement with the self-consistent values shown in Fig. 5.18.

The values of I are larger in the 3d- than in the 4d-series and, for Fe and Ni, $I=67$ and 73 mRy respectively. The spontaneous magnetizations obtained from self-consistent LSD calculations (Wang & Callaway 1977, Andersen *et al.* 1977) are 2.17 and 0.58 μ_B/atom for bcc Fe and fcc Ni respectively, and these compare well with the experimental values (appropriately corrected for the deviation of the g-factor from 2) of 2.12 and 0.55. The magnetic metals have abnormally large atomic volumes and abnormally small bulk moduli (Gschneidner 1964; Pearson 1964); for bcc Fe, $\Delta s = 4.5$ per cent and $\Delta B = -1.0$ Mbar. In a band description, this is interpreted as an effect of the increase in conduction-electron kinetic energy following the exchange splitting, and the resulting positive contribution to the pressure is given by

$$(3PV)_{\text{magn}} = -\frac{Im^2}{\mu_d} + \frac{\delta \ln \mu_d s^2}{\delta \ln s} \int_0^m \frac{m'\, dm'}{2\bar{N}(n, m')} \propto m^2 \tag{86}$$

in the Stoner and atomic sphere approximations. The increase of the magnetization with atomic volume is described approximately by (85) and the fact that the density of states is proportional to $\mu_d s^2$, hence increasing like s^5. Estimates by Janak & Williams (1976) and self-consistent calculations by Poulsen *et al.* (1976) give somewhat smaller anomalies than those found experimentally; in bcc Fe, for instance, $\Delta s = 1.9$ per cent and $\Delta B = -0.7$ Mbar. Although large on a relative scale this discrepancy between theory and experiment is small on an absolute scale, and strong support for the band description is given by calculations for the early actinides (5f-metals) where the observed anomalies are much larger (Skriver, Andersen & Johansson 1978a).

The crystal structures of the magnetic 3d-metals are generally different from those of their non-magnetic 4d- and 5d-counterparts. Fe, for instance, is bcc while Ru and Os are hcp. Moreover, by reducing slightly the magnetic moment in bcc Fe by the application of pressure, it is possible to induce a phase transition to the hcp structure, at which point the moment vanishes. The fact that a transition metal with spin-up and spin-down d-band occupancies of respectively $6.50+2.2$ and $6.5-2.2$ is bcc follows immediately from Fig. 5.17 and so does the fact that the hcp structure becomes stable when the moment is just slightly reduced. The vanishing of the moment, once the hcp phase has formed, follows from (85) and the form of the hcp density-of-states shown in Fig. 5.7.

5.4 Band structures and Fermi surfaces

In this section, we shall review our present experimental and theoretical knowledge of the electronic band structures of transition metals. In general, the most complete description of any particular metal is arrived at by a judicious combination of measurement and calculation, since precise experimental evidence on the band structure away from the Fermi level is relatively difficult to obtain, and calculations which are not subjected to experimental verification may be unreliable. In this respect, Fermi surface studies have been of crucial value, both in establishing the validity of the one-electron picture and in testing the approximations used in constructing potentials and in performing band calculations.

We will not review here the experimental determination of the Fermi surface; a compact discussion of the various methods has been given, for example, by Mackintosh (1968). In practice, only the dHvA effect, discussed extensively elsewhere in this volume, gives a complete quantitative specification of the Fermi surface, although other experiments may give useful additional information on its topology and dimensions. The dHvA effect may also be used in favourable cases to derive detailed information on effective masses, relaxation times and g-factors over the Fermi surface. Our discussion will therefore largely be based on experimental results obtained with this technique.

The band structure away from the Fermi level may be explored by studying the absorption of electromagnetic radiation. The absorption at a frequency ω, due to direct transitions of electrons from filled to empty bands with conservation of crystal momentum, is closely related to the imaginary part of the corresponding dielectric constant, being given by the well-known expression

$$\omega^2\varepsilon_2(\omega)=\frac{e^2\hbar^2}{3\pi m_0^2}\sum_{\mathrm{i,f}}\int \mathrm{d}^3k\;|M_{\mathrm{if}}|^2 \times\delta(\mathscr{E}_{\mathrm{f},\mathbf{k}}-\mathscr{E}_{\mathrm{i},\mathbf{k}}-\hbar\omega)f(\mathscr{E}_{\mathrm{i}})[1-f(\mathscr{E}_{\mathrm{f}})], \tag{87}$$

where M_{if} is a matrix element of the electric dipole operator between the initial state with energy $\mathscr{E}_{\mathrm{i}}$ and the final state $\mathscr{E}_{\mathrm{f}}$, and f is the Fermi–Dirac function. This expression may readily be evaluated with modern computational techniques; in the approximation in which the matrix elements are taken as constant the summation is known as the joint-density of states (JDOS). The absorption in the visible and ultraviolet regions reflects the structure in the JDOS for the occupied and empty conduction band states, although in practice much of the structure arising from individual bands or critical points is heavily damped by the summation and integration in (87). Such structure may, however, be enhanced by modulating the sample with periodic strains, electric fields or heat pulses. In X-ray absorption, the initial state lies deep within the ion-core and the fine-structure near the absorption edge reflects the structure of the energy bands up to several Rydbergs above the Fermi level.

The contribution to the photoelectron spectrum from bulk photoemission, with direct transitions induced by photons of energy $\hbar\omega$, is proportional to

$$N(\mathscr{E}+\hbar\omega, \hbar\omega)=\sum_{\mathrm{i,f}}\int \mathrm{d}^3k\, |M_{\mathrm{if}}|^2 P[\omega, \mathscr{E}_{\mathrm{f,k}}] \times \delta(\mathscr{E}_{\mathrm{f,k}}-\mathscr{E}_{\mathrm{i,k}}-\hbar\omega)\delta(\mathscr{E}-\mathscr{E}_{\mathrm{i}})f(\mathscr{E}_{\mathrm{i}})[1-f(\mathscr{E}_{\mathrm{f}})]. \quad (88)$$

The escape function $P[\omega, \mathscr{E}_{\mathrm{f,k}}]$ represents the average probability that an electron excited to a final state $\mathscr{E}_{\mathrm{f,k}}$ travels to the surface and escapes through it, in which case its energy can be measured. A knowledge of the work function then allows a determination of $\mathscr{E}_{\mathrm{i}}$ and $\mathscr{E}_{\mathrm{f}}$ relative to the Fermi level. If P and M are both taken as constant, (88) reduces to the energy distribution of the joint-density of states (EDJDOS). In practice, the energy distribution curves for ultraviolet photoemission show substantial structure, which is directly related to that of the energy bands, especially if higher derivatives of the photocurrent are measured by modulating the potential on the collector. The energy range of such experiments may be enhanced by cesiating the surface, and hence reducing the work function, and the information content increased by using single crystals.

In the following, we will briefly discuss the electronic structures of the individual transition metals which, for convenience, we will classify by crystal structure, though considering the noble and alkaline earth metals separately. For each structure, we will outline the main features of the energy bands and the constant energy contours in the energy region of interest for Fermi surface studies, using the 4d-metals as examples. The different metals will then be considered and the extent of the correspondence between theoretical band structures and experimental Fermi surfaces and excitation energies will be evaluated. We shall generally refer only to the most recent or complete investigations, since these usually contain extensive references to earlier work, and will not discuss the ferromagnetic transition metals, which are treated in detail in the next chapter. Much useful information on recent developments is contained in the proceedings of the international conference on the Physics of Transition Metals, held in Toronto in 1977. An attempt will be made to clarify which aspects of our present knowledge of the band structures are inadequate.

5.4.1 The noble metals

The noble metals, Cu, Ag and Au, occupy a unique position among the metallic elements by virtue of the fact that their electronic structures have been, by a large margin, the most extensively studied, both experimentally and theoretically. In particular, as mentioned in Section 5.1, Cu is notable as the first metal for which realistic calculations of the d-band states were performed, and also the first whose Fermi surface was fully determined.

The noble metals crystallize in the fcc structure, for which the canonical bands are shown in Fig. 5.4. Their electronic structures, especially that of Cu, have recently been discussed in some detail by O. Jepsen, D. Glötzel & A. R. Mackintosh (unpublished data), and we will summarize some of their conclusions in the following. The Cu band structure is shown in Fig. 5.19, and the important effects of hybridization may be appreciated by comparing the two halves of this figure. Along the $\langle 100 \rangle$ directions, for example, the Δ_1-bands hybridize strongly, giving rise to a gap, and a corresponding dip in the density of states, and raising the Fermi level. Since the $L_{2'}$ level is very slightly depressed by weak hybridization with the high-lying 4f-band, this elevation of the Fermi level results in the formation of the neck in the Fermi surface near L. The states at the Fermi level are strongly mixed and have pronounced d-character. According to the calculations of O. Jepsen, D. Glötzel & A. R. Mackintosh (unpublished data), the partial d-density of states at the Fermi surface is about 50 per cent of the total for both Cu and Au, and about 30 per cent for Ag, where the d-band lies much lower. In this sense it is therefore very reasonable to classify the noble metals as transition metals. Many states are affected by weak hybridization, notably the predominantly s-like L_1

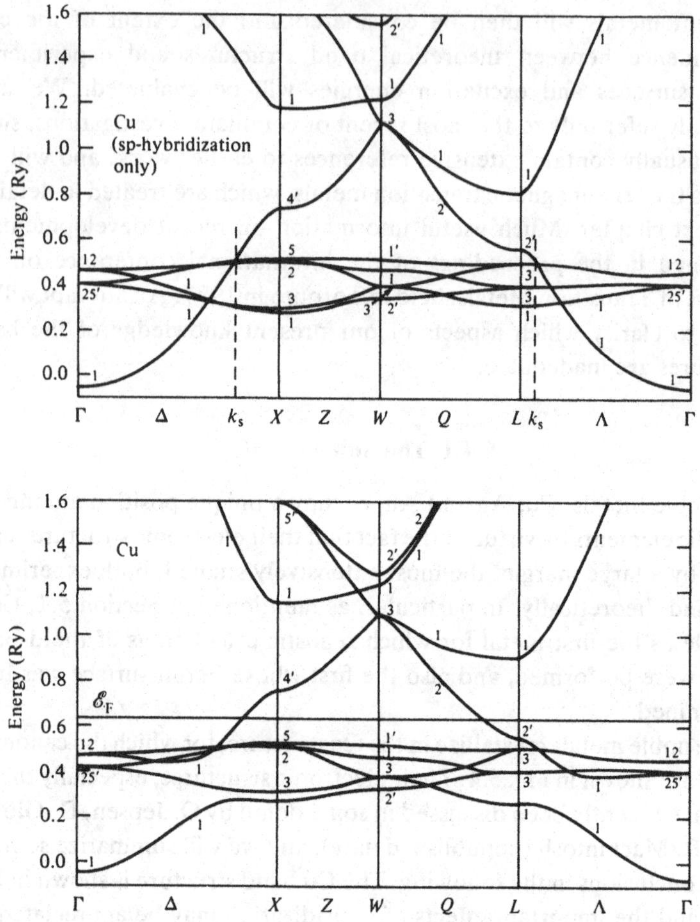

Fig. 5.19. The band structure of Cu. In the upper figure, the hybridization between the sp-bands and the d-bands has been omitted. k_s is the radius of the free-electron sphere. The lower figure includes the d-hybridization.

level, which is raised by mixing with the lowest d-level, thereby increasing $L_1 - L_{2'}$, the 'L-gap', substantially.

The pioneering work of Pippard, Shoenberg, and Alekseevskii and Gaidukov, mentioned earlier, and that of Kip (1960) and his colleagues on cyclotron resonance, and Morse (1960) and his colleagues on the magnetoacoustic effect, was followed by a steady effort to increase the accuracy of measurements of the dHvA frequencies in the noble metals. As described in Chapter 9, this has resulted in determinations of extremal areas in all three metals with an absolute accuracy of better than 2 parts in

10^5, and a relative accuracy of about 1 in 10^6. In addition, a steady improvement has taken place in measurements of the effective mass, culminating in the work of Lengeler *et al.* (1977), who have determined the masses in all three metals to within 0.3 per cent. By performing difficult measurements of absolute dHvA amplitudes, Alles, Higgins & Lowndes (1975) have shown that the *g*-factor in Au varies for different orbits.

These very precise data on extremal orbits contain detailed information on the variation of the radius vector and electron velocities over the Fermi surfaces, and hence implicitly on the crystal potential and electron–phonon mass enhancement. Different techniques have been used to extract this information. Roaf (1962) used a Fourier series representation of the Fermi surface to obtain the radius vectors for all three metals, and Halse (1969) refined these calculations, using more recent data. He also inverted the existing cyclotron mass measurements to obtain a map of the Fermi velocities. His results were in turn refined by Lengeler *et al.* (1977), using their own data. They integrated the resulting velocities over the Fermi surfaces to determine densities of states, which were found to agree very well with those deduced from the heat capacity measurements of Martin (1973). By comparing these experimental velocities with those predicted from band calculations they, and a number of other authors, have concluded that the electron–phonon mass enhancement is highly anisotropic over the Fermi surface. Although this anisotropy appears to be well established qualitatively, a quantitative determination would require more reliable one-electron potentials than have yet been constructed.

An alternative approach, originally suggested by Segall & Ham (1968), is to use the logarithmic derivatives at the Fermi energy as an adjustable parameter set for a Fermi surface calculation. A number of people have used this method and Shaw, Ketterson & Windmiller (1972) have made a very precise fit to the extremal areas of the noble metals with adjustable s, p, d and (for Cu) f phase shifts in a least-squares KKR parametrization scheme. They also used the energy derivatives of the phase shifts as parameters to fit the cyclotron masses, including, of course, the electron–phonon renormalization by this procedure. Finally, the dependences of the extremal areas on hydrostatic pressure, measured by Schirber & O'Sullivan (1970b) were parametrized in terms of the derivatives of the phase shifts with respect to lattice constant. A satisfactory fit was obtained for Cu and Ag, but the pressure derivatives for Au cannot be accounted for without explicit inclusion of spin–orbit effects.

This technique has the great advantage of permitting a straightforward comparison between Fermi surface measurements and the predictions of *a priori* calculations, and hence an evaluation of the potentials on which such calculations are based. Because of the insensitivity of the Fermi surface shape to the potential in the interstitial regions, it may be fitted satisfactorily over a wide range of assumed Fermi energy, $\mathscr{E}_F - v_{MTZ}$ by a suitable choice of phase shifts. These phase shifts are highly sensitive to the assumed Fermi energy but, as shown by Andersen (1971b), the logarithmic derivatives are much less sensitive and indeed, if they are evaluated at suitably chosen radii, the energy dependence may be removed. For evaluating potentials, it is probably most convenient to use the logarithmic derivatives at the atomic sphere, as in the work of O. Jepsen, D. Glötzel & A. R. Mackintosh (unpublished data), where the results of Shaw *et al.* (1972) for Cu were compared with various potentials. It is important to include relativistic band shifts when making comparisons at this level of precision. The neck area, which is very sensitive to the relative positions of all the bands, is not a good figure of merit for a potential, since the self-consistent $X\alpha$-potential of Janak, Williams & Moruzzi (1975), which was constructed to fit the neck, does not give a particularly good overall representation of the Fermi surface, while a self-consistent LD potential, which gives too large a neck area accounts otherwise very well for the Fermi surface shape.

Extensive information on the band structures of the noble metals away from the Fermi level has been obtained from optical spectroscopy. Their optical properties have been studied since the time of Drude, and measurements, such as those of Johnson & Christy (1972), of the dielectric constant as a function of frequency give support to the general form of the band structure predicted from theory. The most accurate measurements of energy differences are generally given by modulation experiments from which, for example, the L-gap can be determined with a precision of about 10 mRy. The pioneering piezo-optical work of Gerhardt (1968) on Cu, that of Nilsson & Sandell (1970) on Ag, and the thermoreflectance measurements on Ag and Au by Colavita, Modesti & Rosei (1976) and Olson, Piacentini & Lynch (1974) respectively, are examples of recent studies from which accurate band gaps may be obtained. Such information may also be derived from photoemission experiments, which give the width of the d-bands with an accuracy of roughly 10 per cent (limited by the broadening of the bottom of the band) as shown by Eastman & Cashion (1970), and a very detailed picture of their structure. The wealth of information obtainable from angular resolved photoemission measurements on single crystals is illustrated, for example, by the work of Becker, Dietz, Gerhardt & Angermüller (1975), who give references to earlier investigations. Photoemission also allows an accurate determination of the position of the top of the d-bands, relative to the Fermi level, which may also be obtained from a careful analysis of the interband absorption edge, as in the work of Guerrisi, Rosei & Winsemius (1975) on Au.

The comprehensive data which exist on the Fermi surface and optical transitions, the most important features of which are summarized in Table 5.1, allow a comparison between calculated and experimental band structures. The essential experimental parameters are the L-gap, the interband absorption energy $\mathscr{E}_F - X_5$ (X_{7+} if spin–orbit coupling is taken into account), the width of the d-bands, and the Fermi surface anisotropy, conveniently characterized by the parameter D, the ratio of the ⟨100⟩ and ⟨110⟩ radii. The L-gap is primarily determined by Δ_{ps}, although the position of the d-bands is of importance through the hybridization of the L_1-level, while Δ_{ds} is the main factor in determining the interband energy. An increase of either Δ_{ps} or Δ_{ds} increases D, while the neck is enlarged by an increase of the latter but diminished by an increase of the former. In Cu, the energy surfaces in the lowest unhybridized sp-band are almost perfectly spherical, as may be seen in Fig. 5.19, but the relativistic

TABLE 5.1. *Electronic parameters for noble metals*

	Cu	Ag	Au
$D \equiv k_F[100]/k_F[110]$	1.11 (E) 1.13 (L) 1.10 (S)	1.09 (E) 1.11 (L) 1.06 (S)	1.20 (E) 1.21 (L) 1.14 (S)
$\dfrac{\text{Neck area}}{\text{Area of free electron sphere}}$	0.036 (E)	0.019 (E)	0.032 (E)
L-gap $(L_1 - L_{2'})$	370 (E) 360 (L) 338 (S)	310 (E) 322 (L) 256 (S)	340 (E) 333 (L) 274 (S)
Interband absorption energy $\mathscr{E}_F - X_5(X_{7+})$	150 (E) 113 (L) 118 (S)	280 (E) 183 (L) 274 (S)	115 (E) 76 (L) 112 (S)
d-band width $X_5(X_{7+}) - X_3(X_{7+})$	220 (E) 206 (L) 209 (S)	260 (E) 273 (L) 240 (S)	420 (E) 425 (L) 403 (S)

Energy differences are given in mRy. (E) denotes an experimental value and (L) and (S) the results of calculations based on respectively LD and standard potentials. The sources of these results are given in the text.

lowering of the s-bands progressively increases the distortion in Ag and Au. These relativistic shifts, together with the low-lying d-bands in Ag, provide a qualitative explanation of the relative distortions and neck areas of the Fermi surfaces.

A systematic study of the band structure of the noble metals has been carried out by Christensen, using the RAPW method and the standard potential (Christensen & Seraphin 1971 (Au); Christensen 1972 (Ag); Christensen 1978 (Cu)). His calculations give a good overall account of the Fermi surfaces and optical properties, but there are discrepancies of detail. As may be seen in Table 5.1, the interband absorption energy is in good agreement with experiment for Ag and Au, but the calculated d-bands are somewhat high in Cu. In all three metals the d-band width is correct within the experimental uncertainty, but the calculated L-gap is substantially too small, and the Fermi surface anisotropy is also too small. Both the L-gap and the Fermi surface anisotropy may be increased by raising the d-bands, but this produces a serious discrepancy with the experimental interband absorption energy. Indeed O. Jepsen, D. Glötzel & A. R. Mackintosh (unpublished data) using a self-consistent LD potential for Cu, found both an L-gap and a Fermi surface in good agreement with experiment, but an absorption edge about 40 mRy below the measured value (see Table 5.1). The main difference between the various potentials which have been constructed for Cu lies in the positioning of the d-bands, and it seems that it may be difficult to devise a self-consistent local potential which reproduces all of the experimental results. Chen & Segall (1975) constructed a set of logarithmic derivatives (equivalent to a set of potential parameters) which were designed to fit both the optical and Fermi surface data, but it is not clear whether these can be derived from a local potential. It is possible that the apparent d-band position observed through optical transitions includes many-body relaxation effects not accounted for by the local potential which accounts for the ground state properties. Such correlation effects would be expected to be particularly strong in narrow d-bands, and this question deserves further investigation. As discussed in the next section, the band positions and widths of high-lying bands in the 4d transition metals (including Ag) observed by X-ray absorption are well reproduced with a local potential.

5.4.2 The fcc metals

The electronic structures of the fcc transition metals, Rh, Pd, Ir and Pt bear a close resemblance to those of the noble metals, as may be seen

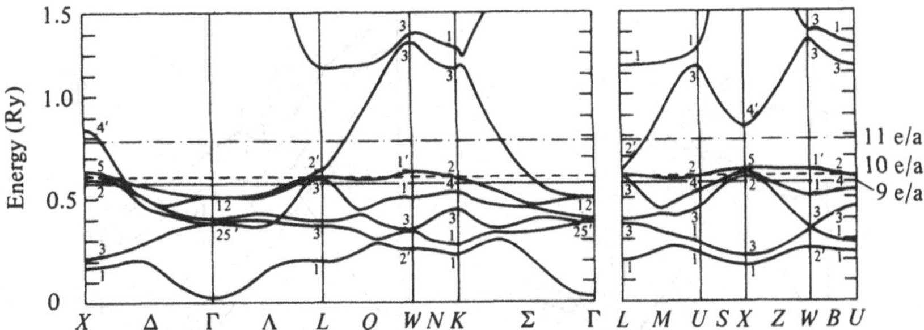

Fig. 5.20. The energy bands for the fcc structure, illustrated by calculations on Pd by H. L. Skriver. Single-group symmetry labels are given at symmetry points. Energies corresponding to 9, 10 and 11 electrons per atom are shown.

from the band structure of Pd, illustrated in Fig. 5.20. However, the d-bands in the fcc metals are broader and higher-lying and, most importantly, the decrease in the number of electrons per atom brings the Fermi level into a region where it intersects the d-bands, giving rise to complex many-sheeted Fermi surfaces. The electronic structures of fcc metals have been discussed in detail by Andersen (1970). The d-band rises relative to the sp-bands in the sequence Pd, Rh, Pt, Ir with a concomitant decrease in the mass, or equivalently, an increase in the d-bandwidth. Both spin–orbit coupling and relativistic band shifts are very important in these metals. For example, the relativistic shift of the $L_{2'}$ p-level, which lies above the d-bands in Pd and Rh, pulls it down into the top of the d-bands in Pt and Ir.

Constant energy surfaces corresponding to the band structure of Fig. 5.20 are illustrated in Fig. 5.21. In the following, it will be convenient to adopt a convention by which a constant-energy surface is labelled by a letter giving the location of the centre of a closed surface, or several letters indicating the open directions of open surfaces, and a number signifying the band with which it is associated, counting the bands in order of increasing energy. For 9 electrons/atom, there are two closed electron surfaces Γ5 and Γ6 and three closed hole surfaces *X*3, *X*4 and *L*4. Spin–orbit coupling separates a number of these surfaces which would otherwise be connected, for example *X*3 and *X*4 along Γ*X*. These surfaces correspond to the observed Fermi surface of Rh, and also that of Ir, except that relativistic effects eliminate *L*4, in the latter. The surfaces associated with the lower bands are predominantly d-like and Γ6 is predominantly s-like, but hybridization is important, especially in Ir. For 10 electrons/atom, corresponding to the Fermi surface in Pd, the *X*3 and *L*4 holes have disappeared and the Γ6 electron sheet has expanded. Most

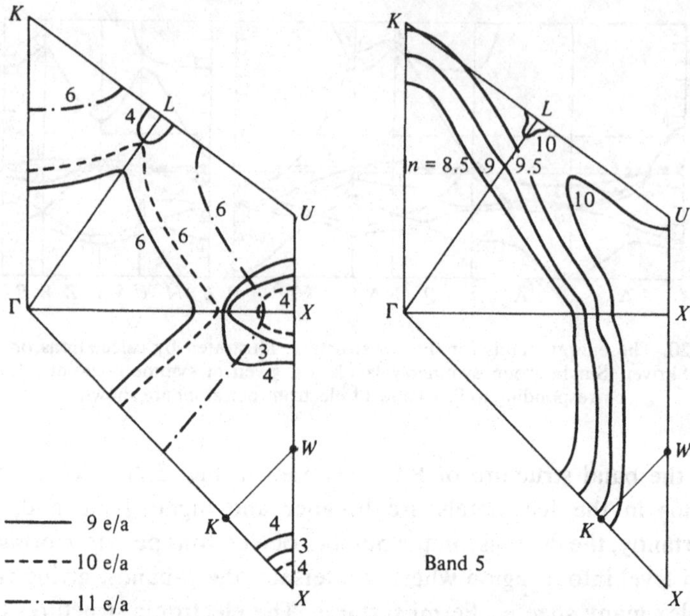

Fig. 5.21. Constant energy surfaces for the fcc structure, derived from the energy bands of Fig. 5.20. In the left figure, the labels on the surfaces indicate the corresponding energy bands, while in the right figure the electron concentrations are shown. The line *KL* does not lie in a symmetry plane, and the energy surface may therefore bulge over the zone boundary.

interesting, however, is the evolution of the fifth band surfaces. As the energy is increased, the closed Γ5 electron surface swells and finally makes contact with neighbouring sheets at P_2 in Fig. 5.22, near the mid-point of the line *LW*, forming a multiply-connected surface *LXW*5. This contact is associated with a saddle point in $\mathscr{E}_{\mathbf{k}}$ along *LW*, and another saddle point along *LX* causes the connection to be broken at P_1 when the energy is further increased, leaving the *XW*5 'scaffolding' surface of Fig. 5.22, and the isolated *L*5 hole pockets, which exist in Pd but not in Pt. These two saddle points give rise to the high peak in the density of states in Pd, and the 'fins' on the scaffolding surface are a manifestation of the *LX* saddle point. These fins are responsible for about half of the total state density on the *XW*5 surface, which in turn accounts for about 90 per cent of $N(\mathscr{E}_F)$ in Pd.

The Fermi surfaces of the fcc metals have been studied in great detail by means of the dHvA effect, most recently by Carrander, Dronjak & Hörnfeldt (1977) (Rh), Windmiller, Ketterson & Hörnfeldt (1971) (Pd), Hörnfeldt, Windmiller & Ketterson (1973b) (Ir), and Dye, Ketterson & Crabtree (1978b) (Pt). In all cases, there is perfect qualitative and

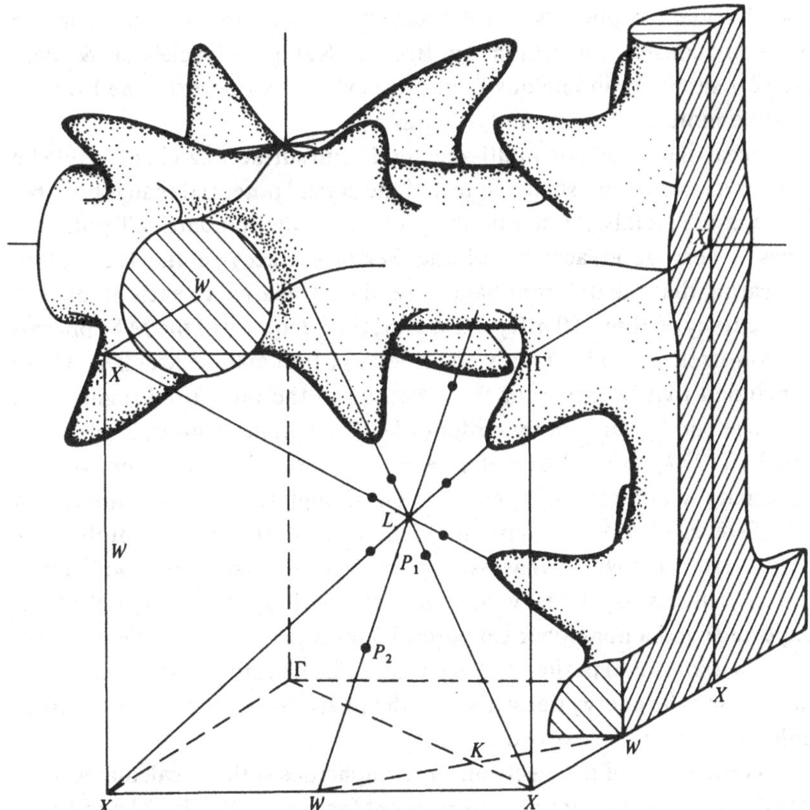

Fig. 5.22. The fifth-band 'scaffolding' surface in Pd. P_1 and P_2 are critical points in this band, as described in the text.

impressive quantitative agreement with the results of calculations based on the standard potential. Windmiller *et al.* (1971) studied the spin-splitting zeros in the dHvA amplitudes in Pd and deduced that the conduction electron *g*-factor differs from 2 and is anisotropic, in qualitative accord with the calculations of Mueller, Freeman, Dimmock & Furdyna (1970a). They also studied the dHvA effect in ferromagnetic alloys containing about 0.1 per cent Co, and deduced an exchange splitting of about 0.1 mRy, from beats in the oscillations associated with the $X4$ pockets. Indirect evidence of exchange splitting was also observed in the disappearance of spin-splitting zeros on the Γ6 electron surface. The effect of hydrostatic pressure on the extremal Fermi surface areas has been measured by Skriver *et al.* (1978a). They find generally good agreement with calculations based on an LD potential and, in particular, find that the $L5$ pockets (first observed as quantum oscillations in the magnetoacoustic absorption by Brown, Kalejs, Manchester & Perz (1972)) show an anomalous behaviour which can be explained by the relative motion of the p- and d-bands.

Precise dHvA data on multi-sheeted Fermi surfaces can in principle be used as an even more stringent test of the crystal potential than is possible in the noble metals. As mentioned previously, the standard MT potential gives a very good account of the Fermi surfaces of the fcc metals, indicating that the different bands are placed relative to each other with an accuracy of about 10 mRy. However, the adequacy of the MT approximation itself may also be tested. Koelling, Freeman & Mueller (1970) concluded that 'warping' of the potential in the interstitial regions can produce energy shifts of the order of 5–10 mRy and Painter, Faulkner & Stocks (1974) showed that this term is greater than that due to non-sphericity within the MT sphere for Pd, though the latter dominates for Nb. Dye *et al.* (1978b) have fitted their accurate and comprehensive dHvA data on Pt with a phase-shift parametrization scheme in which they found it necessary to take into account both spin–orbit coupling (as expected) and a non-spherical potential term in order to fit the extremal areas within the experimental accuracy. The effective masses were also fitted using the energy derivatives of the phase shifts, into which the mass enhancement was absorbed.

A comparison of the electronic heat capacities with the calculated state densities indicates that the enhancement factors for Rh, Ir, Pd and Pt are respectively 1.44, 1.37, 1.66 and 1.63 (Andersen 1970). The value for Ir is in very good agreement with the estimate of the electron–phonon enhancement by McMillan (1968), and the larger values for Pd and Pt may indicate a contribution from spin fluctuations. The mass enhancement appears to be generally higher on surfaces with a strong d-character.

The optical properties of the fcc metals have been extensively studied (Weaver, Olson & Lynch 1977, and references therein). The most detailed information on the band structure has, however, been obtained from photoemission measurements, especially those of Traum & Smith (1974). They found that the complex structure in the second derivative of the photoelectron energy distribution is recognizably similar in all of the fcc and noble metals, and that most of it can be brought into coincidence by a uniform shifting and stretching of the energy scales for the different metals. As they point out, this provides direct evidence for the existence of a d-band with a canonical structure, placed and scaled in energy in the individual metals. The d-band positions and widths which they find, expressed in terms of empirical band structures by Smith (1974), are generally in good agreement with those of Andersen (1970). The energy separations between the Fermi level and the unoccupied L_1 sd-levels may be deduced fairly directly from the optical and photoemission experiments, to an accuracy of about 10 mRy, and have the values (in mRy) 630(640), 530(540), 520(540) and 450(440) for Rh, Pd, Ir and Pt respectively. The calculated values of Andersen (1970) are in parenthesis, and the agreement is seen to be excellent. The same potential has been used by Christensen (1976) to give a satisfactory interpretation of the optical properties of Pd up to photon energies of about 2 Ry, and to identify those regions of the zone contributing to various elements of the structure in the dielectric constant.

The energy bands at even higher energies can be studied through the structure near the *K*-edge in the X-ray absorption (Kostroun, Fairchild, Kukkonen & Wilkins 1976). The peaks and valleys in this structure reflect the positions of the 5p- and 4f-bands several Ry above the Fermi level, even though the energy levels are broadened by as much as 0.4 Ry. Suitably broadened energy bands based upon the standard potential reproduce the structure observed in all of the 4d metals with remarkable fidelity (Muller *et al.* 1978). Further information is contained in the structure near the *L*-edges. In Pt, for example, there is a strong absorption ('white line') near the L_3-edge but not near the L_2-edge (Brown, Peierls & Stern 1977). At the L_2-edge, the electrons originate from a $p_{\frac{1}{2}}$-state and can make transitions to s- or $d_{\frac{3}{2}}$-states, whereas the $p_{\frac{3}{2}}$-electrons responsible for the L_3-edge can only make transitions to the $d_{\frac{5}{2}}$-states. Christensen (1978) has calculated the spin–orbit projected d-state densities for Pt, and finds indeed that only the $5d_{\frac{5}{2}}$ component has a large value near the Fermi level.

For the sake of completeness, we will mention that fcc La is a trivalent 5d transition metal. Similarly Th is a quadrivalent fcc 6d transition metal, though its principal interest, as in all the actinides, lies in the role of the 5f-electrons. The 6d-bands are relatively broad and high, rather as in the alkaline earth metals, with very large spin–orbit coupling, and the Fermi surface is predominantly d-like. According to the calculations of J. P. Jan & H. L. Skriver (private communication), based on an LD potential, an unoccupied 5f-band, with a mass μ_f of about 12, is sufficiently close to the Fermi level that $N_f(\mathscr{E}_F)$ comprises about 25 per cent of the total state density. The Fermi surface consists entirely of closed sheets; $\Gamma 2$ and $L2$ hole surfaces and a set of $\Sigma 3$ electron surfaces on the lines joining Γ and K. Koelling & Freeman (1975) have shown that the influence of the f-states on the Fermi surface is essential for a quantitative explanation of the dHvA results of Thorsen, Joseph & Valby (1967) and Boyle & Gold (1969). The pressure dependence of the extremal areas has been studied by Schirber, Schmidt & Koelling (1977), who find the unusual result that all of the surfaces decrease in size as the lattice constant is reduced. The optical properties have been studied by Weaver & Olson (1977), who find structure which can plausibly be associated with both d- and f-states, although no definite identification has been made.

5.4.3 The bcc metals

The form of the canonical d-bands of Fig. 5.5 is clearly reflected in the band structure of a typical bcc transition metal, such as that of Fig. 5.23. Neither at Γ nor at H are the d-levels hybridized with the s- and p-levels, and the states H_{12} and N_3 which define the bottom and top of the d-band are purely d-like. The deep minimum in the centre of the canonical d-state density, shown in Fig. 5.7, persists in the metals and, as we have seen earlier, plays an important role in determining the stability of their bcc structure. In the ⟨100⟩ directions, the Δ_1 d-band mixes strongly with the sp-band, while Δ_5 hybridizes weakly with the higher-lying p-states. The position of the lowest-lying $N_{1'}$-level of the p-band, which is unoccupied in all the metals, largely determines the sizes of the hole surfaces associated with it.

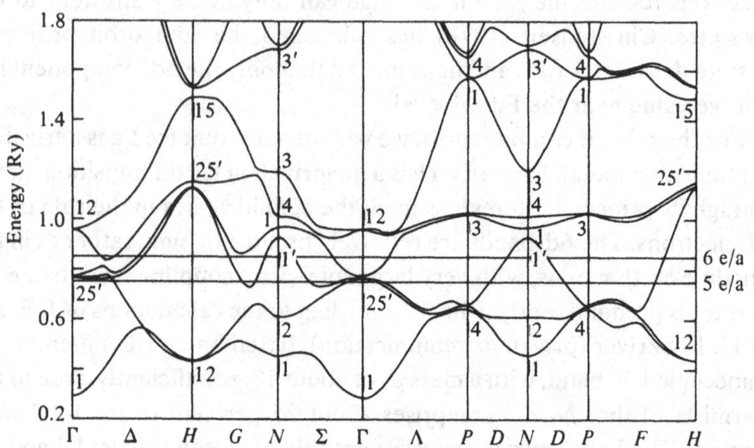

Fig. 5.23. The energy bands for the bcc structure, illustrated by calculations on Mo by N. E. Christensen. Single-group symmetry labels are given at symmetry points.

Constant energy surfaces corresponding to the band structure of Fig. 5.23 are shown in Fig. 5.24. For 5 electrons/atom, as discussed by Mattheiss (1970), there are two closed hole surfaces, $N3$ and the distorted octahedron $\Gamma 2$. If relativistic effects are neglected, the latter makes contact with the open $\Gamma H3$ hole surface along the line ΓP and in the {100} and {110} planes, but these degeneracies are lifted by spin–orbit coupling, as illustrated in Fig. 5.24(*a*). The $\Gamma H3$ surface, usually known as the 'jungle-gym', has the form of an array of non-uniform cylinders in the ⟨100⟩ directions, intersecting at Γ and *H*. The Fermi surfaces of the bcc metals V, Nb and Ta are all believed to have the topology of Fig. 5.24(*a*).

As the energy is raised, the $N3$ and $\Gamma 2$ holes shrink, while the jungle-gym first disconnects along Δ and then contracts into the $H3$ octahedron illustrated in Fig. 5.24(*b*). Simultaneously, the fourth band begins to fill, giving rise to a $\Gamma 4$ octahedron and $\Delta 4$ pockets, which coalesce to form the $\Gamma 4$ electron 'jack' of Fig. 5.24(*b*). The $\Delta 5$ 'lens' surfaces contact the necks of the jack if spin–orbit coupling is neglected. The Fermi surfaces of Mo and W have the general form of Fig. 5.24(*b*), except that the spin–orbit splitting in W is sufficiently great that the electron lens is completely eliminated.

Of the bcc transition metals, the Fermi surface of Nb has been the most precisely determined, by Karim, Ketterson & Crabtree (1978). They observed dHvA oscillations from all the sheets and deduced the corresponding extremal areas and cyclotron masses. Magnetic breakdown

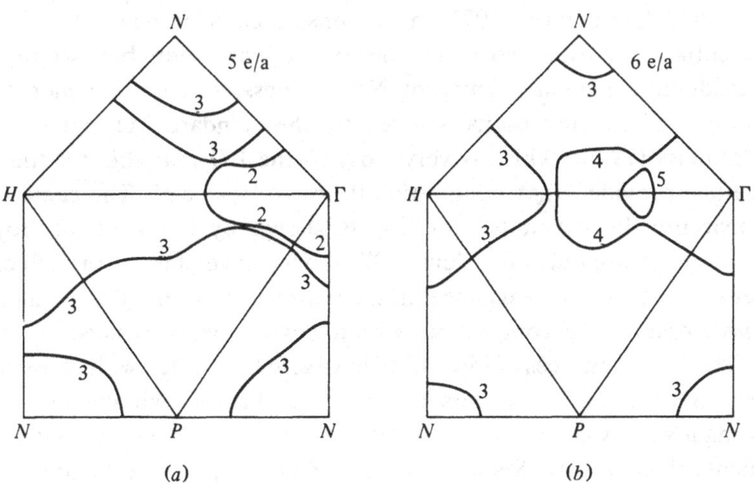

Fig. 5.24. Constant energy surfaces for the bcc structure, derived from the energy bands of Fig. 5.23. The labels on the surfaces indicate the corresponding energy bands.

was observed across the spin–orbit gap between the Γ2 and ΓH3 surfaces. Dye, Karim, Ketterson & Crabtree (1978a) used these results as the basis for a KKR parametrization of the Fermi surface and electron velocities. They find that it is necessary to include both non-spherical terms in the potential and (unusually) the f phase shift to obtain an accurate representation of all the surfaces, but cannot determine the spin–orbit splitting, which has rather little effect on the extremal areas. By comparing the measured velocities with the calculations of Elyashar & Koelling (1977), they deduce a mass enhancement which appears to be larger on the d-like sheets than on the largely p-like N3 holes.

Ta and V have also been studied, in less detail, using a combination of magnetothermal oscillations and the dHvA effect, by Halloran *et al.* (1970) and Parker & Halloran (1974) respectively. The sizes of the N3 holes are primarily determined by the separation Δ_{pd} between the p- and d-bands, and therefore decrease in the sequence Nb, Ta, V. The ΓH3 jungle-gym surface has been studied in some detail in Ta, and some of its smaller orbits have been measured in V, but in neither of these metals has the Γ2 surface been observed, presumably because of its high mass. Parker & Halloran (1974) suggest that the third band in V may be above $\mathscr{E}_F$ along the whole of the line Σ, giving rise to a neck joining ΓH3 and N3, but the evidence is rather unconvincing and not supported by existing band calculations. However, this question cannot yet be considered as finally settled.

Extensive band calculations have been performed for this group. Elyashar & Koelling (1977) have considered a number of different potentials for Nb and shown that the non-MT corrections both within and outside the sphere are significant. Nevertheless, the best agreement with Fermi surface dimensions is given by the standard MT potential of Mattheiss (1970), which is very close to the LD potential for the bcc metals, provided that relativistic effects are included. The same may apparently be concluded for Ta, as shown by the work of Boyer, Papaconstantopoulos & Klein (1977), who have studied the effect of self-consistency and exchange on the Fermi surfaces of all three metals, and present useful comparisons with the experimental results.

The Fermi surfaces of Mo and W have also been fitted with phase-shift parametrizations by Ketterson, Koelling, Shaw & Windmiller (1975) using a variety of experimental data, but primarily the dHvA measurements of Hoekstra & Stanford (1973) and Girvan, Gold & Phillips (1968) respectively. Spin–orbit coupling and non-MT terms are essential for a satisfactory fit in both cases. The dependence of the Mo Fermi surface on uniaxial strain has been studied by Griessen, Lee & Stanley (1977a), who deduced the volume-dependence of the phase shifts from their results. The qualitative and quantitative differences between the Fermi surfaces of Mo and W are primarily due to the large relativistic effects in the latter. The standard potential gives a good quantitative account of these surfaces as shown, for example, by Koelling, Mueller & Veal (1974) and Christensen & Feuerbacher (1974) respectively.

Cr is of especial interest, and complexity, because of its anti-ferromagnetic ordering at low temperatures, which takes the form of a static spin-density wave along a single [100] direction in the crystal, with a period which is incommensurable with that of the lattice. As discussed by Lomer (1964), this gives rise to a series of 'magnetic' energy gaps in the band structure, and the effects of these on the Fermi surface have been extensively studied, principally by Marcus and his colleagues. The paramagnetic Fermi surface, as calculated most recently by Rath & Callaway (1973), resembles that of Mo, but the magnetic energy gaps modify it profoundly, giving rise, for example, to open orbits along the direction of the spin-density wave. The effect of these on the galvanomagnetic properties has been convincingly demonstrated by Arko, Marcus & Reed (1968). The dHvA frequencies in the anti-ferromagnetic phase have been measured in considerable detail by Graebner & Marcus (1968) but, although they have been plausible identifications of a number of the oscillations, making use of the theoretical analysis of Falicov & Zuckermann (1967), they were unable to account for much of the data. By measuring the strain dependence of the dHvA frequencies, Fawcett, Griessen & Stanley (1976) were able to identify a number of them, but there remains much to be done before the electronic structure of this unique metal is fully understood.

Although the optical properties of all the bcc metals have been studied, relatively little explicit information about the band structure away from the Fermi level has generally been deduced from the results. Weaver, Lynch, Culp & Rosei (1976) measured the thermoreflectance of V, Nb and Cr, for example, but concluded that most of the structure originated in large volumes of the Brillouin zone, rather than being associated with well-defined transitions at singularities in the JDOS. The outstanding exception is W, for which the previously mentioned band structure of Christensen & Feuerbacher (1974) gives an excellent account both of their single-crystal photoemission experiments and of the dielectric constants deduced from optical studies by Weaver, Olson & Lynch

(1975). Furthermore, Christensen & Willis (1978) have shown that the structure in the energy distribution of secondary electrons emitted from W single crystals may be explained with the same band structure to a precision of about 15 mRy up to about 2 Ry above the Fermi level. The standard potential thus reproduces the one-electron energy spectrum of W accurately over the whole range of observation. The single-crystal photoemission measurements of Cinti, Al Khoury, Chakraverty & Christensen (1976) on Mo are also generally accounted for by the calculated band structure, but the comparison between theory and experiment has not yet been carried out in detail.

5.4.4 The alkaline earth metals

Ca, Sr and Ba assume a position intermediate between the transition metals proper, in which the d- and s-band centres lie fairly close to each other, and the simple metals, in which the d-band generally lies well above the p-band, as in the free electron gas. The d-band centres in the alkaline earths are between the s- and p-bands, but closer to the latter. In all three metals, the conduction electrons have considerable d-character, as first emphasized by Vasvari, Animalu & Heine (1967). Even Ca, where the d-bands lie highest, has approximately one-half of a d-electron per atom (J. P. Jan & H. L. Skriver, private communication).

The position of the d-band is correlated with the crystal structure. For Ca and Ba, the values of Δ_{dp} calculated with LD potentials are 8.6 and 7.0 and the structures are respectively fcc and bcc. The d-band in Sr also lies fairly high in energy, but its fcc structure may be transformed to bcc by the application of pressure, which lowers the bottom of the d-band (Johansen & Mackintosh 1970). The divalent rare earth metals Eu and Yb, with half- and completely-filled 4f-shells, have d-band positions and structural properties at room temperature analogous to Ba and Sr respectively, although Yb at lower temperatures also has a stable hcp phase.

As may be seen from Fig. 5.17, the fcc structure should be favoured for small numbers of d-electrons, if hybridization is neglected. There is, however, a minimum in the state density near the Fermi level in the bcc structure, principally due to p–d hybridization, which becomes deeper as the d-band is lowered (Y. Glötzel, D. Glötzel & O. K. Andersen, unpublished data). By reducing the total electronic energy, this minimum, which is much less pronounced in the fcc structure, tends to favour the bcc structure as the d-band falls. The hybridization gap is clearly manifested in the p-state density at the Fermi level of Ba, which is only about 20 per cent of that in Ca.

The position of the d-band is also of crucial importance for the Fermi surface topology of the alkaline earths. Johansen (1969) used the standard potential to calculate the band structure of Ba, and found that the lowest H_{15} d-level lies well below the Fermi level. The calculated Fermi surface consists of two closed sheets, an $H2$ electron surface and a $P1$ hole surface, separated along PH by the spin–orbit coupling. McEwen (1971) found that all of his observed dHvA frequencies could be accounted for by orbits on the calculated $P1$ surface, provided that the shape is modified slightly, which can be accomplished by a small distortion and increase of the Fermi energy. The lack of oscillations which could be associated with the $H2$ surface was ascribed to the large mass, which is a consequence of its d-character. It would be interesting to calculate the Fermi surface of Ba with an LD potential, which gives d-bands differing significantly from those derived from the standard potential.

The d-band in Ca, on the other hand, lies sufficiently high that it does not affect the Fermi surface topology as shown by the dHvA experiments of Jenkins & Datars (1973) and Gaertner (1973). They found a Fermi surface consisting of $L2$ electron lenses and a $WK1$ open hole surface which follows the edges of the Brillouin zone, as in the nearly free-electron model. The band structure of Ca has been discussed by many authors, most recently by J. P. Jan & H. L. Skriver (private communication). The position of the d-band affects the Fermi surface principally through the first-band spd K_1-level. When Δ_{ds} is reduced, this level falls beneath $\mathscr{E}_F$ and the open hole surface is dismembered. The standard potential gives d-bands which are much too low and narrow, and hence a qualitatively incorrect Fermi surface. Both the LD potential and a non-self-consistent Xα-potential, with $\alpha = \frac{2}{3}$, give slightly too low values for this K_1-energy, but Nilsson & Forssell (1977) found that the latter potential gives a good account of their measured optical properties. With a self-consistent, $\alpha = \frac{2}{3}$ potential, McCaffrey, Anderson & Papaconstantopoulos (1973) found both the correct Fermi surface topology, and a d-band position and width in good agreement with the estimates of Kress & Lapeyre (1971), based on photoemission experiments. They also calculated the energy bands as a function of lattice spacing and found a transition to semiconducting behaviour, when account is taken of the spin–orbit splitting of accidental degeneracies along Q. The behaviour of the band structure of the alkaline earths under pressure deserves further study, both theoretically and experimentally.

5.4.5 The hcp metals

The hcp structure is the most common among the transition metals, and the combination of the uniaxial symmetry and the two atoms per unit cell results in a band structure which is more complex than that of the cubic metals. This complexity is reflected in the Fermi surfaces and optical properties, which have not generally been studied as intensively as those of more straightforward metals. However, enough has been done to demonstrate that there are no obvious obstacles to elucidating their electronic structure as completely as those of the cubic metals, if a sufficient effort is made.

The hcp canonical bands are shown in Fig. 5.6, and the way in which hybridization and relativistic effects influence the band structure is illustrated in Fig. 5.8. The hcp transition metals fall into two groups, separated by the bcc metals in the periodic table, and typical band structures for each of these are shown in Figs. 5.25 and 5.27. The essential difference between the two groups stems from the lowering of the d-bands as the number of conduction electrons is increased. The band structures of hcp transition metals have been discussed in some detail by Jepsen, Andersen & Mackintosh (1975), and we will restrict ourselves to summarizing some of the general features.

On account of the screw axis along [0001] in the hcp structure, all bands on the hexagonal face of the Brillouin zone of Fig. 5.1 are at least fourfold degenerate (including spin), when spin–orbit coupling is neglected. It is therefore possible to adopt a double-zone representation, in which the number of bands is reduced by a factor 2, by allowing each to extend over twice the distance to the zone face along the hexagonal axis. However, spin–orbit coupling lifts this degeneracy, except along the line joining A and L, and it is therefore generally most convenient to use the primitive

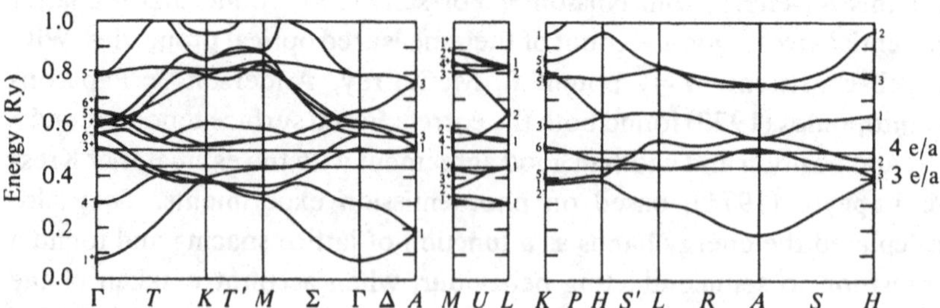

Fig. 5.25. Energy bands for hcp metals with a low electron concentration, illustrated by calculations on Y by H. L. Skriver. The spin–orbit coupling, which is relatively small in Y, has been omitted for clarity. Single-group symmetry labels are given at symmetry points.

zone, bearing in mind that magnetic breakdown may be important for orbits intersecting the hexagonal face of the zone, especially near R.

The d-bands in the hcp metals are strongly affected by hybridization, as illustrated in Fig. 5.8, especially at low energies. Except at certain symmetry points, all levels hybridize either strongly or weakly, in contrast to the cubic metals, in which bands of purely d-character exist along certain symmetry lines. The lowest-lying Γ_4-, M_1+ and L_1 d-states all hybridize with s-states, leaving K_2 as the lowest unhybridized d-level. Γ_3+ is a pure p-level and, as we shall see, its position, which is strongly influenced by relativistic shifts, is of importance for the Fermi surface topology in a number of the hcp metals.

A typical band structure for the hcp transition metals at the beginning of the series is shown in Fig. 5.25, and corresponding constant energy surfaces are illustrated in Fig. 5.26. For 3 electrons/atom, there are only two surfaces, a $\Gamma AL3$ open hole surface and an $HKM4$ open electron surface, which are separated by spin–orbit coupling near H on the hexagonal face of the zone. As the energy is raised, the third-band surface shrinks and disconnects at L, and the resulting $\Gamma A3$ surface is dissected into closed $\Gamma 3$ and $A3$ surfaces, at an electron concentration just less than 4 electrons/atom. The fourth band surface turns into a $\Gamma A4$ open hole surface, and closed $H5$ and $H6$ electron pockets appear in higher bands.

Only one trivalent and one quadrivalent hcp transition metal have been studied in detail. Mattocks & Young (1978) have obtained rather complete dHvA data for Y, and explained their results on the basis of a Fermi surface with the form of Fig. 5.26(*a*). The spin–orbit coupling is sufficiently large that magnetic breakdown effects were not observed. Good quantitative agreement with most of the experimental areas was obtained in the calculations of H. L. Skriver & A. R. Mackintosh (unpublished data), who used an LD potential. Similarly extensive dHvA measurements have been made on Ti by Kamm & Anderson (1974), and successfully interpreted in terms of energy bands calculated with a standard potential by Jepsen (1975). The Fermi surface topology is that of Fig. 5.26(*b*). In contrast to Y, magnetic breakdown is of crucial importance for the interpretation and leads to a series of coupled orbits between the $H5$ and $H6$ surfaces. The relatively sparse dHvA and magnetoresistance data on Zr (Everett 1972) indicate strongly that its Fermi surface has the same general form, which is in disagreement with the calculations of Jepsen *et al.* (1975), with a standard potential, in which the third band surface was found to be open along ΓA. However, they point out that a small raising of the d-band will produce a Fermi surface of

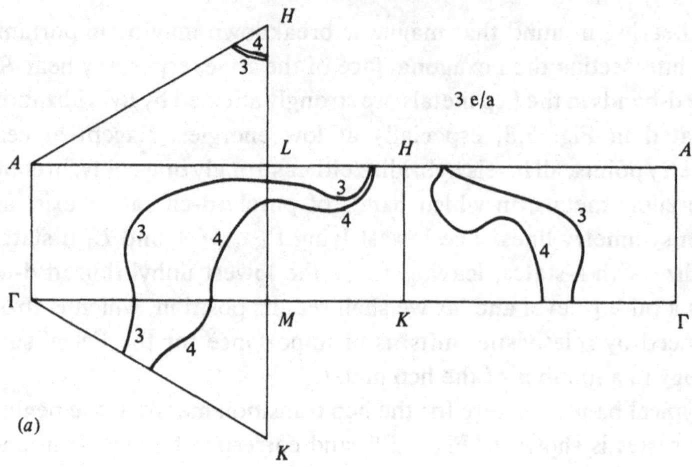

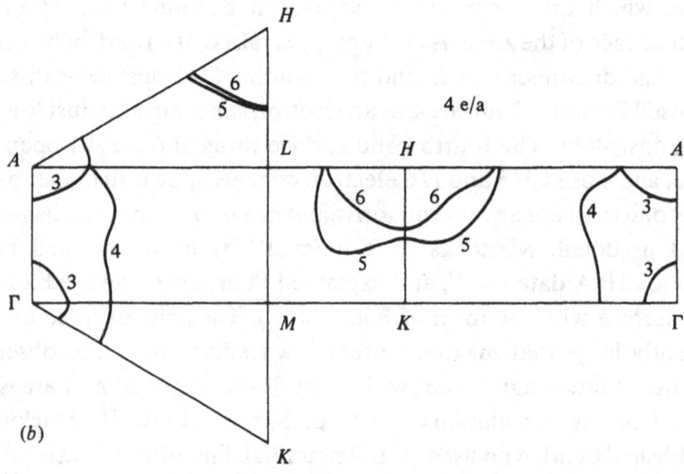

Fig. 5.26. Constant energy surfaces for the hcp structure, derived from the energy bands of Fig. 5.25. The labels on the surfaces indicate the corresponding bands.

the experimentally observed form, and it is noteworthy that the LD potential does give a greater value of Δ_{ds}, as shown in Fig. 5.12.

Towards the end of the transition series, the centre of the d-band is close to that of the s-band, and the effect of this downward shift may be appreciated by comparing Figs. 5.25 and 5.27. The constant energy surfaces are rather complicated. For 7 electrons/atom, the largest sheets are the $\Gamma A8$ open electron surface and the $L7$ hole surface, which touch at a point on the line AL, as illustrated in Fig. 5.28(a). There are also two

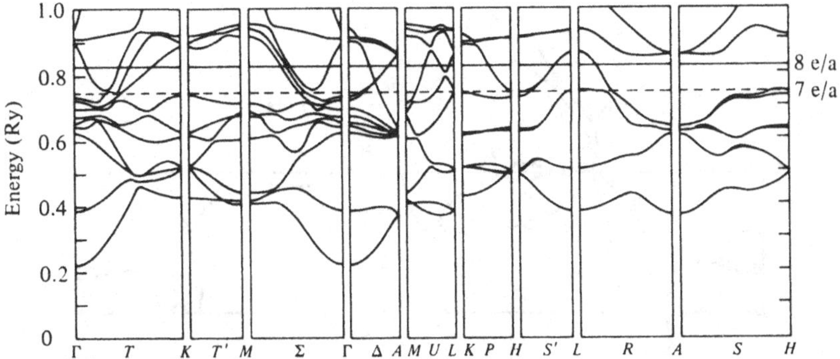

Fig. 5.27. Energy bands for hcp metals with a high electron concentration, illustrated by calculations on Ru by O. Jepsen. The symmetry of the states at the symmetry points may be deduced from Figs. 5.6, 5.8 and 5.25.

further hole pockets, $L5$ and $L6$, centred on L, a $\Gamma 9$ electron surface of complex form, and a lens-shaped $\Delta 8$ hole surface on the line joining Γ and A. As the energy is raised, the energy surfaces are modified with bewildering rapidity, but at 8 electrons/atom the result is as shown in Fig. 5.28(*b*). The eighth band open surface has turned into a $KML8$ open hole surface, the $\Gamma 9$ electron surface has expanded, and the $L7$ hole surface has contracted and split into two distinct $L7$ and $U7$ pockets. The tenth band is partially populated and supports a $\Gamma 10$ electron surface enclosing a small $\Gamma 10$ hole surface.

The dHvA data of Arko *et al.* (1974) on Tc are not complete but are consistent with the topology of Fig. 5.28(*a*) which, apart from the effect of spin–orbit coupling, is similar to that derived from a non-relativistic calculation by Faulkner (1977). Of particular interest is the observation of oscillations which may be ascribed to the $\Delta 8$ surface. The greater relativistic shift of the p-band in Re decreases the energy of the Γ_3+ level and, according to the calculations of Mattheiss (1966), who used a standard potential, eliminates this surface, leaving a small $\Gamma 8$ hole pocket. This has been observed in the magnetoacoustic measurements of Testardi & Soden (1967), and the ninth-band surface, which takes the form of a series of $\Sigma 9$ pockets in Re, has recently been studied by Holroyd, Fawcett & Perz (1978) as part of their investigation of the stress-dependence of the Fermi surface. The rest of the surfaces have the general form of Fig. 5.28(*a*) and dHvA oscillations from all of them have been observed by Thorsen, Joseph & Valby (1966). Magnetic breakdown occurs between the $\Gamma A8$ and $L7$ surfaces and, when this is taken into account, the agreement with the calculations is generally very satisfactory.

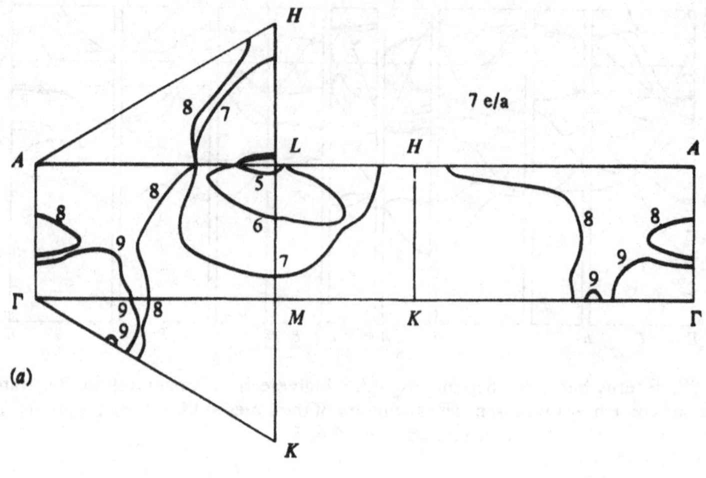

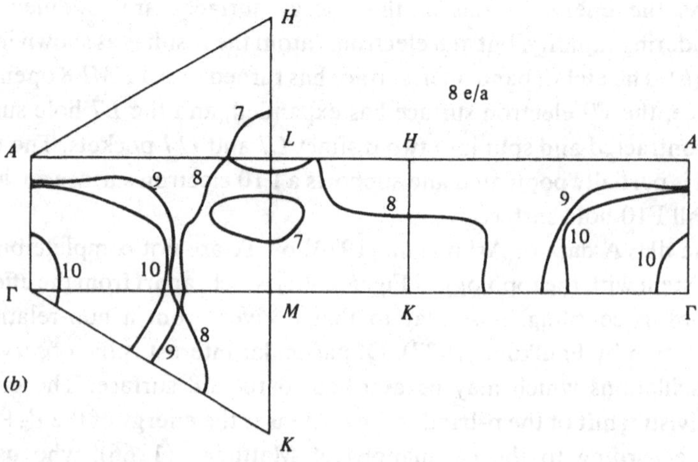

Fig. 5.28. Constant energy surfaces for the hcp structure, derived from the energy bands of Fig. 5.27. The labels on the surfaces indicate the corresponding bands.

All of the sheets of the Fermi surface of Ru shown in Fig. 5.28(*b*) have been observed by Coleridge (1969) in his dHvA experiments, and the quantitative agreement with the calculations of Jepsen *et al.* (1975) is very good. The same may be said of Os, except that Kamm & Anderson (1970) have not detected the *L*7 and Γ10 hole pockets predicted from the calculations. The existence of the Γ10 surface depends on the position of the Γ_{3+} p-level, and the *L*7 pockets must exist if the *KML*8 surface is open in the [0001] direction, as the galvanomagnetic measurements of Alekseevskii, Dubrovin, Karstens & Mikhailov (1968) indicate. Jepsen *et al.* (1975) have given arguments for supposing that these surfaces should exist, and it would be of interest to search for them further, since they would give a precise determination of the band positions near the Fermi level.

Extensive measurements have been made of the optical properties of some of the hcp transition metals and a number of structural features and trends have been observed, but few transitions have yet been convincingly identified. These studies have recently been reviewed by Weaver (1978). Single-crystal photoemission experiments would clearly be a valuable source of information on the band structure of these metals.

Although the main interest in the hcp heavy rare earth metals lies in the magnetic behaviour associated with the 4f-electrons, the conduction electrons also play an important role. Their band structures are similar to Fig. 5.25, with a relatively high-lying d-band, although the spin–orbit coupling is larger. The Fermi surface in divalent Yb has been studied by Tanuma, Datars, Doi & Dunsworth (1970) and their dHvA frequencies interpreted by Jepsen & Andersen (1971), by means of a calculation based on a standard potential. The Fermi surface consists of a number of small closed sheets on symmetry lines in the zone, and the experiment and theory can be brought into satisfactory agreement by a small shift of the d-bands. The lowest d-band is extremely flat, giving rise to a very high peak in the state density just above $\mathscr{E}_F$, and Jepsen & Andersen (1971) suggest that this might be the cause of the instability of the hcp structure when the temperature is increased.

The band structures of the trivalent metals have been calculated by Keeton & Loucks (1968). Experimental evidence on most of them, however, is rather sparse, although dHvA oscillations have been observed in non-magnetic Lu (Hoekstra & Phillips 1971) and ferromagnetic Tb (Mattocks & Young 1977a). The only detailed study is that of Mattocks & Young (1977b) on Gd, which is, however, of particular interest by virtue of its simple ferromagnetic structure and small anisotropy. They observed a great variety of oscillations and were able to give a convincing interpretation of all the high frequencies in terms of extremal areas derived from a rigid spin-splitting of the paramagnetic energy bands. The description of the small Fermi surface sheets requires the correct treatment of the spin–orbit coupling in the presence of the external and exchange fields, and this has not yet been accomplished, although Harmon, Schirber & Koelling (1978) have performed simplified calculations. Gd differs from the ferromagnetic transition metals both in its strong spin–orbit coupling and in the exchange field, which is derived predominantly from localized 4f-electrons, as distinct from the conduction electrons. It is therefore unique in Fermi surface studies, and of great importance in understanding the magnetism of rare earth metals.

5.5 Conclusion

In this chapter, we have attempted to review our knowledge and understanding of the role of the d-electrons in determining the properties of pure transition metals. The single-particle model and the resulting band structure picture are indeed very successful in this respect. The concept of canonical bands allows a concise description of those aspects of the electronic structure which depend only on the symmetry of the lattice, while the potential parameters, especially the d-band position and mass, account for the variations among the individual metals. The trends in the potential parameters, which are almost independent of crystal symmetry for the closely packed structures, can largely be explained in terms of the atomic properties, and the inclusion of the relativistic effects is essential for a quantitative description, especially of the heavier metals.

A great deal of effort has been devoted to constructing potentials for accurate calculations and to checking them against experiment. Fermi surface studies have been decisive in this context since, with suitable interpretation and parametrization, they give precise information about the logarithmic derivatives, and hence the wave-functions, at the Fermi energy, and furthermore indicate the extent to which the MT approximation is valid. The Fermi surfaces of the great majority of the transition metals have been studied to some extent, and its form in some of the cubic metals is known with a precision which is well beyond the ability of calculations to reproduce. Because of their complexity, the hcp metals are generally less well characterized, and it would be useful to select one of them for intensive study and a phase-shift parametrization. The magnetic metals, which are discussed in Chapter 6, are of particular interest and it would again be valuable to attempt to fit the extensive data available on, for example, Fe or Ni with spin-dependent logarithmic derivatives, and to explore the extent to which these parameters can be derived from a first-principles LSD potential. The importance of understanding the results on Gd, from the viewpoint of elucidating the magnetism of the rare earth metals, has already been emphasized. As described elsewhere in this volume, many other properties of the electrons at the Fermi surface have also been measured. The effective masses may be compared with the calculated values and, despite the uncertainties in the latter, it appears that the many-body enhancement is generally greater for surfaces with strong d-character than for the predominantly sp-sheets.

The great majority of accurate band-structure calculations on transition metals have been made utilizing the non-self-consistent standard potential construction, in which the Slater approximation to exchange and correlation is employed. Experience has shown that this type of potential places the different bands relative to each other with an accuracy of 10–20 mRy in the transition metals proper, where the d-band is significantly but incompletely filled. It appears that non-MT effects can cause distortions of the bands of roughly the same magnitude. Similar accuracy is obtained for these metals with the more systematically constructed self-consistent LD potential. At the ends of the transition series, particularly in the alkaline earths and the noble metals, it appears, however, that this potential gives a generally superior account of the Fermi surface geometry to that provided by the standard potential.

This LD potential also gives a good quantitative description of the bulk ground-state properties. The principal features which lead to the stability of the different closely packed structures are understood, although there is a persistent difficulty with the fcc metals at the end of the series. Similarly, calculations of the electronic pressure give lattice constants and bulk moduli which agree remarkably well with experiment. The separation of the pressure into the different angular momentum components gives a useful insight into the factors which determine the bulk properties. With the use of a LSD potential, it is possible to give a satisfactory first-principles account of the magnetic properties and to clarify their connection with the lattice structure in the magnetic transition metals.

Although the LD description is not immediately applicable to excited-state properties, local potentials and the energy bands derived from them can also give a good quantitative account of the energies of electromagnetic excitations. The standard potential leads to electronic energy differences which generally agree with those derived from optical and ultraviolet spectroscopy to within the usual uncertainty of 10–20 mRy. The same potential can also be used for an accurate prediction of the positions of the broadened energy bands, up to several Rydbergs above $\mathscr{E}_F$, observed in X-ray spectroscopy and secondary electron emission. The important exception is the noble metals, where the standard potential underestimates the Fermi surface anisotropy and the L-gap, due to p–s transitions, while apparently placing the d-bands correctly, and the

LD potential, which accounts well for both the Fermi surface and the *L*-gap, apparently places the d-bands much too high in energy. It is tempting to suppose that many-body effects in the narrow d-bands may be responsible for this apparent discrepancy.

The development of self-consistent potentials which, when combined with efficient and physically transparent methods for performing band-structure calculations, give a quantitative description of the electronic properties of transition metals, opens up wide perspectives for further studies of these important materials. Such potentials may be systematically improved to give an even more accurate account of simple properties, and may also be used to extend our understanding of more complex phenomena. Some progress has already been made in calculating, for example, elastic constants, phonon spectra and electron–phonon interactions, and their relation to superconductivity, and we may expect rapid developments in these areas, as well as in the treatment of lattice defects and lattice stability at finite temperatures. A start has also been made on a quantitative description of the electronic structure of transition metal films and surfaces, a topic of both fundamental and technological importance. The availability of prescriptions for constructing self-consistent potentials also allows an improved treatment of the electronic, energetic and structural, and magnetic properties of ordered transition metal compounds and, allied with appropriate statistical techniques, disordered alloys. Although such extensions in our knowledge will require the development of both new experimental and theoretical techniques, they will inevitably be based on the sound understanding which we have now achieved of the behaviour of the electrons in the periodic lattice of a pure, perfect and infinite transition metal at absolute zero.

Acknowledgements

We are grateful to our colleagues N. E. Christensen, D. Glötzel, O. Jepsen, J. Kollár and H. L. Skriver for supplying us with the results of their calculations before publication, and for many useful discussions. The cooperation of many others in providing us with information which we have used in this review is also gratefully acknowledged.

6

Fermi surface studies of ground-state and magnetic excitations in itinerant electron ferromagnets

G. G. LONZARICH

6.1 Introduction

The development and refinement of certain experimental and theoretical methods in solid state physics in recent years has led to major advances in the understanding of itinerant-electron magnetism in the 3d transition metals and alloys. Despite considerable progress, however, a number of formidable long-standing as well as novel problems, which will be reviewed briefly in the following section, remain unresolved. This chapter is concerned in particular with recent Fermi surface studies of the ground-state and low-lying magnetic excitations in iron and nickel which have helped to shed light on some of these problems. The chapter is organized as follows. A brief review (in Section 6.2) of our present understanding of ferromagnetic phenomena in transition metals is followed by a detailed discussion (in Section 6.3) of the type of information which may be derived, at least in principle, from experimental studies of the Fermi surface, regarding both the ground-state and thermal properties of ferromagnetic metals. In Sections 6.4 and 6.5 the results of such experimental studies in iron and nickel are presented and discussed in the light of current theoretical models.

6.2 Itinerant magnetism in 3d transition metals

Magnetic metals and alloys based primarily on elements of the first transition series such as iron, nickel and cobalt have been the subjects of intensive experimental and theoretical investigations for many years (see e.g. Herring 1966; and recent papers in Foner 1976; Lowde & Wohlfarth 1977; Lee, Perz & Fawcett 1978). The relatively slow progress in attaining a detailed understanding of the complex properties of these materials is to a large extent due to the fact that the 3d-electrons responsible for the magnetic behaviour cannot be treated as if they were fully localized, as, for example, the f-electrons in the rare earth metals, nor as essentially fully itinerant, as is often assumed for the conduction electrons in the simple metals. The width of the energy bands, and thus the hopping energy, of the 3d-electrons is thought to be of the same general magnitude as the relevant intra-atomic Coulomb interaction energy, so that dynamical correlations in the motion of the electrons do not necessarily lead to localization as is required in the former approximation, nor can they be effectively ignored as assumed in the latter approach.

The treatment of the many-electron systems in this intermediate coupling regime has been a controversial subject for many years (see e.g. Herring 1966; Edwards 1977; Moriya 1977 and references cited therein). Perhaps the most successful treatment of at least the *ground-state* properties of transition metals has been based on the spin density function (SDF) formalism originating from the work of Hohenberg, Kohn and Sham, which in principle treats exchange and Coulomb correlations exactly (Gunnarsson & Lundqvist 1976). According to this theory the spin densities and quasi-particle energies associated with the ground state may be obtained from the eigenfunctions $\psi(\mathbf{k}, \sigma)$ and eigenvalues $\mathscr{E}(\mathbf{k}, \sigma)$, respectively, of a one-particle Hamiltonian containing a complicated self-consistent potential which takes into account the effect of the electron–ion and electron–electron interactions. The latter is taken into account in the well-known Hartree *and* the so-called exchange-correlation parts of the potential. It is known that there exists a *local* potential, a functional of the spin densities, which leads to the exact particle and spin densities when these are calculated by summing the amplitudes $|\psi(\mathbf{k}, \sigma)|^2$ over the occupied states as in the usual one-electron theory. However, a non-local potential, which depends self-consistently on the excitation energies, must also be included if the eigenvalues of the one-particle Schrödinger equation are to reproduce the true quasi-particle energies needed to construct, for example, the Fermi surface (defined formally from the position of the Midgal–Luttinger discontinuities of the occupation numbers of the bare electrons in wave-vector space (Luttinger 1960)).

The exact forms of the local and non-local potentials are unfortunately not known and almost all SDF calculations carried out thus far have been based on the so-called local spin density (LSD) approximation in which the non-local potential is ignored and the local potential is assumed to have the same form as that calculated for a homogeneous electron gas (von Barth & Hedin 1972; Gunnarsson & Lundqvist 1976). Though a more realistic treatment of electron correlations than the earlier well-known $X\alpha$ method derived from a Hartree–Fock approximation, this scheme nevertheless ignores certain dynamical effects which could be important in the determination of the Fermi surface and in particular the quasi-particle density of states (Rasolt & Vosko 1974; Edwards 1977).

The SDF theory in the LSD approximation provides the framework for a band theory of magnetism. In the spin-polarized or ferromagnetic state the self-consistent LSD potential is, on account of exchange correlations which help to reduce the interaction energy of parallel spins, lower for the majority (↑) than the minority (↓) spin quasi-particles. Thus the energy bands for the majority carriers are in general shifted down in energy relative to the minority bands. This exchange splitting of the energy bands is found to be more or less uniform in **k**-space, at least for pure d-states, and approximately proportional to the magnetization (Gunnarsson 1977). The equilibrium magnetization is determined self-consistently by requiring that the ↑ and ↓ spin quasi-particles moving in their ↑ or ↓ spin LSD potentials, have the same chemical potential. The SDF–LSD scheme has provided at least partial theoretical support for the well-known Stoner model at $T = 0$ (see e.g. Wohlfarth 1976 for a recent review). According to this model the exchange splitting of the energy bands is assumed to be the same for all states and proportional to the magnetization at all temperatures. The validity of the latter assumption in particular is discussed further on.

Energy-band calculations based on the SDF–LSD or SDF–$X\alpha$ approximations have on the whole been reasonably successful in accounting for many of the ground-state properties of the 3d-ferromagnets iron, nickel and cobalt. For example, the calculated magnetic moment per atom (and its pressure dependence), the charge and spin density form factors, the high field galvanomagnetic properties, and the Fermi surfaces are all more or less in general accord with experimental findings (Section 6.4 and references cited therein). On the other hand the observed electronic heat capacities and cyclotron effective masses in iron and nickel, for example, are typically nearly twice as large as the calculated values (Sections 6.3.3 and 6.4 and references given therein). Part of the enhancement can be attributed, as in normal metals, to the electron–phonon interaction; however, it has been suggested that a large contribution arises from the electron–magnon interaction, one aspect of dynamical electron correlations which is ignored in the LSD approximation. The apparent difference between the calculated and observed separation between the top of the majority d-bands and the Fermi level in Ni has also been explained in these terms (Edwards 1977 and references given therein; but see also Kleinman 1978).

Apart from these difficulties, which may perhaps be overcome in more realistic treatments of the exchange-correlation potential, the SDF formalism has been criticized for not making explicit the precise nature of Coulomb and exchange-correlation effects which appear to be implicitly contained in the local and non-local potentials. A number of attempts have been made over the years to arrive at a more microscopic or physical understanding of the magnetic ground state in itinerant systems (Anderson 1961; Friedel, Leman & Olszewski 1961; Kanamori 1963; Hubbard 1963; Edwards 1970; Stearns 1977; Matsumoto, Umezawa, Seki & Tachiki 1978); however, formidable mathematical difficulties have thus far precluded a completely satisfactory and general solution.

On the whole, however, the $T=0$ properties of transition metal magnets are far better understood than the thermodynamic properties. The earliest attempt at understanding the magnetic behaviour of itinerant ferromagnets at finite temperatures was based on the Stoner model defined earlier. This model considers only independent particle–hole excitations above the ground state, and includes interaction effects solely through the assumption that the exchange splitting Δ of $\uparrow$ and $\downarrow$ spin states is proportional to the total magnetization; i.e. $\Delta = I(n_\uparrow - n_\downarrow)$, where n_σ is the total number of σ-spin electrons per atom and I is an interaction parameter. The theory is appealing for its simplicity, requiring only a knowledge of the quasi-particle density of states, the zero temperature exchange splitting and the well-known Fermi function. However, in this form the theory predicts unrealistically high Curie temperatures for iron, nickel and cobalt and it cannot, for example, reasonably account for the detailed temperature dependence of the magnetization and the high-temperature susceptibility in these ferromagnets. The overall agreement with experiment can be improved if it is assumed that the interaction parameter I varies substantially with temperature (Shimizu 1977). However, the precise temperature dependence required to obtain this agreement cannot yet be confirmed from first principles calculations.

As is well known, the central weakness of the above Stoner theory lies in its oversimplified picture of the magnetic (spin-flip) excitation spectrum. In this theory thermally excited quasi-particles and holes of opposite spin (so-called Stoner excitations) move independently over their common exchange-correlation potential. However, in the ferromagnetic state the electron–electron interaction is expected to bring about a considerable degree of correlation in the motion of these particle–hole pairs. Magnetic excitations are then expected to be in the nature of spin density fluctuations which, at long wavelengths and low temperatures, correspond to classical spin waves. The excitation energy spectrum, and thus the expected thermodynamic behaviour of a ferromagnet should, for these correlated modes of particle–hole pairs, be considerably different from that in the Stoner theory (see e.g. Herring 1966; Edwards 1977; Moriya 1977 and references cited therein).

Spin fluctuations can be conveniently discussed in terms of the dynamical wave-vector and frequency-dependent susceptibility tensor $\chi_{\alpha\beta}(\mathbf{q}, \omega)$, which measures the magnetic response of a system to an impressed time and space varying magnetic field characterized by a wave-vector $\mathbf{q}$ and a frequency ω. The singularities or resonances in the imaginary part of the transverse component $\chi_{-+}(\mathbf{q}, \omega)$ of the generalized susceptibility, which can be measured directly by means of neutron inelastic scattering, reveal the spectrum of magnetic (spin-flip) excitations. In Stoner theory the assumption that the exchange-correlation potential is essentially proportional to the time and space average magnetization leads to an independent-particle expression for $\chi_{-+}(\mathbf{q}, \omega)$, whose singularities correspond to the Stoner excitations discussed above. In the next level of approximation for $\chi_{-+}(\mathbf{q}, \omega)$, referred to as the time-dependent Hartree–Fock or random phase approximation (RPA), the exchange-correlation potential is essentially allowed to depend on the *local dynamical* spin polarization. This approximation is treated by means of several theoretical techniques, most of which are based on a Stoner ground state. For this Stoner–RPA theory the singularities of $\chi_{-+}(\mathbf{q}, \omega)$ are found to consist of magnons for low q and ω and weakly correlated Stoner-like modes at higher momenta and energies (see e.g. Izuyama, Kim & Kubo 1963, Cooke 1973, Edwards 1977, Moriya 1977 and references given therein).

Calculations of $\chi_{-+}(\mathbf{q}, \omega)$ based on the RPA method and the Stoner or more general SDF approximations of the ground states yield generally good agreement, within the accuracy of the assumed ground state, with inelastic neutron scattering data in nickel and iron well below the Curie temperature (Cooke, Lynn & Davis 1974; see also Callaway & Wang 1977a for a somewhat different approach). In addition, this type of model has been used to explain at least qualitatively (Izuyama *et al.* 1963) certain aspects of critical neutron scattering near T_c and paramagnetic scattering in nearly ferromagnetic metals or in ferromagnets above T_c. The Stoner–RPA model is also capable of accounting for the observed temperature dependence of the magnetization at relatively low temperatures. However, certain important ambiguities, discussed in detail under Sections 6.3.6 and 6.5, remain regarding the relative importance of Stoner and collective spin excitations in certain ferromagnetic metals.

Though a considerable improvement over simple Stoner theory, the Stoner–RPA model is still inadequate in several important respects. For instance, in this approximation (in its simplest form) the exchange splitting between the ↑ and ↓ quasi-particle bands is found to be proportional to the difference in the total number of ↑ and ↓ spin carriers and thus to the *total* magnetization (Izuyama *et al.* 1963; Thompson, Wohlfarth & Bryan 1964). However, it is evident (Herring 1966) that this cannot be valid if the magnetization is reduced by means of relatively long wavelength spin-waves. In the limit $q \to 0$, for instance, such excitations merely give rise to a rotation of the entire magnetization vector away from the spin quantization axis, and, at least in the absence of magnetic anisotropy, cannot modify the quasi-particle energies or exchange splitting of the energy bands. The insensitivity of the exchange splitting to thermally excited spin-waves at low temperatures has been demonstrated in several theoretical analyses (Izuyama & Kubo 1964; Edwards 1974; Lonzarich 1974; Prange & Korenman 1975; Liu 1976) and recently confirmed experimentally (see Section 6.5 and references cited therein). The theoretical studies referred to above indicate that the RPA theory is improved if it is assumed that, at least for low temperatures, the exchange splitting is approximately proportional to the total *single-particle* magnetization; i.e. the $T = 0$ magnetization plus the change in magnetization induced by the thermal excitation of Stoner particle–hole pairs, the magnon contribution being excluded.

For temperatures approaching T_c and beyond, the RPA approach (even with this refinement) is expected to be a rather poor approximation since it does not take proper account of the coupling among the different wave-vector components of the spin fluctuations, or the mode–mode coupling, which should play an increasingly important role in the thermodynamic behaviour as the temperature is raised. Indeed a number of striking properties of the itinerant ferromagnets such as iron and nickel at high temperatures cannot be accounted for in terms of the RPA theory. For example, (i) the magnitude of the Curie temperature (which is found to be considerably lower than that calculated in the Stoner–RPA models), (ii) the persistence of spin-wave-like modes of relatively short wavelength above T_c, (iii) the apparent absence of any appreciable decrease in the exchange splitting with increasing temperatures (for temperatures well above T_c), (iv) the Curie–Weiss behaviour of the paramagnetic susceptibility, as well as other high-temperature properties of iron and nickel (see e.g. Herring 1966; Fadley & Wohlfarth 1972; Wohlfarth 1975; Cooke 1976; Edwards 1977 for reviews of some of the experimental information) are evidence of the importance of electron correlations in these systems which cannot be adequately treated in terms of the theoretical models discussed thus far. It has been suggested that such correlation effects may be less important in certain weak itinerant ferromagnets such as $ZrZn_2$ and Ni_3Al (Edwards & Wohlfarth 1968); however, the situation in these materials is still uncertain (see e.g. Murata & Doniach 1972; Moriya 1977).

A number of theoretical models which go well beyond the RPA method have been proposed in recent years in an effort to account for the complex high temperature phenomena described above (Wang, Evenson & Schrieffer 1969; Murata & Doniach 1972; Moriya 1977; Edwards 1977; Hertz & Klenin 1977; Korenman, Murray & Prange 1977; Sokoloff 1978; Liu 1978). These theories have been shown to yield results which are in qualitative agreement with some finite temperature measurements, but they do not as yet provide a complete and unified picture. The development of a realistic formalism for the treatment of the finite temperature properties of itinerant ferromagnets must therefore await future study.

6.3 Application of the de Haas–van Alphen effect to the study of itinerant ferromagnetism

6.3.1 General background

The brief introduction and review given in the last section indicate that, despite recent advances, certain aspects of itinerant ferromagnetism are still relatively poorly understood and require further study. In this section information concerning these systems, which may be derived from experimental studies of the Fermi surface will be considered.

For comparison and perspective we begin with a brief discussion of the information which may be obtained by what is thought to be perhaps the most powerful technique for investigating magnetic phenomena in solids, namely thermal neutron scattering. As is well known, the usefulness of the scattering technique in general lies in the fact that the response of a many-particle system to a weak external probe (such as a beam of weakly scattered particles) is closely linked with certain equilibrium correlation functions of that system (see e.g. Pines & Nozières 1966). These in turn are connected with important properties of the ground state and with the spectrum of a certain type of elementary excitations. In part the neutron itself interacts – via the magnetic dipole–dipole interaction – with the spins of the electrons in a system and thus probes the spatial and temporal correlations in the *spin-density*. These provide information about the static spin-density distribution and the spin-fluctuation or magnetic excitation spectrum.

More specifically, if the orbital angular momentum of the electrons can be ignored, the differential cross-section for magnetic scattering of a neutron from a state of wave-vector $\mathbf{k}$ to one of wave-vector $\mathbf{k}+\mathbf{q}$, with energy change $\hbar\omega$, is in general proportional to $|\mathbf{k}+\mathbf{q}|/|\mathbf{k}|$ and the components perpendicular to $\mathbf{q}$ of the tensor (see e.g. White 1970)

$$S_{\alpha\beta}(\mathbf{q}, \omega) = \int \mathrm{d}^3\mathbf{r} \int \mathrm{d}^3\mathbf{r}' \int_{-\infty}^{\infty} \mathrm{d}t\, \mathrm{e}^{-\mathrm{i}(\mathbf{q}\cdot\mathbf{r}'+\omega t)} \langle S_\alpha(\mathbf{r}+\mathbf{r}', t) S_\beta(\mathbf{r}, 0)\rangle. \quad (1)$$

This is the Fourier transform of a function describing the correlation between the β component $S_\beta(\mathbf{r}, 0)$ of the spin-density at position $\mathbf{r}$ and time 0 and the α component at a different position $\mathbf{r}+\mathbf{r}'$ and time t. The angular brackets signify a quantum-mechanical ensemble average. The components of the local magnetization may be expressed in the usual manner as $M_\alpha(\mathbf{r}, t) = -g\mu_\mathrm{B} S_\alpha(\mathbf{r}, t)$.

From (1) and the convolution theorem it is evident that in the zero frequency or elastic scattering limit the differential cross-section yields the modulus square of the Fourier transform of the static spin density. In a periodic system the scattering is concentrated in Bragg peaks centred on values of $\mathbf{q}$ equal to a wave-vector in the magnetic reciprocal lattice. The intensities of these peaks provide the time-average total spin per atom and the atomic spin-density form factor for wave-vectors in the reciprocal lattice.

It also follows from (1) and the fluctuation dissipation theorem that the differential cross-section for *inelastic* scattering is related to components of the imaginary part $\chi''_{\alpha\beta}(\mathbf{q}, \omega)$ of the dynamical susceptibility tensor discussed in the introduction, which reveal the nature of the spectrum of magnetic excitations. At high temperatures in the spin-disordered state $S_{\alpha\beta}(\mathbf{q}, \omega)$ may be used to infer the correlation length and time of short-range spin fluctuations.

The wealth of information available from neutron scattering is remarkable. However, a number of equilibrium properties associated especially with the energy spectrum of single particles cannot readily be explored by this technique. Properties in this category discussed in the following sections may on the other hand be investigated by means of detailed studies of the Fermi surface and its variation with temperature and pressure. Thus the neutron scattering and Fermi surface methods are complementary techniques which can jointly provide a detailed picture of the equilibrium state of magnetic metals.

Of the various experimental probes of the Fermi surface, the most useful thus far in the study of ferromagnetic metals have been the high-field galvanomagnetic and de Haas–van Alphen (dHvA) effects, the rapidly developing positron annihilation technique which has recently been applied to the study of iron and nickel both at low and high temperatures (Mijnarends 1973; Stachowiak 1976, and angle-resolved photoemission). We shall confine our attention to the technique which has thus far proved to be most useful for these materials at low temperatures, namely the dHvA effect which has been studied extensively over the years by a number of investigators and in particular by D. Shoenberg and his students and colleagues (see Gold 1968; and D. Shoenberg in a forthcoming book).

The theory of the dHvA effect, which consists of an oscillatory variation as a function of the magnetic induction of the diamagnetic susceptibility, has been worked out in considerable detail for normal and nearly ferromagnetic metals and it has been assumed that the basic results carry over to the ferromagnetic systems. Detailed analyses which have included the effects of electron–electron as well as electron–phonon interactions have indicated that the oscillatory magnetization $\tilde{\mathbf{M}}$ in a real metal should be virtually the same as that expected for a system of independent quasi-particles as in the Lifshitz–Kosevich theory, provided the correct quasi-particle energies are used and certain parameters are suitably renormalized (Lifshitz & Kosevich 1956; Luttinger 1961; Bychkov & Gorkov 1962; Prange & Sachs 1967; Engelsberg & Simpson 1970; Isihara & Tsai 1974). According to the quasi-particle model, an oscillatory magnetization $\tilde{\mathbf{M}}$ is associated with each extremal cross-sectional area $\mathscr{A}(\hat{\mathbf{B}})$ of the Fermi surface lying in a plane normal to the magnetic induction $\mathbf{B}$. The frequency of the oscillations in B^{-1} is given by the relation

$$F = \frac{\hbar c}{2\pi|e|}\mathscr{A}, \tag{2}$$

and the component of $\tilde{\mathbf{M}}$ parallel to $\mathbf{B}$ for a given area $\mathcal{A}$ of the ↑ or ↓ spin part of the Fermi surface is (at least for sufficiently low fields so that F/B is large and magnetic breakdown can be ignored) given by the formula

$$\tilde{M} = \frac{aTF}{(B|\mathcal{A}''|)^{\frac{1}{2}}} \sum_{r=1}^{\infty} \frac{\exp\left(-\dfrac{rKm_c^*x}{B}\right)}{r^{\frac{1}{2}}\sinh\left(\dfrac{rKm_c^*T}{B}\right)} \sin\left[2\pi r\left(\frac{F}{B}+\eta-\gamma\right)\mp\frac{\pi}{4}\right]. \quad (3)$$

In this expression T is the temperature, x is an effective (Dingle) temperature which describes approximately the effects of the finite lifetime of the quasi-particles and of the phase cancellation due to inhomogeneities in F/B over the crystal, $m_c = (\hbar^2/2\pi)|\partial\mathcal{A}/\partial\mathscr{E}|_{\mathscr{E}_F}$ is the cyclotron effective mass as determined from the quasi-particle energies calculated excluding the electron–phonon and electron–magnon interactions, and an asterisk designates enhancement due to these interactions. The quantity $|\mathcal{A}''|$ is a dimensionless curvature factor equal to 2π for a spherical Fermi surface, and K and a are related to fundamental constants and numerically given as $K = 1.469\times10^5$ G K^{-1} m^{-1} and $a = -3.261\times10^{-6}$ $G^{\frac{1}{2}}$ K^{-1}. (A many-body correction to the value of the latter parameter has recently been predicted by Isihara & Tsai (1974).) It has been shown that at least for the simple metals (i) the volume of the Fermi surface is unaffected by the electron–electron interaction (Luttinger 1961), (ii) the shape and volume of the Fermi surface and the product m_c^*x in (3) are unaffected by the electron–phonon interaction (Prange & Sachs 1967) and (iii) the electron–phonon enhancement in the product m_c^*T in (3) should be essentially that at $T = 0$ (Engelsberg & Simpson 1970). Corresponding results for the electron–magnon interaction in ferromagnetic metals have not yet been rigorously established.

It is useful to consider briefly the orders of magnitude of the parameters in (3). For a cross-sectional area of the Fermi surface of iron of intermediate size, typically $F \sim 5\times10^7$ G, $m_c^*/m \sim 2$ and $\mathcal{A}'' \sim 6$, so that, assuming $T \sim 1$ K, $xm_c^*/m \sim 2$ and $B \sim 5\times10^4$ G, the amplitude of the fundamental component in (3) is found to be $|\tilde{M}| \sim 5\times10^{-6}$ G. This value (which depends critically on the choice of m_c^*, T, x and B is approximately 9 orders of magnitude *smaller* than the steady ferromagnetic magnetization in Fe but still several orders of magnitude *greater* than that which can be measured with the most sensitive present detection techniques.

In the argument of the sinusoidal function in (3) γ is a constant, equal to $\frac{1}{2}$ for an isotropic system, and η is a spin-dependent phase term given by the formula

$$\eta = \frac{\sigma \nu g \Gamma m_c}{2m}\left(\frac{N_{-\sigma}}{N_\sigma + N_{-\sigma}}\right), \tag{4}$$

where g is the band g-factor and Γ is the many-body exchange (or Stoner) enhancement, both of which are here assumed to be isotropic. In addition N_σ is the density of states at the Fermi level for the σ-spin quasi-particles, σ as a factor is $+1$ or -1 for $\uparrow$ or $\downarrow$ spins, respectively, and ν is $+1$ for an electron surface and -1 for a h le surface. This phase term measures the rate of change of the given cross-sectional area with magnetic induction and is closely connected with the Pauli susceptibility of the conduction electrons. When $N_\uparrow$ equals $N_\downarrow$, (4) reduces to the conventional expression for η in paramagnetic metals which leads to the so-called Dingle-cosine factor when the $\uparrow$ and $\downarrow$ components of the magnetization are combined. Note than the mass enhancement defined earlier has not been included in the parameters appearing in (4) (see Engelsberg & Simpson 1970 for nearly ferromagnetic metals).

There is compelling evidence that the Lifshitz–Kosevich theory as outlined above provides an excellent description of the dHvA effect in normal metals (see e.g. Palin, Randles, Shoenberg & Vanderkooy 1971). The situation in the ferromagnetic systems is perhaps less certain; however, a recent study (Vuillemin & Mori 1977) has demonstrated that the absolute amplitude of the dHvA magnetization arising from the majority necks in Ni *could* be described with a precision of about 7 per cent in terms of the Lifshitz–Kosevich formula (3). The parameters m_c^* and x were obtained from the temperature and field dependences of the amplitude, respectively, and $\mathscr{A}''$ was derived from an empirical model of the Fermi surface.

With this brief background as a guide we consider in general the type of information which the dHvA effect is capable of providing regarding the ferromagnetic state. Detailed experimental results obtained thus far will be presented and discussed in Sections 6.4 and 6.5.

6.3.2 Majority and minority spin sheets of the Fermi surface

The magnitudes and orientation dependences of the dHvA frequencies may be used in conjunction with inversion schemes, geometrical models or empirical-interpolation band models (which are of particular value when the experimental data are incomplete), to determine the overall ↑ and ↓ spin Fermi surface. Although certain important discrepancies and ambiguities remain, investigations of this kind in Fe and Ni have produced Fermi surface models which are in general accord with recent first-principles band-structure calculations based on the SDF–LSD or SDF–Xα methods (see Section 6.4). The central conclusion which follows from this overall agreement between experiment and band theory is that the magnetic electrons in Fe and Ni are itinerant, at least in the sense that the difference in the volumes enclosed by the ↑ and ↓ spin surfaces is roughly consistent with that expected from the observed spin magnetization, and the sum with the known number of valence electrons. As discussed in Section 6.4, these conclusions have not been rigorously established, particularly in the case of iron, on the basis of the experimental evidence alone and further studies would be of considerable interest, particularly in view of recent suggestions that, due to many-body effects ignored in the SDF–LSD calculations, the former of the two conditions mentioned above may not be satisfied exactly (Herring 1978; Matsumoto *et al.* 1978; but see also Kondratenko 1965).

Certain difficulties which must be faced in any precise determination of the volumes of the separate ↑ and ↓ spin parts of the Fermi surface are noted below. In the first instance it is recalled that the dHvA effect itself does not provide completely unambiguous information regarding the spin or charge character of an individual cross-sectional area. This information can usually be safely inferred from theoretical models; however, in some cases involving the smaller parts of complex Fermi surfaces such a procedure is not necessarily reliable. In principle, measurements of the infinite field phase (see (3) and (4)), or the pressure dependence of the phase (see Section 6.3.5) would help in this identification in ambiguous cases.

Apart from this, a completely clear-cut separation of the Fermi surface in terms of ↑ and ↓ spin sections is not always possible even in principle, on account of spin-hybridization, or the mixing of ↑ and ↓ spin states induced by the spin–orbit interaction (see Gold 1974 and Singh, Wang & Callaway 1975 for the most recent discussions). The blurring of the distinction between ↑ and ↓ spin states on the Fermi surface will only be serious in those regions of **k**-space where ↑ and ↓ bands are nearly degenerate at the Fermi level. In these regions the spin–orbit interaction (which is off-diagonal in a basis of ↑ and ↓ spin Bloch states since the orbital angular momentum is quenched by the crystal field) can give rise to first-order energy shifts and the one-particle wave-functions can be strong admixtures of both ↑ and ↓ spin Bloch states. Since these regions generally involve only very small volumes in **k**-space the uncertainty in the determination of the volumes of the ↑ and ↓ spin parts of the Fermi surface may not be important. Furthermore it is noted that in sufficiently high applied magnetic fields, magnetic breakdown (Stark & Falicov 1967) may ultimately permit a determination of the *unhybridized* ↑ and ↓ spin parts of the Fermi surface directly.

6.3.3 Majority and minority spin densities of states

Although the Fermi surface results mentioned in the last section point to the itinerant character of the magnetic carriers in Fe and Ni, they do not clarify completely the role of dynamical electron correlations in these systems. As indicated in Section 6.2, additional information may be obtained from measurements of the enhanced cyclotron effective mass m_c^* which appears in the temperature-dependent factor in (3). The partial information presently available (Gold, Hodges, Panousis & Stone 1971; Goy & Grimes 1973) suggests that the mass enhancement factor is as large as 1.5–2 in Fe and Ni and thus considerably larger than that in most simple metals and comparable to that found in nearly ferromagnetic metals such as Pt and Pd where the electron–paramagnon interaction is known to be important (Dye, Karim, Ketterson & Crabtree 1978a).

Provided spin hybridization is not too extensive, a detailed study of m_c^* over the entire ↑ and ↓ Fermi surface would enable for the first time a reliable determination of the separate (enhanced) ↑ and ↓ spin densities of states $N_\uparrow^*$ and $N_\downarrow^*$ at the Fermi level. This information would be of particular interest in view of recent calculations which reveal a considerable spin dependence in the electron–magnon enhancements in $N_\uparrow^*$ and $N_\downarrow^*$ (Edwards 1977 and references cited therein). Such a study may also be useful in the interpretation of recent extensive investigations of the spin polarization of electrons photoemitted from the vicinity of the Fermi levels in Fe, Ni and Co (Siegmann 1978 and references cited therein).

6.3.4 Magnetic anisotropy

An interesting and rather far reaching consequence of the combined presence of the exchange and spin–orbit interactions in a ferromagnetic metal is a dependence of the energy-band structure and thus the detailed features of the Fermi surface on the *direction* of the spin magnetization **M** or essentially the magnetic induction **B** (Gold 1974; Singh *et al.* 1975, and references cited therein). Some insight into the origin of this effect may be obtained by noting that, at least in the limit of full spin polarization of the valence electrons, the orientation of the spin **s** of any occupied one-particle state is essentially determined by **B** whereas the orbital angular momentum **l** points along certain directions in the crystal dictated primarily by the symmetry of those Bloch states which are admixed by the spin–orbit interaction (to produce the spatial part of the one-particle state in question). Thus the inner product $\mathbf{s} \cdot \mathbf{l}$ and consequently the spin–orbit contribution to the individual quasi-particle energies can depend dramatically on the direction of **B** relative to the crystal axes. When the valence electrons are only partially spin polarized this argument must be extended to include admixtures of both ↑ and ↓ spin Bloch states in regions where ↑ and ↓ spin bands are nearly degenerate. In the non-magnetic state this admixture becomes all important and leads to spin–orbit energy shifts which are essentially independent of the direction of the magnetic induction (in normal laboratory applied fields).

The energy shifts produced by the spin–orbit interaction are quite small compared with the average width of the 3d-bands; however, their influence on a number of physical properties is quite remarkable. By lifting the degeneracies at crossing-points of different sheets of the Fermi surface the spin–orbit interaction can greatly alter the Fermi surface topology and thus the dHvA and galvanomagnetic effects. The dependence of these changes in topology on the orientation of **B** in ferromagnets can greatly complicate the interpretation of the experimental dHvA data (and, for example, the empirical determination, discussed under Section 6.3.2, of the precise volumes of the ↑ and ↓ spin parts of the Fermi surface).

Closely related to these spin–orbit effects are the anomalous damping of low-energy spin-waves and more generally the existence of a preferred axis of magnetization, or the intrinsic magnetocrystalline anisotropy in metallic ferromagnets. The former is at least partly due to energy losses associated with a dynamic repopulation of states near the Fermi surface driven by the precession of the magnetization vector, which gives rise to periodic variations in the dimensions of the Fermi surface (Korenman & Prange 1972). The second effect arises from the dependence of the total spin–orbit energy, summed over all one-particle states, on the orientation of **M**. It is of interest to note, in this connection, that an unusual orientation dependence of the magnetic anisotropy energy in Ni at low temperatures was recently attributed to the existence of very small pockets of the Fermi surface which undergo dramatic changes in size as a function of $\hat{\mathbf{M}}$ (Gersdorf 1978). Evidence for the existence of these pockets has *not* been reported in the dHvA or galvanomagnetic effects (see Section 6.4) and further studies are needed to confirm or refute this conjecture. A temperature modulation technique (see e.g. Legkostupov 1971) may be more suitable for the detection of the very low frequency and high cyclotron mass dHvA oscillations predicted for these small pockets, than the conventional field modulation method which has been used thus far.

6.3.5 Information from the pressure dependence of the Fermi surface

The dHvA effect has been an invaluable tool in the investigation of the volume and strain dependence of energy-band structures in normal metals (Templeton 1966; Bosacchi, Ketterson & Windmiller 1970; Fawcett, Griessen, Joss, Lee & Perz in Chapter 7). In this section the application of this technique in the study of volume effects in ferromagnetic metals is considered. A new and important feature in these systems is the volume dependence of the spin magnetization.

The parameter measured in the dHvA method is essentially the pressure derivative (at constant temperature and applied magnetic field) of the phase Φ of the fundamental component of $\tilde{M}$ in (3), which may be expressed as

$$\frac{\mathrm{d}\Phi}{\mathrm{d}P}=2\pi\frac{\mathrm{d}(\eta-\gamma)}{\mathrm{d}P}+\frac{2\pi F}{B}\left[\frac{\mathrm{d}\ln\mathscr{A}}{\mathrm{d}P}-\frac{4\pi(1-\mathscr{D})M}{B}\frac{\mathrm{d}\ln M}{\mathrm{d}P}\right], \tag{5}$$

where $B=H_{\mathrm{ext}}+4\pi(1-\mathscr{D})M$, H_{ext} is the external magnetic field, $\mathscr{D}$ is the demagnetizing factor, M is the total magnetization and η is defined in (4). A least-squares fit of a set of experimental values of $\mathrm{d}\Phi/\mathrm{d}P$ versus B, to a function of the form $a_0+a_1B^{-1}+a_2B^{-2}$ would yield the pressure derivatives of essentially $N_{-\sigma}/(N_{-\sigma}+N_\sigma)$ (from (4)), $\mathscr{A}$ and M from the coefficients a_0, a_1 and a_2, respectively. The last two parameters, which are of considerable theoretical interest, can in principle be accurately measured with available dHvA detection methods. Since under suitable conditions phase changes as small as 10^{-4} cycles can now be detected, it follows from (5) that for $F/B\sim10^3$, the changes in the cross-sectional area $\mathscr{A}$ or in the magnetization M of one part in 10^7 can be observed. It is expected that shifts of this magnitude, at least in Fe, can be induced with applied pressures of the order of bars (i.e. in the helium fluid-pressure range).

We now relate the pressure derivative of a cross-sectional area $\mathcal{A}$ associated with band l' of the σ-spin part of the Fermi surface to the pressure derivatives of the relevant band-structure parameters, and in particular the spin magnetization. If $\mathscr{E}(l, \mathbf{k}, \sigma)$ is the energy of a quasi-particle associated with band l, wave-vector $\mathbf{k}$ and spin σ, and $\mathscr{E}_F$ is the Fermi energy, then the cross-sectional area $\mathcal{A}$ and the number n_σ of the σ-spin particles per atom can be expressed as

$$\mathcal{A} = \int \mathrm{d}^2\mathbf{k}\, \theta\{\nu[\mathscr{E}_F - \mathscr{E}(l', \mathbf{k}, \sigma)]\} \tag{6}$$

and

$$n_\sigma = \frac{V}{(2\pi)^3 N_a} \sum_l \int \mathrm{d}^3\mathbf{k}\, f[\mathscr{E}(l, \mathbf{k}, \sigma)], \tag{7}$$

where

$$f(\mathscr{E}) = \{\exp[\beta(\mathscr{E} - \mathscr{E}_F)] + 1\}^{-1}$$

is the Fermi function, θ is the step function, V the volume, N_a the number of atoms, and ν is $+1$ if $\mathcal{A}$ is associated with an electron surface and -1 if with a hole surface. The domain of integration in (6) is an area containing $\mathcal{A}$ in the extremal section and in (7) the volume of the Brillouin zone. From (6) and (7) the pressure derivatives of $\mathcal{A}$ and n_σ at $T = 0$ are given as

$$\begin{aligned}\frac{\mathrm{d}\mathcal{A}}{\mathrm{d}P} &= \int \mathrm{d}^2\mathbf{k}\, \nu\left[\frac{\mathrm{d}\mathscr{E}_F}{\mathrm{d}P} - \frac{\mathrm{d}\mathscr{E}(l', \mathbf{k}, \sigma)}{\mathrm{d}P}\right]\delta\{\nu[\mathscr{E}_F - \mathscr{E}(l', \mathbf{k}, \sigma)]\} \\ &= \frac{2\pi\nu m_c}{\hbar^2}\left[\frac{\mathrm{d}\mathscr{E}_F}{\mathrm{d}P} - \left\langle\frac{\mathrm{d}\mathscr{E}(l', \mathbf{k}, \sigma)}{\mathrm{d}P}\right\rangle_{\mathcal{A}}\right]\end{aligned} \tag{8}$$

and

$$\begin{aligned}\frac{\mathrm{d}n_\sigma}{\mathrm{d}P} &= \frac{V}{(2\pi)^3 N_a}\sum_l \int \mathrm{d}^3\mathbf{k}\left[\frac{\mathrm{d}\mathscr{E}_F}{\mathrm{d}P} - \frac{\mathrm{d}\mathscr{E}(l, \mathbf{k}, \sigma)}{\mathrm{d}P}\right]\delta[\mathscr{E}_F - \mathscr{E}(l, \mathbf{k}, \sigma)] + n_\sigma^{\mathrm{eff}}\frac{\mathrm{d}\ln V}{\mathrm{d}P} \\ &= N_\sigma\left[\frac{\mathrm{d}\mathscr{E}_F}{\mathrm{d}P} - \left\langle\frac{\mathrm{d}\mathscr{E}(l, \mathbf{k}, \sigma)}{\mathrm{d}P}\right\rangle_\sigma\right] + n_\sigma^{\mathrm{eff}}\frac{\mathrm{d}\ln V}{\mathrm{d}P},\end{aligned} \tag{9}$$

where m_c is the cyclotron effective mass associated with area $\mathcal{A}$, N_σ is the σ-spin density of states per atom at $\mathscr{E}_F$ and $\langle\ \rangle_{\mathcal{A}}$, $\langle\ \rangle_\sigma$ signify orbital and σ-spin Fermi surface averages, respectively. It has been assumed that the electron–phonon and electron–magnon enhancement factors can be ignored in the band-structure parameters entering (8) and (9). The parameter n_σ^{eff} would be equal to n_σ if the pressure derivatives of the limits of integration (i.e. the boundaries of the Brillouin zone) could be ignored. A general explicit expression for n_σ^{eff}, which is defined implicitly through (7) and (9), will not be needed in the present analysis. Eliminating $d\mathscr{E}_F/dP$ from (8) and (9) and making use of the fact that $2\,dn_\sigma = \sigma(dn_\uparrow - dn_\downarrow)$, from the conservation of the total number of particles $n = n_\uparrow + n_\downarrow$, we obtain

$$\frac{d\ln\mathscr{A}}{dP} = \frac{2\pi\nu m}{\hbar^2\mathscr{A}}\left[\frac{\sigma(n_\uparrow - n_\downarrow)}{2N_\sigma}\frac{d\ln\zeta}{dP} - \frac{n_\sigma^{\text{eff}}}{N_\sigma}\frac{d\ln V}{dP} + \left\langle\frac{d\mathscr{E}(l,\mathbf{k},\sigma)}{dP}\right\rangle_\sigma - \left\langle\frac{d\mathscr{E}(l',\mathbf{k},\sigma)}{dP}\right\rangle_{\mathscr{A}}\right], \tag{10}$$

where $\zeta = (n_\uparrow - n_\downarrow)/n$ is the spin polarization.

To make further progress it is now assumed that the quasi-particle energies in the ferromagnetic state may be derived approximately from the paramagnetic band structure (i.e. $\zeta = 0, T = 0$) by introducing an appropriate rigid exchange splitting proportional to the spin polarization, i.e. $\Delta = (n_\uparrow - n_\downarrow)I$, for the d(LCAO)-states and a zero splitting for the sp(OPW)-states, before sp–d hybridization (Hodges, Ehrenreich & Lang 1966; Ehrenreich & Hodges 1968; Zornberg 1970). This model, which is considerably more realistic than the simple Stoner theory for $T = 0$ defined in Section 6.2, is reasonably consistent with the results of recent SDF–LSD band-structure calculations in Fe and Ni (see Section 6.4). In our assumed model the quasi-particle energies $\mathscr{E}(l, \mathbf{k}, \sigma)$ may be regarded as functions of the volume V and the d-exchange splitting $\Delta = n\zeta I$ so that

$$\frac{d\mathscr{E}(l,\mathbf{k},\sigma)}{dP} = \frac{\partial\mathscr{E}(l,\mathbf{k},\sigma)}{\partial\ln V}\frac{d\ln V}{dP} + \frac{\partial\mathscr{E}(l,\mathbf{k},\sigma)}{\partial\ln\Delta}\left(\frac{d\ln I}{dP} + \frac{d\ln\zeta}{dP}\right). \tag{11}$$

Combining (10) and (11) the pressure derivative of $\mathscr{A}$ can then be expressed, in terms of the present band model, in the following form

$$\frac{d\ln\mathscr{A}}{dP} = \left(\frac{d\ln\mathscr{A}}{dP}\right)_{\text{nm}} + \left(\frac{d\ln\mathscr{A}}{dP}\right)_{\text{m}}, \tag{12a}$$

where

$$\left(\frac{d\ln\mathscr{A}}{dP}\right)_{\text{nm}} = \frac{2\pi\nu m_c}{\hbar^2\mathscr{A}}\frac{d\ln V}{dP}\left[\left\langle\frac{\partial\mathscr{E}(l,\mathbf{k},\sigma)}{\partial\ln V}\right\rangle_\sigma - \left\langle\frac{\partial\mathscr{E}(l',\mathbf{k},\sigma)}{\partial\ln V}\right\rangle_{\mathscr{A}} - \frac{n_\sigma^{\text{eff}}}{N_\sigma}\right], \tag{12b}$$

$$\left(\frac{d\ln\mathscr{A}}{dP}\right)_{\text{m}} = \frac{\pi\nu m_c\sigma(n_\uparrow - n_\downarrow)}{\hbar^2\mathscr{A}N_\sigma}\left[\frac{d\ln\zeta}{dP} - N_\sigma I\Lambda\left(\frac{d\ln\zeta}{dP} + \frac{d\ln I}{dP}\right)\right], \tag{12c}$$

$$\Lambda = -2\sigma\left[\left\langle\frac{\partial\mathscr{E}(l,\mathbf{k},\sigma)}{\partial\Delta}\right\rangle_\sigma - \left\langle\frac{\partial\mathscr{E}(l',\mathbf{k},\sigma)}{\partial\Delta}\right\rangle_{\mathscr{A}}\right]. \tag{12d}$$

The first term on the rhs of (12a) is that expected for a non-magnetic metal having a quasi-particle spectrum $\mathscr{E}_{l,\mathbf{k}} = \mathscr{E}(l, \mathbf{k}, \sigma)$ and Fermi level $\mathscr{E}_F$. The experimental evidence available thus far in the simple as well as the transition metals (Griesson, Lee & Stanley 1977a) indicates that the values of $(\mathrm{d}\ln\mathscr{A}/\mathrm{d}P)_{\mathrm{nm}}$ for corresponding sheets of the Fermi surface for two metals (1 and 2) having similar overall Fermi surfaces, scale approximately according to the expression

$$\left(\frac{\mathrm{d}\ln\mathscr{A}}{\mathrm{d}P}\right)_{\mathrm{nm}2} \Big/ \left(\frac{\mathrm{d}\ln\mathscr{A}}{\mathrm{d}P}\right)_{\mathrm{nm}1} = \frac{m_2\mathscr{A}_1}{m_1\mathscr{A}_2}\left(\frac{\partial\ln V}{\partial P}\right)_2 \Big/ \left(\frac{\partial\ln V}{\partial P}\right)_1, \tag{13}$$

a result which is consistent with (12b) if the quantity in the square brackets in (12b) is the same in both metals.

The last term in (12a) is unique to the ferromagnetic metals and depends on the pressure dependence of the spin polarization and interaction parameter I. The term containing Λ in (12c) is usually small compared to the first and vanishes if the orbital and σ-spin Fermi surface averages in (12d) are the same. The condition $\Lambda = 0$ is always true if the exchange splitting near the Fermi surface is uniform (i.e. as in the simple Stoner theory) but can hold in more general cases. It is noted that since $\partial\mathscr{E}(l, \mathbf{k}, \sigma)/\partial\Delta$ is equal to $(-\sigma/2)$ for pure d-states and zero for pure s–p-states then $|\Lambda|$ is in general smaller than unity.

Our analysis given above provides a basis for understanding the recent studies of the pressure dependence of parts of the Fermi surface of Ni (Anderson, Heimann, Schirber & Stone 1976; Vinokurova, Gaputchenko & Itskevich 1977) and Fe (Section 6.5). In Ni the magnitude of $\mathrm{d}\ln\mathscr{A}/\mathrm{d}P$ for the copper-like necks of the $\uparrow$-spin Fermi surface (which are of particular interest in Section 6.5.2) is found to be unusually small, two or three times smaller than that for the corresponding necks in Cu (see Table 6.3, p. 268). This result may be understood at least qualitatively in terms of a partial cancellation in Ni of the *positive* non-magnetic term by the *negative* magnetic term (which is of course absent in Cu) in equation (12a). From the parameters given in Tables 6.2 and 6.3 (pp. 267, 268) the magnitude of the magnetic term (12c) is estimated to be $(\mathrm{d}\ln\mathscr{A}/\mathrm{d}P)_{\mathrm{m}} \approx -(1.9\text{–}2.2)\times10^{-6}\ \mathrm{bar}^{-1}$ (the range in values corresponds to the range $\Lambda = 0$ to 1 in (12c)) and to account for the observed value of $\mathrm{d}\ln\mathscr{A}/\mathrm{d}P$, given in Table 6.3, we require $(\mathrm{d}\ln\mathscr{A}/\mathrm{d}P)_{\mathrm{nm}} \approx (2.5\text{–}3.0)\times10^{-6}\ \mathrm{bar}^{-1}$. This is similar in magnitude and sign to the value $(\mathrm{d}\ln\mathscr{A}/\mathrm{d}P)_{\mathrm{nm}2} \approx 3.6\times10^{-6}\ \mathrm{bar}^{-1}$ obtained from (13) taking 1 = Cu, 2 = Ni and parameters m_c and $\mathscr{A}$ for the corresponding neck surfaces in the two metals. A more quantitative analysis of the experimental data in Ni must await a detailed band-structure calculation, in particular of the non-magnetic term in (12a), as well as a more accurate determination of the pressure derivative of the spin polarization.

A study of the pressure dependence of a part of the Fermi surface of Fe, which can be understood quantitatively in terms of the present analysis, will be discussed under Section 6.5 in connection with an investigation of the variation of the Fermi surface with temperature.

6.3.6 Information from the temperature dependence of the Fermi surface

The studies described thus far have been concerned with properties of the ferromagnetic ground state. In this section we consider the information regarding the nature of magnetic excitations, which may be obtained from studies of the temperature dependence of the Fermi surface.

6.3.6.1 Thermodynamic properties of itinerant ferromagnets at low temperatures

The thermal properties of ferromagnetic metals at low temperatures have been described in terms of three types of elementary excitations from the ground state, single-particle (as in the Landau theory of normal Fermi liquids), phonon, and magnon excitations (Herring & Kittel 1951; Edwards 1962; Izuyama & Kubo 1964; Kondratenko 1965; Cornwell 1965; Herring 1966). Treating interactions between excitations in a self-consistent field approximation (i.e. in terms of Landau interaction functions) the energies of the quasi-particles and of (acoustic) spin-waves, can be expressed, respectively, in the forms (Izuyama & Kubo 1964; Herring 1966; Edwards 1974)

$$\mathscr{E}(l, \mathbf{k}, \sigma) = \mathscr{E}_0(l, \mathbf{k}, \sigma) + \sum_{l', \mathbf{k}', \sigma'} f_{\text{ee}}(l, \mathbf{k}, \sigma; l', \mathbf{k}', \sigma')\delta n(l', \mathbf{k}', \sigma')$$

$$+ \sum_{\alpha, \mathbf{q}} f_{\text{e}\alpha}(l, \mathbf{k}, \sigma; \mathbf{q})\upsilon_\alpha(\mathbf{q}), \tag{14a}$$

$$\hbar\omega(\mathbf{q}) = \hbar\omega_0(\mathbf{q}) + \sum_{l, \mathbf{k}, \sigma} f_{\text{em}}(l, \mathbf{k}, \sigma; \mathbf{q})\delta n(l, \mathbf{k}, \sigma)$$

$$+ \sum_{\alpha, \mathbf{q}'} f_{\text{m}\alpha}(\mathbf{q}, \mathbf{q}')\upsilon_\alpha(\mathbf{q}'), \tag{14b}$$

where $\mathscr{E}_0(l, \mathbf{k}, \sigma)$ and $\hbar\omega_0(\mathbf{q})$ are the corresponding energies at $T = 0$, $\delta n(l, \mathbf{k}, \sigma)$ is the deviation in the quasi-particle occupation number $n(l, \mathbf{k}, \sigma)$ from the ground state, $\upsilon_\alpha(\mathbf{q})$ gives the number of excitations present of wave-vector $\mathbf{q}$ and type α, where α designates one of the possible acoustic phonon or magnon modes, and f_{ee}, $f_{e\alpha}$, and $f_{m\alpha}$ are Landau-interaction functions describing interactions between the various types of excitations (e, m and p will designate electron-quasi-particle (or simply particle), magnon, and phonon excitations, respectively). From (14a,b), certain general properties of the interaction functions, the form of $\omega_0(\mathbf{q})$, and the known statistical nature of the excitations, it has been shown that in thermal equilibrium at a low temperature T, the quasi-particle and magnon energies are given, respectively, by the relations (Izuyama & Kubo 1964, Herring 1966, Edwards 1974)

$$\mathscr{E}(l, \mathbf{k}, \sigma) = \mathscr{E}_0(l, \mathbf{k}, \sigma) + \mathscr{E}_{ee}T^2 + \mathscr{E}_{em}T^{\frac{5}{2}} + \mathscr{E}_{ep}T^4, \tag{15a}$$

$$\hbar\omega(\mathbf{q}) = \mathscr{E}_g + D(q^2 + C_1q^4 + \ldots), \tag{15b}$$

where

$$D = D_0 + D_{me}T^2 + D_{mm}T^{\frac{5}{2}} + D_{mp}T^4. \tag{15c}$$

The temperature-dependent terms in (15a) arise from the particle–particle, particle–magnon and particle–phonon interactions, respectively, while the correction terms in (15c) to the $T = 0$ spin-wave stiffness D_0 arise from corresponding magnon–particle, magnon–magnon and magnon–phonon interactions. In (15b) $\mathscr{E}_g$ is an energy gap defined below (see (17b)) and the weakly temperature-dependent parameter C_1 measures the strength of the leading correction term to the quadratic spin-wave dispersion curve for small q. In arriving at (15a) and (15c) the effects of interband transitions and magnetic anisotropy have been essentially ignored. The latter can be shown to give rise to a very weak $T^{\frac{3}{2}}$ term in $\mathscr{E}(l, \mathbf{k}, \sigma)$ which can be ignored in the present discussion (Lonzarich 1974). The effect of interband transitions, which may be important if there is a significant overlap of ↑ and ↓ spin sheets of the Fermi surface, has not been analysed in detail. It is also noted that a large $T^{\frac{3}{2}}$ term in $\mathscr{E}(l, \mathbf{k}, \sigma)$ was originally obtained in simple RPA theory in which the exchange splitting is found to be proportional to the total magnetization at finite temperatures. The incorrectness of this result was discussed under Section 6.2..

Of particular interest in the following discussion are those elementary excitations which involve a change in the total spin of the many-particle system; i.e. magnons and Stoner (spin-flip single-particle) excitations. The magnetic excitation spectrum for a ferromagnet characterized by a single parabolic energy band (for each spin), a uniform exchange splitting Δ, and a partial polarization ($\zeta < 1$), is illustrated in Fig. 6.1. Magnons are well-defined excitations up to the boundary of the spectrum of Stoner excitations (which forms a continuum since for each $\mathbf{q}$ there exists a large number of wave-vectors $\mathbf{k}$ for which states $\psi(\mathbf{k}, \uparrow)$ and $\psi(\mathbf{k}+\mathbf{q}, \downarrow)$ are occupied and empty, respectively, at $T=0$). Within the Stoner continuum, collective modes are expected to be in the nature of short-lifetime spin-density fluctuations similar to paramagnons in nearly ferromagnetic metals. At low temperatures only well-defined spin-waves of low energy and small q and Stoner excitations of low energy and wave-vectors in the range $q_{min} < q < q_{max}$ (which correspond to transfers of particles from the vicinity of the $\uparrow$-spin to the vicinity of the $\downarrow$-spin part of the Fermi surface) are important. It is noted that if more than one energy band (for each spin) is considered, Stoner excitations of both low energy *and* low momentum can arise from interband transitions. in addition 'optical' as well as 'acoustic' magnon branches are possible. The former are not expected to be important at very low temperatures and are therefore ignored in the following discussions (see, however, the end of Section 6.5).

In terms of our present model of undamped spin-wave and single-particle excitations (whose mutual interactions are treated in terms of Landau interaction functions) the deviation of the spin polarization ζ

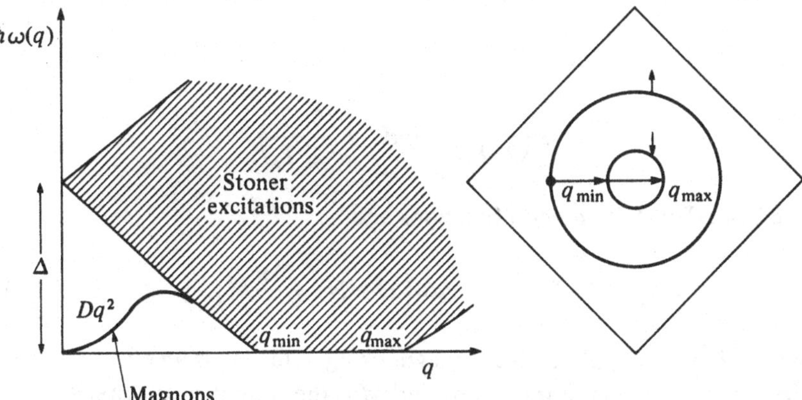

Fig. 6.1. Magnetic excitation spectrum for an itinerant ferromagnet in the random phase approximation. A single parabolic band (for each spin) and a partial spin polarization ($\zeta < 1$) is assumed. Δ is the exchange splitting of the energy bands and q_{max}, q_{min} are the sum and difference, respectively, of the $\uparrow$ and $\downarrow$ spin Fermi radii (as indicated in the drawing of a central section of the Fermi surface on the right).

from its $T = 0$ value ζ_0 may be expressed as

$$\delta\zeta = \delta\zeta_{sw} + \delta\zeta_{sp}, \tag{16a}$$

where

$$\delta\zeta_{sw} = -\frac{2V}{(2\pi)^3 nN_a} \int d^3\mathbf{q}\, \upsilon(\mathbf{q}), \tag{16b}$$

$$\delta\zeta_{sp} = \frac{1}{n}\delta(n_\uparrow - n_\downarrow). \tag{16c}$$

The components $\delta\zeta_{sw}$ and $\delta\zeta_{sp}$ arise from spin-wave and single-particle excitations, respectively, $\upsilon(\mathbf{q})$ is the Bose function for magnons of energy $\hbar\omega(\mathbf{q})$ and n_σ is defined in (7). Note that n_σ in this definition is only equal to the *total* number of σ-spin *electrons* per atom if spin-fluctuations in the ground state can be ignored and if $\delta\zeta_{sw} = 0$.

The calculation of $\delta\zeta$ from (16a,b,c and 7) must, strictly speaking, be carried out self-consistently since the quasi-particle and magnon energies required to evaluate $\delta\zeta_{sw}$ and $\delta\zeta_{sp}$ depend through (14a,b) on the actual magnitudes of $\delta\zeta_{sw}$ and $\delta\zeta_{sp}$ themselves. However, if the spin-wave parameters $D, C_1, \ldots$ are assumed known at a temperature T then from (15b) and (16b) it is readily shown that (see e.g. Argyle, Charap & Pugh, 1963; Aldred & Froehle 1972)

$$\frac{\delta\zeta_{sw}}{\zeta_0} = -\frac{2V}{nN_a\zeta_0}\left(\frac{k_BT}{4\pi D}\right)^{\frac{3}{2}}\left[F\left(\frac{3}{2}, \frac{T}{T_g}\right) - 15\pi C_1\left(\frac{k_BT}{4\pi D}\right)F\left(\frac{5}{2}, \frac{T}{T_g}\right) + \ldots\right], \tag{17a}$$

where

$$F\left(\frac{3}{2}, \frac{T}{T_g}\right) \quad \text{and} \quad F\left(\frac{5}{2}, \frac{T}{T_g}\right)$$

are Bose–Einstein integral functions and

$$k_BT_g = \mathscr{E}_g \approx g\mu_B\left(H + H_A + \frac{4\pi}{3}M\right), \tag{17b}$$

where H is the applied plus demagnetizing field, H_A is an effective field arising from magnetic anisotropy and M is the saturation magnetization. In the detailed analysis given in Section 6.5.3 it is assumed (following Aldred 1975) that D has the form

$$D = D_0 - D_1T^2, \tag{17c}$$

where D_0 is the $T=0$ spin-wave stiffness and the term D_1T^2 takes into account approximately the effects of many-body interactions at low temperatures (see (15c)).

We now consider the single-particle contribution $\delta\zeta_{sp}$ (see (16c)). It follows from (7) that the deviation of n_σ from its $T=0$ value can at a low temperature T be expressed in the form

$$\delta n_\sigma = \frac{\pi^2}{6}N'_\sigma(k_BT)^2 + N_\sigma[\delta\mathscr{E}_F - \langle\delta\mathscr{E}(l, \mathbf{k}, \sigma)\rangle_\sigma], \tag{18a}$$

where N'_σ is the energy derivative of the σ-spin density of states N_σ at the Fermi level (N_σ is assumed to be non-zero for both spins), $\langle\ \rangle_\sigma$ designates a σ-spin Fermi surface average as defined in Section 6.3.5 and $\delta\mathscr{E}(l, \mathbf{k}, \sigma)$ is the deviation of the quasi-particle energy from its $T=0$ value (see (14a)) which can be expressed in terms of the particle–particle (ee) and particle–magnon–phonon (emp) contributions as

$$\delta\mathscr{E}(l, \mathbf{k}, \sigma) = \delta_{ee}\mathscr{E}(l, \mathbf{k}, \sigma) + \delta_{emp}\mathscr{E}(l, \mathbf{k}, \sigma). \tag{18b}$$

From (18a) and the requirement $\delta(n_\uparrow + n_\downarrow) = 0$ we obtain

$$\frac{\delta\zeta_{sp}}{\zeta_0} = \frac{N_{\uparrow\downarrow}}{(n_\uparrow - n_\downarrow)}\left[\frac{\pi^2}{6}(k_BT)^2\left(\frac{N'_\uparrow}{N_\uparrow} - \frac{N'_\downarrow}{N_\downarrow}\right)\right.$$

$$\left. + \langle\delta\mathscr{E}(l, \mathbf{k}, \downarrow)\rangle_\downarrow - \langle\delta\mathscr{E}(l, \mathbf{k}, \uparrow)\rangle_\uparrow\right], \tag{19}$$

where

$$N_{\uparrow\downarrow}^{-1} = (N_\uparrow^{-1} + N_\downarrow^{-1})/2.$$

To make further progress we now treat the particle–particle contribution on the rhs of (18b) approximately in terms of the band-structure model introduced in Section 6.3.5, according to which

$$\delta_{ee}\mathscr{E}(l, \mathbf{k}, \sigma) \simeq \frac{\partial\mathscr{E}(l, \mathbf{k}, \sigma)}{\partial\Delta} nI\delta\zeta_{sp}, \tag{20a}$$

where $\Delta = I(n_\uparrow - n_\downarrow)$ is the exchange splitting of the d-bands before s–p–d hybridization. If in addition the T^4 particle–phonon contribution in (18b) is ignored and the particle–magnon term in (18b) is taken to be of the order of magnitude $k_BT\delta\zeta_{sw}/\zeta_0$ (ignoring interband transition; Edwards 1974; Lonzarich 1974; Liu 1978), then from (19) the self-consistent solution for $\delta\zeta_{sp}$ is given by the relation

$$\frac{\delta\zeta_{sp}}{\zeta_0} = \frac{N_{\uparrow\downarrow}\Gamma}{(n_\uparrow - n_\downarrow)}\left[\frac{\pi^2}{6}(k_BT)^2\left(\frac{N'_\uparrow}{N_\uparrow} - \frac{N'_\downarrow}{N_\downarrow}\right) + \mathscr{K}k_BT\frac{\delta\zeta_{sw}}{\zeta_0}\right], \tag{20b}$$

where Γ is the exchange or Stoner enhancement factor,

$$\Gamma^{-1} = 1 - N_{\uparrow\downarrow} I \left\langle -\frac{2\sigma\, \partial \mathscr{E}(l, \mathbf{k}, \sigma)}{\partial \Delta} \right\rangle, \tag{20c}$$

and $\langle\ \rangle$ designates an average over the entire $\uparrow$ and $\downarrow$ spin parts of the Fermi surface. The parameter $\mathscr{K}$ in (20b) is of order unity and $\langle -2\sigma\, \partial\mathscr{E}(l, \mathbf{k}, \sigma)/\partial\Delta \rangle$ in (20c) is in general between zero and one and is equal to one if the quasi-particle states on the Fermi surface are d-like in character, or in simple Stoner theory. It can be readily shown within the present theoretical framework that the quantity $(N_{\uparrow\downarrow}\Gamma)$ may be determined from the Pauli spin susceptibility χ_P according to the relation

$$\chi_P = \frac{N_a}{2V} g^2 \mu_B^2 N_{\uparrow\downarrow} \Gamma, \tag{21}$$

where g is the Fermi surface average of the band g-factor.

If $\langle -2\sigma\, \partial\mathscr{E}(l, \mathbf{k}, \sigma)/\partial\Delta \rangle$ is set equal to one and the particle–magnon interaction is ignored ($\mathscr{K} = 0$ in (20b)), then (20b) reduces to the well-known Stoner T^2 term in the temperature dependence of the magnetization (Thompson *et al.* 1964). In Fe and Ni the magnon induced term in $\delta\zeta_{sp}/\zeta_0$ in (20b) is much smaller than $\delta\zeta_{sw}/\zeta_0$ itself at liquid helium temperatures, but is not necessarily very much smaller than the Stoner T^2 term. Furthermore it is noted that in weak itinerant ferromagnets in which the factor $\Gamma/(n_\uparrow - n_\downarrow)$ can be very large, the spin-wave induced term in (20b) can be of the same order of magnitude as $\delta\zeta_{sw}/\zeta_0$ in the liquid helium temperature range.

6.3.6.2 Temperature dependence of a cross-sectional area of the Fermi surface

In this section it is shown that the temperature dependence of a cross-sectional area of the Fermi surface provides information about the single-particle component $\delta\zeta_{sp}$ in $\delta\zeta$ without interference from the spin-wave contribution $\delta\zeta_{sw}$ (see (16a), (17a), (20b); Edwards 1974 and Lonzarich & Gold 1974a). This result, which provides the basis of a new technique for investigating magnetic excitations in ferromagnetic metals (Section 6.5), is a consequence of the following points: (i) at low temperatures $\delta\zeta_{sp}$ is associated with a repopulation of $\uparrow$- and $\downarrow$-spin quasi-particle states precisely at the Fermi surface (which leads to changes in the dimensions of the Fermi surface), whereas $\delta\zeta_{sw}$ is connected with an infinitesimal variation of the occupation numbers for states extending over the entire singly occupied volume of **k**-space (see e.g. Herring 1966), with a negligible repopulation precisely on the Fermi surface itself; (ii) the effect (beyond that already taken into account in the derivation of $\delta\zeta_{sp}$, see (20b)) of thermally excited magnons on the energies of the quasi-particles and thus, indirectly, on the Fermi surface is usually very small and ignorable at low temperatures.

From (6), (18a) and (19) the variation $\delta\mathscr{A}$, of a cross-sectional area $\mathscr{A}$ associated with band l' of the σ-spin part of the Fermi surface, from its $T=0$ value $\mathscr{A}_0$ can be expressed as

$$\frac{\delta\mathscr{A}}{\mathscr{A}_0}=\frac{2\pi\nu m_c}{\hbar^2\mathscr{A}_0}\left[\frac{\sigma(n_\uparrow-n_\downarrow)}{2N_\sigma}\frac{\delta\zeta_{sp}}{\zeta_0}-\frac{\pi^2N'_\sigma(k_BT)^2}{6N_\sigma}\right.$$
$$\left.+\langle\delta\mathscr{E}(l,\mathbf{k},\sigma)\rangle_\sigma-\langle\delta\mathscr{E}(l',\mathbf{k},\sigma)\rangle_{\mathscr{A}}\right],\tag{22}$$

where the various parameters are defined in Sections 6.3.5 and 6.3.6.1. If interactions between excitations are treated according to the approximations leading to (20a,b,c) then (22) becomes

$$\frac{\delta\mathscr{A}}{\mathscr{A}_0}=\frac{2\pi\nu m_c}{\hbar^2\mathscr{A}_0}\left[\frac{\sigma(n_\uparrow-n_\downarrow)(1-\Lambda N_\sigma I)}{2N_\sigma}\frac{\delta\zeta_{sp}}{\zeta_0}\right.$$
$$\left.-\frac{\pi^2N'_\sigma(k_BT)^2}{6N_\sigma}+\mathscr{K}_Ak_BT\frac{\delta\zeta_{sw}}{\zeta_0}\right],\tag{23}$$

where Λ is defined by (12d) and $\mathscr{K}_A$ is a parameter of order unity closely associated with $\mathscr{K}$ in (20b). If $\mathscr{K}$, Λ and $\mathscr{K}_A$ are set to zero in (20b) and (23), then (23) reduces to the result given by Edwards (1974).

In the analysis leading to our main results (20b) and (23) the temperature dependence of the volume and of the interaction parameter I, and the effects of the quasi-particle–phonon interaction have been ignored. Under the experimental conditions described in Section 6.5 these effects lead to area changes which are smaller than the experimental resolution. The latter also holds for the last two terms in (23), so that for the purposes of the analysis given under Section 6.5 we may write simply

$$\frac{\delta\mathscr{A}}{\mathscr{A}_0}=\frac{\pi\nu\sigma m_c(n_\uparrow-n_\downarrow)(1-\Lambda N_\sigma I)}{\hbar^2\mathscr{A}_0N_\sigma}\frac{\delta\zeta_{sp}}{\zeta_0}\tag{24}$$

This result, which provides the direct relationship between $\delta\mathscr{A}/\mathscr{A}_0$ and $\delta\zeta_{sp}/\zeta_0$, is similar to that obtained for the magnetic term (12c) for the *pressure* derivative of the cross-sectional area $\mathscr{A}$.

6.3.6.3 Temperature dependence of dHvA phase

The variation of a cross-sectional area $\mathscr{A}$ with temperature may be determined from the variation (at constant pressure and applied magnetic field) of the phase Φ of the fundamental component of $\tilde{M}$ in (3), which may be expressed as

$$\delta\Phi = \delta[2\pi(\eta-\gamma)+\psi(T)] + \frac{2\pi F}{B}\left[\frac{\delta\mathscr{A}}{\mathscr{A}_0} - \frac{4\pi(1-\mathscr{D})M}{B}\frac{\delta M}{M_0}\right], \quad (25)$$

a result analogous to (5) for the pressure derivative of the phase, except for the inclusion of a new temperature- (and field-) dependent phase term $\psi(T)$ which arises from higher order corrections to the conventional theory of the dHvA effect. The most important terms entering $\psi(T)$ (Lonzarich 1974; Lonzarich & Gold 1974b) have a linear temperature dependence over the experimental temperature range of interest and can therefore be easily distinguished from any T^2 term in $\delta\mathscr{A}/\mathscr{A}_0$ arising from the single particle component $\delta\zeta_{sp}/\zeta_0$ (see (24)). Apart from this, the various terms in (25) can be separated in principle by virtue of their different dependences on B. Thus from a consideration of both the temperature and field dependence of $\delta\Phi$ the variation of $\mathscr{A}$ with temperature can be determined unambiguously.

The effect of quasi-particle–magnon interactions on the dimensions of the Fermi surface is discussed in some detail under Sections 6.3.6.1 and 6.3.6.2; however, no corresponding theoretical analysis of the influence of magnons on the dHvA phase Φ itself has been carried out thus far. The following points, based on the fact that the cyclotron frequency is somewhat greater than the mean spin-wave frequency in the experimental temperature range (Section 6.5), should be noted: (i) the magnitude of the quantity $\delta M/M_0$ in (20) is expected to be somewhat different from the bulk-average change in magnetization; and (ii) for a non-spherical Fermi surface the area $\mathscr{A}$ appearing in the phase Φ may be expected to vary slightly with temperature by virtue of the fact that the effective orientation of **B** seen by the electrons in their cyclotron orbit will be slightly temperature dependent. The latter effect in particular can be shown to be ignorable under the experimental conditions described in Section 6.5.

We end this section with some brief comments regarding the special experimental precautions which are required to carry out reliable measurements of the temperature dependence of the dHvA phase. In contrast to the more conventional measurements of the *pressure* dependence of the phase, in the former the *amplitude* of the dHvA oscillations can vary by several orders of magnitude over the course of the experiment (i.e. between the lowest and highest temperatures studied). This difference leads to a number of experimental complications. For example, if the observed dHvA magnetization is in reality a superposition of two or more components having essentially the same frequency but different phases and amplitudes with different temperature dependence, then the effective phase of the total oscillatory magnetization as measured over a limited field range would appear to vary with temperature. Phase shifts of this kind can be quite important in practice and have been included symbolically in the function $\psi(T)$ in (25). Contributions to $\psi(T)$ arise from a variety of effects which have been investigated in detail and can now be taken into account, as discussed earlier. The main contributions originate from higher order corrections to the Lifshitz–Kosevich theory and to the conventional treatment of the field-modulation method for detecting $\tilde{M}$. Additional contributions arise from incomplete penetration of the a.c. and slowly varying d.c. fields, inhomogeneities of the dHvA frequency and magnetic induction over the sample, Shoenberg magnetic interactions, and effects connected with the approach to the quantum limit. Beyond this, to measure phase shifts reliably it is necessary to prevent movements of the specimen as the temperature is varied, to take proper account, by digital techniques, of the effect of interfering signals (whose temperature dependences are usually much different from that of the dHvA signal of interest), and to minimize noise and inaccuracies in the measurements of the magnetic field. With suitable precautions and refinements in detection technique and data analysis, it has recently been possible to measure shifts in the phase of the oscillatory magnetization arising from the majority neck in Ni to a precision of about 0.0003 cycles up to temperatures as high as 15 K, for $B \approx 70$ kG (Cooper & Lonzarich, unpublished). The temperature dependence of the Fermi surface of ferromagnetic metals may be investigated to higher temperatures in the future by means of quantum interference effects (Stark & Friedberg 1971), positron annihilation (see Kontrym–Sznajd *et al.* 1975 for a recent study in Ni below and above T_c) and angle-resolved photoemission.

6.4 The band structures and Fermi surfaces of iron and nickel

The band structures and Fermi surfaces of the iron-group ferromagnets have recently been discussed in comprehensive reviews by Gold (Fe, Ni and Co, 1974) and Stark (Ni, unpublished). In this section we summarize briefly the published theoretical and experimental information relating to the Fermi surfaces of Fe and Ni and then consider in greater detail the results of more recent studies which have helped to shed light on the existing ambiguities regarding in particular the Fermi surface of Fe.

The interpretation of the complex Fermi surface data in Fe and Ni has been guided to a considerable extent by the results of first-principles energy-band calculations, the most recent of which are those of Callaway & Wang (1977b) for Fe and Wang & Callaway (1977) for Ni. The essential features of these SDF self-consistent-field calculations are the use of (i) the Kohn, Sham and Gaspar (KSG; i.e. Xα with $\alpha = \frac{2}{3}$) or the von Barth–Hedin (vBH; i.e. essentially the LSD) approximations for the local exchange-correlation potential discussed in Section 6.2, and (ii) the tight-binding procedure for finding the eigenvalues of the one-particle Hamiltonian.

The energy bands along the principal symmetry directions obtained by these authors for Ni (fcc) and Fe (bcc) are illustrated in Figs. 6.2 and 6.3, respectively. The bold solid curves and lighter dashed curves correspond to the majority ($\uparrow$)- and minority ($\downarrow$)-spin carriers, respectively. The main features of the results are the more or less uniform exchange splitting of states having a strong d-character (which fall in the intermediate energy range in the figures) and a relatively small and even negative splitting of states with dominant s–p character (which appear primarily in the lower and upper portions of the figures). Callaway and Wang have used their band-structure models to calculate the magneton number ($n_\uparrow - n_\downarrow$), the

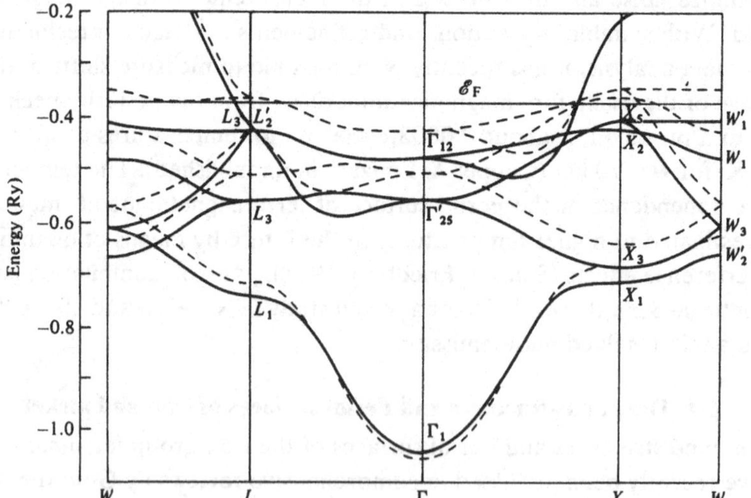

Fig. 6.2. The majority (solid lines) and minority (dashed lines) energy bands for fcc nickel calculated by means of the vBH approximation for the exchange-correlation potential and a tight-binding computational procedure. (After Wang & Callaway, 1977.)

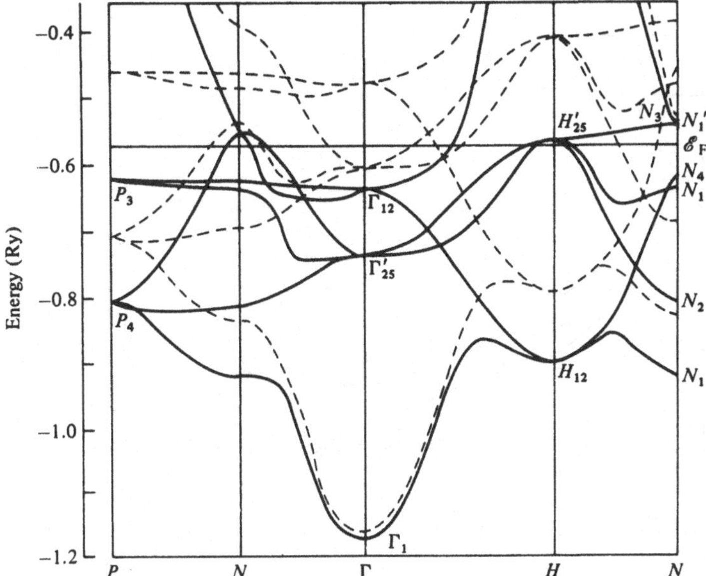

Fig. 6.3. The majority (solid lines) and minority (dashed lines) energy bands for bcc iron calculated by means of the KSG approximation for the exchange-correlation potential and a tight-binding computational technique. (After Callaway & Wang, 1977b.)

characteristic exchange splitting Δ, the total ↑ and ↓ spin density of states $N(\mathscr{E}_F)$ at the Fermi energy, the spin-density form factors, and the Fermi surface. Some of their numerical results for the KSG and vBH potentials are compared with experimental values in Table 6.1. It is seen that the calculated magneton numbers are in very reasonable agreement with experiment, particularly for the vBH potential. On the other hand the calculated densities of states at the Fermi level are much smaller than the experimental values, presumably because of the omission of the electron–phonon and electron–magnon interactions in the SDF–Xα or the SDF–LSD approximations, as discussed in Section 6.2.

The central (110) section of the Fermi surface of Ni calculated by Wang & Callaway (1977) using the vBH potential is illustrated in Fig. 6.4(*a*). The majority sheet (solid curve labelled sp↑) is very similar to the Fermi surfaces of the noble metals with the characteristic necks which contact the {111} zone faces near the points *L*. The minority surface (dashed curves) consists of an sp↓ sheet, which is a contracted version of the sp↑ surface, a major d↓ sheet and two d-hole pockets centred at *X*, labelled $X_{5\downarrow}$ and $X_{2\downarrow}$, respectively. Wang and Callaway have compared their theoretical Fermi surface model with Stark's unpublished experimental results illustrated in Fig. 6.4(*b*). When the effect of the spin–orbit interaction in lifting certain degeneracies is taken into account (Wang &

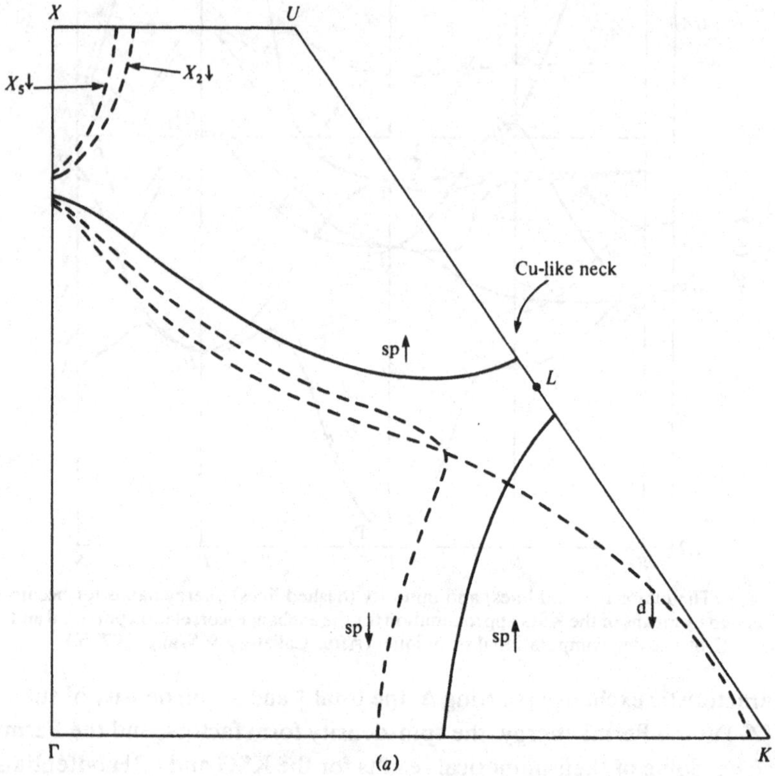

Fig. 6.4. Central (110) sections through the Fermi surface of nickel (*a*) as computed by Wang & Callaway (1977) using the vBH potential, and (*b*) as determined experimentally by Stark (see Wang & Callaway, 1977). The solid and broken lines correspond to the majority and minority spin carriers, respectively. The dotted line and the position of the thin solid line in (*b*) are dictated by the results of Tsui's (1967) dHvA study.

Callaway 1974), they find good overall agreement except for the apparent absence of the large $X_{2\downarrow}$ pocket predicted in their calculation. The existence of very tiny $X_{2\downarrow}$ pockets (see Gersdorf 1978 or Section 6.3.4) would not significantly alter the seriousness of this discrepancy which may arise (Callaway & Wang 1977a) from a significant defect in the SDF–Xα or SDF–LSD approximations for the description of the band structures of ferromagnetic metals.

We now consider the Fermi surface of Fe which is considerably less well understood. The (110) and (100) central sections of the Fermi surface calculated by Callaway & Wang (1977b) with the vBH potential, are illustrated in Fig. 6.5(*a*). The topology of the Fermi surface, particularly

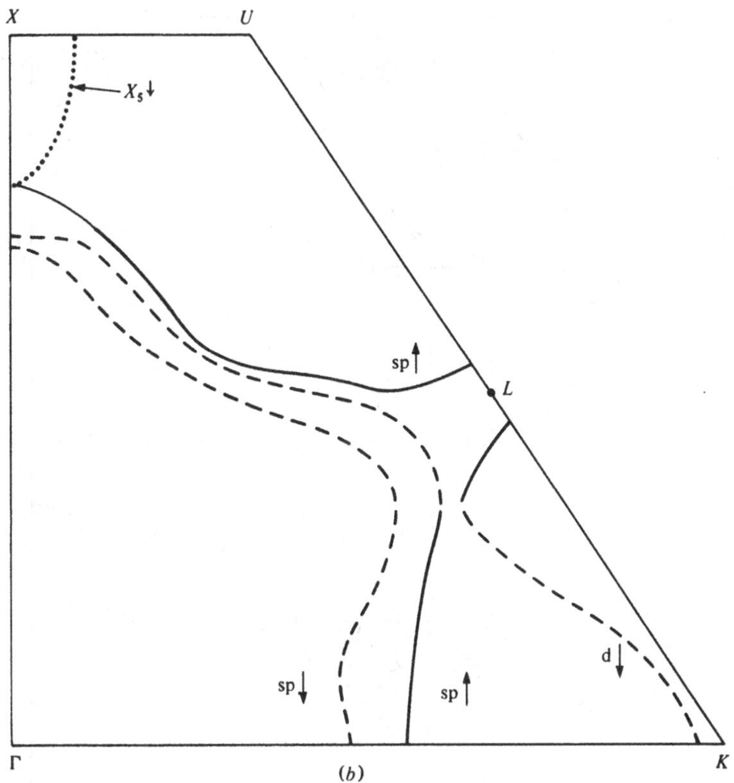

(b)

TABLE 6.1. *Comparison of experimental and band-structure values of the magneton number* $(n_\uparrow - n_\downarrow)$ *(in electrons atm*$^{-1}$*), the exchange splitting of states at the top of the* d*-band* Δ *(in eV), the total density of states at the Fermi energy (in states atm*$^{-1}$ *Ry*$^{-1}$*), and the difference between majority and minority spin densities at the nuclear site expressed as an effective hyperfine field (in* 10^3 *G). The calculated values refer to the KSG potential (i.e. the* Xα *approximation with* $\alpha = \frac{2}{3}$*) and the vBH potential (i.e. the LSD approximation). (After Callaway & Wang 1977a and references cited therein)*

	$(n_\uparrow - n_\downarrow)$	Δ	$N(\mathscr{E}_F)$	Hyperfine field
Fe KSG	2.30	2.68	15.37	−343
vBH	2.25	2.21	15.97	−237
Exp	2.12	(1.2–2.0)	27.37	−339.0(3)
Ni KSG	0.65	0.88	22.92	−69.7
vBH	0.58	0.63	25.45	−57.9
Exp	0.56	(0.3–0.5)	40.41	−76(1)

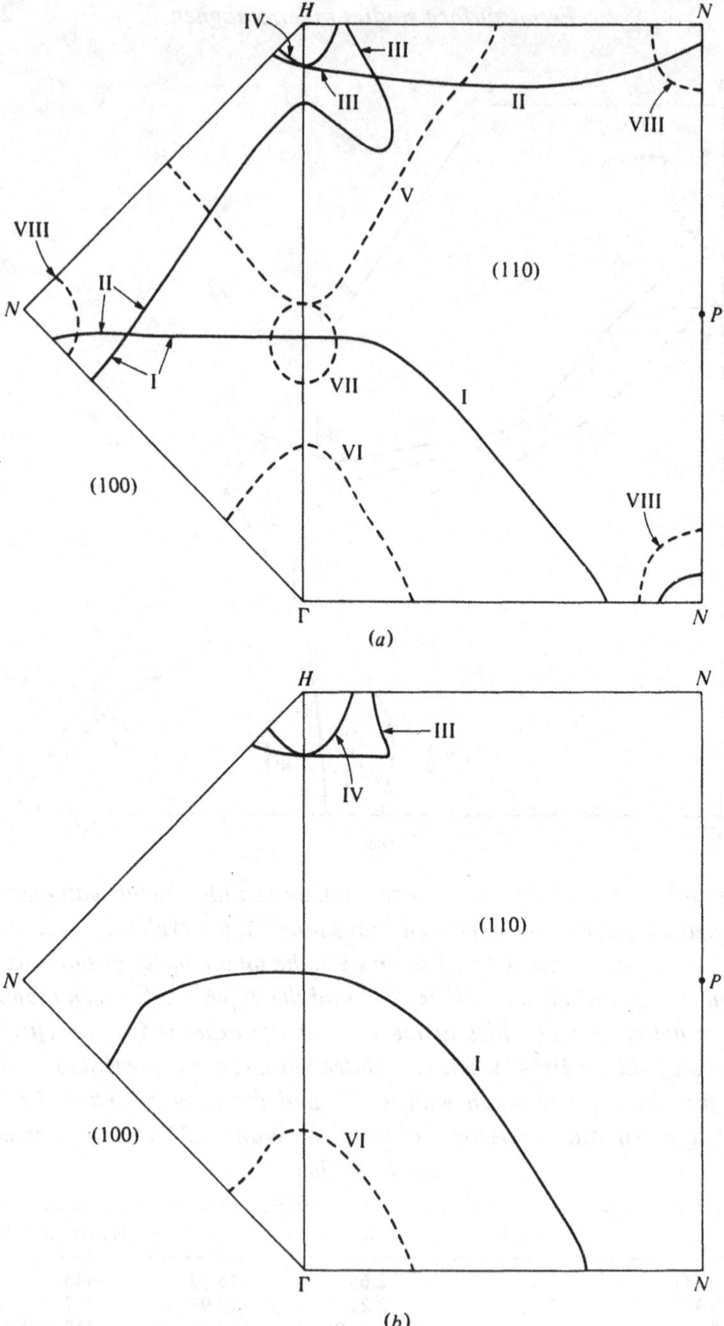

Fig. 6.5. Central (100) and (110) sections through the Fermi surface of iron (*a*) as computed by Callaway & Wang (1977b) using the vBH potential, and (*b*) as determined by Baraff (1973) from the unambiguous and complete dHvA frequency branches. The solid and broken lines correspond to the majority and minority spin sheets of the Fermi surface, respectively.

when the effect of the spin–orbit interaction is taken into account (Singh *et al.* 1975), is evidently quite complex. The majority surface consists of a large electron sheet (I), an arm-like hole surface extending along *HNH* (II), and two small hole pockets centred on *H* (III, IV). The minority surface on the other hand consists of a large hole octahedron centred on *H* (V), an electron octahedron centred on Γ (VI), electron balls along Γ*H* (VII) and finally a set of hole ellipsoids centred on the points *N* of the Brillouin zone (VIII). The minority surface is quite similar to the Fermi surface of molybdenum or tungsten; the principal difference being that the Γ*H* electron balls in Mo and W are larger and intersect the electron octahedron centred on Γ. Previous band-structure calculations for Fe (see references cited in Callaway & Wang 1977b) have yielded a variety of topological models which are significantly different from the one described above. On the whole there has been little general theoretical agreement regarding in particular the ordering of the energy bands in the vicinity of *N* and (to a lesser extent) the Γ*H* line. A discussion of a number of topological models which are plausible on the basis of past band-structure calculations has been given by Gold *et al.* (1971). In the following only those models which appear to be in closest accord with the present experimental information will be considered.

The Fermi surface of Fe has been investigated by the dHvA technique by Gold *et al.* (1971) and more recently by Baraff (1973) who discovered several new frequency branches and followed those previously reported over the entire $(1\bar{1}0)$ and (in some cases) (001) planes. Some of the observed frequency branches were sufficiently complete to permit an unambiguous determination of the Fermi surface radii and those parts of the Fermi surface whose dimensions may now be regarded as being firmly established are illustrated in Fig. 6.5(*b*). The remaining Fermi surface data, which are still in some respects ambiguous, will be discussed in terms of the three topological models A_1, A_2 and B illustrated in Figs. 6.6(*a*) and (*b*). Models A_1 and A_2 differ only in the size of the minority hole ellipsoids (surface VIII, see Fig. 6.6(*a*)) and models A_2 and B differ essentially in the sizes of the minority electron balls (VII) and the hole octahedron (V). In Figs. 6.6(*a*) and (*b*) the bold solid and dashed curves represent the unambiguous sheets of the Fermi surface (Fig. 6.5(*b*)) which serve as valuable constraints in the interpretation of the remaining experimental data. Models A_1 and A_2 are closest to those proposed somewhat tentatively by Baraff (1973) and Gold *et al.* (1971), respectively, on the basis of the information available to them, while model B is closest, in so far as the ordering along Γ*H* is concerned, to the theoretical model of Callaway & Wang (1977b). In the following it is shown that, of

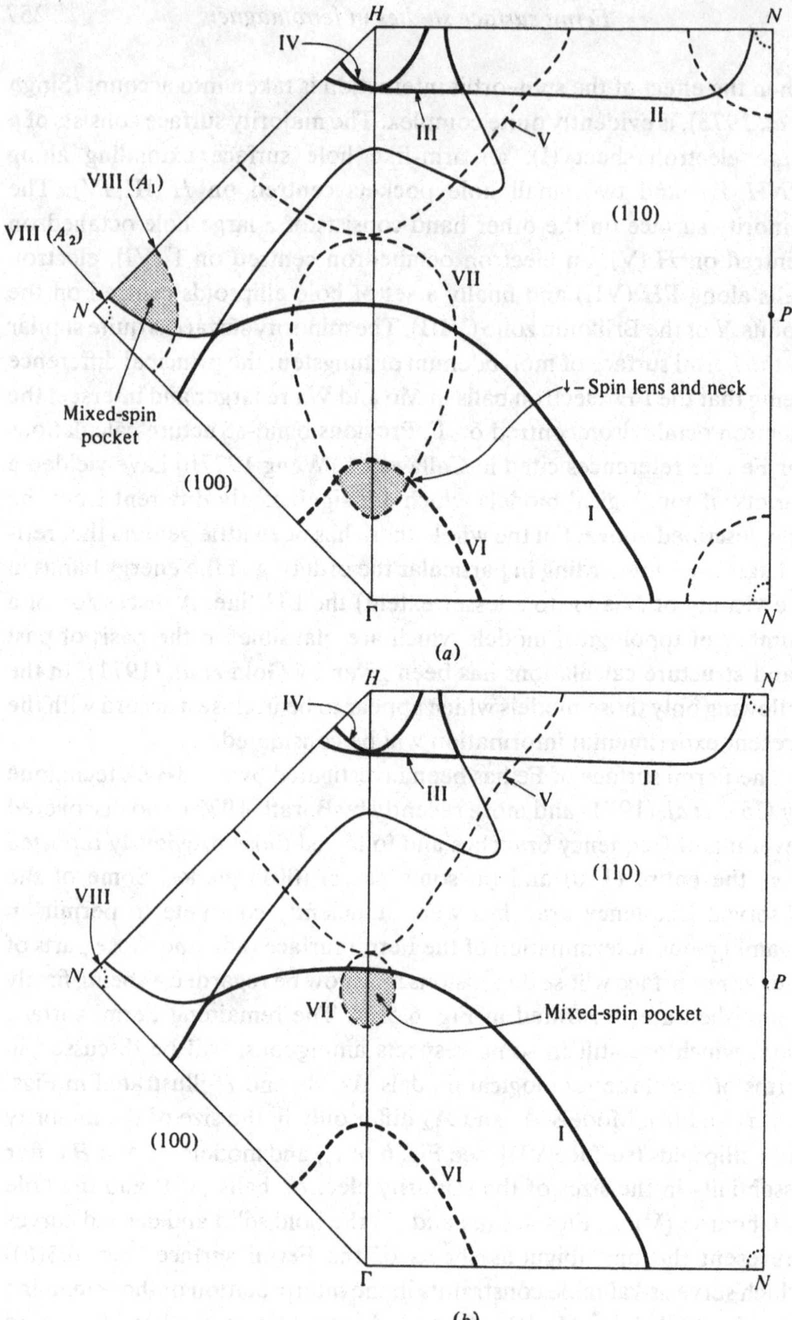

Fig. 6.6. Central (100) and (110) sections through the Fermi surface of iron (*a*) for models A_1 and A_2 and (*b*) for model *B*. Models A_1 and A_2 differ only in the size of the minority hole ellipsoid at *N* (i.e. the surfaces labelled VIII). The solid and dashed lines are for majority and minority sheets of the Fermi surface, respectively. The bold solid and dashed lines are those of Fig. 6.5(*b*).

the three, model B is on the whole in closest agreement with all the experimental and theoretical information available at this time.

Models A_1, A_2 and B provide alternative descriptions for the origins of the low-frequency ν, ε and η branches observed by several investigators in the dHvA and de Haas–Shubnikov (dHS) effects, and the very weak intermediate and high-frequency ζ, γ, ι and κ branches observed by Baraff below 0.3 K, for field orientations over narrow angular ranges near the [001] (for ζ) and the [110] (for γ, ι and κ) symmetry directions. To help shed light on the true origins of these branches a careful investigation has been carried out of the variation of the frequencies and amplitudes of the ν, ε and η oscillations for field orientations in the $(1\bar{1}0)$ plane (see Figs. 6.7–6.9). The ν-oscillations have been observed by several investigators (Gold 1965 and Lonzarich & Gold 1974b in the dHvA effect; Angadi & Fawcett 1973 and Coleman, Morris & Sellmyer 1973 in the dHS effect); however, due to experimental difficulties their frequency components remained until now unresolved. The ε and η branches were first investigated by Gold *et al.* (1971), whose results for the orientation dependence of the frequencies in the $(1\bar{1}0)$ plane are in good agreement with the present higher resolution study. The behaviour of the ν, ε and η frequency branches in the $(1\bar{1}0)$ plane (Figs. 6.7 and 6.8) as well as the peculiar orientation dependence of the amplitudes of the ε and η oscillations near [001] (Figs. 6.9(a), (b)) will be discussed in terms of models A_1, A_2 and B (Figs. 6.6(a), (b)). The possible interpretation of Baraff's ζ, γ, ι and κ branches are also discussed briefly in terms of these models.

The orientation dependence and symmetry of the ν branches are consistent with that expected for small and nearly spherical pockets of the Fermi surface in the vicinity of the points N of the Brillouin zone. The solid curves in Fig. 6.7 represent a three-parameter fit of the experimental data assuming the ν oscillations arise from ellipsoidal surfaces centred at N. The three parameters correspond to the three principal axes of the ellipsoids. In terms of model A_1, Baraff's ζ branch (~35 MG), which was observed within 26° of the [110], would arise from large $\downarrow$ hole ellipsoids (VIII) at N, and the ν branches from smaller mixed spin pockets which arise (when the effect of the spin–orbit interaction is taken into account) from the intersections near N of the $\uparrow$ hole arms (II) and the $\downarrow$ hole ellipsoids (see Fig. 6.6(a)). The mixed spin pockets obtained in this model are, however, much larger than required to account for the magnitudes of the ν frequencies and any attempt to reduce their size by shifting the ends of the $\uparrow$ hole arms further from N would produce a model inconsistent with the known orientation dependence of the amplitude of the ζ

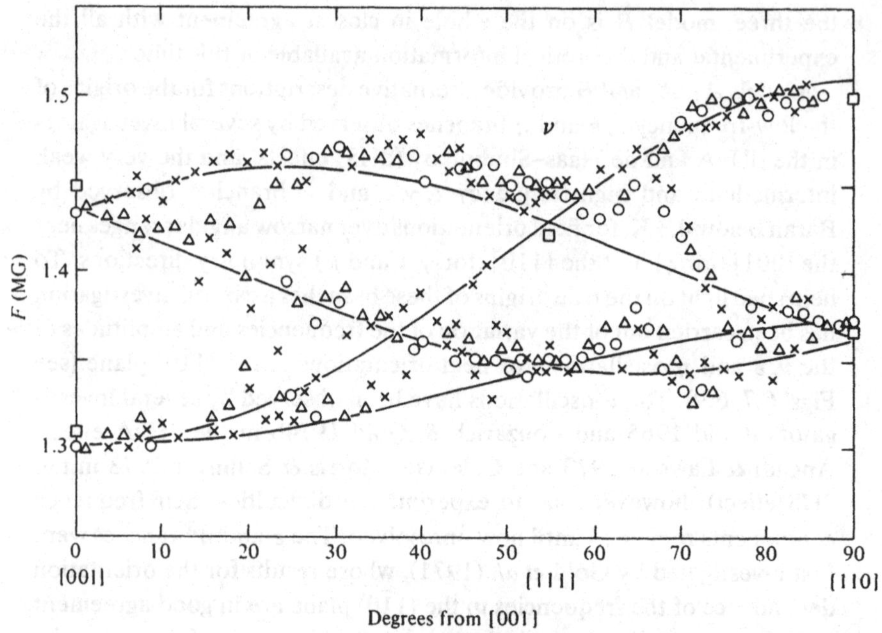

Fig. 6.7. Variation of the ν frequency branches in iron for **B** in the $(1\bar{1}0)$ plane, as measured in a $(1\bar{1}0)$ disc (×, ○, △) and in [001], [111] and [110] whisker samples (□) in the field range 23 kG < B < 55 kG. The full curves are frequency branches expected for ellipsoids centred on the points ν of the Brillouin zone, with principal axes $P(NH) = 0.0317$, $P(N\Gamma) = 0.0349$, $P(NP) = 0.0368$, in atomic units (a.u.). Conversion of F in MG to area $\mathscr{A}$ in a.u. is given by the relation $\mathscr{A}$ (a.u.) $= 2.673 \times 10^{-3}$ F (MG).

oscillations (Baraff 1973). Model A_1 suffers from a number of other difficulties, perhaps the most important of which is the fact that a [110] dHvA frequency expected to arise from the necks surrounding the mixed spin pockets in this model has not been observed, despite an intensive search.

These difficulties are avoided in models A_2 or B, in which Baraff's ζ branch would arise from the ↑ hole arms and the ν branches from very small ↓ ellipsoids at N which do not intersect the ↑ arm sheets. Since the ↑ arm surfaces are, in contrast to the ↓ ellipsoids, expected to have a high cyclotron mass, the present interpretation also leads to a natural explanation for the unusually small amplitude of the ζ branch. Furthermore it is noted that the observed increase of the ζ frequency with angle from [110] (Baraff 1973) is consistent with the behaviour expected for the ↑ arms, but in conflict with that predicted for the ↓ hole ellipsoids in all band-structure calculations carried out thus far.

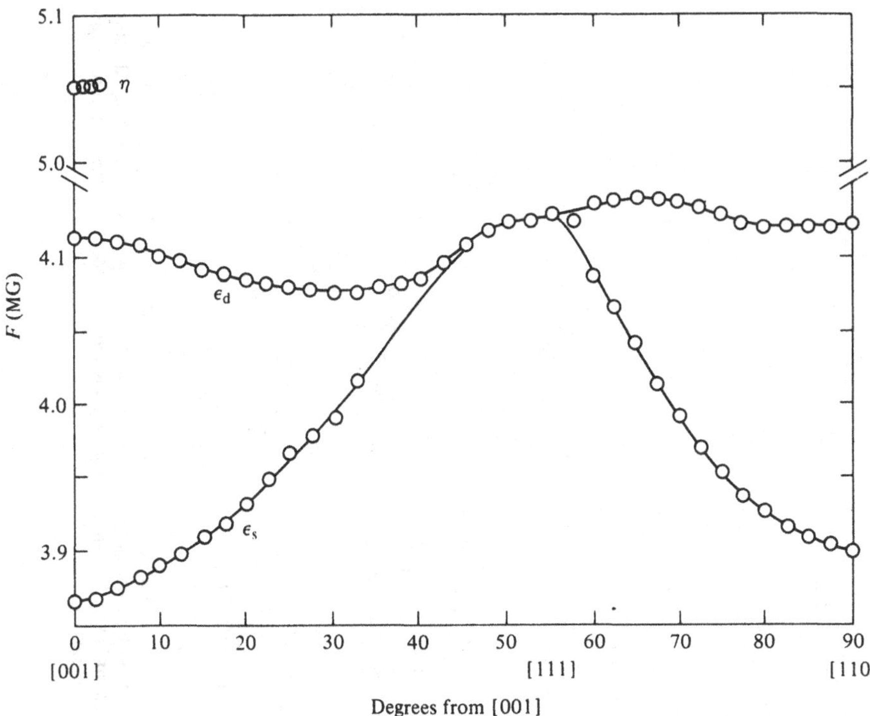

Fig. 6.8. Variation of the ε and η frequency branches in iron for **B** in the $(1\bar{1}0)$ plane, as measured in a $(1\bar{1}0)$ disc sample in the field range $23\ \text{kG} < B < 55\ \text{kG}$. The absolute accuracy in the frequencies is ±0.5 per cent. ($\mathscr{A}$ (a.u.) $= 2.673 \times 10^{-3}\ F$ (MG)).

The experimental and theoretical evidence therefore strongly favours models A_2 or B over A_1. Models A_2 and B provide alternative descriptions in particular of the origins of the ε and η branches. The orientation dependence and symmetry of the ε-frequency branches (the doublet ε_d and singlet ε_s in Fig. 6.8) are consistent with that expected for small nearly spherical pockets centred somewhere along the ΓH lines of the Brillouin zone. In model A_2 these would be the lenses produced by the intersection of the $\downarrow$ electron balls (VII) with the $\downarrow$ electron octahedron (VI, Fig. 6.6(*a*)). The η branch in this model would then arise from the necks which surround the lenses and which are separated from these by the effect of the spin–orbit interaction. In terms of model B, on the other hand, the ε branches would arise from the mixed spin pockets produced by the hybridization of the small $\downarrow$ electron balls (VII) and the large $\uparrow$ electron sheet (I; see Fig. 6.6(*b*)). The η branch which is observed only in the immediate vicinity of [001] would then arise from the basic $\downarrow$ electron

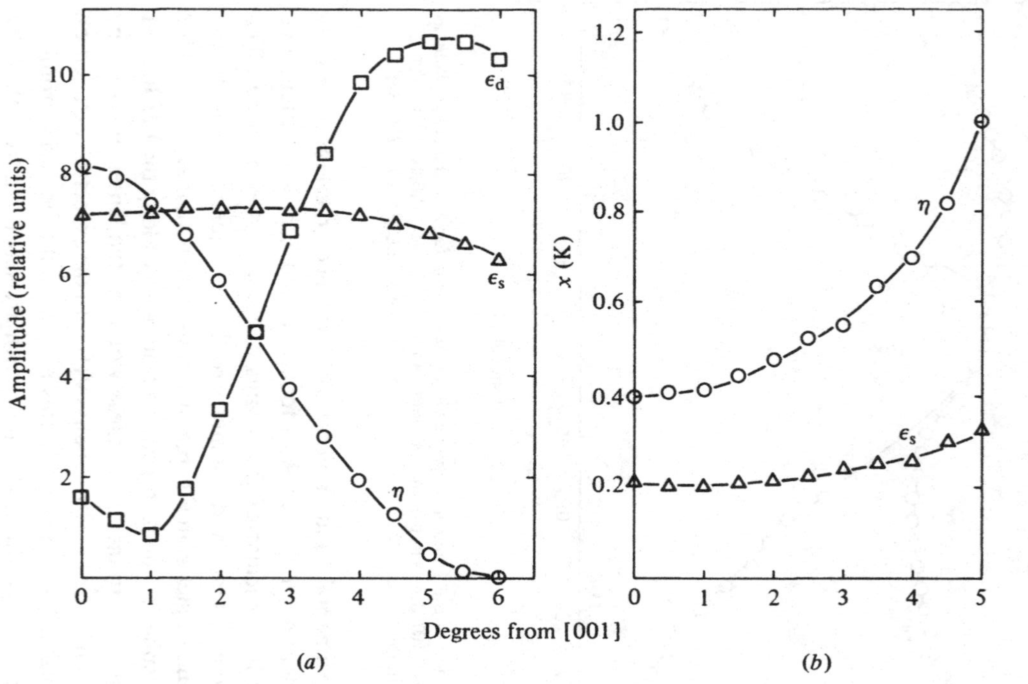

Fig. 6.9. Variation of the amplitudes and the Dingle temperatures (x) of the ε and η oscillations in iron for **B** in the $(1\bar{1}0)$ plane near [001]. Measurements were carried out in the field range 25 kG $< B <$ 55 kG in a very high quality [001] whisker sample closely resembling a long ellipsoid of revolution. The effect of shape anisotropy was carefully taken into account in the determination of the angular dependence of the amplitudes and Dingle temperatures.

balls which are not expected to hybridize with the majority sheet I in the central (001) plane when **B** is along [001].

To help in the identification of the correct model the effect of the spin–orbit interaction on the relevant points of degeneracy has been calculated by means of an interpolation band-model similar to that described by Gold *et al.* (1971; see also Ehrenreich & Hodges 1968). The spin–orbit parameter was taken to be $\xi = 5$ mRy, which is the same as that used by Maglic & Mueller (1971) and slightly greater than the value 4.3 mRy recently calculated by Singh *et al.* (1975). Where comparisons can be made, the results of the present calculations for the (100) and (110) sections, illustrated in Figs. 6.10(*a*) and (*b*) for models A_2 and B, respectively, are in qualitative agreement with those published by Singh *et al.* and are in all cases consistent with the symmetry requirements given by Ruvalds & Falicov (1968).

Referring to the results for model B (Fig. 6.10(*b*)), it is seen that the spin–orbit gap and thus the magnetic breakdown field (Stark & Falicov

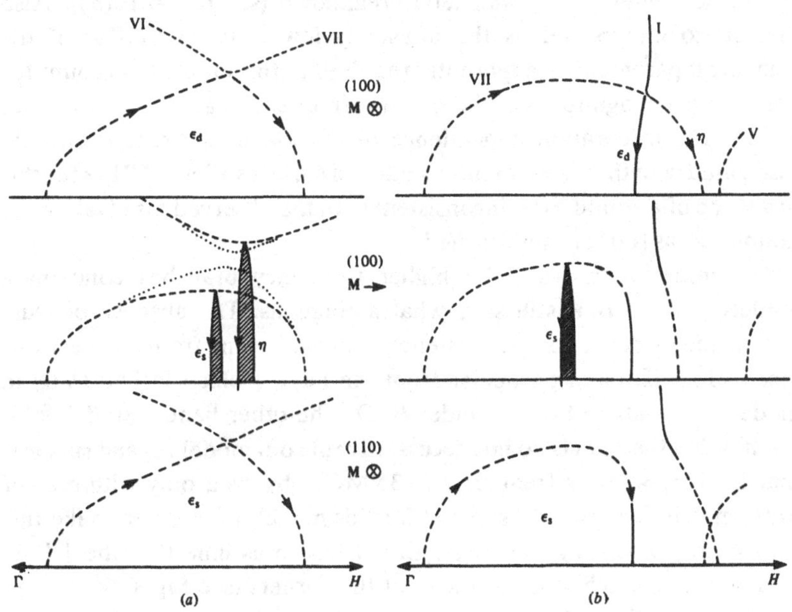

Fig. 6.10. Central (100) and (110) sections of the Fermi surface of iron in the vicinity of (*a*) the minority electron lenses of models A_1 and A_2 (Fig. 6.6(*a*)) and (*b*) the mixed spin pockets of model B (Fig. 6.6(*b*)), as computed by means of the band-structure model (including the spin–orbit interaction) defined in the text. The dotted lines in (*a*) are obtained using a spin–orbit parameter $\xi = 1.5$ mRy; all other results are for $\xi = 5.0$ mRy. **M** is the magnetization vector.

1967) associated with the intersections of Sheets VII and I in the (001) plane nearly vanish for **B** along [001]. Thus for this orientation the η oscillations (associated with the unhybridized balls in the (001) plane) should be strongest and the ε_d oscillations (associated with the hybrid balls in the (001) plane) weakest. As **B** is tilted from the [001] direction (in any plane) the magnetic breakdown field rises and the amplitudes of the η and ε_d oscillations are expected to fall and rise, respectively, in a complementary manner. It is seen from Fig. 6.9 that this is indeed the observed behaviour of the amplitudes of these oscillations, and in particular the orientation dependence of the magnetic breakdown field estimated from the data in Fig. 6.9(*b*) is consistent with that estimated (Chambers 1966) from the results of the present band-structure calculation.

The overall behaviour of the amplitude of the ε–η branches cannot on the other hand be as readily understood in terms of model A_2. In particular the latter does not seem to offer a straightforward explanation for the rapid fall of the amplitude of the η oscillations near [001], nor for the strong amplitude of the ε_s oscillations near [110] which should in this model be attenuated by magnetic breakdown (see Fig. 6.10(*a*)). Also difficult to understand, is the unusually low value (1.5 mRy) of the spin–orbit parameter ξ apparently required in this model to account for the observed magnitudes of the η and ε_s frequencies at [001] (Gold *et al.* 1971). The orientation dependence of the magnetic breakdown field associated with the points of intersection of surfaces VI and VII, is for this low value of ξ, found to be inconsistent with the observed rapid fall of the doublet ε_d as [001] is approached.

The information from the higher frequency branches concerning models A_2 and B is still somewhat ambiguous. The absence of truly convincing evidence for the existence of orbits arising from a ↓ electron-jack (VI+VII) or in particular from the large ↓ electron ball (VII) in model A_2, tends to favour model B. On the other hand Baraff (1973) implies that these facts do not necessarily rule out model A_2 and suggests that his very weak γ frequency (~35 MG) observed only within 8° of [001] may in fact provide support for this model. In order to make this conjecture convincing, however, Baraff has to assume that the ↓ hole octahedron is much more rounded at the corners (see Fig. 6.6(*a*)) than predicted by all band-structure calculations carried out thus far, and to this end he assigns his ι(~198 MG) rather than the higher frequency κ(~240 MG) branch near [001] to the unhybridized ↓ hole octahedron (V). Since these branches are observed only over a very narrow angular range near [001], their correct assignment is still very much in doubt. In terms of model B the unhybridized ↓ hole octahedron would be associated with the κ frequency at [001], whose magnitude is in close agreement with that calculated for this surface by Callaway & Wang (1977b). (These authors obtain the correct accepted value, within experimental error, for the central [111] frequency for this surface.) The ι branch would then be associated with a non-central orbit in the hybrid surfaces (II+V) centred on H.

Other sources of information relating to the Fermi surface of Fe have not, thus far, helped to distinguish between models A_2 and B. In particular both models appear to be capable of sustaining the narrow bands of open orbits along the [100] and [110] directions (Gold *et al.* 1971) which are needed to account for the high-field galvanomagnetic properties of Fe (Fawcett & Reed 1963; Coleman *et al.* 1973). In addition, positron annihilation studies in Fe (Mijnarends 1973), while providing support for models A_2 or B over A_1, appear to give little information concerning the ordering of the energy bands in the vicinity of the ΓH line of the Brillouin zone.

We conclude that the available experimental information relating to the Fermi surface of Fe can on the whole be most readily understood in terms of our model B (Fig. 6.6(*b*)). However, more detailed investigations especially of the ambiguous ζ, γ, ι and κ branches as well as of the lower frequency η branch (which should be a singlet in model A_2 but a doublet in model B near [001] are required to achieve a complete understanding of this complex Fermi surface. Additional Fermi surface studies of the ground states of Fe and Ni, as indicated in Section 6.3, would also be of considerable interest.

From Figs. 6.6(*b*) and 6.5(*a*) it is seen that the discrepancies between theory and experiment are substantial even in the case of model B. In contrast to a number of non-ferromagnetic transition metals, no single first-principles band calculation has yet come very close to reproducing the experimental Fermi surface data for Fe.

6.5 Magnetic (spin-flip) excitations in iron and nickel at low temperatures

The temperature dependence of a cross-sectional area of the Fermi surface is known to provide information concerning the contribution of Stoner single-particle excitations to the temperature dependence of the magnetization, without interference from spin-waves at low temperatures (see Section 6.3.6, (16a), (17a), (20b) and (24). The precise role of

Stoner excitations in the thermodynamic behaviour of ferromagnetic metals, particularly the weak itinerant ferromagnets such as $ZrZn_2$ and Ni_3Al but also certain strong ferromagnets such as Ni, has been a controversial subject for a number of years. In this section the results of recent Fermi surface studies of the Stoner contribution ($\delta\zeta_{sp} = -a_{sp}\zeta_0 T^2$) in Fe and Ni at low temperatures, are presented. The findings are compared with theoretical estimates based on parameters obtained in recent band-structure calculations, and with the results of analyses based on the known spin-wave parameters and the temperature dependence of the total magnetization.

6.5.1 Fermi surface study in iron

Information concerning $\delta\zeta_{sp}$ in Fe at low temperatures has been obtained in a high-resolution study (Lonzarich & Gold 1974a) of the temperature dependence of the [111] extremal cross-sectional area $\mathscr{A}$ of the minority ΓH pockets discussed in Section 6.4 (i.e. the lenses in model A_2 in Fig. 6.6(a) or the mixed spin pockets in model B in Fig. 6.6(b)). As a check on the validity of the general theory presented in Sections 6.3.5 and 6.3.6 which relates $\delta\mathscr{A}/\mathscr{A}_0$ to $\delta\zeta_{sp}/\zeta_0$ as a function of temperature, (24), or pressure, (12), the pressure dependence of the [111] cross-sectional area has also been investigated.

For comparison the temperature measurements $\delta\mathscr{A}/\mathscr{A}$ versus T ($1.3 < T < 4.0$ K, $P \approx 0$) and the pressure measurements $\delta\mathscr{A}/\mathscr{A}$ versus P ($T =$ 1.3 K, $1 < P < 22.5$ bars) are plotted on the same graph as a function of the *total* change in polarization $\delta\zeta/\zeta_0$ arising fron the specified changes in temperature or pressure, respectively. The temperature dependence of ζ was determined from Legkostupov's (1971) low-temperature measurements of $\mathrm{d}\zeta/\zeta_0\,\mathrm{d}T$ (for $H = 17$ kG), and the pressure dependence from the value of $\mathrm{d}\zeta/\zeta_0\,\mathrm{d}P$ given in Table 6.3.

From (12), (13) and the parameters given in Tables 6.2 and 6.3, the value of the ratio ($-\delta \ln \mathscr{A}/\delta \ln \zeta$) is calculated to be 42 for model A_2 and 38 for model B of the Fermi surface, assuming the pressure P is the independent variable (at constant temperature). Considering the present uncertainties in the magnitude of some of the parameters used in this calculation, these results are consistent with the experimental value of 33.5 ± 1 obtained from the slope of the line passing through the triangular data points in Fig. 6.11.

From (24), on the other hand, the value of ($-\delta \ln \mathscr{A}/\delta \ln \zeta$), assuming T is the independent variable and $\delta\zeta_{sp} = \delta\zeta$, is calculated to be 32 for model

TABLE 6.2. *Experimental or band structure values for certain parameters of the ground states of iron and nickel used in the analyses in the text. Quantities in parentheses are order of magnitude estimates based on the available band-structure data*

Parameter	Fe	Ni
a (Å)	2.86[a]	3.51[f]
$(n_\uparrow - n_\downarrow)$ (electrons atm^{-1})	2.12[a]	0.56[f]
g	2.09[a]	2.22[f]
H_A (10^3 G)	0.6[c]	2.0[c]
χ_P (10^{-5})	2.5[d]	0.5[d]
D_0 (meV Å^2)	314[e]	450[e]
D_1 (10^{-4} meV Å^2 K^{-2})	6.4[e]	7.0[e]
Δ (eV)	2.2[a]	0.6[f]
$N_\uparrow$ (states atm^{-1} Ry^{-1})	11.3[a]	2.3[f]
$N_\downarrow$ (states atm^{-1} Ry^{-1})	3.4[a]	21.3[f]
$N'_\uparrow$ (states atm^{-1} Ry^{-2})	(10^2)[b]	(-10)[f]
$N'_\downarrow$ (states atm^{-1} Ry^{-2})	(± 10)[b]	(10^3)[f]

[a] See e.g. Callaway & Wang (1977b).
[b] See e.g. Singh *et al.* (1975).
[c] See e.g. Argyle *et al.* (1963).
[d] Foner *et al.* (1969).
[e] See Table 6.4.
[f] See e.g. Wang & Callaway (1974, 1977).

A_2 and 25 for model B. Both are at least 20 times larger than the upper limit ($\lesssim 1$) derived from the circular data points in Fig. 6.11 (see Lonzarich & Gold 1974a for a discussion of the probable origin of the small apparent negative shift of the cross-sectional area with temperature). This result can be understood if the *change in polarization with temperature in Fe is almost entirely due to spin-waves,* i.e. $\delta\zeta \approx \delta\zeta_{sw}$, in the temperature and field range of the experiment. More precisely, from the experimental data and (24) an upper limit for the Stoner coefficient a_{sp} of the T^2 term in $\delta\zeta_{sp}/\zeta_0$ is found to be $|a_{sp}| \lesssim 3\times10^{-8}\,\mathrm{K}^{-2}$ (for models A_2 *or* B of the Fermi surface). This limit is consistent with the value $|a_{sp}| \approx 1.0\times10^{-8}\,\mathrm{K}^{-2}$ obtained from the theoretical expressions (20b), (21) and the band-structure parameters given in Table 6.2. Since the derivatives of the densities of states entering (20b) are rather poorly known, this theoretical value is only an order of magnitude estimate. Estimates of a_{sp} obtained by means of other experimental techniques will be presented and discussed in Section 6.5.3.

TABLE 6.3. *Experimental or theoretical values of parameters (for $T \approx 0$ K) used in the interpretation of the pressure and temperature dependences of the [111] cross-sectional areas of the ΓH pockets of the Fermi surface of iron and of the majority necks of the Fermi surface of nickel. The quantities* $d \ln \mathcal{A}/dP$, *F*, m_c^*/m *and* m_c/m *are those associated with the [111] ΓH pockets in Fe (see models* A_2 *and B in Section 6.4), the [111] lenses in Mo (corresponding to model* A_2*), the [111] ΓH balls in W (corresponding to model B), the [111] majority necks in Ni, and the [111] necks in Cu, under columns 1, 2, 3, 4 and 5, respectively. Pressure derivatives are given in* (10^{-7} bar^{-1}) *and F in* (10^6 G)

Parameters	Fe	Mo[A]	W[B]	Ni	Cu
$d \ln \zeta/dP$	−3.0(2)[a]	—	—	−2.7[b]	—
$d \ln I/dP$	—	—	—	5[c]	—
$d \ln V/dP$	−6.0(1)[d]	−3.8(1)[e]	−3.2(1)[e]	−5.4(1)[d]	−7.0(1)[f]
$d \ln \mathcal{A}/dP$	100(3)[g,s]	14(1)[g]	7(1)[o]	6(1)[i], 8(1)[j]	18(2)[f]
F	4.15(3)[g]	5.41(3)[g]	23.6(4)[h]	2.68(3)[k]	21.9(3)[f]
m_c^*/m	0.71(2)[g]	0.30(2)[g]	0.65(2)[h]	0.25(2)[k]	0.46(2)[l]
m_c/m	0.36[g]	0.30[m]	0.65[m]	0.13[n]	0.41[l]
$\Lambda N_\sigma I$	0.0[p], 0.2[p]	—	—	0–0.2[r]	—

[a] Bloch & Pavlovic (1969) and references cited therein.
[b] Tange & Tokunaga (1969); the uncertainty in this parameter may be large (see e.g. Fujiwara, Okamoto & Tatsumoto 1965; Bloch & Pavlovic 1969 and references cited therein).
[c] Mathon (1972).
[d] Gschneidner (1964).
[e] Featherston & Neighbours (1963).
[f] Bosacchi *et al.* (1970).
[g] Present work.
[h] Girvan, Gold & Phillips (1968).
[i] Anderson *et al.* (1976) (high-pressure results only).
[j] Vinokurova *et al.* (1977).
[k] Tsui (1967).
[l] O'Sullivan, Switendick, & Schirber (1970).
[m] The mass enhancement has been assumed to be small and ignorable.
[n] Zornberg (1970).
[o] Svechkarev & Pluzhnikov (1973) ($d \ln \mathcal{A}/dP$ is given for the [100] orientation only).
[p] Estimated using the band-structure model described in Section 6.4. The first and second values correspond to models A_2 and B, respectively, of the Fermi surface.
[r] Range of values for $\Lambda = 0$–1. The analysis in the text assumes $\Lambda = 0$.
[s] Measurements at high pressures (between 1 and 10 k bars), since this work was completed, have yielded $d \ln \mathcal{A}|dp = (80 \pm 5) \times 10^{-7}$ bar^{-1} (Vinokurova, L. I., Gaputchenko, A. G. & Itskevich, E. S. (1978) *JETP Lett.* **28**, 256; *Zh. Eksp. Teor. Fiz.* **28**, 280).

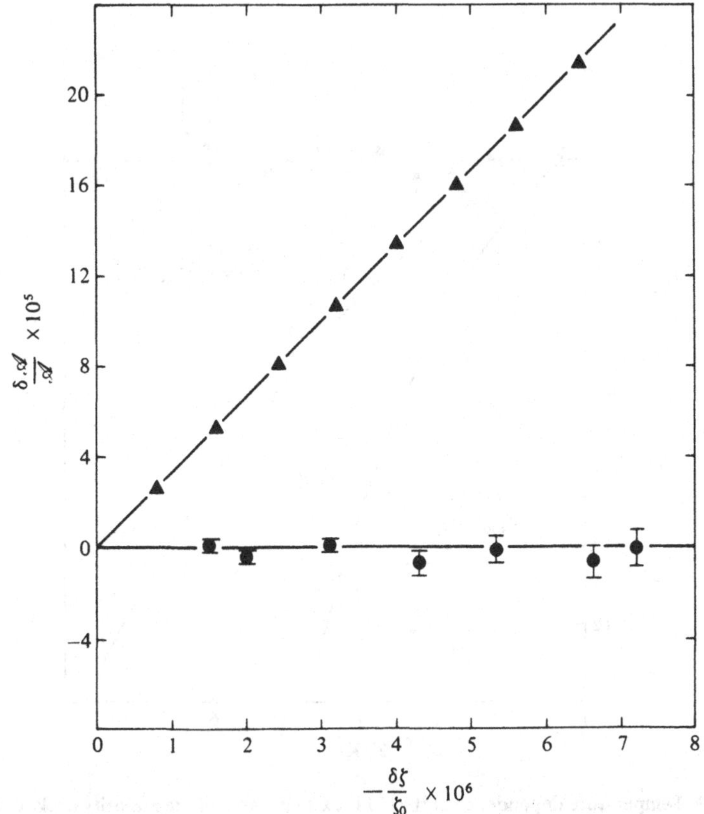

Fig. 6.11. Temperature (●) and pressure (▲) dependence of the [111] extremal area of the ΓH-pockets of the Fermi surface of iron; (●) $T = 1.3$–4.0 K, $P \approx 0$, $H = 17.0$ to 21.0 kG (after Lonzarich & Gold, 1974a); (▲) $T = 1.3$ K, $P = 1$–22.5 bars, $H = 20.0$ kG. $\delta\zeta/\zeta_0$ is the total change in the spin-polarization as a function of temperature or pressure.

6.5.2 Fermi surface study in nickel

The Stoner component $\delta\zeta_{sp}$ in Ni has been investigated in a recent study of the temperature dependence of the [111] cross-sectional area $\mathscr{A}$ associated with the majority-spin necks, in the temperature range 1.2–6.7 K (Lonzarich 1978). The experimental results $\delta\mathscr{A}/\mathscr{A}$ versus T are given in Fig. 6.12. Each point represents an average of up to 20 separate measurements for H (i.e. the applied plus demagnetizing field) in the range 30–34 kG. The temperature dependence of $\mathscr{A}$ is derived from the temperature dependence of the dHvA phase according to the procedure

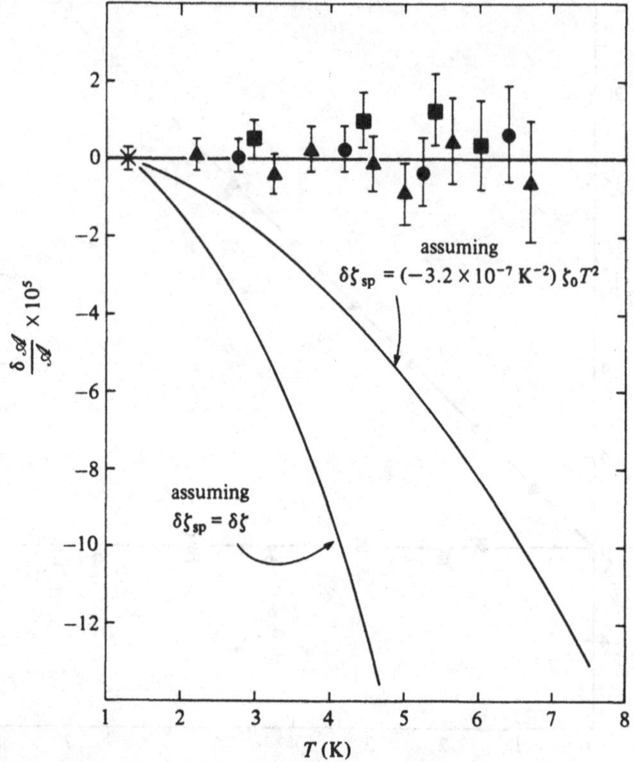

Fig. 6.12. Temperature dependence of the [111] extremal area of the majority necks of the Fermi surface of nickel ($H = 30.0$–34.0 kG). The lower full curve shows the variation in $\mathscr{A}$ expected if the total observed change in the spin polarization, under the conditions of the experiment, were due to Stoner excitations. The upper full curve shows the area change expected if Riedi's T^2 term in the spin-polarization were due to Stoner excitations.

outlined in Section 6.3.6.3 (in which any linear variation of the phase with temperature is ignored). The pressure dependence of $\mathscr{A}$ has also been investigated (Anderson *et al.* 1976 and Vinokurova *et al.* 1977) and the results are presènted and satisfactorily interpreted in terms of (12) and (13) in Section 6.3.5.

The lower curve in Fig. 6.12 gives the variation of $\delta\mathscr{A}/\mathscr{A}$ with temperature predicted from (24) and the relevant parameters given in Tables 6.2 and 6.3, *if* $\delta\zeta/\zeta_0 = \delta\zeta_{sp}/\zeta_0$. The temperature dependence of ζ was obtained from Legkostupov's (1971) low-temperature values of $d\zeta/\zeta_0\,dT$ (for $H = 32$ kG). A comparison of the observed and calculated variation of $\delta\mathscr{A}/\mathscr{A}$ with T leads to the conclusion that in Ni, as in Fe, the *change in polarization with temperature is almost entirely due to spin-wave excitations* in the field and temperature range of the experiment. In particular, from the experimental data and (24) an upper limit for the Stoner coefficient a_{sp} is found to be $|a_{sp}| \lesssim 5 \times 10^{-8}\,\mathrm{K}^{-2}$. This limit is consistent with the theoretical estimate $a_{sp} \approx 3.1 \times 10^{-8}\,\mathrm{K}^{-2}$ obtained from (20b), (21) and the relevant parameters given in Table 6.2. Since the derivatives of the densities of states required in (20b) are poorly known the latter is an order of magnitude estimate only.

The experimental and theoretical estimates of a_{sp} given above are 6–10 times *smaller* than that recently determined by Riedi (1977) on the basis of his analysis of the temperature dependence of the total magnetization between 4.2 K and room temperature. From (24) and the relevant parameters in Tables 6.2 and 6.3, the expected variation $\delta\mathcal{A}/\mathcal{A}$ versus T for Riedi's value $a_{sp} = 3.2 \times 10^{-7}\,\mathrm{K}^{-2}$, is given by the upper curve in Fig. 6.12. In the next section we consider (i) the detailed procedure used to obtain this value and (ii) the possible significance of the large difference between this value and the experimental and theoretical estimates given above.

6.5.3 Information from other experimental techniques

In principle the Stoner and spin-wave components $\delta\zeta_{sp}$ and $\delta\zeta_{sw}$ may be obtained directly from the temperature and field dependences of $\delta\zeta$ at very low temperatures, where only the leading terms in the expansions given in Section 6.3.6 are important. Unfortunately at these low temperatures (i.e. $T \lesssim 20$ K for Fe and Ni) both random and systematic errors in the existing data for $\delta\zeta/\zeta_0$ versus T are quite large, so that a reliable separation of the magnon and Stoner terms in $\delta\zeta$ has thus far not been possible for Fe and Ni (Argyle *et al.* 1963; Tsarev & Zavaritskii 1965; Legkostupov 1971; Aldred & Froehle 1972; Aldred 1975; Riedi 1977).

In the intermediate temperature regime (i.e. $20\,\mathrm{K} \lesssim T \lesssim 300\,\mathrm{K}$) magnetization measurements have been more reliable and reproducible (see e.g. Argyle *et al.* 1963; Aldred & Froehle 1972; Aldred 1975; Riedi 1977). However, as emphasized by Aldred (1975) and Riedi (1977), the necessity of including higher order terms in the expansions of $\delta\zeta_{sw}$ and $\delta\zeta_{sp}$ versus T in this temperature range, very seriously complicates the detailed analysis of the magnetization data.

In an effort to reduce the number of independent variables appearing in such an analysis Aldred (1975) and Riedi (1977) investigate essentially the 'residual' polarization defined as the difference between $\delta\zeta/\zeta_0$ and the first spin-wave term in (17a), i.e.

$$\frac{\delta\zeta_x}{\zeta_0}=\frac{\delta\zeta}{\zeta_0}+\frac{2VF(\frac{3}{2},T/T_g)}{n\zeta_0N_a}\left[\frac{k_BT}{4\pi(D_0-D_1T^2)}\right]^{\frac{3}{2}}. \tag{26}$$

The spin-wave parameters D_0 and D_1 are obtained from spin-wave resonance or inelastic neutron scattering experiments (see Tables 6.2 and 6.4). If the values of the parameters inserted in this expression, in particular $\delta\zeta/\zeta_0$ and D_0, are accurately known then $\delta\zeta_x/\zeta_0$ is a measure of $\delta\zeta_{sp}/\zeta_0$ plus the spin-wave terms beyond the first in (17a) which may be ignored at sufficiently low temperatures. In principle higher order spin-wave components could also be subtracted from $\delta\zeta$ in (26) if the precise form of the magnon dispersion curve were known as a function of temperature (and if the effect of spin-wave damping, which has been ignored in the calculation leading to (17a), were taken into account).

The quantity $\delta\zeta_x/\zeta_0$ in the range $1\text{ K}\lesssim T\lesssim 300\text{ K}$ has been calculated for Fe and Ni (see Figs. 6.13 and 6.14), using the parameters given in Table 6.2 and the values of $\delta\zeta/\zeta_0$ versus T published by Argyle *et al.* (1963; Fe, Ni), Legkostupov (1971; Fe, Ni), Aldred & Froehle (1972; Fe), Aldred (1975; Ni) and Riedi (1977; Fe). The spin-wave parameters

TABLE 6.4. *Experimental values of spin-wave stiffness parameters at room temperature and low temperature in iron and nickel*

	Technique[a]	D (295 K)	D (4.2 K)
Fe	TFR	281(17)[b], 312(13)[b]	
	DM	266(15)[b]	
	SAS	256(8)[b], 260(20)[c]	314(10)[c]
	TA	281(10)[b], 285(10)[e]	
Ni	TFR[i]	354[b], 387(12)[b]	412[b]
	DM	374(20)[b], 391(20)[b]	
	SAS	403(7)[b], 407(10)[c]	450(10)[c], 422(20)[g]
	TA	458(9)[d,f], 433[f], 400[f], 454(9)[f], 400(15)[h]	555(6)[d], 544(6)[f], 530(15)[h]

[a] TFR: thin film resonance; DM: diffraction method; SAS: small-angle scattering; TA: triple-axis spectrometry; DM, SAS and TA are neutron scattering techniques.
[b] Shirane, Minkiewicz & Nathans (1968) and references cited therein.
[c] Stringfellow (1968).
[d] Mook, Lynn & Nicklow (1973).
[e] Lynn (1975).
[f] Aldred (1975) and references cited therein.
[g] Menshikov, Kazantsev, Kuzmin & Sidorov (1975).
[h] Hennion & Hennion (1978) and unpublished data.
[i] Uncertainties in TFR results for Ni may be large (see e.g. Mitra, D. P. & Whiting, J. S. S. (1978), *J. Phys.* F **8**, 2401).

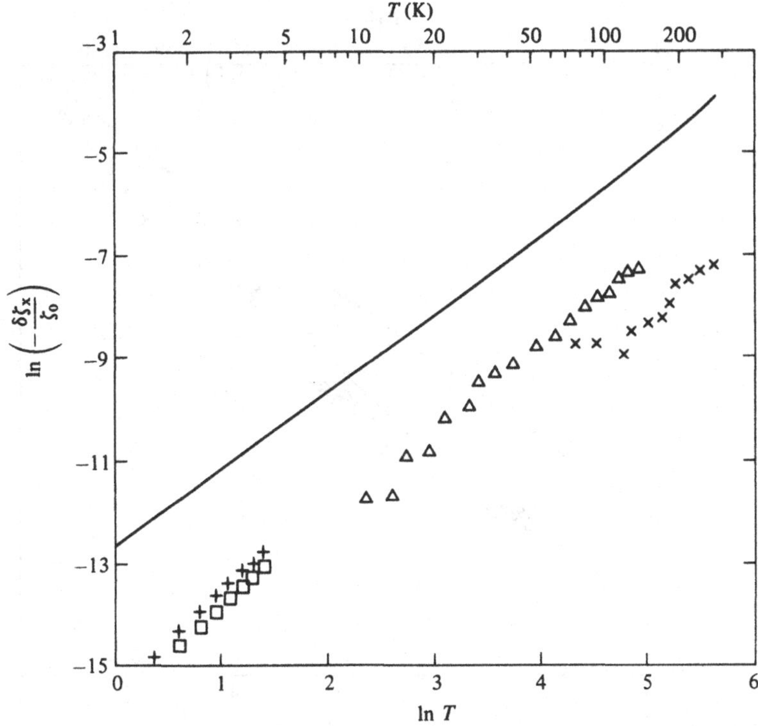

Fig. 6.13. Temperature dependence of the residual polarization in iron calculated from (26), the parameters given in Table 6.2, and the data $\delta\zeta/\zeta_0$ versus T of Argyle *et al.* (1963; △, $H = 6.67$ kG), Legkostupov (1971; +, $H = 22.0$ kG, □, $H = 34.0$ kG) and Aldred & Froehle (1972; ×, $H = 10.2$ kG). The full curve is a log plot of the second term on the rhs of (26) (i.e. the spin-wave term) for $T_g = 0$.

D_0 and D_1 were obtained from the published experimental values of D at 4.2 K and room temperature. It is seen from Table 6.4 that the various experimental estimates of D are in reasonable agreement with one another for Fe. On the other hand the values of D for Ni found by the triple axis (TA) method are significantly higher than those obtained by thin-film resonance (TFR), the diffraction method (DM) or small angle scattering (SAS). In the present analysis the values of D_0 and D_1 were estimated from the results of the DM and SAS techniques which probe the spin-wave spectrum at smaller wave-vectors q than the TA method. Aldred (1975) and Riedi (1977), on the other hand, base most of their analyses on the higher values of D obtained in the TA studies. Their values of $|\delta\zeta_x/\zeta_0|$ are thus somewhat greater than those presented in Figure 6.14. (The calculation of the precise magnitude of $\delta\zeta_{sw}/\zeta_0$ versus temperature must await a more accurate determination of the spin-wave spectrum at low energies and wave-vectors.)

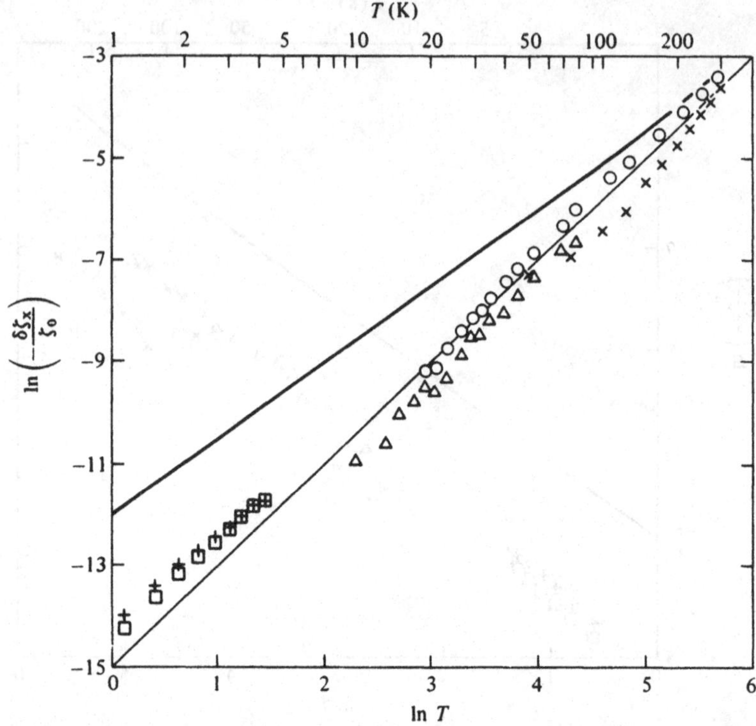

Fig. 6.14. Temperature dependence of the residual polarization in nickel calculated from (26), the parameters given in Table 6.2, and the data of argyle *et al.* (1963; △, $H = 17.0$ kG), Legkostupov (1971; +, $H = 22.0$ kG, □, $H = 34.0$ kG), Aldred (1975; ×, $H = 12.8$ kG) and Riedi (1977; ○, $H \approx 0$). The bold solid curve is a log plot of the second term on the rhs of (26) (i.e. the spin-wave term) for $T_g = 0$. The thin full line is a log plot of Riedi's residual polarization $(-\delta\zeta_x/\zeta_0) = (3.2 \times 10^{-7}\ \mathrm{K}^{-2})T^2$.

The bold solid curve in Fig. 6.13 or 6.14 represents the contribution of the last (i.e. the spin-wave) term in (26) assuming $T_g = 0$. From Fig. 6.13 it is seen that $(-\delta\zeta_x/\zeta_0)$ is much smaller than the spin-wave term in Fe over the entire temperature range investigated. As pointed out by Aldred & Froehle (1972) this suggests that the Stoner term $\delta\zeta_{sp}/\zeta_0$ and higher spin-wave components in $\delta\zeta/\zeta_0$ are very small in Fe below room temperature (an accidental cancellation of these components is highly unlikely, especially since the leading spin-wave $T^{\frac{5}{2}}$ and Stoner T^2 terms in $\delta\zeta_x/\zeta_0$ are expected to have the same sign (see e.g. Aldred & Froehle 1972)). From the room temperature magnitude of $\delta\zeta_x/\zeta_0$ we obtain an upper limit $|a_{sp}| \lesssim 1 \times 10^{-8}\ \mathrm{K}^{-2}$ for Fe. This limit is consistent with the experimental and theoretical estimates given in Section 6.5.1 and with the value $a_{sp} = (1 \pm 3) \times 10^{-8}\ \mathrm{K}^{-2}$ obtained by Argyle *et al.* (1963) in a detailed curve-fitting analysis of their magnetization data in the range $4.2\,K < T < 140\,K$. On the other hand Legkostupov's (1971) data for $\delta\zeta/\zeta_0$ versus T in the range $0.7\,K \lesssim T \lesssim 4.2\,K$ yields a value $a_{sp} = (11 \pm 4) \times 10^{-8}\,K^{-2}$ (see Fig. 6.13 or Legkostupov's analysis based on a different approach), which is considerably greater than all the experimental and theoretical limits given above. This high value of a_{sp} is, however, in doubt for the following reasons. (i) Measurements of $\delta\zeta/\zeta_0$ at low temperatures are difficult and susceptible to large systematic errors; for example, the values of $\delta\zeta/\zeta_0$ measured by Legkostupov (1971) differ from those of Tsarev & Zavaritskii (1965) by as much as 40 per cent over the same temperature and field range. (ii) A detailed examination of Legkostupov's data shows that $\delta\zeta_x/\zeta_0$ varies quite dramatically with magnetic field particularly below 20 kG, a behaviour which can be shown to be inconsistent with the high value of the Stoner coefficient a_{sp} given above. (iii) Finally it is noted that for $a_{sp} \approx 11 \times 10^{-8}\,K^{-2}$, the Stoner T^2 and spin-wave $T^{\frac{3}{2}}$ terms in $\delta\zeta/\zeta_0$ would have comparable magnitudes at room temperature, a result which cannot be reconciled in particular with the well-known values of $\delta\zeta/\zeta_0$ and D at $T \approx 300\,K$ (Argyle *et al.* 1963, Aldred & Froehle 1972).

It is concluded that, with the sole exception of the low-temperature measurements of Legkostupov which require further study, all experimental techniques and methods of analysis of the experimental data lead to a similar upper limit for a_{sp} in Fe (i.e. $|a_{sp}| \lesssim 3 \times 10^{-8}\,K^{-2}$). Furthermore this limit is consistent with the theoretical estimate given in Section 6.5.1 (i.e. $|a_{sp}| \approx 1 \times 10^{-8}\,K^{-2}$).

As first pointed out by Aldred (1975) and confirmed by Riedi (1977) the residual polarization $\delta\zeta_x/\zeta_0$ is generally much larger in Ni than in Fe and is comparable to the $T^{\frac{3}{2}}$ spin-wave term itself at room temperature (see Fig. 6.14). Arguing that higher order spin-wave terms in $\delta\zeta_x/\zeta_0$ are ignorable below room temperature, Aldred and Riedi have proposed that $\delta\zeta_x/\zeta_0 \approx \delta\zeta_{sp}/\zeta_0$. On the basis of his high resolution NMR measurements of $\delta\zeta/\zeta_0$ in the range $4.2\,K \lesssim T \lesssim 300\,K$, Riedi showed that $\delta\zeta_x/\zeta_0$ could be fitted approximately to a T^2 law and for the value of D_0 given in Table 6.2 he finds $(-\delta\zeta_x/\zeta_0) = (3.2 \times 10^{-7}\,K^{-2})T^2$, a result represented by the thin solid line in Fig. 6.14. The coefficient of the T^2 term is over six times larger than the upper limit for $|a_{sp}|$ determined from the Fermi surface measurements in Section 6.5.2 and over ten times larger than the theoretical estimate given in the same section. An examination of Fig. 6.14, however, shows that, although Riedi's result does indeed provide a best overall fit of the cumulative data to a T^2 law, the precise temperature dependence of $\delta\zeta_x/\zeta_0$ in the important low and intermediate temperature region is still quite uncertain. Thus, although there is little doubt that the residual $\delta\zeta_x/\zeta_0$ is important in Ni, its precise temperature and field dependence and thus its physical origin has not been firmly established on the basis of the present analysis of the magnetization data alone. In particular it is suggested that, on the basis of the Fermi surface study discussed in Section 6.5.2, the *residual $\delta\zeta_x/\zeta_0$ in Ni cannot be attributed to a T^2 term in $\delta\zeta/\zeta_0$ arising from Stoner excitations*, as defined in the present theoretical framework (Section 6.3.6).

In the intermediate temperature range and above, the following contributions to $\delta\zeta_x$ may be important: (i) higher order spin-wave components in $\delta\zeta$ arising essentially from the damped plateau region of the spin-wave spectrum which is expected to exist over a limited region in $\mathbf{q}-\omega$ space in the energy range 110–160 meV (see e.g. Cooke 1976); (ii) higher order single-particle repopulation arising from the high density of Stoner excitations thought to exist in the vicinity of the spin-wave plateau in $\mathbf{q}-\omega$ space; and (iii) single-particle repopulation driven by the effects of thermally excited magnons (see e.g. the second term on the rhs of (20b) in the case of no interband transitions). The importance of the damped magnons in (i) and the corresponding Stoner excitations described in (ii), which give rise to terms in $\delta\zeta_{sp}$ which are higher order in T than the T^2 term retained in the low temperature analysis presented under section 6.3.6, may be enhanced by the effects of the electron–magnon interaction which is expected to reduce the separation between the top of the majority d-density of states and $\mathscr{E}_F$, and thus lower the energy of the high density region of the Stoner excitation spectrum and of the spin-wave plateau in $\mathbf{q}-\omega$ space (Edwards 1977 and private communications).

The above effects are, however, unlikely to be important in the low temperature regime. In particular the Fermi surface study presented above as well as a more recent investigation (Cooper & Lonzarich unpublished) suggest that single-particle repopulation of any kind cannot account for the magnitude of $\delta\zeta_x$ at least for $T \lesssim 20$ K. It is noted that if, as is suggested by some TFR measurements (see e.g. Mitra & Whiting 1978 and Table 6.4), the effective spin-wave stiffness D_0 at low q is as low as approximately 400 meV Å^2 in Ni, a value about 30 per cent lower than the effective D_0 determined by the TA technique in the wave-vector range $0.1 \lesssim q \lesssim 0.4$ Å^{-1} (Table 6.4), most of the low temperature magnetization data discussed in this section (with the exception of the very low temperature data of Legkostupov) can be reasonably accounted for in terms of spin-wave theory alone. More reliable measurements of the magnon spectrum at very low energies and wave-vectors, as well as of the temperature and field dependence of the magnetization at low temperatures and high fields may help to shed further light on this problem.

Finally we wish to point out that in recent theoretical analyses of the temperature dependence of the magnetization in weak itinerant ferromagnets Yamada (1974) and Makoshi & Moriya (1975) obtain an enhancement of the T^2 term in $\delta\zeta$ at low temperatures due to the influence of correlated spin-fluctuations. Though probably insignificant in Ni, this enhancement may perhaps provide further insight into the detailed behaviour of the magnetization as a function of temperature in some very weak itinerant ferromagnets.

Acknowledgements

It is a pleasure to acknowledge the invaluable support of A. V. Gold and D. Shoenberg during the course of this work, which was initiated at the University of British Columbia under a grant from the National Research Council, and completed in the Cavendish Laboratory, Cambridge. I am also grateful to R. V. Coleman for the loan of a [110] iron whisker used in a part of this investigation and to the Battelle Memorial Institute, Columbus, Ohio, for providing high-purity bulk iron from which single-crystal discs were prepared. Finally, I wish to thank a number of other investigators, in particular D. M. Edwards, E. Fawcett, N. Cooper and E. P. Wohlfarth, for informative communications.

7

The effect of strain on the Fermi surface

E. FAWCETT, R. GRIESSEN, W. JOSS,
M. J. G. LEE AND J. M. PERZ

7.1 Introduction

Experimental measurements of the pressure dependence of the Fermi surfaces of most elemental metals have been made in the last decade or so, in order to obtain information about the dependence of the energy-band structure on the lattice parameter. The response of the Fermi surface to uniaxial strain, which can be determined either directly by applying an external stress to the sample, or indirectly from the oscillatory magnetostriction or sound velocity, has also been measured in many metals. David Shoenberg and his students and co-workers have made substantial contributions to this work.

Work on the hydrostatic pressure dependence of the Fermi surfaces of metals has been reviewed by Brandt, Itskevich & Minina (1971), by Svechkarev & Panfilov (1974) (d-band transition metals), and most recently and comprehensively by Schirber (1974, 1978). Schirber has given a full account of experimental techniques for measuring the hydrostatic pressure dependence of the Fermi surface, together with comprehensive references. Early oscillatory magnetostriction work was surveyed in a review by Chandrasekhar & Fawcett (1971), but much more complete studies are now available. Griessen (1978) has given a brief account of work on the uniaxial strain dependence of the Fermi surfaces of transition metals.

The earliest measurements of the strain dependence of the Fermi surface were performed by Verkin, Dmitrenko & Lazarev (1956) on bismuth using the pressure bomb technique (see also Dmitrenko, Verkin & Lazarev 1958). The development by Shoenberg & Stiles (1964) of a technique for observing the de Haas–van Alphen (dHvA) effect, using field modulation in a superconducting solenoid operating in the persistent mode, made it possible to measure the small variations in the phase of the oscillations induced by a stress applied to the sample. Shoenberg & Watts (1967) made the first direct measurements of the uniaxial stress dependence of Fermi surfaces, on the noble metals, while Templeton (1966) developed the fluid helium hydrostatic pressure technique and also applied it to the noble metals, which were subsequently investigated further by the solid helium technique by Schirber & O'Sullivan (1970a).

Green & Chandrasekhar (1963) made the first measurements of oscillatory magnetostriction, on bismuth. The measurement of oscillatory magnetostriction and magnetic torque was developed by Griessen & Olsen (1971) to measure the stress dependence of the Fermi surface of aluminium, and was later used to study several other free-electron-like metals (Griessen & Sorbello 1974), as well as to provide results on the uniaxial stress dependence of the Fermi surface of a d-band transition metal, molybdenum (Posternak *et al.* 1975). Testardi & Condon (1970) combined measurements of oscillatory sound velocity and magnetization to deduce the strain dependence of the Fermi surface of beryllium, while Lee, Perz & Stanley (1976b) combined sound velocity and torque measurements to determine the angular shear strain dependence of the Fermi surface of tungsten.

In this chapter we describe the experimental investigation of the effects of stress and strain on the Fermi surfaces of some representative metals. Most of the experimental data have been obtained from studies of quantum oscillations, to which we confine our discussion. Direct uniaxial stress techniques and indirect techniques based on measurements of magnetostriction or sound velocity will be described and compared. To illustrate the use of these techniques and to indicate the type of information they provide, we conclude by describing work on three representative systems, the nearly-free-electron metal lead, the noble metal copper, and the group VI transition metal tungsten. For a comprehensive summary of the stress and strain dependences of the Fermi surfaces of metals, the new series of Landolt–Börnstein tables should be consulted.

7.2 Characterization of lattice strains and stresses

Measurements of the way in which the Fermi surface is distorted by the application of hydrostatic pressure or uniaxial tension yield information about the response of the electronic structure to homogeneous lattice strain. The lattice strain is related to the applied stress through the elastic stiffness constants c_{ij} of the metal. We shall restrict this discussion to cubic crystals, which have three independent stiffness constants, c_{11}, c_{12} and c_{44}.

The lattice strain in a cubic crystal can be specified by the associated changes in the translation vectors of the conventional unit cell. An arbitrary state of strain may involve the elongation of the basis vectors so that, for example, the basis vector $\mathbf{x} = a\hat{\mathbf{x}}$ takes the form $\mathbf{x}' = a\hat{\mathbf{x}}(1+\varepsilon_{xx})$; ε_{xx} represents the fractional elongation of the x-axis. It may also involve a relative rotation of the basis vectors so that, for example, the basis vector $\mathbf{x} = a\hat{\mathbf{x}}$ takes the form $\mathbf{x}' = a(\hat{\mathbf{x}} + \varepsilon_{xy}\hat{\mathbf{y}})$. Thus ε_{xy} represents the angle (in radians) through which the x-axis is rotated towards the direction of the y-axis.

It will prove convenient to express the state of strain of the lattice in terms of the dilation, together with six volume-conserving shears. The dilation Δ is defined as the fractional change in the volume Ω of the unit cell,

$$\Delta = \frac{\delta\Omega}{\Omega} = \varepsilon_{xx} + \varepsilon_{yy} + \varepsilon_{zz}. \tag{1}$$

The tetragonal shears γ_x, γ_y and γ_z are volume-conserving combinations of the elongations, ε_{xx}, ε_{yy} and ε_{zz}. For example, the strains associated with a pure tetragonal shear γ_x along the x-axis are

$$\varepsilon_{xx} = \gamma_x;\ \varepsilon_{yy} = \varepsilon_{zz} = -\tfrac{1}{2}\gamma_x. \tag{2}$$

The angular shears γ_{xy}, γ_{yz} and γ_{zx} are combinations of the axis rotations ε_{xy}, ε_{yx}, ε_{xz}, ε_{zx}, ε_{yz}, ε_{zy}. For example γ_{xy}, the angular shear in the x–y plane, is defined by

$$\gamma_{xy} \equiv \gamma_{yx} \equiv (\varepsilon_{xy} + \varepsilon_{yx}). \tag{3}$$

It follows that γ_{xy} is equal to the shear-induced decrease in the angle between the x- and y-axes of the real space lattice. The different components of strain are illustrated in Fig. 7.1.

When a hydrostatic pressure P is applied to a cubic crystal, the resulting strain is a pure dilation given by

$$\Delta = -\frac{3P}{c_{11} + 2c_{12}}. \tag{4}$$

The strain field induced by a uniaxial tension is more complicated. If a stress σ is applied in a direction whose direction cosines with respect to

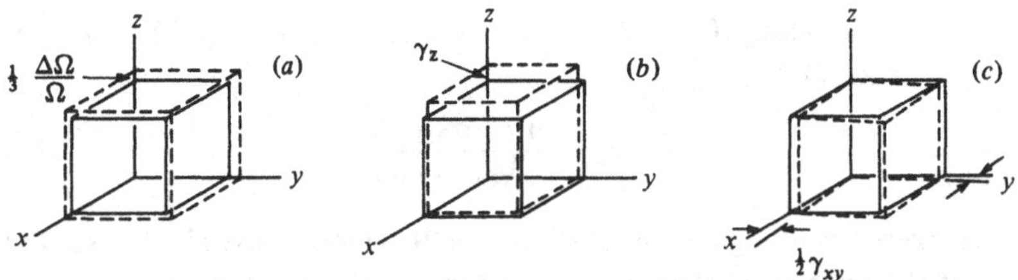

Fig. 7.1. Distortions (dashed lines) of a unit cube (solid lines) by (*a*) isotropic dilation $\Delta\Omega/\Omega$, (*b*) tetragonal shear γ_z, (*c*) angular shear γ_{xy}.

the x, y and z directions are l_x, l_y and l_z, then there is a dilation

$$\Delta = \frac{\sigma}{c_{11}+2c_{12}}, \tag{5}$$

whose magnitude is independent of the direction in which the stress is applied. In addition, elongations occur along the three axes. They are given by

$$\tilde{\varepsilon}_{ii} = \frac{\sigma}{c_{11}-c_{12}}(l_i^2 - \tfrac{1}{3}), \tag{6}$$

where we have introduced $\tilde{\varepsilon}_{ii}$ to denote the elongation which remains after the dilation contribution has been subtracted out. The elongations $\tilde{\varepsilon}_{ii}$ are related to the tetragonal shears γ_i by

$$\begin{aligned} \tilde{\varepsilon}_{xx} &= \gamma_x - \tfrac{1}{2}(\gamma_y + \gamma_z), \\ \tilde{\varepsilon}_{yy} &= \gamma_y - \tfrac{1}{2}(\gamma_z + \gamma_x), \\ \tilde{\varepsilon}_{zz} &= \gamma_z - \tfrac{1}{2}(\gamma_x + \gamma_y). \end{aligned} \tag{7}$$

The tetragonal shears are not uniquely determined by these equations, because the elongations must satisfy the volume conservation condition,

$$\tilde{\varepsilon}_{xx} + \tilde{\varepsilon}_{yy} + \tilde{\varepsilon}_{zz} = 0. \tag{8}$$

If a stress is applied along the [001] direction, $\tilde{\varepsilon}_{xx} = \tilde{\varepsilon}_{yy}$, and hence, from (7), $\gamma_x = \gamma_y$. It is natural to take the tetragonal shear to lie along the [001] direction, so that $\gamma_x = \gamma_y = 0$. Then the tetragonal shear is given by

$$\gamma_z = \frac{2}{3}\left(\frac{\sigma_{001}}{c_{11}-c_{12}}\right). \tag{9}$$

If the stress lies along [011], $\tilde{\varepsilon}_{yy} = \tilde{\varepsilon}_{zz}$. It is natural to take $\gamma_x = 0$, and then, from (6) and (7),

$$\gamma_y = \gamma_z = \frac{1}{3}\left(\frac{\sigma_{011}}{c_{11}-c_{12}}\right). \tag{10}$$

If the stress lies along any other direction, it is most convenient to set the sum of the tetragonal shears equal to zero. It follows that

$$\gamma_i = \frac{2\sigma}{3(c_{11}-c_{12})}(l_i^2 - \tfrac{1}{3}) \qquad (i = x, y, z). \tag{11}$$

If the stress lies along [111], all tetragonal shears must vanish by symmetry.

Angular shear is intrinsically a volume-conserving shear. When a uniaxial stress σ is applied along a direction whose direction cosines are l_x, l_y and l_z, the angular deformations of a cubic crystal are given by

$$\varepsilon_{ij} = \frac{l_i l_j \sigma}{2c_{44}}. \tag{12}$$

Hence the component of angular shear in the (i, j) plane is, from (3),

$$\gamma_{ij} = \frac{l_i l_j \sigma}{c_{44}}. \tag{13}$$

From this result it is easily seen that all angular shear components vanish when the uniaxial stress is applied along any one of the ⟨100⟩ directions. If the stress is applied along [011], the angular shears are

$$\gamma_{xy} = \gamma_{xz} = 0; \qquad \gamma_{yz} = \frac{\sigma_{011}}{2c_{44}}. \tag{14}$$

If the tension is applied along the [111] direction, the angular shears are

$$\gamma_{xy} = \gamma_{xz} = \gamma_{yz} = \frac{\sigma_{111}}{3c_{44}}. \tag{15}$$

To summarize, uniaxial tension causes a dilation that is independent of the direction in which it is applied. In addition, a tension applied along a ⟨100⟩ direction produces a tetragonal shear in that direction, whereas a tension applied along a ⟨111⟩ direction produces equal angular shears about all three axes. A tension applied along a general direction produces both tetragonal shears and also angular shears about all three axes.

Measurements of the pressure and uniaxial tension dependence of cross-sectional areas of the Fermi surface can in principle yield a complete description of the response to the six independent components of homogeneous lattice strain. It is clear from the preceding discussion that the dilation response is most easily obtained from pressure data. If the dilation response is known, the response to tetragonal shear can be most easily obtained from the response to tension applied along two of the ⟨100⟩ directions. Alternatively, the dilation and tetragonal shear responses can be deduced from the response to tension applied along each of the three ⟨100⟩ directions. The response to a symmetric combination of angular shears can be deduced by subtracting the dilation component from the response to tension along ⟨111⟩. The response to individual angular shears can be obtained by subtracting the dilation and tetragonal shear components from the response to uniaxial tension along the three ⟨110⟩ directions.

The response of the Fermi surface to an individual component of lattice strain can be deduced much more easily by propagating ultrasonic waves in the crystal. A longitudinal wave propagating along one of the ⟨100⟩ axes, i, produces a pure elongational strain ε_{ii} along that axis, while a transverse wave propagating along the axis i with polarization along another cube axis j produces an angular shear strain γ_{ij} together with a rotation $\theta_{ij}=\frac{1}{2}\gamma_{ij}$ about the third cube axis. The response of the Fermi surface cross-sectional area to θ_{ij} can be determined by measuring the angular dependence of the dHvA frequency, so the rotation term can be subtracted. The response of the Fermi surface to all ε_{ii} and γ_{ij} can be determined from at most six wave propagation measurements. The dilation and tetragonal shear responses are readily deduced from the response to elongation.

Unusual circumstances, such as high ultrasonic absorption for propagation along the cube axes, might dictate that waves be propagated along other crystallographic directions. The accompanying lattice strains are combinations of elongations, angular shears and rotations, and the ultrasonic approach loses some of its simplicity; we shall not pursue the details here.

The strain response of the Fermi surface of an anisotropic metal may be expressed in terms of the generalized deformation tensor K_{ij}, which is defined at every point on the Fermi surface by writing the normal displacement Δk_n at that point in response to the homogeneous strain component ε_{ij} in the form

$$\Delta k_n = K_{ij}\varepsilon_{ij}. \tag{16}$$

In general, the deformation Δk_n occurs in two stages. The change in the structure of the lattice associated with the strain brings about an adiabatic change in the wave-vector of each electron state on the Fermi surface, given by

$$(\Delta k_n)_1 = -k_i \varepsilon_{ij} n_j. \tag{17}$$

Usually the displaced wave-vectors do not constitute a constant energy surface of the distorted lattice. If this is so, the displacement of (17) is followed by a further displacement

$$(\Delta k_n)_2 = D_{ij} \varepsilon_{ij}, \tag{18}$$

as the distribution of occupied electronic states relaxes towards the equilibrium Fermi surface of the strained lattice. By writing

$$\Delta k_n = (\Delta k_n)_1 + (\Delta k_n)_2, \tag{19}$$

we obtain an expression,

$$K_{ij} = -k_i n_j + D_{ij}. \tag{20}$$

Equation (20) relates the deformation tensor K_{ij}, which is determined experimentally by measuring the response of the Fermi surface to homogeneous strain, to the tensor D_{ij}, the dissipative component of the deformation tensor, which is of fundamental importance in the theory of the attenuation of ultrasonic waves in the metals (Pippard 1960).

In practice, the deformation tensor K_{ij} has not yet been measured directly as a point function over the Fermi surface of any metal. We describe in this chapter how experimental studies of the strain dependence of extremal orbits on the Fermi surface can be interpreted in terms of orbital averages of K_{ij} calculated on the basis of appropriate theoretical models.

7.3 Experimental methods

7.3.1 Direct methods

The effects of stress and strain on the Fermi surface can be measured by the direct application of stress, either uniaxial or hydrostatic. The strains must be limited to about 1 part in 10^4 in order to ensure that no plastic deformation occurs. This in turn means that the fractional changes in Fermi surface areas will be of the order of 1 part in 10^4. Fortunately it is possible to measure such small changes with sufficient precision by following the shifts of phase of quantum oscillations in a constant magnetic field.

The free energy density $\mathscr{F}$ of a metal in a magnetic field **H** includes an oscillatory term contributed by each extremal cross-sectional area $\mathscr{A}$ of the Fermi surface which lies in a plane normal to **H**. Following Lifshitz & Kosevich (1955) we may write this oscillatory term as

$$\mathscr{F}_{\text{osc}} = \sum_{r=1}^{\infty} G_r \cos\left(\frac{2\pi rF}{H} + \psi_r\right), \tag{21}$$

where G_r, a slowly-varying function of temperature and magnetic field, depends on the electronic structure of the metal near the Fermi surface and on the Dingle temperature of the sample. We shall simplify the analysis by considering only the fundamental term, $r = 1$, in (21), since the fundamental is usually larger than the harmonics and hence can be measured more accurately.

The quantum oscillations in various physical properties can be calculated by taking derivatives of $\mathscr{F}_{\text{osc}}$ with respect to magnetic field, stress, etc. We need retain only the term involving the derivative of the cosine function, which is the dominant term when the phase $2\pi F/H$ is large. For example, the dHvA oscillations in the component of the magnetization parallel to **H** are given by

$$\tilde{M} = -\frac{\partial \mathscr{F}_{\text{osc}}}{\partial H} = -\frac{G}{H}\left(\frac{2\pi F}{H}\right) \sin\left(\frac{2\pi F}{H} + \psi\right), \tag{22}$$

where we have rewritten $G_1 \equiv G$ and $\psi_1 \equiv \psi$.

The change $\delta\zeta$ in the phase of the quantum oscillations at constant field H is given by

$$\delta\zeta = 2\pi \delta F/H + \delta\psi. \tag{23}$$

If we assume that $\delta F/F$ and $\delta\psi/\psi$ are of the same order, while the phase ζ is typically 10^4, the variation in ψ may be neglected. Therefore, we obtain

$$\delta\zeta = \frac{2\pi F}{H}\,\delta \ln F = \frac{2\pi F}{H}\,\delta \ln \mathscr{A}. \tag{24}$$

It follows that a change in area of 1 part in 10^4 yields a phase change of order unity, which can be measured with reasonable accuracy. This approach, which has been extremely useful for measuring hydrostatic pressure effects, has been reviewed recently by Schirber (1974, 1978). The main difficulty, particularly with non-cubic metals, is to ensure that the stress is truly isotropic. One must also take great care to ensure that the strain does not cause any change in the orientation of the sample, because the frequency F and the area $\mathscr{A}$ depend on orientation. In fact, precise measurements can be made only along symmetry directions, where the first derivative of the frequency with respect to sample rotation vanishes.

Direct uniaxial stress experiments are subject to severe difficulties of sample damage and sample rotation. In their pioneering work on the effects of tension on the noble metals, Shoenberg & Watts (1965) experienced difficulty in applying a measured stress through a vacuum seal; their subsequent work (Shoenberg & Watts 1967) improved on this (but see Gamble & Watts 1972). In work on white tin, Perz & Hum (1971) measured the tension by observing the extension of a calibrated spring in the upper part of the vacuum chamber, but they still experienced difficulties associated with friction in the mechanical pulling device, which must have only a very small clearance to minimize variations in the sample orientation. Spurgeon & Lazarus (1972), and Gamble & Watts (1973), applied a uniaxial compression to their samples and measured the resulting shift in the phase of the magnetization oscillations. Gerstein & Elbaum (1973) measured the effect of uniaxial compression on the phase of quantum oscillations in ultrasonic absorption in lead. The direct uniaxial stress experiments have the added disadvantage that, for practical reasons, the stress must lie essentially along the field direction in a solenoid, or perpendicular to it in a transverse magnet. The greater flexibility of indirect techniques, and the advantage that the sample need not be macroscopically strained, have led to their ascendency over direct uniaxial stress methods in recent years.

7.3.2 Indirect methods

The amplitudes of quantum oscillations in the linear dimensions and in the elastic stiffness constants of metallic single crystals depend on the stress and strain derivatives of the cross-sectional areas of the Fermi surface. The oscillatory strain (i.e. the magnetostriction in the i-direction) is equal to the derivative of the free energy with respect to stress σ_i,

$$\varepsilon_i = -\frac{\partial \mathscr{F}_{\text{osc}}}{\partial \sigma_i} = G\left(\frac{2\pi F}{H}\right)\frac{\partial \ln F}{\partial \sigma_i}\sin\left(\frac{2\pi F}{H}+\psi\right), \tag{25}$$

and the elastic stiffness tensor (which determines the sound velocity) contains oscillatory components,

$$c_{ij} = \frac{\partial^2 \mathscr{F}_{\text{osc}}}{\partial \varepsilon_i \partial \varepsilon_j} = -G\left(\frac{2\pi F}{H}\right)^2 \frac{\partial \ln F}{\partial \varepsilon_i} \frac{\partial \ln F}{\partial \varepsilon_j} \cos\left(\frac{2\pi F}{H} + \psi\right). \tag{26}$$

The magnetic torque about the i-direction is

$$\tau_i = \frac{\partial \mathscr{F}_{\text{osc}}}{\partial \theta_i} = -G\left(\frac{2\pi F}{H}\right) \frac{\partial \ln F}{\partial \theta_i} \sin\left(\frac{2\pi F}{H} + \psi\right), \tag{27}$$

where θ_i measures rotation of the field about the axis of torque. In (25), (26) and (27) the slowly-varying envelope function G is regarded as constant, since its logarithmic derivative may be neglected relative to the derivative of the circular function when the phase $2\pi F/H$ of the quantum oscillations is large.

The stress or strain dependence of the frequency F, and therefore of the corresponding area $\mathscr{A}$, can be determined by combining measurements of the oscillatory magnetization, or alternatively the torque, (equations (22) or (27)) with either the oscillatory magnetostriction or an oscillatory component of the elastic tensor (equations (25) or (26)), or by combining measurements of the latter two quantities and using the static elastic constants to relate stress and strain. The advantage of this approach is that, by dividing any pair of these expressions, the unknown amplitude G can be eliminated.

To measure the absolute magnetization, one needs to know the degree of coupling between the pick-up coil and the sample. To accompany their ultrasonic velocity measurements in beryllium, Testardi & Condon (1970) measured the magnetization of their sample by using a single-layer coil wound on a close-fitting former, but with this arrangement the magnetization signal proved to be too weak to measure with high accuracy. Recently de Wilde & Meredith (1976) measured magnetization with a superconducting flux transformer and ferrite core flux-gate magnetometer, and determined some aspects of the strain dependence of the Fermi surfaces of aluminium and indium by measuring simultaneously oscillations in magnetization and ultrasonic velocity. Finkelstein (1974) has combined direct measurements of magnetization with magnetostriction measurements in tin. In their analyses of magnetostriction data for the noble metals, Aron (1972) and Slavin (1973) calculated the amplitude of the oscillatory magnetization using known values of the

Fermi surface parameters, but for most metals the parameters are not known with sufficient accuracy to allow use of this method.

The magnetic torque associated with the component of magnetization perpendicular to the applied field can generally be measured more accurately and conveniently than the magnetization. The simultaneous measurement of torque and magnetostriction was pioneered by Griessen, Krugmann & Ott (1974), and has been used extensively since. Lee *et al.* (1976b) and Stanley, Perz, Lee & Griessen (1977) first combined simultaneous torque and ultrasonic velocity measurements in a study of tungsten. A limitation of the torque technique is that the torque vanishes when the magnetic field direction lies along a crystallographic direction of high symmetry.

The combination of an oscillatory component of the elastic stiffness tensor c_{ij}, which can be deduced from an ultrasonic velocity measurement, with torque or magnetization has the disadvantage that a product of strain derivatives is obtained. For example, for wave propagation along a cube axis, oscillations in the longitudinal sound velocity give $(\partial \ln \mathscr{A}/\partial \varepsilon_1)^2$, so that the sign of a strain derivative is not determined by the experiment. However, if the sign can be determined from other measurements, the ultrasonic technique gives directly the magnitudes of the strain derivatives. Although the strain derivatives can often be determined from a combination of two magnetostriction measurements along mutually perpendicular directions, the extraction of the individual shear dependences often involves a significant loss of accuracy. The ultrasonic technique also selects preferentially those oscillations that correspond to cross-sections with large strain derivatives, because a product or square factor appears in the amplitude. Furthermore, studies of the velocity of ultrasonic shear waves can provide a measure of the dependence of the Fermi surface on individual components of angular shear (Lee *et al.* 1976b) which cannot be determined directly from magnetostriction experiments.

The combination of magnetostriction and sound velocity measurements has the advantages that both effects have finite amplitude when the field direction lies along a symmetry direction, and that the signs of the strain and stress derivatives can be determined. The main disadvantage is that a single amplitude ratio determines a combination of strain or stress derivatives, whereas the techniques described earlier yield individual stress or strain derivatives. This may be no disadvantage in a comprehensive study, as two pairs of measurements yield two individual strain derivatives and stress derivatives. To give a specific example, if in a cubic crystal magnetostriction is measured, first along a cube axis parallel, then along a cube axis perpendicular to the magnetic field, while the longitudinal sound velocity parallel to the field is measured, the two experiments yield the ratios,

$$\left(\frac{\partial \ln \mathscr{A}}{\partial \varepsilon_1}\right)^2 \Big/ \frac{\partial \ln \mathscr{A}}{\partial \sigma_1} = \left(\frac{\partial \ln \mathscr{A}}{\partial \varepsilon_1}\right)^2 \Big/ \left(s_{11}\frac{\partial \ln \mathscr{A}}{\partial \varepsilon_1} + 2s_{12}\frac{\partial \ln \mathscr{A}}{\partial \varepsilon_2}\right)$$

and (28)

$$\left(\frac{\partial \ln \mathscr{A}}{\partial \varepsilon_1}\right)^2 \Big/ \frac{\partial \ln \mathscr{A}}{\partial \sigma_2} = \left(\frac{\partial \ln \mathscr{A}}{\partial \varepsilon_1}\right)^2 \Big/ \left[2s_{12}\frac{\partial \ln \mathscr{A}}{\partial \varepsilon_1} + (s_{11}+s_{12})\frac{\partial \ln \mathscr{A}}{\partial \varepsilon_2}\right].$$

The individual strain derivatives $\partial \ln \mathscr{A}/\partial \varepsilon_1$ and $\partial \ln \mathscr{A}/\partial \varepsilon_2$ can be deduced with the aid of the static elastic compliance constants s_{11} and s_{12}. It should be emphasized that, although (28) is quadratic, one of the solutions corresponds to zero strain and stress derivatives, so that the remaining solution is unique. This was not the case with the geometry used by Stanley, Perz & Au (1976), who combined measurements in tungsten of magnetostriction and sound velocity parallel to the field with measurements of both perpendicular to the field. A quartic equation resulted, and the correct roots could be selected only by reference to other experiments.

In earlier experimental work the two physical quantities needed to determine the strain dependence of the Fermi surface were measured separately, but Griessen *et al.* (1974) first demonstrated the great advantage of measuring them simultaneously, which is that their relative phase, and hence the sign of the strain dependence, can be determined. Moreover, difficulties of reproducing the same temperature, field and orientation, and even the same sample condition for those materials which suffer damage from thermal cycling, are avoided by making the two measurements simultaneously. The dilatorquemeter described by Posternak *et al.* (1975), and in a modified version by Griessen, Lee & Stanley (1977a), combines the dilatometer of Brändli & Griessen (1973) with the torquemeter of Griessen (1973). The precision of the measurements is limited in practice by such factors as possible coupling between the longitudinal and transverse magnetostriction due to the rigid mounting of the sample (Slavin 1973), and possible coupling between the magnetostriction and torque due to relaxation of the glue attaching the sample to the base plate of the magnetostriction cell.

Other experimental arrangements worthy of note include that of Thompson, Aron, Chandrasekhar & Langenberg (1971), who combined measurements of oscillations in Young's modulus with magnetostriction, and that of Pudalov & Khaĭkin (1974), who measured the magnetostriction in very pure tin, and determined the amplitude of the oscillatory magnetization from the Shoenberg effect (see Chapter 4).

Finally, we should point out the circumstances in which the preceding analysis is inadequate. First, when the phase $2\pi F/H$ of the quantum oscillations is small, the variation of G_r in (21) may not be neglected. In particular, ε_i and c_{ij} will contain contributions from the strain dependence of the effective mass m_c^*. Moreover, c_{ij} will contain a term,

$$-G\frac{\partial^2 \ln \mathscr{F}}{\partial\varepsilon_i\partial\varepsilon_j}\left(\frac{2\pi F}{H}\right)\sin\left(\frac{2\pi F}{H}+\psi\right), \tag{29}$$

which is normally neglected in comparison with the term in (26), since it is smaller by a factor $H/2\pi F$. For extremal orbits for which gm_c^*/m is an odd integer, the fundamental oscillations vanish in both the magnetization and torque. However, fundamental oscillations may be observed in magnetostriction and sound velocity measurements because of the strain dependence of gm_c^* (Holroyd, Fawcett & Perz, 1978).

In a ferromagnetic metal the magnetic induction $\mathbf{B}=\mathbf{H}_{ext}+4\pi\mathbf{M}+\mathbf{H}_d$ should replace $\mathbf{H}$ in (21); $\mathbf{H}_{ext}$ is the applied field and $\mathbf{H}_d$ the demagnetizing field. The analysis remains straightforward provided that the sample shape is ellipsoidal so that $\mathbf{H}_d$ is constant and parallel to $\mathbf{H}_{ext}$ (see Chapter 6).

When the oscillatory magnetization is sufficiently large, the magnetic interaction effect becomes important (see Chapter 4). When $d\tilde{M}/dB$ is of order unity, the distribution of magnetization across the sample breaks up into diamagnetic domains (Condon 1966a). Testardi & Condon (1970) have shown that the domain walls in a sample of beryllium are unable to follow the strains associated with ultrasonic waves of frequency 20 MHz, and intricate analysis is required to extract the strain dependence of the Fermi surface. Furthermore, even when the sample does not break up into diamagnetic domains (Condon 1966a). Testardi & Condon (1970) fundamental and harmonic amplitudes are altered. Even though $d\tilde{M}/dB$ has the same line-shape as c_{ij}, the preceding simplified analysis cannot be used.

Finally, at ultrasonic frequencies which are sufficiently low that the wavelength is large relative to the classical skin depth, the lines of $\mathbf{B}$ follow the particle motion, giving rise to a coupling between the magnetic induction and the lattice strain (Alpher & Rubin 1954). Testardi & Condon (1970) discussed the propagation of transverse and longitudinal waves parallel and perpendicular to the magnetic field; a treatment generalized to include arbitrary field direction has recently been given by de Wilde (1978).

7.4 Theoretical models

The very precise information that is obtainable from Fermi surface studies has stimulated the development of theoretical models whose purpose is to extract from the experimental data parameters which describe various aspects of the interaction between the conduction electrons and the lattice. A plane-wave representation of the conduction electron states, in combination with a weakly-scattering pseudopotential, has been very successful in accounting for the strain dependence of the Fermi surfaces of simple metals (Heine 1970; Cohen & Heine 1970). However, this approach is valid only if the core is highly localized, and if there are states in the core of the same symmetry character as the electron states at the Fermi surface, so that the lattice potential is largely 'cancelled' by orthogonalization to the core. The electron states at the Fermi surfaces of the noble metals, although of predominantly s- and p-wave character, are strongly perturbed by the d-bands which lie only one or two electron-volts below the Fermi level, and the small core approximation does not hold. Thus the assumptions of plane-wave pseudopotential theory are not strictly valid for the noble metals. For metals of the d-transition series, where large regions of the Fermi surface are derived from electronic states of d-wave character, the combination of a plane-wave representation with a weakly-scattering pseudopotential fails completely.

In order to describe the Fermi surfaces of the noble and transition metals, it is necessary to develop a model which takes into account the strong interaction between the conduction electrons and the lattice. Perhaps the most successful approach has been that of the phase-shift pseudopotential, in which the lattice potential is approximated by an array of non-overlapping spherically symmetric potentials, each centred on a lattice site. The phase shifts which characterize the interaction between the conduction electrons and the lattice are evaluated by fitting experimental Fermi surface cross-sections. Phase-shift pseudopotential calculations based on the multiple-scattering formalism of Korringa, Kohn and Rostoker (KKR) give a satisfactory account of many aspects of the electronic structure of the noble metals, including the shape of the Fermi surface (Lee 1969; Shaw, Ketterson & Windmiller 1972), the anisotropy of the Fermi velocity (Lee 1970) and the anisotropy of impurity scattering (Coleridge, Holzwarth & Lee 1974). They have also been applied, to a more limited extent but with comparable success, to various transition metals (see, for example, Ketterson, Koelling, Shaw & Windmiller 1975).

In this section we outline the different versions of pseudopotential theory which have been found to be appropriate for various metals. In Section 7.5 we consider lead, copper and tungsten as illustrative of the simple, noble and transition metals respectively. We present comprehensive experimental data on the strain dependence of the Fermi surfaces of these metals, and make a comparison with the results of pseudopotential calculations.

7.4.1 Free-electron scaling

The simplest treatment of the strain dependence of the Fermi surface is based on free-electron theory, in which the Fermi surface is spherical, and its volume is inversely proportional to the volume of the real space unit cell. It follows that cross-sectional areas of extremal orbits are inversely proportional to the two-thirds power of the volume, i.e. $d \ln \mathscr{A}/d \ln \Omega = -\frac{2}{3}$. Free-electron scaling is often a useful approximation to the dilation dependence of the Fermi surfaces of real metals, even when the free-electron model is not strictly applicable.

7.4.2 The OPW pseudopotential

In the OPW method of band-structure calculation, the shape of the Fermi surface of a homogeneously strained lattice can be determined by finding the locus of wave-vectors **k** that satisfy an implicit equation of the form,

$$\lambda(\mathbf{k}, \mathscr{E}_F, V_s, \varepsilon) = 0. \tag{30}$$

λ is an eigenvalue of the transformed Hamiltonian matrix, which depends parametrically on the Fermi energy $\mathscr{E}_F$, the pseudopotential matrix elements V_s, and the lattice strains ε. However, it is more accurate to calculate strain derivatives of the areas of orbits on the Fermi surface by evaluating directly the derivative of the wave-vector at each point on the orbit. This can be done by evaluating the derivative of the Hamiltonian matrix with respect to each of the parameters analytically, and using the Hellman–Feynman theorem to calculate the corresponding derivatives of the eigenvalue. In order to remain on the constant energy surface $\lambda = 0$, the variation in a parameter X_r must be accompanied by a variation in the Fermi wave-vector whose component normal to the Fermi surface of the unstrained lattice at the point **k** is given by

$$(\partial k_F/\partial X_r) = -\frac{(\partial\lambda/\partial X_r)}{\nabla_{\mathbf{k}}\lambda}. \tag{31}$$

A calculation of the response of the cross-sectional area of an orbit on the Fermi surface to an arbitrary homogeneous lattice strain involves evaluating orbital integrals of the form

$$\frac{\partial\mathcal{A}}{\partial X_r} = -\oint \frac{(\partial\lambda/\partial X_r)}{|(\nabla_{\mathbf{k}}\lambda)|}\,\mathrm{d}l, \tag{32}$$

where $(\nabla_{\mathbf{k}}\lambda)$ is the projection of the gradient of the eigenvalue onto the plane of the orbit, and $\mathrm{d}l$ is an element of length along the path of the orbit.

The response to a general strain ε is given by

$$\frac{\mathrm{d}\ln\mathcal{A}}{\mathrm{d}\varepsilon} = \frac{\partial\ln\mathcal{A}}{\partial\varepsilon} + \frac{\partial\ln\mathcal{A}}{\partial\ln\mathscr{E}_F}\left(\frac{\mathrm{d}\ln\mathscr{E}_F}{\mathrm{d}\varepsilon}\right) + \sum_s \frac{\partial\ln\mathcal{A}}{\partial\ln V_s}\left(\frac{\mathrm{d}\ln V_s}{\mathrm{d}\varepsilon}\right). \tag{33}$$

All the partial derivatives of $\mathcal{A}$ can be calculated from integrals of the form of (32). The strain dependence of $\mathscr{E}_F$ can be assumed to be free-electron-like (Griessen & Sorbello 1974) and therefore to depend only on the dilation. The strain derivatives of $\mathscr{E}_F$ and V_s can be evaluated either by *a priori* calculations or by considering the derivatives as parameters to be determined by fitting to experimental strain-dependence data.

In an *a priori* calculation, the volume derivative of $\mathscr{E}_F$ can be determined from the equation,

$$n = \int_0^{\mathscr{E}_F} N(\mathscr{E})\,\mathrm{d}\mathscr{E}, \tag{34}$$

where n is the number of electrons per atom, which remains invariant under deformation, and $N(\mathscr{E})$ is the density of states defined by

$$N(\mathscr{E}) = \sum_i \frac{2\Omega}{(2\pi)^3}\int_{BZ} \delta(\mathscr{E} - \lambda_i)\,\mathrm{d}^3k, \tag{35}$$

where the index i denotes the energy band. For the dilation dependence of the Fermi energy we find,

$$\frac{\mathrm{d}\mathscr{E}_F}{\mathrm{d}\ln\Omega}=\frac{\sum_i \frac{2\Omega}{(2\pi)^3}\int_{BZ}\frac{\mathrm{d}\lambda_i}{\mathrm{d}\ln\Omega}\delta(\mathscr{E}_F-\lambda_i)\,\mathrm{d}^3k}{N(\mathscr{E}_F)}. \tag{36}$$

The pseudopotential matrix element V_s that corresponds to the reciprocal lattice vector $\mathbf{G}_s$ is the product of a structure factor $S(\mathbf{G}_s)$ and a form factor $v_s = v(G_s)$. For a Bravais lattice with a basis of one atom, the structure factor is unity, and the strain derivative of V_s equals that of the form factor v_s itself. Assuming that the strain dependence of the form factor $v(q)$ is local in nature, we can write

$$\frac{\mathrm{d}V_s}{\mathrm{d}\varepsilon}=\frac{\mathrm{d}v_s}{\mathrm{d}\varepsilon}=\left.\frac{\partial v}{\partial\ln q}\right|_{q=G_s}\frac{\mathrm{d}\ln G_s}{\mathrm{d}\varepsilon}+\frac{\partial v_s}{\partial\ln k_F}\frac{\mathrm{d}\ln k_F}{\mathrm{d}\varepsilon}, \tag{37}$$

where k_F is the radius of the free-electron sphere. For a cubic crystal, the volume derivative of v_s is

$$\frac{\mathrm{d}v_s}{\mathrm{d}\ln\Omega}=-\frac{1}{3}\left(\left.\frac{\partial v_s}{\partial\ln q}\right|_{q=G_s}+\frac{\partial v_s}{\partial\ln k_F}\right), \tag{38}$$

since both k_F and G_s are proportional $\Omega^{-\frac{1}{3}}$; the shear derivatives are simply

$$\frac{\mathrm{d}v_s}{\mathrm{d}\gamma}=\left.\frac{\partial v}{\partial\ln q}\right|_{q=G_s}\frac{\mathrm{d}\ln G_s}{\mathrm{d}\gamma}, \tag{39}$$

as k_F is constant for volume-conserving shears.

The dependence of the form factor on k_F in the second term of (38) can be evaluated with sufficient accuracy (Griessen & Sorbello 1974) by noticing that $v(q, k_F)$ can be approximated by the functional form

$$v(q, k_F)=f(q)/(\Omega\varepsilon(q, k_F)), \tag{40}$$

where $\varepsilon(q, k_F)$ is the Hartree dielectric function and $f(q)$ is a function of q.

7.4.3 The phase-shift pseudopotential

In the KKR method of band-structure calculation the shape of the Fermi surface is obtained from an implicit equation of the form

$$\lambda(\mathbf{k}, \mathscr{E}_F, \eta_l)=0, \tag{41}$$

where λ is an eigenvalue of the KKR secular matrix. The phase shifts η_l are adjusted to bring the shape of the Fermi surface calculated from (41) into coincidence with the experimental data. They constitute a very economical parameter set, apparently because they have a direct physical interpretation.

The Fermi energy $\mathscr{E}_F$ is also a parameter in a phase-shift pseudopotential. The fact that its value cannot be determined by fitting Fermi surface data can be explained as follows. The value of $\mathscr{E}_F$ governs the wave-vector, and hence the curvature, of the wave-function in the interstitial region of the lattice. To the extent that the interstitial volume is small, the curvature in this region can vary over a wide range without changing the physical character of the wave-function. Therefore, the Fermi energy parameter $\mathscr{E}_F$ is best regarded as a free parameter in a phase-shift fit to Fermi surface data. As $\mathscr{E}_F$ is varied, the phase shifts required to fit the data vary, and so (to a much smaller extent) do the logarithmic derivatives of the wave-function at the surface of the muffin-tin (MT) sphere, but physical quantities derived from the fit to the data are relatively insensitive to the choice of $\mathscr{E}_F$. In practice, it is usual to estimate the value of $\mathscr{E}_F$ by carrying out a first-principles band-structure calculation, and to adopt the resulting value in fitting the Fermi surface data.

The phase-shift pseudopotential method has recently been extended in order to calculate the homogeneous strain dependence of the Fermi surfaces of transition metals (Griessen *et al.* 1977a). In principle, the shape of the Fermi surface of a homogeneously strained lattice can be determined by finding the locus of wave-vectors **k** that satisfy the generalized KKR secular equation

$$\lambda(\mathbf{k}, \mathscr{E}_F, \eta_l, \varepsilon) = 0, \tag{42}$$

where the lattice strains ε are included as parameters.

This approach to the strain dependence of the Fermi surface assumes a 'rigid MT-sphere' model of the lattice potential – that is to say, the non-overlapping potential associated with each lattice site is supposed to remain spherically symmetric, and to be displaced rigidly, as the lattice is strained. The discussion of a cubic metal in this approximation allows an important simplification, for Gray & Gray (1976) have shown by symmetry considerations that scalar parameters such as the Fermi energy, the lattice potential, and the phase shifts, must remain invariant in any volume-conserving lattice deformation. It follows that the shear response of the Fermi surface is completely determined by the phase shifts that characterize the electron–ion interaction in the unstrained lattice, which can be evaluated by fitting Fermi surface data for the unstrained metal. For each individual component of shear, a single partial derivative with respect to shear strain determines the distortion of the Fermi surface. That is why it is advantageous to decompose the strain into shear components and a dilation.

In the rigid MT-sphere approximation, the phase shifts and the Fermi energy depend on the dilation of the crystal. Several partial derivatives of the form of (32) each contribute to the response of the Fermi surface to dilational strain. The total derivative of the cross-sectional area $\mathscr{A}$ of an extremal orbit on the Fermi surface with respect to the volume of the cubic unit cell can be written

$$\frac{\mathrm{d}\ln\mathscr{A}}{\mathrm{d}\ln\Omega}=\left(\frac{\partial\ln\mathscr{A}}{\partial\ln\Omega}\right)_{\eta,\mathscr{E}_\mathrm{F}}+\frac{\mathrm{d}\mathscr{E}_\mathrm{F}}{\mathrm{d}\ln\Omega}\left(\frac{\partial\ln\mathscr{A}}{\partial\mathscr{E}_\mathrm{F}}\right)_{\eta,\Omega}+\sum_l\frac{\mathrm{d}\eta_l}{\mathrm{d}\ln\Omega}\left(\frac{\partial\ln\mathscr{A}}{\partial\eta_l}\right)_{\Omega,\mathscr{E}_\mathrm{F}}. \tag{43}$$

As we noted above, the Fermi energy parameter $\mathscr{E}_\mathrm{F}$ can be treated as a free parameter in a phase-shift pseudopotential fit to Fermi surface data. It follows that $(\mathrm{d}\mathscr{E}_\mathrm{F}/\mathrm{d}\ln\Omega)$ can be assigned a value that varies within rather wide limits, without altering significantly the quality of the fit to the data, although the numerical values of the phase-shift derivatives $(\mathrm{d}\eta_l/\mathrm{d}\ln\Omega)$ do, of course, depend on the numerical value assigned to $(\mathrm{d}\mathscr{E}_\mathrm{F}/\mathrm{d}\ln\Omega)$. A common procedure is to take $(\mathrm{d}\mathscr{E}_\mathrm{F}/\mathrm{d}\ln\Omega)=-1/N(\mathscr{E}_\mathrm{F})$, where $N(\mathscr{E}_\mathrm{F})$ is the density of states at the Fermi energy, and to estimate $N(\mathscr{E}_\mathrm{F})$ from a first-principles band-structure calculation. Alternatively, perhaps the most convenient procedure is to take $(\mathrm{d}\mathscr{E}_\mathrm{F}/\mathrm{d}\ln\Omega)=-\frac{2}{3}\mathscr{E}_\mathrm{F}$. This has the advantage that the expression for the total derivative of the cross-sectional area of the Fermi surface with respect to volume reduces to the simplified form

$$\frac{\mathrm{d}\ln\mathscr{A}}{\mathrm{d}\ln\Omega}=-\tfrac{2}{3}+\sum_l\frac{\mathrm{d}\eta_l}{\mathrm{d}\ln\Omega}\left(\frac{\partial\ln\mathscr{A}}{\partial\eta_l}\right)_{\Omega,\mathscr{E}_\mathrm{F}}, \tag{44}$$

where the departures from free-electron scaling are expressed explicitly in terms of phase shift derivatives. This is the form adopted by Templeton (1974) in his analyses of the dilation dependences of the Fermi surfaces of the noble metals. The coefficients $(\mathrm{d}\eta_l/\mathrm{d}\ln\Omega)$ can be deduced by making a least squares fit to the experimental dilation dependences in the form of (44).

7.5 Representative results

7.5.1 Lead (an example of the simple metals)

The Fermi surface of lead, and its dependence on strain, have been the subjects of extensive study. Anderson, O'Sullivan & Schirber (1967) showed in an early investigation that the pressure dependence of the Fermi surface of lead is influenced strongly by the pressure dependence of the pseudopotential form factor. In a subsequent paper (Anderson *et al.* 1972a) they compared their experimental pressure derivatives, $d \ln \mathcal{A}/dP$ (where P is the hydrostatic pressure), to the predictions of various Fermi surface models. They pointed out that, although the nearly-free-electron model accounts for the general topological features of the Fermi surface of lead, it fails to predict correctly its pressure dependence. They found it necessary to use a non-local pseudopotential model, with spin–orbit interaction, and to scale the pseudopotential core radius R_M in proportion to the lattice constant in order to reproduce the hydrostatic pressure data.

The source of the failure of simple 4-OPW calculations such as that of Anderson *et al.* (1972a) using a local pseudopotential is discussed by Van Dyke (1973). According to Van Dyke, the local pseudopotential results of Anderson *et al.* (1972) do not deviate significantly from free-electron scaling because they use only a few orthogonalized plane waves. Van Dyke obtained reasonably good agreement with the experimental values of $d \ln \mathcal{A}/dP$ of Anderson *et al.* (1972a) with a local pseudopotential calculation involving 90 OPWs.

From the analyses of Anderson *et al.* (1972a) and Van Dyke (1973) one can conclude that either a non-local pseudopotential with variable core radius and a small number of OPWs, or a local fully converged pseudopotential calculation, are required to reproduce the pressure dependence of the Fermi surface of lead. Sorbello & Griessen (1973) showed, however, that fair agreement with the experimental pressure derivatives $d \ln \mathcal{A}/dP$ can be obtained by a simple local 4-OPW model without spin–orbit interaction (see Table 7.2). The root-mean-square deviation for the ζ, ξ, ν, ψ [110] and ψ [100] orbits (0.69×10^{-6} bar^{-1}) is the same as that of Van Dyke's (1973) 90-OPW calculation (0.67×10^{-6} bar^{-1}) and better than the value (0.82×10^{-6} bar^{-1}) that Anderson *et al.* (1972a) obtained using a non-local model including spin–orbit coupling but with constant core radius. Anderson *et al.* (1972a) obtained significantly better agreement by scaling the core radius R_M in a proportion to the lattice constant. As pointed out by Van Dyke (1973), a scaled ion core radius is inconsistent with the basic assumptions underlying the OPW pseudopotential method.

We show here that a good description of the response of the Fermi surface of lead to a general homogeneous strain can be achieved by a pseudopotential calculation using a small number of OPWs in a local approximation, without including a spin–orbit parameter, and keeping the core radius constant. For our calculation of the Fermi surface we have used an 8-OPW matrix and neglected all the matrix elements $v(q)$ with $q > 2k_F$.

The minimum number of reciprocal lattice vectors which have to be included in a band-structure calculation for a cubic metal subjected to a general homogeneous strain is 8. Calculations in one-eighth of the Brillouin zone have been carried out using the reciprocal lattice vectors (0, 0, 0), (1, 1, 1), $(\bar{1}, 1, 1)$, $(1, \bar{1}, 1)$, $(1, 1, \bar{1})$, (2, 0, 0), (0, 2, 0) and (0, 0, 2). A fit of the three parameters $\mathscr{E}_F$, V_{111} and V_{200} (the spin–orbit interaction is neglected) to experimental zero pressure cross-sections of the Fermi surface leads to the values (1 a.u. = 27.2 eV)

$$\left.\begin{aligned} \mathscr{E}_F &= 0.35085 \text{ a.u.}, \\ V_{111} &= -0.04769 \text{ a.u.}, \\ V_{200} &= -0.02409 \text{ a.u.} \end{aligned}\right\} \qquad (45)$$

The fitted values of the matrix elements are close to the theoretical values given by Appapillai & Heine (1972) ($V_{111} = -0.04678$ a.u. and $V_{200} = -0.01993$ a.u.) for an optimized Shaw potential. $\mathscr{E}_F$ is slightly lower than the free-electron value, $\mathscr{E}_F^0 = 0.35227$ a.u. (our zero temperature lattice constant is 9.2874 a.u.), as expected from a perturbation calculation (see equation 3-55 in Harrison (1966)). The value of $\mathscr{E}_F$ obtained from the fit to the cross-sectional areas is in good agreement with the value $\mathscr{E}_F = 0.3512$ a.u. obtained from the integrated density of states calculated with our 8-OPW model by the method of Gilat & Raubenheimer (1966). Our calculation fits the experimental areas of eight different orbits with a root-mean-square deviation of 2.9 per cent, which is not as good as the results of Anderson *et al.* (1972) or of Van Dyke (1973). This is not surprising, as only three fitting parameters are used in our model, compared to four for the local models with spin–orbit interaction of Anderson *et al.* (1972a) and Van Dyke (1973). As the main objective of this work is to compare the predictions of various band-structure models for the strain response of the Fermi surface, we have not tried to improve the quality of our zero pressure fit. This seems justified because it has been found that in other simple metals (Griessen & Sorbello 1974) the quality of the zero pressure fit has relatively little influence on the values of the strain derivatives.

We shall now proceed along three different lines to show that our simple 8-OPW model gives an adequate description of the response of the Fermi surface to a general strain. First, we shall compare the results of an *a priori* calculation based on this model with hydrostatic pressure data. Secondly, the 8-OPW model will be used to calculate the partial derivatives in (33). It is then possible to determine the volume dependence of $\mathscr{E}_F$, V_{111} and V_{200} from a fit to experimental pressure data. Finally we shall compare theoretical values to experimental data on the uniaxial stress dependence of the Fermi surface.

Let us first consider the response to a dilational strain. From (33), the volume dependence of the area $\mathscr{A}$ of an extremal cross-section of the Fermi surface is given by

$$\frac{\mathrm{d}\ln\mathscr{A}}{\mathrm{d}\ln\Omega}=\left(\frac{\partial\ln\mathscr{A}}{\partial\ln\Omega}\right)_{\mathscr{E}_F,V_s}+\left(\frac{\partial\ln\mathscr{A}}{\partial\ln\mathscr{E}_F}\right)_{\Omega,V_s}\frac{\mathrm{d}\ln\mathscr{E}_F}{\mathrm{d}\ln\Omega}+\sum_s\left(\frac{\partial\ln\mathscr{A}}{\partial V_s}\right)_{\Omega,\mathscr{E}_F,V_s}\frac{\mathrm{d}V_s}{\mathrm{d}\ln\Omega}. \tag{46}$$

All the partial derivatives of $\mathscr{A}$ can be calculated by means of (32); they depend only on the zero pressure parameters $\mathscr{E}_F$, V_{111} and V_{200}. We have calculated the volume derivative of $\mathscr{E}_F$ from the relation

$$\frac{\mathrm{d}\ln\mathscr{E}_F}{\mathrm{d}\ln\Omega}=\left(\frac{\partial\ln\mathscr{E}_F}{\partial\ln\Omega}\right)_{V_s}+\sum_s\left(\frac{\partial\ln\mathscr{E}_F}{\partial V_s}\right)_{\Omega}\frac{\mathrm{d}V_s}{\mathrm{d}\ln\Omega}. \tag{47}$$

The partial derivatives have been determined from (36) by computing the density of states and its derivatives by the method of Gilat & Raubenheimer (1966). This calculation gives

$$\left.\begin{aligned}\frac{\partial\ln\mathscr{E}_F}{\partial\ln\Omega}&=-0.668\text{ a.u.},\\ \frac{\partial\ln\mathscr{E}_F}{\partial V_{111}}&=0.282\text{ a.u.},\\ \frac{\partial\ln\mathscr{E}_F}{\partial V_{200}}&=-0.532\text{ a.u.}\end{aligned}\right\} \tag{48}$$

We have evaluated $\mathrm{d}V_s/\mathrm{d}\ln\Omega$ using (37). The derivatives $\partial v/\partial \ln q$, at the appropriate reciprocal lattice vectors, have been taken from the pseudopotential form factor of Appapillai & Heine (1972). Equation (40) has been used to calculate

$$\left.\begin{aligned}\frac{\partial v_{111}}{\partial \ln k_F} &= -0.1150\ \text{a.u.},\\ \frac{\partial v_{200}}{\partial \ln k_F} &= -0.0587\ \text{a.u.}\end{aligned}\right\} \tag{49}$$

The volume derivatives $\mathrm{d}\ln\mathscr{E}_F/\mathrm{d}\ln\Omega$ and $\mathrm{d}V_s/\mathrm{d}\ln\Omega$, and the derivatives of the form factor $\partial v/\partial \ln q$, are summarized in Table 7.1, and the corresponding area derivatives $\mathrm{d}\ln\mathscr{A}/\mathrm{d}\ln\Omega$ are given in Table 7.2. The calculated area derivatives agree well with the measured values, significantly better than the results of all the previous model calculations, except those of Anderson *et al.* (1972a), who used non-local pseudopotential and spin–orbit coupling, and scaled the core radius with the lattice constant.

TABLE 7.1. *Volume derivatives* $\mathrm{d}\ln\mathscr{E}_F/\mathrm{d}\ln\Omega$, $\mathrm{d}V_s/\mathrm{d}\ln\Omega$ *and form factor derivatives* $\mathrm{d}v/\mathrm{d}\ln q$ *used in the determination of* $\mathrm{d}\ln\mathscr{A}/\mathrm{d}\ln\Omega$ *in lead. All quantities are expressed in a.u.*

	a	*b*	*c*	*d*
$\dfrac{\mathrm{d}\ln\mathscr{E}_F}{\mathrm{d}\ln\Omega}$	−0.649	−0.641	$-\frac{2}{3}$	−0.870
$\left.\dfrac{\partial v}{\partial \ln q}\right\rvert_{q=G_{111}}$	0.180	0.162	0.117	—
$\left.\dfrac{\partial v}{\partial \ln q}\right\rvert_{q=G_{200}}$	0.198	0.162	0.135	—
$\dfrac{\mathrm{d}V_{111}}{\mathrm{d}\ln\Omega}$	−0.0215	−0.0157	—	−0.0561
$\dfrac{\mathrm{d}V_{200}}{\mathrm{d}\ln\Omega}$	−0.0464	−0.0345	—	−0.0707

[a] This work: $\mathrm{d}\ln\mathscr{E}_F/\mathrm{d}\ln\Omega$ obtained from equation (47) and the slopes of the form factor from Appapillai & Heine (1972).
[b] This work: fitted values.
[c] Sorbello & Griessen (1973): 4-OPW pseudopotential calculation and fitted form-factor slopes.
[d] Van Dyke (1973): 90-OPW pseudopotential calculation.

TABLE 7.2. *Dilation dependence,* $(d \ln \mathscr{A}/d \ln \Omega)$, *of areas of extremal orbits on the Fermi surface of lead*

Orbit	*a*	*b*	*c*	*d*	*e*	*f*	*g*	*h*	*i*
$\zeta[110]$	-1.22 ± 0.10	-1.51	-1.26	-1.36	-0.68	-1.07	-1.02	-1.36	-1.41
$\zeta[111]$	-1.32 ± 0.10	-1.53	-1.27		-0.68	-1.07	-1.02	-1.36	-1.41
$\xi[100]$	0 ± 0.15	0.20	0.03	0.10	-0.97	-1.32	0	0.05	0.44
$\nu[100]$	-1.12 ± 0.05	-1.37	-1.11	-1.27	-0.73	-1.02	-0.78	-1.22	-0.93
$\theta[111]$	-0.58 ± 0.05	-0.46	-0.56		-0.54	-0.19	-0.68	-0.49	-0.68
$\psi[111]$	-1.02 ± 0.10	-1.21	-1.16		-0.68	-0.78	-0.63	-1.02	-1.36
$\psi[110]$	-1.41 ± 0.10	-1.17	-1.13	-0.83	-0.63	-0.73	-0.68	-1.22	-1.41
$\psi[100]$	-1.17 ± 0.19	-1.34	-1.28	-0.93	-0.63	-0.78	-0.73	-1.17	-1.75
RMS deviation		0.23	0.13	0.33	0.63	0.62	0.40	0.11	0.32

[a] Experimental values Anderson *et al.* (1972a).
[b] This work: obtained with the computed value, $d \ln \mathscr{E}_F/d \ln \Omega = -0.649$, and the derivatives of the Appapillai & Heine (1972) form factors, $|\partial V/\partial \ln q|_{q=G_{111}} = 0.180$ a.u., $|\partial V/\partial \ln q|_{q=G_{200}} = 0.198$ a.u.
[c] This work: obtained with the fitted values, $d \ln \mathscr{E}_F/d \ln \Omega = -0.641$, $|\partial V/\partial \ln q|_{q=G_{111}} = 0.162$ a.u., $|\partial V/\partial \ln q|_{q=G_{200}} = 0.162$ a.u.
[d] Sorbello & Griessen (1973), 4-OPW calculation with local pseudopotential.
(*e*), (*f*), (*g*), (*h*): Anderson *et al.* (1972a) 4-OPW pseudopotential calculation including spin–orbit interaction.
[e] Local potential.
[f] Local potential with scaled core radius.
[g] Non-local potential.
[h] Non-local potential with scaled core radius.
[i] Van Dyke (1973): 90-OPW pseudopotential calculation including spin–orbit interaction.

A second way to test our band-structure model is to use the experimental volume derivatives (obtained by dividing the measured pressure derivative by the negative of the isothermal compressibility, -2.05×10^{-6} bar^{-1}) to determine the volume derivatives of $\mathscr{E}_F$, V_{111} and V_{200}. The fitted derivatives given in Table 7.1 are found to be in good agreement with the results of the 'first principles' calculation from the Appapillai & Heine (1972) form factor. In contrast to Van Dyke's (1973) results it is found that the Fermi energy scales with volume almost in the way predicted for free electrons. It must be realized that, in contrast to Van Dyke's (1973) approach, our fitting procedure places no constraint whatever on the derivatives of the form factor $\partial v/\partial \ln q$. The good agreement with the Appapillai & Heine (1972) form-factor derivatives is not accidental; similar results have been obtained for other simple metals, such as aluminium, indium, zinc and magnesium (Griessen & Sorbello 1974). The agreement between the experimental derivatives and the derivatives of the pseudopotential form factor, shown in Fig. 7.2, can be

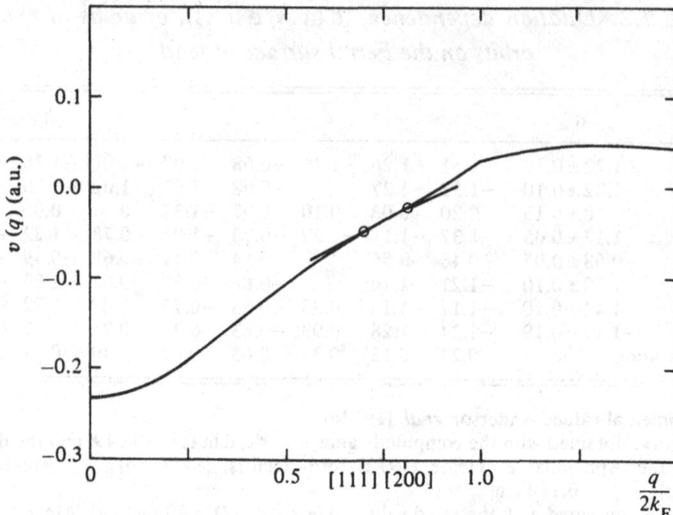

Fig. 7.2. Appapillai and Heine form factor $v(q)$ for lead and experimental tangents to the phenomenological form factor.

regarded as convincing *a posteriori* justification of the various approximations entering our simple model.

The volume derivatives $d \ln \mathscr{A}/d \ln \Omega$ calculated with the fitted volume derivatives of $\mathscr{E}_F$, V_{111} and V_{200} are listed in Table 7.2, and are found to give a good description of the volume dependence of the Fermi surface. The influence of dilation on the shape of the Fermi surface of lead in the (010) and (1$\bar{1}$0) planes, calculated with our fitted values, is illustrated in Fig. 7.3.

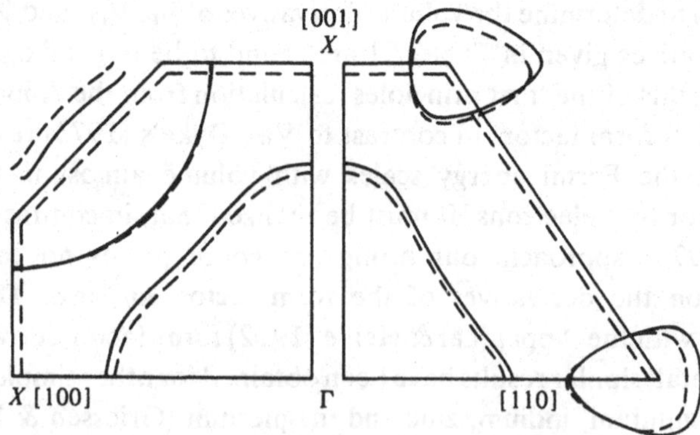

Fig. 7.3. Calculated deformation of the Fermi surface of lead in the (010) plane and in the (1$\bar{1}$0) plane for a dilation $\Delta\Omega/\Omega = 0.10$ (dashed line).

The tetragonal or angular shear dependence of a cross-sectional area $\mathscr{A}$ of the Fermi surface can be calculated from (33), with $\mathrm{d}\mathscr{E}_F/\mathrm{d}\gamma = 0$. Then we obtain

$$\frac{\mathrm{d}\ln\mathscr{A}}{\mathrm{d}\gamma} = \frac{\partial\ln\mathscr{A}}{\partial\gamma} + \sum_s \frac{\partial\ln\mathscr{A}}{\partial V_s}\frac{\mathrm{d}V_s}{\mathrm{d}\gamma}, \tag{50}$$

where $\partial\ln\mathscr{A}/\partial\gamma$ is a partial derivative taken at constant V_s. The shear dependence of the matrix elements can be evaluated from (39). For a tetragonal shear $\mathrm{d}\ln G_{111}/\mathrm{d}\gamma_i$ vanishes, and the only parameter entering our 8-OPW calculation of the shear dependence is the derivative of the form factor at $q = G_{200}$. In contrast, $\mathrm{d}\ln G_{200}/\mathrm{d}\gamma_{ij}$ vanishes for an angular shear, and that shear dependence is determined by $(\partial v/\partial\ln q)|_{q=G_{111}}$.

The tetragonal shear response of the Fermi surface of lead can be deduced from experimental data on the uniaxial stress dependence by using the relation

$$\frac{\mathrm{d}\ln\mathscr{A}}{\mathrm{d}\gamma_i} = (c_{11}-c_{12})\left[\frac{\mathrm{d}\ln\mathscr{A}}{\mathrm{d}\sigma_i} - \frac{1}{2}\left(\frac{\mathrm{d}\ln\mathscr{A}}{\mathrm{d}\sigma_j} + \frac{\mathrm{d}\ln\mathscr{A}}{\mathrm{d}\sigma_k}\right)\right]. \tag{51}$$

We have studied the uniaxial stress dependence of some extremal cross-sectional areas of the Fermi surface of lead by measuring the oscillatory magnetostriction and the dHvA torque. The experimentally determined tetragonal shear derivatives of the areas of three orbits are given in Table 7.3, and are compared with results of a 'first principles' calculation by using the derivative of the Appapillai & Heine (1972) form factor. The value of the form-factor derivative determined to give the best fit to the

TABLE 7.3. *Tetragonal shear strain dependence of areas of three extremal orbits on the Fermi surface of lead*

	$\mathrm{d}\ln\mathscr{A}/\mathrm{d}\gamma_{001}$		
Orbit	*a*	*b*	*c*
$\xi[100]$	-2.07 ± 0.015	-1.94	-2.04
$\nu[100]$	0.86 ± 0.015	0.73	0.83
$\psi[110]$	-0.96 ± 0.015	-0.95	-0.98

[a] Experimental values.
[b] Obtained with the values $(\partial v/\partial\ln q)|_{q=G_{200}}$ from the Appapillai & Heine form factor.
[c] Obtained with the fitted value, $(\partial v/\partial\ln q)|_{q=G_{200}} = 0.174$ a.u.

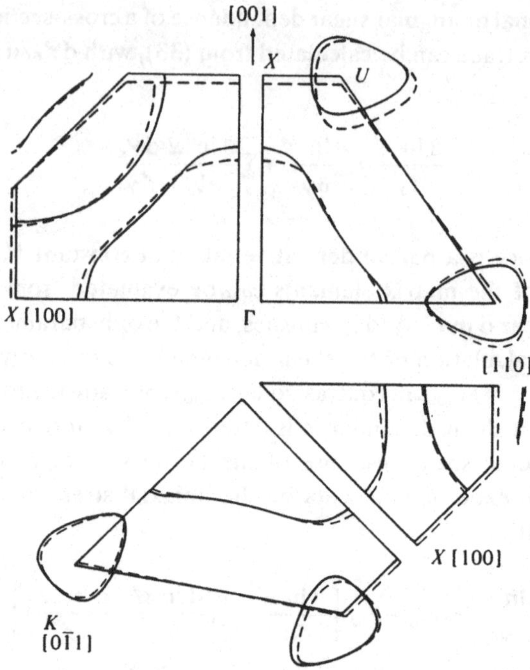

Fig. 7.4. Tetragonal shear response of the Fermi surface of lead in the (010), (1$\bar{1}$0), (001) and (011) planes, for a tetragonal shear $\gamma_z = 0.04$ (dashed line).

experimental tetragonal shear data is

$$\left.\frac{\partial v}{\partial \ln q}\right|_{q=G_{200}} = 0.174 \text{ a.u.} \tag{52}$$

This is in fair agreement with the Appapillai & Heine (1972) form-factor derivative of 0.198 at $q = G_{200}$. The influence of a tetragonal and an angular shear on the shape of the Fermi surface of lead in different planes is illustrated in Figs. 7.4 and 7.5.

7.5.2 Copper (an example of the noble metals)

The noble metals are of special importance as a bridge between the simple metals and the transition metals. The relative simplicity of their Fermi surfaces, coupled with their sensitivity to perturbation by the d-bands, makes them ideal systems in which to test techniques of calculation which are designed for the transition metals. In this sense, the noble metals are the simplest of the transition metals. In this section we

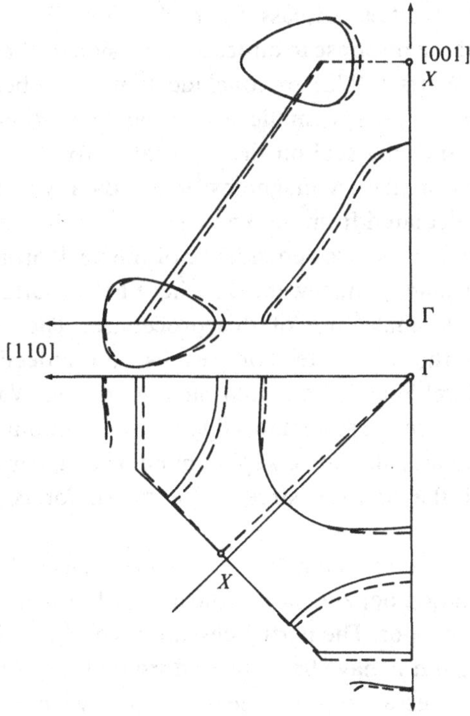

Fig. 7.5. Angular shear dependence of the Fermi surface of lead in the $(1\bar{1}0)$ and (001) planes. The dashed curve corresponds to an angular shear γ_{xy} of 3 degrees.

attempt a consistent explanation of the pressure and uniaxial tension data on the basis of a phase-shift pseudopotential model of the electronic structure of copper. The pressure data are interpreted to determine the volume dependence of the phase shifts that describe the scattering of conduction electrons by the ion cores, and the deformation at points around principal orbits on the Fermi surface is predicted. The shear dependence of the areas of the principal orbits is calculated, and their response to uniaxial tension is deduced.

Templeton's (1974) measurements of the hydrostatic pressure dependence of the Fermi surfaces of the noble metals exploited the high precision of the fluid helium technique. Shoenberg & Watts (1967) measured the uniaxial stress dependence of the Fermi surfaces of the noble metals using a direct-tension technique. The random errors associated with direct measurements of the uniaxial stress dependence are an order of magnitude higher than for pressure measurements because of experimental difficulties discussed in Section 7.3. Furthermore, measurements of the response to direct compression (rather than tension) led Gamble & Watts (1972) to conclude that Shoenberg and Watts' results were subject to systematic error due to friction between the tension rod and the ring seal on the cryostat. Aron (1972) and Slavin (1973) combined oscillatory magnetostriction data with the oscillatory magnetization calculated from the known shape of the Fermi surface to determine the uniaxial stress dependence of the neck orbits in the noble metals; these are among the few metals whose Fermi surfaces are known sufficiently well to make use of this procedure. Their results, when compared with the direct tension results, do indeed support the conclusion of Gamble and Watts that Shoenberg and Watts' values for the stress dependence are systematically too low by amounts up to 50 per cent. This discrepancy should be kept in mind when comparing theoretical estimates of the uniaxial stress dependence for copper with the experimental data.

The phase-shift pseudopotential method has been applied to calculate the shear and dilation dependence of the areas of principal orbits on the Fermi surface of copper. The partial derivatives of the orbital areas with respect to Ω, $\mathscr{E}_F$, and η_l have been derived from (31) and (32). For these calculations it is necessary to know the phase shifts which characterize the electron–ion interaction. These have been determined by fitting experimental Fermi surface data for the unstrained metal.

A least-squares fit to the dilation dependence of the cross-sectional areas of principal orbits on the Fermi surface of copper has been carried out within the rigid MT-sphere approximation. The experimental data were taken from the work of Templeton (1974), and the density of states at the Fermi surface was estimated from a potential constructed by Chodorow, with the result $(\mathrm{d}\mathscr{E}_F/\mathrm{d}\ln\Omega)\approx-0.2719$ Ry. The best fit to the data yields the following values for the volume derivatives of the phase shifts:

$$\left.\begin{aligned}(\mathrm{d}\eta_0/\mathrm{d}\ln\Omega)&=0.129\pm0.032\text{ rad},\\(\mathrm{d}\eta_1/\mathrm{d}\ln\Omega)&=-0.233\pm0.004\text{ rad},\\(\mathrm{d}\eta_2/\mathrm{d}\ln\Omega)&=0.0022\pm0.0039\text{ rad}.\end{aligned}\right\}\qquad(53)$$

The results of the fit are compared with the experimental data in Table 7.4. As a check, the dilation dependence of the volume V of the Fermi surface was calculated, and found to satisfy the condition $(\mathrm{d}\ln V/\mathrm{d}\ln\Omega)=-1$, to better than 0.4 per cent.

TABLE 7.4. *Pressure and dilation dependence of principal orbits on the Fermi surface of copper*

Orbit		$(\mathrm{d}\ln\mathscr{A}/\mathrm{d}P)_{\mathrm{expt}}$ [a] $[10^{-7}\ \mathrm{bar}^{-1}]$	$(\mathrm{d}\ln\mathscr{A}/\mathrm{d}\ln\Omega)_{\mathrm{expt}}$ [b]	$(\mathrm{d}\ln\mathscr{A}/\mathrm{d}\ln\Omega)_{\mathrm{fit}}$ [c]
Belly	⟨001⟩	4.42 (3)	−0.628 (4)	−0.627
Belly	⟨111⟩	4.21 (3)	−0.598 (4)	−0.602
Neck	⟨111⟩	19.8 (5)	−2.81 (7)	−2.85
Rosette	⟨001⟩	4.42 (3)	−0.628 (4)	−0.630
Dog's bone	⟨011⟩	4.04 (2)	−0.574 (3)	−0.574

[a] Templeton (1974).
[b] Derived from data of Templeton using elastic constant data of Overton & Gaffney (1955).
[c] Result of calculations described in the text.

Templeton (1974) obtained somewhat different values for the phase-shift derivatives from a phase-shift analysis of his pressure data, because he used free-electron scaling to estimate the first two terms of (43). This is equivalent to taking $\mathrm{d}\mathscr{E}_\mathrm{F}/\mathrm{d}\ln\Omega = -\frac{2}{3}\mathscr{E}_\mathrm{F} \approx -0.3667$ Ry. With this choice of $\mathrm{d}\mathscr{E}_\mathrm{F}/\mathrm{d}\ln\Omega$, the present calculation yields phase-shift derivatives identical to those of Templeton. As we have shown above, either parameter set provides an equally valid description of the data.

It is usual to express the experimental results as a fraction of the homogeneous strain dependence of the Fermi surface calculated from the free-electron model. According to the free-electron model, the radius of the Fermi sphere depends only on the electron density, and hence only on the dilation component of the homogeneous strain. Let $\mathscr{A}_0$ denote the cross-sectional area of an orbit on the free-electron Fermi sphere. Then

$$\frac{\delta\mathscr{A}_0}{\mathscr{A}_0} = -\frac{2}{3}\frac{\delta\Omega}{\Omega} \equiv -\frac{2}{3}\Delta, \tag{54}$$

and from (5) it will be seen that the relationship,

$$\frac{\mathrm{d}\ln\mathscr{A}}{\mathrm{d}\ln\mathscr{A}_0} = -\frac{3(c_{11}+2c_{12})}{2}\frac{\mathrm{d}\ln\mathscr{A}}{\mathrm{d}\sigma}, \tag{55}$$

relates the usual form in which the data are presented to the usual form in which the stress dependence is derived from band-structure theory.

The responses of the area of an orbit to tetragonal and angular shear, and to dilation, can be combined to predict the response to uniaxial tension. The response to stress in the [001] direction is composed of a dilation term and a tetragonal shear term. Expressing the stress dependence as a fraction of the uniaxial stress dependence of the free-electron sphere, gives

$$\left(\frac{\mathrm{d}\ln\mathscr{A}}{\mathrm{d}\ln\mathscr{A}_0}\right)_{\sigma_{001}}=-\frac{3}{2}\left(\frac{\mathrm{d}\ln\mathscr{A}}{\mathrm{d}\ln\Omega}\right)-\left(\frac{c_{11}+2c_{12}}{c_{11}-c_{12}}\right)\frac{\mathrm{d}\ln\mathscr{A}}{\mathrm{d}\gamma_z}. \tag{56}$$

Similarly, the response to stress in the [011] direction is composed of a tetragonal shear term and an angular shear term,

$$\left(\frac{\mathrm{d}\ln\mathscr{A}}{\mathrm{d}\ln\mathscr{A}_0}\right)_{\sigma_{011}}=-\frac{3}{2}\frac{\mathrm{d}\ln\mathscr{A}}{\mathrm{d}\ln\Omega}-\frac{(c_{11}+2c_{12})}{2(c_{11}-c_{12})}\left(\frac{\mathrm{d}\ln\mathscr{A}}{\mathrm{d}\gamma_y}+\frac{\mathrm{d}\ln\mathscr{A}}{\mathrm{d}\gamma_z}\right)-\frac{3(c_{11}+2c_{12})}{4c_{44}}\frac{\mathrm{d}\ln\mathscr{A}}{\mathrm{d}\gamma_{yz}}, \tag{57}$$

and the response to stress in the [111] direction is composed of a dilation term and an angular shear term,

$$\left(\frac{\mathrm{d}\ln\mathscr{A}}{\mathrm{d}\ln\mathscr{A}_0}\right)_{\sigma_{111}}=-\frac{3}{2}\frac{\mathrm{d}\ln\mathscr{A}}{\mathrm{d}\ln\Omega}-\frac{(c_{11}+2c_{12})}{2c_{44}}\left(\frac{\mathrm{d}\ln\mathscr{A}}{\mathrm{d}\gamma_{xy}}+\frac{\mathrm{d}\ln\mathscr{A}}{\mathrm{d}\gamma_{xz}}+\frac{\mathrm{d}\ln\mathscr{A}}{\mathrm{d}\gamma_{yz}}\right). \tag{58}$$

Using the values for the elastic stiffness constant given by Overton & Gaffney (1955), we deduce the uniaxial tension dependence of principal orbits on the Fermi surface of copper. Our results for a tension applied along the direction of the magnetic field, and therefore normal to the plane of orbit, are presented in Table 7.5.

TABLE 7.5. *Uniaxial tension dependence of cross-sectional areas of principal orbits on the Fermi surface of copper. In each case the tension is applied normal to the plane of the orbit, and the tension dependence is expressed relative to that of the free-electron sphere*

Orbit		d ln $\mathscr{A}$/d ln $\mathscr{A}_0$				
		Dilation[a]	Tetragonal shear[a]	Angular shear[a]	Total[a]	Experimental[b]
Belly	⟨001⟩	0.940	2.04	0	2.98	2.4 (5)
Belly	⟨111⟩	0.903	0	0.118	1.02	0.6 (2)
Neck	⟨111⟩	4.27	0	−50.5	−46.2	−44 (10)
Rosette	⟨001⟩	0.945	−2.65	0	−1.70	−2.1 (8)
Dog's bone	⟨011⟩	0.860	−6.87	7.32	1.31	—

[a] Result of calculations described in the text.
[b] Shoenberg & Watts (1967).

The results of our calculations are generally in very satisfactory agreement with the experimental data of Shoenberg and Watts, although the agreement may be fortuitous in view of the sizeable experimental uncertainties. The results show that the very large uniaxial tension dependence of the neck frequency is predominantly associated with the shear component of lattice strain. The physical interpretation of this result is clear. Under an applied tension, the lattice is elongated in the [111] direction. Correspondingly, the separation between the (111) and $(\bar{1}\bar{1}\bar{1})$ Brillouin zone planes is reduced, thereby increasing the area of contact with the Fermi surface and enlarging the neck orbit. This effect is slightly compensated by an overall decrease in the volume of the Fermi sphere, which is caused by the dilation component of strain associated with the tension. Our calculations for the dog's bone orbit illustrate a difficulty that may arise in interpreting data for a direction in which both tetragonal and angular shear contribute to the strain. While each gives a sizeable contribution, the cancellation between the shear terms is almost complete, and the resulting stress dependence of the Fermi surface is dominated by dilation of the lattice.

The phase-shift pseudopotential approach can also be used to calculate the stress dependence of the Fermi surface at each point around an orbit. Some of our results are plotted in Figs. 7.6 and 7.7. It is clear from these results that the stress dependence of the orbital areas is a consequence of substantially anisotropic distortion of the orbits. Unfortunately it would seem to be prohibitively difficult, with present experimental techniques, to measure the stress dependence from point to point around the orbit.

7.5.3 Tungsten (an example of the group VI transition metals)

The strain dependence of the Fermi surfaces of the group VI bcc transition metals has recently been determined experimentally (chromium – Fawcett, Griessen & Stanley 1976; molybdenum – Posternack *et al.* 1975, Griessen *et al.* 1977a; tungsten – Lee *et al.* 1976b, Stanley *et al.* 1976, Stanley *et al.* 1977). We shall summarize the results for tungsten, which is the only one of these metals for which comprehensive data are available on the pressure derivatives, the uniaxial stress and strain derivatives, and the angular shear derivatives of cross-sectional areas of orbits on all sheets of the Fermi surface.

The Fermi surface of tungsten consists of an electron jack centred on the zone origin Γ, a hole octahedron centred at *H*, and small ellipsoidal hole pockets centred at the points *N*. Girvan, Gold & Phillips (1968)

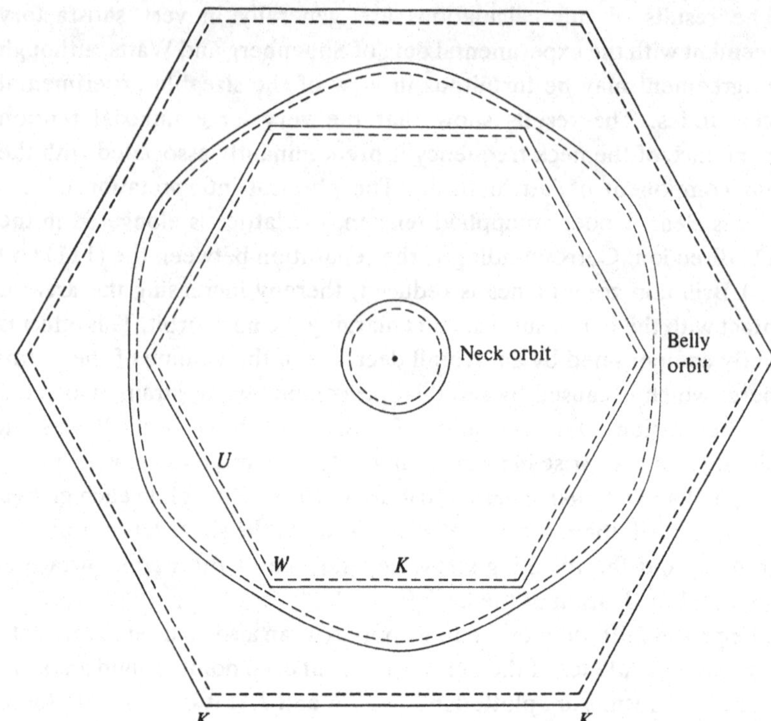

Fig. 7.6. Deformation of the Fermi surface of copper in a (111) plane centred on Γ (outer) and centred on L (inner) corresponding to a dilation, $\Delta\Omega/\Omega = 0.10$.

made an extensive study of the Fermi surface geometry, using the dHvA effect, and their notation is used in labelling orbits here. In Fig. 7.8 we show five orbits whose strain derivatives have been studied in detail. The volume derivatives of the areas of these orbits were measured independently by Schirber (1971), and Svechkarev & Pluzhnikov (1973), by observing the pressure-induced shift in the phase of dHvA oscillations. The uniaxial strain and stress derivatives were studied by Stanley *et al.* (1977), who made simultaneous measurements both of magnetostriction and torque and of ultrasonic velocity and torque. They found that the stress derivatives of the five orbits shown in Fig. 7.8 are typical of orbits for general field directions. The uniaxial strain derivatives for the ρ_1 and π orbits have also been deduced by Stanley *et al.* (1976) from measurements of ultrasonic velocity and magnetostriction. The dilation dependence of the orbit areas can be deduced from the derivatives with respect

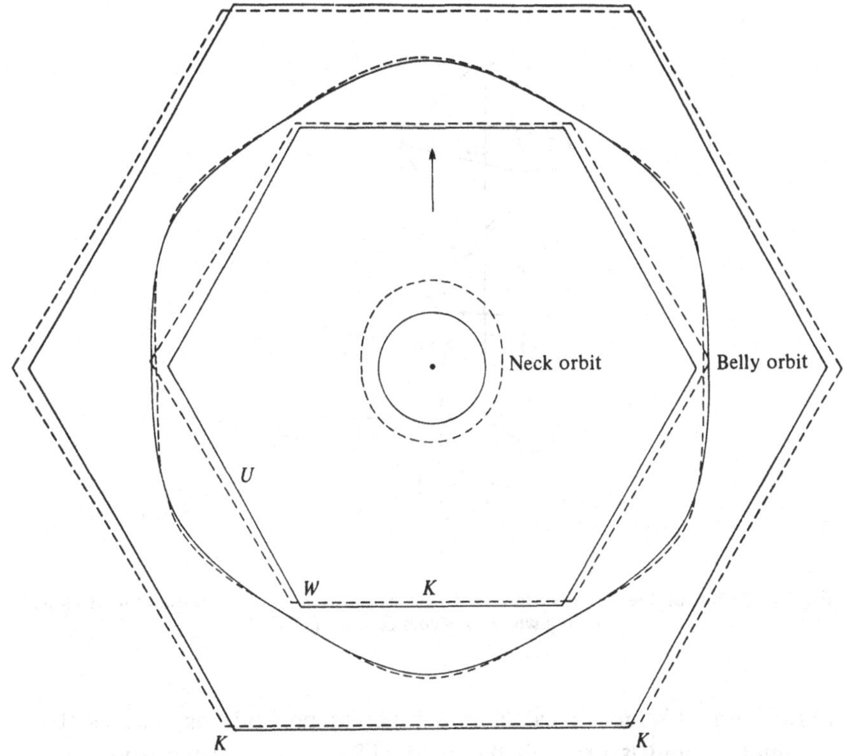

Fig. 7.7. Angular shear response of the Fermi surface of copper in a (111) plane centred on Γ (outer) and centred on L (inner) corresponding to an angular shear, $\gamma_{xy} = 0.10$. Note that the (111) plane centred on L is the positive octant, with the arrow indicating the projection of the z-axis into this plane. Note also that uniaxial strain along [111] corresponds to the sum of three such angular shear, γ_{xy}, γ_{xz} and γ_{yz}.

to strain parallel and perpendicular to the magnetic field. The results obtained from the different uniaxial experiments given in Table 7.6 are mutually consistent within the experimental accuracy. Stanley *et al.* (1977), using the methods described in Section 7.4, determined the point-by-point deformation of the Fermi surface due to uniform dilation, as illustrated in Fig. 7.9.

The response of the Fermi surface to volume-conserving tetragonal shear was also deduced from the experimentally determined uniaxial stress and strain derivatives, and the results are summarized in Table 7.7. The area derivatives with respect to tetragonal shear were computed by use of the KKR formalism from phase shifts fitted to the zero-strain

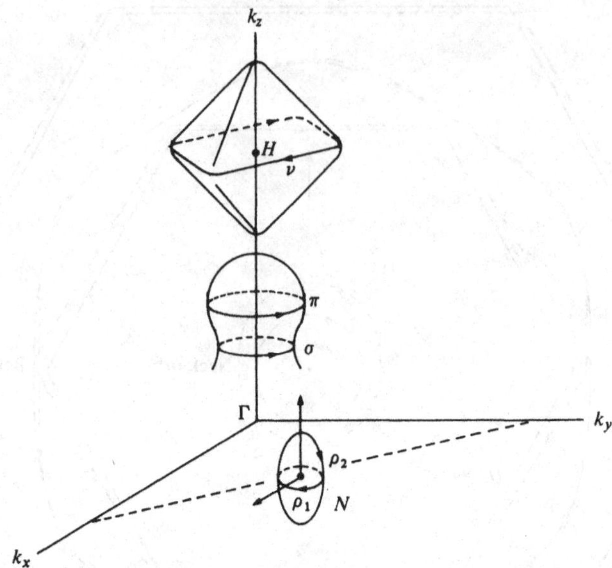

Fig. 7.8. Orbits on the Fermi surface of tungsten; ρ_1, π, σ and ν correspond to **H** along [001], while ρ_2 occurs **H** along [100].

Fermi surface. Such a calculation involves no additional adjustable parameters, and is exact in the rigid MT-sphere approximation. The results of these calculations are included in Table 7.7, and the point-by-point response of the Fermi surface to a tetragonal shear is shown in Fig. 7.10. The results agree well with the experimentally-determined derivatives of the large area orbits π and ν, but somewhat overestimate the

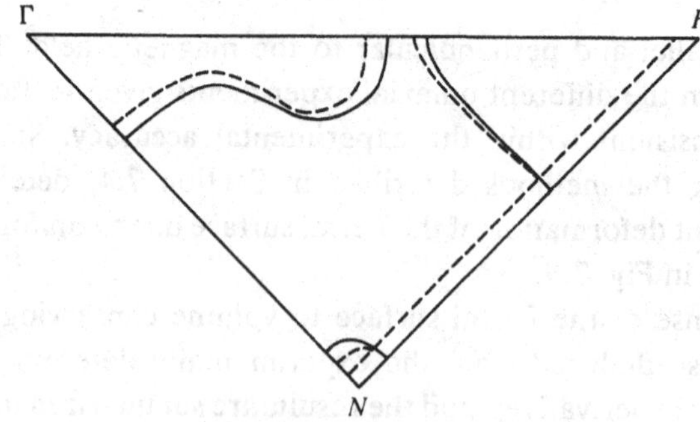

Fig. 7.9. Deformation of the Fermi surface of tungsten in the (100) plane corresponding to a dilation of $\delta\Omega/\Omega = 0.10$.

TABLE 7.6. *Volume dependence of the areas of extremal orbits on the Fermi surface of tungsten*

	Orbit	$\mathscr{A}$ (a.u.)	d ln $\mathscr{A}$/d ln Ω						
			a	*b*	*c*	*d*	*e*	*f*	*g*
σ	neck	0.0164	−5.7 (21)	−4.9 (4)		−4.1 (9)	−4.1 (4)	−4.5 (4)	−4.87
π	ball	0.0583	−2.1 (3)	−2.1 (3)	−2.4 (2)	−2.4 (5)	−2.4 (5)	−2.2 (2)	−2.00
ν	octahedron	0.3836	−1.4 (3)	−1.7 (2)		−0.9 (1)		−1.2 (3)	−1.08
ρ_1	ellipsoid	0.0158	−9.0 (15)	−8.0 (17)	−7.8 (11)	−9.7 (9)	−9.4 (9)	−8.9 (9)	−8.78
ρ_2	ellipsoid	0.0228	−8.8 (26)	−10.0 (12)			−9.6 (12)		

[a] Magnetostriction and torque experiments (Stanley *et al.* 1977).
[b] Elastic constant and torque experiments (Stanley *et al.* 1977).
[c] Elastic constant and magnetostriction experiments (Stanley *et al.* 1976).
[d] Schirber (1971).
[e] Svechkarev & Pluzhnikov (1973).
[f] Weighted average of (*a*)–(*e*).
[g] KKR calculation (Stanley *et al.* 1977). The calculated volume dependence is determined from the set of phase-shift derivatives (radians) that gives the best fit to the experimental data. The results that correspond to $\mathscr{E}_F = 0.85$ Ry are:

$$\frac{d\eta_0}{d\ln\Omega} = 7.7015, \quad \frac{d\eta_{1,j}}{d\ln\Omega} = -0.5873, \quad \frac{d\eta_{2,j}}{d\ln\Omega} = -1.2251.$$

TABLE 7.7. *Tetragonal shear dependence of areas of extremal orbits on the Fermi surface of tungsten*

Orbit	$\mathcal{A}$ (a.u.)	i	d ln $\mathcal{A}$/dγ_i				
			a	b	c	Av. expt.	Calcn.
σ neck orbit [001]	0.0164	x, y	1.0 (7)	0.80 (35)		0.85 (35)	0.27
		z	−2.0 (14)	−1.6 (7)		−1.70 (70)	−0.55
π ball orbit [001]	0.0583	x, y	1.65 (15)	1.65 (30)	2.05 (25)	1.75 (15)	2.05
		z	−3.30 (30)	−3.30 (60)	−4.10 (50)	−3.50 (30)	−4.10
ν octahedron orbit [001]	0.3836	x, y	0.75 (15)	1.0 (2)		0.85 (15)	0.91
		z	−1.50 (3)	−2.0 (4)		−1.70 (15)	−1.83
ρ_1 ellipsoid at (1, 1, 0) π/a	0.0158	x, y	−9.1 (5)	−9.75 (130)	−9.2 (7)	−9.25 (50)	−13.8
orbit ⊥ [001]		z	18.2 (10)	19.50 (260)	18.4 (15)	18.50 (100)	27.6
ρ_2 ellipsoid at (1, 1, 0) π/a	0.0228	x	−7.2 (7)	−10.10 (230)		−7.9 (7)	−11.9
orbit ⊥ [100]		y	−7.2 (7)	−6.35 (230)		−7.0 (7)	−15.2
		z	14.4 (15)	16.45 (320)		15.0 (15)	27.1

[a] Magnetostriction and torque (Stanley *et al.* 1977).
[b] Elastic constant and torque (Stanley *et al.* 1977).
[c] Elastic constant and magnetostriction (Stanley *et al.* 1976).

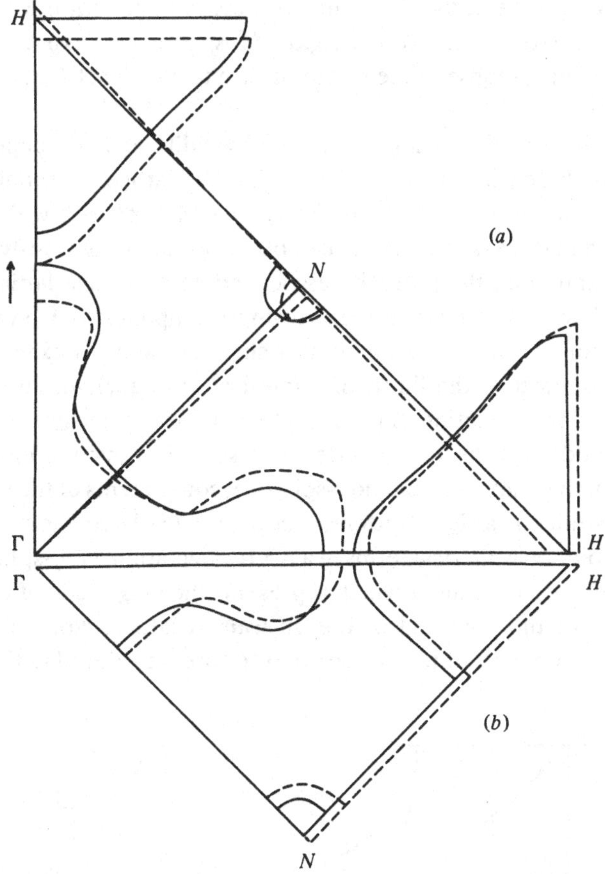

Fig. 7.10. Tetragonal shear response of the Fermi surface of tungsten (*a*) in the (100) plane containing the shear axis, and (*b*) in the (001) plane perpendicular to the shear axis, for $\gamma_z = 0.04$. The arrow denotes the shear axis.

response of the ellipsoid orbits ρ, and underestimate the response of the neck orbit σ.

The angular shear response of the ellipsoidal hole pockets in tungsten was obtained from simultaneous measurements of quantum oscillations in ultrasonic velocity and torque (Lee *et al.* 1976b). A feature of the results is that, while torque oscillations were observed from all sheets of the Fermi surface, the only sound velocity oscillations observed correspond to ellipsoids whose centres N lie in the plane, $k_z = 0$, normal to the [001] axis of shear. This is dictated by symmetry, as discussed by Griessen *et al.* (1977a). The variation with magnetic field direction of the non-vanishing area derivative is in good agreement with the results of a KKR calculation based on a rigid MT-sphere approximation. The point-by-point response of the Fermi surface to angular shear is depicted in Fig. 7.11.

We conclude by discussing the limits of validity of strain dependence calculations based on our assumption that the lattice potential can be represented by an array of non-intersecting MT spheres, which move rigidly as the lattice is strained. It can be shown quite rigorously that, in this approximation, the potential associated with a single lattice site is unchanged by any but a lattice strain having a component of Γ_1 symmetry, i.e. a dilation. Thus our calculation of shear response is exact (to first order and subject to the limits of numerical accuracy) within the rigid MT-sphere approximation. It is known that in order to construct the most accurate models of the Fermi surfaces of several transition metals it is necessary to take into account non-spherical components of the potential of the unstrained lattice. Ketterson *et al.* (1975) have shown how to include cubic non-spherical terms in a KKR calculation. Non-sphericity is represented by a parameter which represents the magnitude of the cubic component of the potential of the unstrained lattice. Non-cubic non-spherical components of the lattice potential are introduced as the lattice

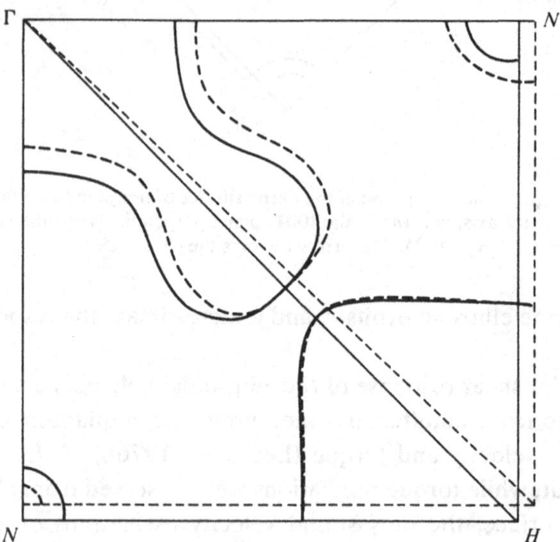

Fig. 7.11. Angular shear response of the Fermi surface of tungsten in the (001) plane perpendicular to the shear axis. The broken curve corresponds to an angular shear γ_{xy} of 3 degrees.

is sheared, and one may suppose that these will contribute significantly to the first-order shear dependence of the Fermi surface cross-sections. It is also likely that non-spherical components of the lattice potential will contribute in a different way to the response to tetragonal shear and to angular shear, since the form of the non-spherical components is governed by overlap between neighbouring cells, and the form of the overlap depends on the character of the shear.

Just as the dilation dependence of the MT potential was estimated by fitting the pressure data, so also one might attempt to estimate the shear dependence of the non-spherical components of the crystal potential by fitting any discrepancy between the results of the rigid MT-sphere calculation and the experimental data. What is needed for such calculations is an estimate of the dependence of the Fermi surface areas on the non-spherical terms in the potential. One can in principle introduce parameters to represent the non-cubic crystal field of the strained lattice. Such calculations have not yet been carried out because of their complexity. They would probably not be justified for copper because the agreement between experiment and calculation based on the spherical potential model is already quite satisfactory. However, in the transition metals the strongly anisotropic d-like states which make up a large part of the shear-induced non-spherical terms in the crystal potential, may make important contributions to the shear dependence of the orbital cross-sections.

7.6 Conclusion

The strain dependence of the Fermi surface is directly related to important properties of metals, including the electronic contribution to acoustic attenuation and thermal expansion, and the scattering by phonons of electrons contributing to transport. It is also related indirectly to properties such as cohesive energy, structure and elastic constants, which depend upon the whole energy-band structure.

The acoustic attenuation was shown by Pippard (1960) to be closely related to the tensor D_{ij} introduced in Section 7.2 (equation (18)). He showed that the electronic contribution to the attenuation of an ultrasonic wave in a metal single crystal may be expressed as a line integral around the Fermi surface. The integrand involves D_{ij}, and also the radii of curvature of the Fermi surface, and the electronic relaxation time. Rayne & Jones (1970) have reviewed experimental and theoretical work in this area, and have shown that the general form of the anisotropy of the attenuation of longitudinal ultrasonic waves in copper can be explained in terms of the Fermi surface geometry, assuming an isotropic deformation tensor and constant relaxation time. Clearly this work should be extended and applied to other metals for which the deformation tensor has been determined from measurements of the strain dependence of the Fermi surface, and for which the anisotropy of the relaxation time is known, so as to check the consistency of the theory for this important physical property of metals.

The electronic Gruneisen coefficient is defined by the equation

$$\gamma_e = \beta_e \Omega / C_e K_T, \tag{59}$$

where β_e is the electronic volume thermal expansion coefficient, Ω the atomic volume, C_e the electronic specific heat and K_T the isothermal compressibility. The Gruneisen coefficient can be expressed in terms of $N_0(\mathscr{E}_F)$, the bare density of states per atom at the Fermi surface, and the electron–phonon enhancement factor λ in the form,

$$\gamma_e = \frac{\mathrm{d} \ln [N_0(\mathscr{E}_F)(1+\lambda)]}{\mathrm{d} \ln \Omega}. \tag{60}$$

Thus a knowledge of the volume dependence of $N_0(\mathscr{E}_F)$ obtained from measurements of the strain dependence of the Fermi surface, combined with the measured value of γ_e, provides an estimate of the volume dependence of λ. Griessen (unpublished) has shown that in molybdenum and tungsten, the volume dependence of λ determined in this way is in agreement with estimates obtained by Bennemann & Garland (1972) from an analysis of superconductivity data.

The tensor D_{ij} plays an essential role in the electron–phonon interaction responsible for scattering of electrons by phonons, and therefore is an important factor in determining metallic transport properties. As we have seen in Section 7.2, D_{ij} is closely related (equation (20)) to the strain response of the Fermi surface. We anticipate that much progress in understanding the transport properties of real metals, especially as they are influenced by the details of the electron–phonon interaction, will be stimulated by the wide range of experimental data now becoming available concerning the strain response of the Fermi surfaces of metals.

Part III Aspects of the de Haas–van Alphen effect

8

The de Haas–van Alphen effect in dilute alloys

P. T. COLERIDGE

8.1 Introduction

The considerable success of the de Haas–van Alphen (dHvA) effect in determining the Fermi surface of pure metals encourages the application of the technique to alloys, so that quantitative comparisons with theory can provide the tests that have been so productive in the case of pure metals. Experimentally, the most obvious effect of adding impurities to a metal is a substantial and highly inconvenient degradation of signal strength caused by the scattering out of the Landau levels of the conduction electrons, so that it is only in the most dilute alloys that dHvA oscillations can be seen. However, when signals are visible and dHvA frequencies can be measured the language of the pure metal can be retained and the dimensions of a Fermi surface measured. It is found that the changes of Fermi surface dimensions are of the order of magnitude of the impurity concentration but vary from alloy to alloy with an anisotropy that depends both on the host and the type of impurity. Furthermore, it is found that in carefully prepared alloys the scattering of the conduction electrons, far from just being an experimental inconvenience, can be measured with some precision and shows the same general kind of behaviour. It is obvious, therefore, that, provided a satisfactory theory can be constructed to explain them, the magnitude and anisotropy of both these experimental quantities provide detailed information on the state of the impurities in the host metal.

A theoretical treatment of the dHvA effect in alloys must address two problems. First, there is the question of how the Landau quantum oscillations are affected by the impurities and whether the Lifshitz–Kosevich (1955) theory, which explains so well dHvA oscillations in pure metals, can be satisfactorily modified to include the effects of the impurities (and other lattice defects). This is equivalent to asking more fundamental questions about the concept of a Fermi surface in alloys, whether the parameters derived from the dHvA effect in alloys define an alloy Fermi surface in a meaningful sense and how such a Fermi surface, if it exists, is related to the pure metal Fermi surface. Secondly, there is the much more general problem of developing a satisfactory theory to explain quantitatively the different behaviour observed in different types of alloys. This is a somewhat less challenging problem than producing a general alloy theory valid for all concentrations because in all but a few exceptional cases it is only necessary to consider the dilute limit of the theory, which is equivalent to considering a single impurity and then arguing that the effects of a low concentration (c) of impurities are additive without interaction, or at least the number of impurities that do interact is proportional to c^2 and is negligible.

8.2 The dilute alloy Fermi surface

By considering the nature of the dHvA signal in an alloy and interpreting the results on the basis of the pure metal theory it is possible to obtain, formally at least, a definition of an alloy Fermi surface. Considering a single harmonic component of a dHvA frequency the oscillatory part of the magnetization takes the form

$$\tilde{M} = M_0(B, T) \cos\left(2\pi\frac{F}{B} + \psi_\infty\right) \exp\left[-K\left(\frac{m_c}{m}\right)\frac{x}{B}\right]. \tag{1}$$

In an alloy the measured frequency F might be written as $F_0 + \Delta F$, where F_0 is the pure metal value and ΔF the change induced by alloying. The Dingle temperature x expresses the scattering effect of the impurity, m_c is the cyclotron mass of the orbit and K is a constant, $2\pi^2 k_B m/e\hbar = 146.9$ kG/K. Experimentally both ΔF and x are linear in impurity concentration so it appears that the criterion of isolated impurities is satisfied.

Formally the effect of alloying can be expressed in the form

$$\tilde{M} = M_0 \operatorname{Re}\{\exp[2\pi i(F_0 + \Delta F_c)/B + i\psi_0]\}, \tag{2}$$

where $\Delta F_c = \Delta F + (iK/2\pi)(m_c/m)x$ is a complex frequency change, proportional to concentration. Corresponding to ΔF_c, the Onsager relation, between frequencies and extremal areas, defines a (complex) change in Fermi surface cross-section $\Delta\mathscr{A}_c$. The imaginary part of $\Delta\mathscr{A}_c$ can be considered as a measure of the average blurring of the Fermi surface produced by the scattering of the quasi-particles as they move round the extremal orbit. Using the normal geometric expression for the cyclotron mass the effect of alloying on the Landau levels can therefore be expressed in terms of a real energy shift Δ and a broadening of half-width Γ (see Fig. 8.1), i.e.

$$\Delta\mathscr{A}_c=\left(\frac{\partial\mathscr{A}}{\partial\mathscr{E}}\right)(\Delta+i\Gamma)=\frac{2\pi}{\hbar^2}m_c(\Delta+i\Gamma). \quad (3)$$

An expression for Γ in terms of a quasi-particle lifetime was first given by Dingle & Shoenberg (1950), Dingle (1952) and Robinson (1950). Explicitly

$$x=\frac{\hbar}{2\pi k_B}\langle 1/\tau(\mathbf{k})\rangle=\frac{\Gamma}{\pi k_B}, \quad (4)$$

where $\langle 1/\tau(\mathbf{k})\rangle$ is an average round the orbit of the reciprocal lifetime of the quasi-particles and a factor 2 has been changed from the original treatment (see Brailsford 1966) so that $\tau(\mathbf{k})$ corresponds to the mean time between collisions.

In pure metals the concept of the Fermi surface follows from the periodic lattice structure. When Fourier transformed into reciprocal space this results in a discrete set of **k**-states at any given energy, and in particular at the Fermi energy, so that a sharp Fermi surface is formed.

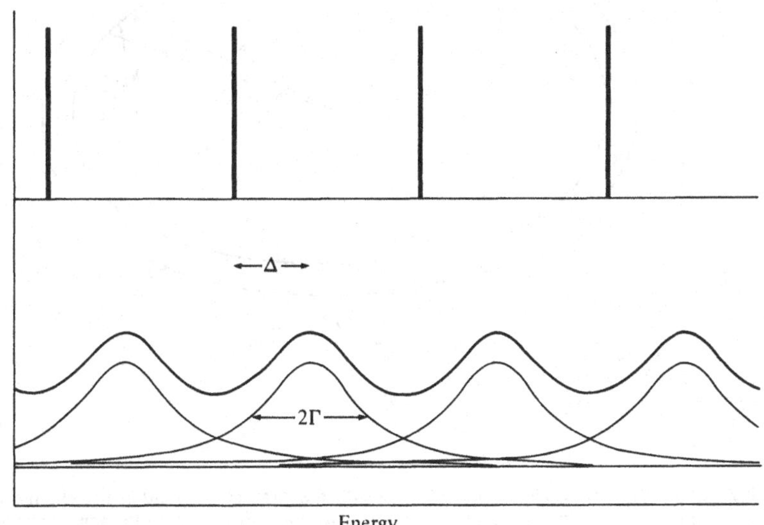

Fig. 8.1. In an alloy the Landau levels, which are delta functions in the pure metal, can be considered as shifted in energy by an amount Δ and broadened by a Lorentzian of half-width Γ.

When the lattice periodicity is interrupted by impurities the sharp Fermi surface no longer exists but a form of blurred Fermi surface can still be defined in terms of the measured (complex) area change ($\Delta\mathscr{A}_c = \Delta\mathscr{A}_r + i\Delta\mathscr{A}_i$).

Considering first the real part of the cross-sectional area change; this is related in a purely geometric fashion to a (real) **k**-vector change $\Delta\mathbf{k}_r$ by the orbital integral

$$\Delta\mathscr{A}_r = \oint \frac{k_\perp \Delta k_r \sec\alpha \, d\eta}{\hat{\mathbf{v}}_\perp \cdot \hat{\mathbf{k}}_\perp}, \tag{5}$$

where $\Delta\mathbf{k}_r$ is taken normal to the Fermi surface, parallel to the velocity, η is the angle round the orbit, $\mathbf{k}_\perp$ is the radius vector in the plane of the orbit, i.e. normal to the applied magnetic field, $\mathbf{v}_\perp$ is the velocity component in the same plane, and α is the angle between $\mathbf{v}_\perp$ and $\mathbf{v}_\mathbf{k}$ (see Fig. 8.2(*a*)). An exactly equivalent integral defines an imaginary **k**-vector $\Delta\mathbf{k}_i$ so the effect of alloying can be formally specified in terms of a complex **k**-vector change ($\Delta\mathbf{k}_c = \Delta\mathbf{k}_r + i\Delta\mathbf{k}_i$) with a direction chosen

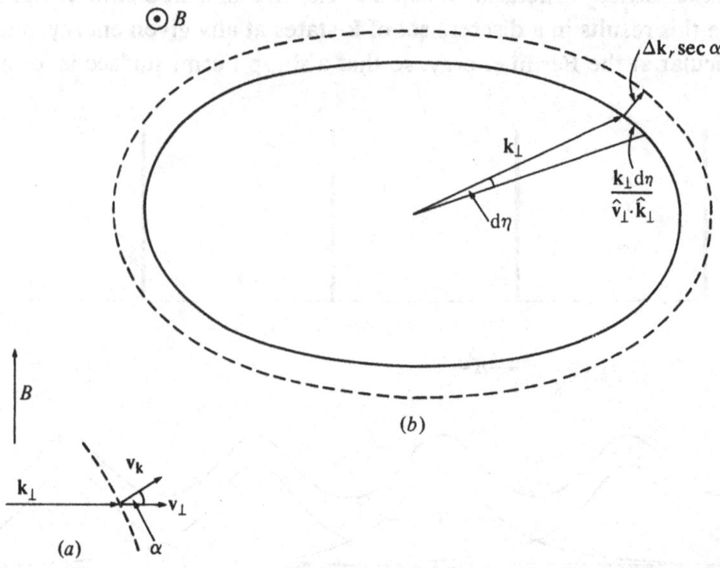

Fig. 8.2. Calculation of the change in area of an extremal orbit corresponding to a change in **k**-vector on the Fermi surface $\Delta\mathbf{k}_r$ which is parallel to the velocity $\mathbf{v}_\mathbf{k}$. (*a*) Perpendicular to the field direction; this shows angle α between $\mathbf{v}_\mathbf{k}$ and $\mathbf{v}_\perp$, which is the projection onto the plane of the orbit of $\mathbf{v}_\mathbf{k}$. (*b*) In the plane of the orbit; $\mathbf{k}_\perp$, the radius vector, is the projection onto the plane of **k** and $\Delta k_r \sec\alpha$ is the projection of $\Delta\mathbf{k}_r$. An element of length round the orbit is $k_\perp d\eta/\hat{\mathbf{v}}_\perp \cdot \hat{\mathbf{k}}_\perp$.

parallel to the velocity. Corresponding to the energies Δ and Γ, which are orbital averages, it is further possible to introduce the equivalent point quantities $\Delta(\mathbf{k})$ and $\Gamma(\mathbf{k})$ defined by

$$\Delta k_c = (\hbar v_{\mathbf{k}})^{-1}(\Delta(\mathbf{k}) + \mathrm{i}\Gamma(\mathbf{k})) \tag{6a}$$

$$= (\hbar v_{\mathbf{k}})^{-1}\{\Delta(\mathbf{k}) + \mathrm{i}\hbar[2\tau(\mathbf{k})]^{-1}\}. \tag{6b}$$

The experimental quantities, ΔF and $m_c x$, can therefore be inverted to get the point quantities by the two exactly parallel deconvolutions

$$\Delta F = \frac{1}{2\pi e}\oint W(\eta)\Delta(\mathbf{k})\,\mathrm{d}\eta \tag{7}$$

and

$$m_c x = \frac{\hbar}{2\pi^2 k_B}\oint W(\eta)\Gamma(\mathbf{k})\,\mathrm{d}\eta \tag{8a}$$

$$= \frac{\hbar^2}{4\pi^2 k_B}\oint W(\eta)[\tau(\mathbf{k})]^{-1}\,\mathrm{d}\eta. \tag{8b}$$

The weight factor $W(\mathbf{k}) = k_\perp^2/\mathbf{v}_{\mathbf{k}}\cdot\mathbf{k}_\perp$, was introduced for the scattering expression by Springford (1971) on the basis of the time an electron spends at each **k**-point on the orbit, here it follows from a purely geometric argument. In terms of this inversion procedure an alloy Fermi surface can therefore be derived from dHvA data provided only that the quantities $\Delta(\mathbf{k})$ and $\Gamma(\mathbf{k})$ are independent of the magnetic field in magnitude and direction. At each **k**-point the alloy Fermi surface is shifted from the pure metal surface by an amount $\Delta\mathbf{k}_r$ and is blurred by an amount $\Delta\mathbf{k}_i$.

It is sometimes stated that under strong scattering conditions the Fermi surface 'does not exist', even in dilute alloys, (Stern 1968, 1969; Sorbello 1977). Strong scattering in this sense means that the broadening of the Landau levels (Γ) is large compared with the energy shift (Δ) and corresponds to a potential that is sufficiently strong for the Born approximation not to be valid, i.e. that one of the phase shifts is comparable to $\pi/2$. This is not true of an alloy Fermi surface as defined above because the uncertainty with which the dimensions of the Fermi surface can be determined depends only on the broadening Γ and, in principle at least, this can always be made sufficiently small by reducing the impurity concentration. This is *not* the same as saying that the energy shift Δ can be determined with arbitrary precision. In the strong scattering limit Δ is small compared with Γ so the error in Δ may be comparable to or larger than itself. It is possible to make an experimental estimate of this error.

The phase of a dHvA oscillation can be measured with a precision of about 0.002 cycles (see Chapter 11) and a reasonable criterion for the signal to be above the noise is for the exponent in equation (1) to be greater than about −10. Using these values the shift can be determined with a typical accuracy of order 0.0015Γ in any alloy in which the dHvA signals can be observed, i.e. provided the alloy is sufficiently dilute.

8.3 The electronic structure of alloys in a magnetic field

Corresponding to the energies $\Delta(\mathbf{k})$ and $\Gamma(\mathbf{k})$ derived from experiment it should be possible to calculate energies describing the changes to the electronic structure of a metal when it is alloyed. They are most conveniently obtained in terms of a spectral density which replaces the single valued $\mathscr{E}$ versus $\mathbf{k}$ curves in a pure metal. This should be calculated using the Green's function of the system in the presence both of the impurities and the magnetic field. A complete solution of this problem in a realistic case has yet to be obtained so various approximations must be discussed. First the theory of dilute alloys in zero magnetic field is summarized and the concept of self-energy introduced; then there is a discussion of how the standard (Lifshitz–Kosevich 1955) theory is extended to alloys by assuming an alloy self-energy that is unchanged by the application of the magnetic field, and finally a simple model calculation is described in which both the magnetic field and the impurities are treated in a self-consistent fashion.

8.3.1 The self-energy in zero magnetic field

This simple treatment of the dilute alloy theory follows Taylor (1970) but more detailed accounts have recently been given by Ehrenreich & Schwartz (1976) and Rennie (1977). If the Bloch states in a pure host crystal are described by

$$\mathscr{H}_0\psi^0_{\mathbf{k}} = \mathscr{E}^0_{\mathbf{k}}\psi^0_{\mathbf{k}} \tag{9}$$

then the introduction of the impurity potential V requires a new Hamiltonian $\mathscr{H} = \mathscr{H}_0 + V$ with eigenstates ψ_i and a Green's function operator $\mathscr{G}$ defined by

$$\mathscr{G}(\mathscr{E}) = (\mathscr{E} - \mathscr{H})^{-1} = (\mathscr{E} - \mathscr{H}_0 - V)^{-1}. \tag{10}$$

The probability of an alloy eigenstate ψ_i being a Bloch state $\psi^0_{\mathbf{k}}$ is given by $|\langle\psi^0_{\mathbf{k}}|\psi_i\rangle|^2$ so that total probability of finding an electron of energy $\mathscr{E}$ in a

Bloch state **k** is given, in terms of a spectral density, by the sum of all such probabilities, i.e.

$$A(\mathbf{k}, \mathscr{E}) = \sum_i |\langle\psi^0_\mathbf{k}|\psi_i\rangle|^2 \delta(\mathscr{E} - \mathscr{E}_i). \tag{11}$$

Taking diagonal matrix elements of the alloy Green's function gives

$$\langle\psi^0_\mathbf{k}|\mathscr{G}(\mathscr{E})|\psi^0_\mathbf{k}\rangle = \sum_i \frac{|\langle\psi^0_\mathbf{k}|\psi_i\rangle|^2}{\mathscr{E} - \mathscr{E}_i + i0^+}, \tag{12}$$

where 0^+ is a vanishingly small positive number. The spectral density is therefore given in terms of the Green's function by

$$A(\mathbf{k}, \mathscr{E}) = -\frac{1}{\pi}\langle\psi^0_\mathbf{k}|\mathscr{G}(\mathscr{E})|\psi^0_\mathbf{k}\rangle. \tag{13}$$

To obtain an expression for $\mathscr{G}$ one approach is to write down a perturbation expansion in terms of the host Green's function $\mathscr{G}_0$ i.e.

$$\mathscr{G} = \mathscr{G}_0 + \mathscr{G}_0 V \mathscr{G}_0 + \mathscr{G}_0 V \mathscr{G}_0 V \mathscr{G}_0 + \dots. \tag{14}$$

If the self-energy operator Σ is defined by $\mathscr{G}(\mathscr{E}) = (\mathscr{E} - \mathscr{H}_0 - \Sigma)^{-1}$ then the matrix elements of the alloy Green's function required for the spectral density can be written as

$$\langle\psi^0_\mathbf{k}|\mathscr{G}|\psi^0_\mathbf{k}\rangle = G_0(\mathbf{k}) + G_0(\mathbf{k})\Sigma(\mathbf{k})G_0(\mathbf{k}) + \dots, \tag{15}$$

where $\Sigma(\mathbf{k})$ is the diagonal matrix element of the self-energy and $G_0(\mathbf{k})$, which is the matrix element of the host Green's function with Bloch states, is also diagonal. The right-hand side of equation (15) is made up of only numbers which can be summed to give the Dyson equation

$$\langle\psi^0_\mathbf{k}|\mathscr{G}|\psi^0_\mathbf{k}\rangle = G_0(\mathbf{k})[1 - \Sigma(\mathbf{k})G_0(\mathbf{k})]^{-1}. \tag{16}$$

If the energy dependence of $\Sigma(\mathbf{k}, \mathscr{E})$ is neglected the spectral density can be written explicitly as

$$A(\mathbf{k}, \mathscr{E}) = -\frac{\pi^{-1}\,\mathrm{Im}\,\Sigma(\mathbf{k})}{[\mathscr{E} - \mathscr{E}^0_\mathbf{k} - \mathrm{Re}\,\Sigma(\mathbf{k})]^2 + [\mathrm{Im}\,\Sigma(\mathbf{k})]^2}, \tag{17}$$

i.e. the delta function, characteristic of the spectral density in a pure metal, is shifted by an amount $\mathrm{Re}\,\Sigma(\mathbf{k})$ and broadened into a Lorentzian of half-width $\mathrm{Im}\,\Sigma(\mathbf{k})$. Ignoring any dependence of $\Sigma(\mathbf{k})$ on energy or magnetic field it is therefore possible to identify $\Delta(\mathbf{k})$ and $\Gamma(\mathbf{k})$ with $\Sigma(\mathbf{k})$ directly

$$\Delta(\mathbf{k}) - \mathrm{i}\Gamma(\mathbf{k}) = \Sigma(\mathbf{k}) \tag{18}$$

and, indeed, Ehrenreich and Schwartz show explicitly that the time decay of an alloy state $\psi_{\mathbf{k}}$ is given by†

$$|\langle\psi_{\mathbf{k}}(t)|\psi_{\mathbf{k}}(0)\rangle|^2 = \exp\left[-t/\tau(\mathbf{k})\right]. \tag{19}$$

In the general case a spectral density can still be defined in terms of the Green's function but it no longer necessarily has a simple Lorentzian shape. As an example we show in Fig. 8.3 the spectral density calculated in a Cu(Ni) alloy using the coherent potential approximation (Stocks, Gyorffy, Guiliano & Ruggeri 1977). Although relatively complicated the general pattern of a shift and broadening of the pure metal states is still apparent.

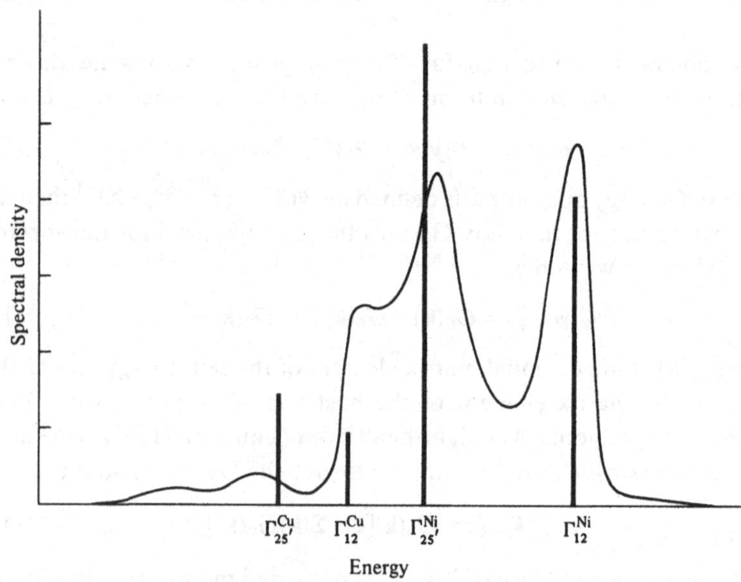

Fig. 8.3. Spectral density at the point Γ for a Cu(0.19) Ni(0.81) alloy calculated by Stocks *et al.* (1977) using the coherent potential approximation. The d-states in pure Cu and pure Ni are shown weighted according to their degeneracy and the concentration.

8.3.2 The T-matrix and level shifts

For a single impurity the zero magnetic field self-energy can be determined by considering the scattering of the conduction electrons off the impurity potential. Simple calculations of the scattering use the Born

† Note, however, the factor of 2 involved in their definition of Γ compared with $\Gamma(\mathbf{k})$ defined by (18).

approximation but this is unsatisfactory when dealing with 'strong' potentials so it is convenient to introduce the T-matrix as a means of treating the higher-order terms exactly. If the scattering of the Bloch states $\psi_{\mathbf{k}}^0$ by the potential V produces a new wave-function $\chi_{\mathbf{k}}$ defined by

$$(\mathcal{H}_0+V)\chi_{\mathbf{k}}=\mathcal{E}_{\mathbf{k}}\chi_{\mathbf{k}}, \tag{20}$$

then time-dependent scattering theory states the probability of scattering between Bloch states $\psi_{\mathbf{k}}^0$ and $\psi_{\mathbf{k}'}^0$, where $\chi_{\mathbf{k}}$ tends to $\psi_{\mathbf{k}}^0$ well outside the range of the potential, is given by

$$P(\mathbf{k},\mathbf{k}')=\frac{2\pi}{\hbar}|\langle\psi_{\mathbf{k}'}^0|V|\chi_{\mathbf{k}}\rangle|^2\delta(\mathcal{E}_{\mathbf{k}}-\mathcal{E}_{\mathbf{k}'}^0), \tag{21}$$

i.e. energy is conserved and the scattering is elastic.

The transition or T-matrix is defined as having matrix elements between *Bloch states* given by the exact scattering, i.e.

$$T_{\mathbf{kk}'}=\langle\psi_{\mathbf{k}'}^0|T|\psi_{\mathbf{k}}^0\rangle=\langle\psi_{\mathbf{k}'}^0|V|\chi_{\mathbf{k}}\rangle. \tag{22}$$

Although the T-matrix is *only* defined for elastic scattering, i.e. for scattering between Bloch states on the energy shell (see e.g. Ziman 1969) this is certainly true for impurity scattering so that

$$P(\mathbf{k},\mathbf{k}')=\frac{2\pi}{\hbar}|T_{\mathbf{kk}'}|^2\delta(\mathcal{E}_{\mathbf{k}}^0-\mathcal{E}_{\mathbf{k}'}^0). \tag{23}$$

The mean scattering time $\tau(\mathbf{k})$ is then given by the total probability of scattering from state $\mathbf{k}$ to all other states $\mathbf{k}'$, i.e.

$$\tau(\mathbf{k})^{-1}=cN\sum_{\mathbf{k}'}P(\mathbf{k},\mathbf{k}'), \tag{24}$$

where we have considered N unit cells in the crystal and a concentration c of impurities which are assumed to be non-interacting.

An alternative expression can be obtained by using the optical theorem which states that the total scattering is given by the imaginary part of the forward scattering as

$$\tau(\mathbf{k})^{-1}=(-2c/\hbar)\,\mathrm{Im}\,T_{\mathbf{kk}}. \tag{25}$$

Comparison with (6) and (18) yields the well-known result that

$$\mathrm{Im}\,\Sigma(\mathbf{k})=c\,\mathrm{Im}\,T_{\mathbf{kk}}. \tag{26}$$

Derivation of a corresponding expression for Re $\Sigma(\mathbf{k})$ involves a rather more subtle argument because it is necessary to consider carefully the effect of the level shifts. When the eigenstates ψ_i of the alloy are formed from the Bloch states of the host there is a difference of energy, $\mathscr{E}_i - \mathscr{E}^0_{\mathbf{k}}$ between the two sets of eigenstates. While an expression for this energy can be written down (see e.g. Jones & March 1973):

$$\mathscr{E}_i - \mathscr{E}^0_{\mathbf{k}} = \frac{\langle \psi^0_{\mathbf{k}} | V | \psi_i \rangle}{\langle \psi^0_{\mathbf{k}} | \psi_i \rangle}, \tag{27}$$

careful consideration shows there might be some difficulties associated with its evaluation. If the eigenstate ψ_i is identified with $\chi_{\mathbf{k}}$, as defined in (20), then the numerator in (27) is just the T-matrix. The T-matrix, however, is only defined between states of the same energy, which is not the case for $\psi^0_{\mathbf{k}}$ and ψ_i. The paradox, which must be resolved by introducing level shift operators (Roman 1965; Fenton 1977), is associated with the normalization of the wave-functions. The normalization is determined by the boundary conditions in the problem and, in particular, any change of boundary conditions that occurs on alloying changes the normalization of the wave-functions. The effect of this has been discussed, among others, by Anderson & McMillan (1967) and Sorbello (1977).

The eigenstates of the alloy form a complete set so that the alloy Green's function can be written as

$$\mathscr{G} = \sum_i (\mathscr{E} - \mathscr{E}_i - \mathrm{i}0^+)^{-1} |\psi_i\rangle\langle\psi_i|, \tag{28}$$

which shows that the poles of $\mathscr{G}$ occur at the real eigenenergies $\mathscr{E}_i$. The self-energy, however, is specified by the poles of the equation

$$\langle \psi^0_{\mathbf{k}} | \mathscr{G} | \psi^0_{\mathbf{k}} \rangle = [\mathscr{E} - \mathscr{E}^0_{\mathbf{k}} - \Sigma(\mathbf{k}, \mathscr{E}) - \mathrm{i}0^+]^{-1} \tag{29}$$

which occur at complex energies $\mathscr{E}^0_{\mathbf{k}} + \Sigma(\mathbf{k}, \mathscr{E})$. For a crystal of finite volume, taking the limit $0^+ \to 0$ in (28) picks out the poles of $\mathscr{G}$ at the real energies $\mathscr{E}_i$ corresponding to the eigenstates ψ_i. To obtain the self-energy, which has an imaginary part, it is necessary to broaden the set of discrete eigenstates ψ_i into a level density by first letting the number of atoms in the crystal (N) tend to infinity and then taking the limit $0^+ \to 0$. In the same fashion there are two ways to evaluate the energy in (27). In terms of the Lipmann–Schwinger scattering theory, ψ_i can be written as (Fenton 1977)

$$\psi_i \equiv \psi^{\pm}_{\mathbf{k}} = |\psi^0_{\mathbf{k}}\rangle + \sum_{\mathbf{k}'} T^{\pm}_{\mathbf{kk}'} (\mathscr{E}^0_{\mathbf{k}} - \mathscr{E}^0_{\mathbf{k}'} \pm \mathrm{i}0^+)^{-1} |\psi^0_{\mathbf{k}'}\rangle. \tag{30}$$

If the limit $0^+ \to 0$ is taken before the limit $N \to \infty$, a divergence occurs in the denominator $\langle \psi^0_{\mathbf{k}} | \psi_i \rangle$ and the equation just reduces to an expression for the level shift (Fenton 1977; Roman 1965). By contrast, if the effect of the impurity potential is considered to be localized and screened so that the perturbation to $\psi^0_{\mathbf{k}}$ is confirmed to the region of the impurity then

$$\langle\psi^0_{\mathbf{k}}|\psi_i\rangle = \langle\psi^0_{\mathbf{k}}|\psi^0_{\mathbf{k}}\rangle[1+O(N^{-1})], \tag{31}$$

so that if the limit $N\to\infty$ is taken before $0^+\to 0$ the divergence in the denominator is suppressed and $\langle\psi^0_{\mathbf{k}}|\psi_i\rangle$ tends to $\langle\psi^0_{\mathbf{k}}|\psi^0_{\mathbf{k}}\rangle$. In this case the energy is complex, and is just the self-energy in the form

$$\Sigma(\mathbf{k}) = \frac{\langle\psi^0_{\mathbf{k}}|V|\chi_{\mathbf{k}}\rangle}{\langle\psi^0_{\mathbf{k}}|\psi^0_{\mathbf{k}}\rangle} = cT_{\mathbf{kk}}, \tag{32}$$

where we have explicitly replaced ψ_i by the correctly normalized wave-function $\chi_{\mathbf{k}}$ and extended the example to a dilute limit of cN non-interacting impurities by normalizing $\psi^0_{\mathbf{k}}$ to a unit volume. Therefore (contrary to the statement of Fenton (1977)), both the real and imaginary parts of $\Sigma(\mathbf{k})$ are given by $T_{\mathbf{kk}}$ *provided the effect of the impurity potential is localized* by screening. This is not necessarily the case when considering spin-flip or Kondo scattering (Fenton 1976).

It is important to note that, when calculating $T_{\mathbf{kk}}$ in the presence of level shifts, the wave-function must be appropriately renormalized (Roman 1965). That is, because the impurity potential must be screened, the boundary conditions are changed on alloying. This can be demonstrated by considering the stationary states for the system of free electrons and a spherical potential confined to a spherical box of radius R (which is not the same problem as considering the scattering of the electrons because then the only boundary condition is that scattered electrons propagate to infinity and are not reflected). Extra stationary states are added, the number of which is given by the Friedel sum rule, and the level shifts take the form $\Delta\mathscr{E}_l = 2\sqrt{\mathscr{E}^0_{\mathbf{k}}}\eta_l/R$ where η_l are the phase shifts of the potential (Anderson & McMillan 1967). In the example, discussed below, of an array of muffin-tin (MT) potentials, the effect of the level shifts and the associated wave-functions renormalization can be explicitly calculated by considering how the backscattering by the surrounding lattice, of the outgoing scattered wave, progressively transforms the wave into a Bloch state.

The reason it is the T-matrix and not the level shift that gives the self-energy has been discussed by Stern (1968). In a momentum-sensitive experiment, such as the dHvA effect or positron annihilation, it is the average over a large number of scattering events that is important. In the multiply-scattered wave it is only the coherent or forward scattered part

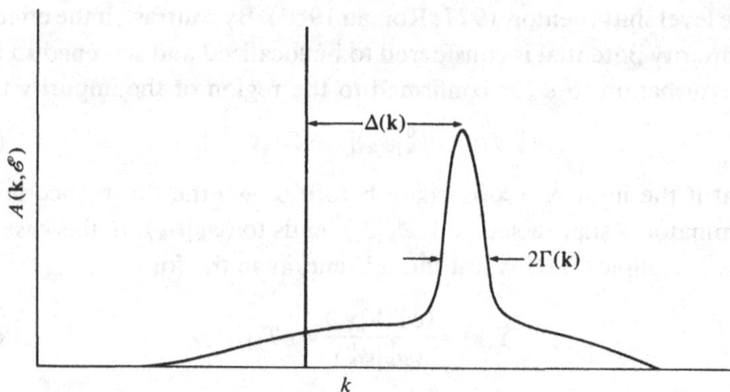

Fig. 8.4. The spectral density of an eigenstate in a dilute alloy (after Stern 1968). In addition to a narrow peak, which contains the 'coherent' states and has a width $2\Gamma(\mathbf{k})$ there is a broad distribution of 'incoherent' states that has a fixed width but a height proportional to the concentration.

that contributes to the momentum and this is given by the forward scattering T-matrix *determined in the presence of the surrounding medium.* The effect of the surrounding medium is to renormalize the wave-functions in such a way as to decrease the weight of that part of the state associated with a particular **k** value. This has the important result that the volume of the Fermi surface in the alloy no longer holds precisely all the electrons in the alloy, and, indeed this can be observed experimentally in the dHvA effect (Coleridge 1975; see also Chapter 11). In terms of the spectral density for the wave-function two alloying effects can be seen (Fig. 8.4). The discrete **k**-state in the pure metal is broadened by an amount $\Gamma(\mathbf{k})$ which is proportional to the impurity concentration but Stern argues that in addition to this relatively sharp peak, which contains the coherent states described by the forward scattering T-matrix, $T_{\mathbf{kk}}$, there is broad distribution of states of fixed width, and a height proportional to concentration, that contains incoherent states which are not measured by momentum-sensitive experiments such as the dHvA effect.

8.3.3 Fermi level shift

An important parameter in calculating the change in shape of the Fermi surface is the change in Fermi level caused by alloying. It will be assumed that in the dilute limit at least there is no change. Friedel (1954) proves this for impurity potentials small compared with the Fermi energy and Stern (1972) argues that if the perturbation is localized, i.e. well shielded, and produces no volume distortion, then the Fermi level is unaltered even in non-dilute alloys. In the dilute case the argument is simple; well away from the impurities the perturbation is screened out so that the conduction electrons have no knowledge of the existence of the impurities and see a 'local band structure' with a Fermi level the same as in the pure crystal. Therefore, because the chemical potential is constant throughout the crystal, it is not changed by alloying until the concentration of impurities is such as to produce overlap of the screening clouds. There is therefore an average or 'coherent' band structure which is moved down (or up) so as to accommodate the excess (or deficit) of electrons carried by the impurity (or at least those that are not localized) and ensure charge neutrality. It is important to realize that the Fermi level in this case is measured absolutely, i.e. with respect to vacuum, *not* relative to the bottom of the band.

In any practical case associated with alloying there is inevitably a change in lattice constant which produces a change in Fermi level. There are few, if any, acceptable theoretical treatments of this problem so that in practice the lattice distortion is usually approximated by some or all of: (i) a uniform expansion of the lattice, (ii) a localized distortion round the impurity and (iii) a change in host potentials appropriate to the new lattice constant. A simple rescaling of experimental results to the new lattice constant accounts for the first term while the second and third can be included as part of the perturbing potential. As discussed below this approximation is frequently inadequate.

8.3.4 The Landau level structure in alloys

Most treatments of the quantum oscillations in alloys ignore any changes to the alloy self-energy that might be induced by the application of the magnetic field and then rederive the Lifshitz–Kosevich expression with various approximations for the electronic structure of the dilute alloy in zero magnetic field. A convenient summary of early work is given by Strěda (1974) and a detailed treatment is contained in the two, similar, analyses by Mann (1971) and Soven (1972). Both authors assume the alloy problem is solved, in zero magnetic field, by determining the self-energy within a coherent potential approximation so the alloy is replaced by a periodic array of potentials each of which is the host potential plus the (complex) self-energy. The Lifshitz–Kosevich expression is then rederived on the assumption that the scattering process and hence the self-energy is not altered by the application of the magnetic field. Further approximations are that the radius of curvature of the electron orbit is large compared with the characteristic diameter of the scattering potential and that the Landau level number at the Fermi surface is large, i.e. that the Fermi energy is large compared with the cyclotron frequency $\hbar\omega_c$. These two approximations are likely to be well satisfied in all cases except maybe in the quantum limit. In the dilute limit both authors also assume, as discussed above, that the self-energy in zero magnetic field, $\Sigma^0(\mathbf{k})$, is given by the forward scattering T-matrix element $T_{\mathbf{kk}}$ and that the Fermi level is unaltered by alloying.

Although Soven considers only a free-electron host, and therefore excludes explicitly any anisotropy effects or the consideration of small sheets of Fermi surface, both authors reach essentially the same conclusion; in accordance with (7), (8) and (18) the frequency change is given by m_c times an average round the orbit of $\text{Re}\,\Sigma^0(\mathbf{k})$ and the broadening by the corresponding average of $\text{Im}\,\Sigma^0(\mathbf{k})$. Contrary to Mann, Soven finds, in the temperature dependence, an extra contribution to the effective cyclotron mass. This term comes from the step where the Fermi distribution function is folded with the energy derivative of the area and results in an effective cyclotron mass

$$m_{\text{eff}} = \frac{\hbar^2}{2\pi}\frac{\partial}{\partial\mathscr{E}}\left[\mathscr{A}_0 - \frac{\partial\mathscr{A}}{\partial\mathscr{E}}\,\text{Re}\,\Sigma(\mathbf{k},\mathscr{E})\right]$$

$$\approx m_c\left[1 - \frac{\partial}{\partial\mathscr{E}}\Sigma^0(\mathbf{k},\mathscr{E})\right], \tag{33}$$

where $\partial^2\mathscr{A}/\partial\mathscr{E}^2$ and higher terms are neglected as small. The observed temperature dependence is then given by $\sinh(-2\pi^2 k_B m_{\text{eff}} T/e\hbar B)$. Mann neglects the second term and therefore finds no change in the temperature dependence. Soven points out that the extra term is *not* related to the zero field density of states and he cites the example of a resonant state at the Fermi level when the zero field density of states has a large increase but m_{eff} can *decrease* because of the energy dependence of $\text{Re}\,\Sigma^0(\mathbf{k},\mathscr{E})$. The change in the cyclotron mass is of course proportional to the concentration of impurities and will therefore be extremely difficult to detect experimentally because measurement of m_c with the required accuracy, of order 1 per cent, is not easy. (See, however, Paton 1971.)

The principle criticism of these two analyses must be the neglect of the fact that the density of states oscillates with a period proportional to the magnetic field. Both the real and imaginary parts of the self-energy and the scattering probability might therefore be expected to oscillate with the field and indeed this is the origin of the Shubnikov–de Haas effect. Most work in which the effect of the magnetic field is explicitly considered (e.g. Bychkov 1960; Skobov 1959; Liu & Toxen 1965) are based on a perturbation approach. The unperturbed states, corresponding to the ideal crystal in a uniform magnetic field show singularities in the density of states at the Landau levels so, as Strěda (1974) argues, it is not immediately obvious that this approach will give a correct answer.

Strěda has solved the problem self-consistently using a simple model calculation. The alloy is represented by a host with parabolic bands and the atomic potentials by zero range δ-functions with an energy cut-off at $E_{\max}$. The temperature is supposed to be zero and the Green's function in a magnetic field is taken as

$$F^0(z) = \frac{1}{N} \sum_{\substack{\alpha \\ E_\alpha < E_{\max}}} \frac{1}{z - E_\alpha}, \tag{34}$$

where α denotes a set of quantum numbers (n, k_y, k_z) and

$$E_\alpha = \frac{\hbar^2 k_z^2}{2m_c} + \hbar\omega_c(n + \tfrac{1}{2}) \tag{35}$$

are the Landau levels.

The 'Soven' equation for the coherent potential is

$$\Sigma = \bar{v} - (\Sigma - v^A)\, F^0(z - \Sigma)\, (\Sigma - v^B), \tag{36}$$

where v^A, v^B are host and impurity potentials and $\bar{v}$, an average potential, is given by

$$\bar{v} = c^A v^A + c^B v^B. \tag{37}$$

This equation was solved numerically and, indeed, the self-energy does oscillate appreciably, particularly for the first few Landau levels. Although these are important when considering the quantum limit (Rode & Lowndes 1977), the more common behaviour is the case of high Landau level numbers and moderate scattering, i.e. fields in the range 30–100 kG and Dingle temperatures of a few kelvin.

This regime was not explicitly considered by Strěda but the calculation has been repeated with new parameters. A typical example in a field of 100 kG, Landau level number 2590, is shown in Fig. 8.5. It is apparent that both the real and imaginary parts of the self-energy oscillate with an amplitude that increases with the impurity concentration but at a rate

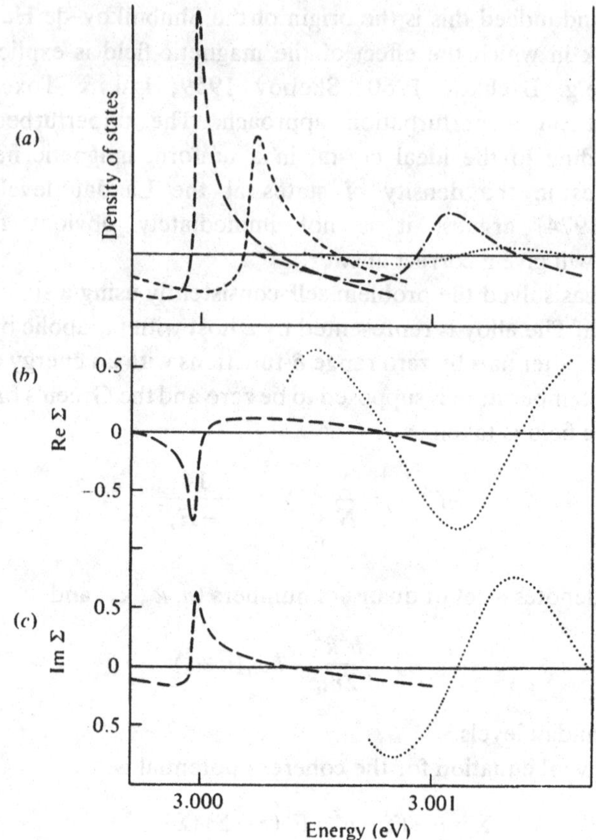

Fig. 8.5. Self-consistent model calculation of dHvA effect in dilute alloys for energies near 3 eV. The δ-function potential is of strength 4 eV and the magnetic field is 100 kG, i.e. $\hbar\omega_c = 1.16 \times 10^{-3}$ eV. For a 1 per cent alloy the non-oscillating part of Re Σ is 3.5×10^{-3} eV (corresponding to a shift in phase of about 3 cycles) and the non-oscillating part of Im Σ is 3.5×10^{-4} eV (corresponding to a Dingle temperature of 1.3 K). (a) Oscillating part of the density of states (in arbitrary units) for various alloy concentrations: ——— 0.03 per cent, ------ 0.1 per cent, —·—·— 0.3 per cent, ······ 1 per cent. For a 3 per cent alloy, in which dHvA signals would be just visible, the amplitude of the oscillations in the density of states cannot be distinguished from the width of the line. (*b*) Oscillating part of Re Σ for 0.03 and 1 per cent alloys in units of 10^{-6} eV (symbols as in (*a*)). (*c*) oscillating part of Im Σ for 0.03 and 1 per cent alloys in units of 10^{-6} eV (symbols as in (*a*)).

considerably slower than linear. Generally the average value of Σ deviates from the zero field value by an extremely small amount, corresponding typically to less than 0.001 cycles or 0.0002 K for the frequency shift and Dingle temperature respectively. This result is somewhat at variance with Strěda's conclusion but his calculation was for a Landau level number of 52, appreciably closer to the quantum limit. For large scattering there is a significant deviation from the simple Dingle expression. The amplitude of both the fundamental and harmonic components is increased. Empirically the requirement for this to occur is that the Dingle temperature exponent for the rth harmonic (rKm_cx/mB, where $K =$ 146.9 kG/K) be larger than about 8. This is close to the limit for a signal to be observable and the effect is not particularly large until the exponent increases to ~10, i.e. for signals an order of magnitude smaller, so it is unlikely that this effect is ever important in any practical circumstances. For example, when using the wave-shape analysis technique (Chapter 10) the exponent is typically ≤ 4 (D. H. Lowndes, private communication) and the signal is two orders of magnitude larger than the critical value.

The calculation was made at $T = 0$ and the effect of a finite temperature is not immediately obvious. It should be noted, however, that the maximum value of $\partial\Sigma/\partial\mathscr{E}$, which occurs for a low concentration of impurities, is only a per cent or so and therefore the effect on the cyclotron mass from (33) is unlikely to be important.

The conclusion from this model calculation is that it is probably a valid approximation to equate the measured values of the self-energy, i.e. those deduced from experimental dHvA data, with those appropriate to zero magnetic field, at least at the present levels of experimental error. Theoretically it is sufficient to calculate the self-energy ignoring the effects of magnetic field and, except near the quantum limit, the results should provide an adequate description of the dHvA oscillations.

8.4 Alloy models

Because the self-energy appearing in the dHvA effect is effectively independent of the magnetic field the results from inverting experimental data can be compared directly with theoretical calculations. Conceptually it is sometimes convenient to retain the language of the perfect crystal and consider alloy band structures which are constructed by shifting the pure metal bands and broadening them (see Fig. 8.6(a)). In the dilute limit the Fermi level in the alloy is unaltered from the pure metal and the shift and broadening are just given by the self-energy.

Simple models for the alloy can be developed in terms of a slightly modified pure metal band structure, i.e. considering only the real part of the energy shift, but the scattering must then be introduced in a rather artificial fashion. Alternatively, if the scattering from a single atom,

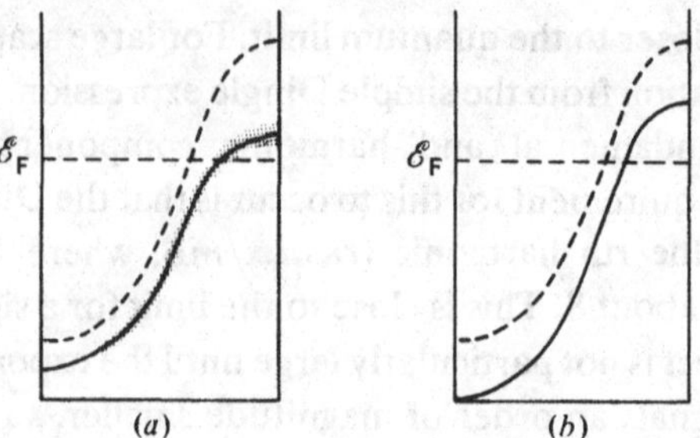

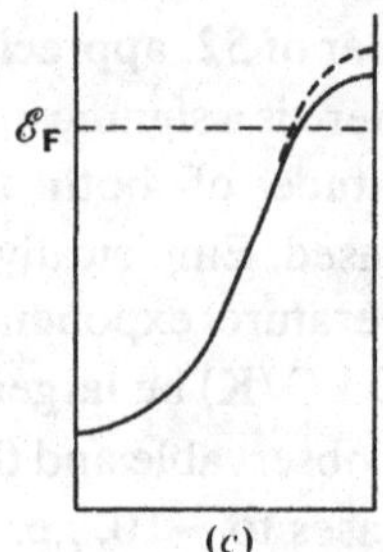

Fig. 8.6. Schematic representation of dilute alloy band structures. (*a*) The pure metal band structure (broken line) is shifted down by Re $\Sigma(\mathbf{k})$ and broadened by Im $\Sigma(\mathbf{k})$, both of which may be anisotropic. The Fermi level is unchanged. (*b*) The rigid-band model where the bands are moved down rigidly and the Fermi level is unchanged. (*c*) A pseudopotential model in which the band structure is unchanged except near the zone boundary but the Fermi level is raised to accommodate extra electrons.

expressed in terms of the 'T-matrix', is generalized to include all the scattering events the self-energy can be obtained rather directly. Some of these approaches are discussed below.

8.4.1 Rigid-band model

The simplest model for estimating the Fermi surface changes produced by alloying is a rigid-band model (Heine 1956; Chollet & Templeton 1968). In the crudest form this consists of neglecting all lattice constant changes and taking the self-energy to be real and given by $-c\Delta Z/N(\mathscr{E})$, where ΔZ is the valence difference between impurity and host and $N(\mathscr{E})$ is the density of states at the Fermi energy. If the bands are moved down rigidly by this amount an extra number of electrons $c\Delta Z$ can be accommodated with the same Fermi level (see Fig. 8.6(*b*)). Expressing the density of states in terms of a thermal mass m_{th} the cross-sectional area change is given by

$$\Delta\mathscr{A} = \frac{\partial\mathscr{A}}{\partial\mathscr{E}}\frac{c\Delta Z}{N(\mathscr{E})} = \frac{2}{3}c\Delta Z\frac{m_c}{m_{\text{th}}}\mathscr{A}_{\text{fe}}, \tag{38}$$

where m_c is the cyclotron mass and $\mathscr{A}_{\text{fe}}$ the free-electron cross-section. This may be an experimental estimate because the density of states can be derived from the electronic specific heat and the cyclotron mass from dHvA measurements. Then both the thermal mass and the cyclotron mass will include electron–phonon enhancement factors so the ratio will be affected only to the extent that the electron–phonon interaction is anisotropic and the two masses involve different averages. However, if one of the masses is taken from a band-structure calculation and one is experimental the electron–phonon enhancement factor must be explicitly included.

Effects of lattice distortion are commonly included in an *ad hoc* fashion by the simple procedure of scaling measured frequencies by the lattice constant appropriate to the alloy. This effect can be large in non-cubic systems, e.g. tetragonal or hexagonal, because the c/a ratio may vary rapidly on alloying even when the volume of the crystal changes little. In some systems, e.g. Cu(Al) (Templeton & Coleridge 1975a) the rigid-band model provides a fair explanation of the data and in other cases it explains the general trends correctly. However, there are cases when it is totally wrong and more accurate experiments always tend to reveal discrepancies with the rigid-band model rather than improve agreement.

8.4.2 Pseudopotential models

When the host metal is nearly free-electron-like an alloy theory can be formulated within the framework of the pseudopotential approximation. We will not discuss this in detail but consider instead two recent examples of this approach applied to Fermi surface changes, the work of Fung & Gordon (1977) on Mg alloys and Holtham & Parsons (1976) on In alloys, pointing out the various steps involved in these alloy calculations. In both examples the crystal structure of the host metal is non-cubic so lattice distortion is particularly important. The first calculation uses a non-local pseudopotential formulation and the second a local 'on the Fermi surface' approximation with spin–orbit effects included.

In zeroth order pseudopotential theory the rigid-band model appears in a slightly different guise (Fig. 8.6(c)); the bottom of the band is fixed and the Fermi level raised to accommodate the extra electrons. This is sometimes called a 'rigid rigid-band model'. Next, lattice distortion is introduced but all changes of potential are ignored. There is a change in the magnitude of the **g**-vectors and therefore in the diagonal energy term $(\mathbf{k}+\mathbf{g})^2$, a change in the cell size and a shift of the Fermi level. Finally changes in the potential are included by decomposing the local part of the pseudopotential into a bare pseudopotential $U^{\text{bare}}(\mathbf{g})$ and a dielectric constant $\varepsilon(\mathbf{g}, k_F)$

$$U(\mathbf{g}) = U^{\text{bare}}(\mathbf{g})/\varepsilon(\mathbf{g}, k_F) \tag{39}$$

and using the virtual crystal approximation to write the alloy potential as

$$U_{\text{alloy}}^{\text{bare}} = (1-c)U_{\text{host}}^{\text{bare}}(\mathbf{g}') + cU_{\text{imp}}^{\text{bare}}(\mathbf{g}'), \tag{40}$$

where $\mathbf{g}'$ are the *alloy* reciprocal lattice vectors. This bare potential is then screened with a dielectric constant appropriate to the alloy, $\varepsilon(\mathbf{g}', k_F')$. In the two examples the non-local part of the pseudopotential and spin–orbit effects are assumed not to change significantly on alloying. The pseudopotential form factors used in the calculations were derived from fitting experimental results in the pure metal but they could have been obtained from model potentials.

The Fermi level is undetermined by the procedure outlined above and is therefore an adjustable parameter of the fit. It may be determined either by fitting a particular frequency or set of frequencies or by integrating over the density of states and selecting the energy needed to accommodate the total number of electrons in the alloy.

In cases where the host metal is nearly free-electron-like and the valence difference small these calculations, which have few adjustable parameters, provide a good fit to experiment. Typically the alloy frequencies are measured with a precision of order 1 per cent and the accuracy of the fit is about the same. The changes in frequency are of order 10 per cent of the pure metal value, so that experiment and calculation agree with an accuracy of only about 10 per cent of the actual change and more accurate measurements may well reveal some discrepancies. The two papers discussed are typical in that although most of the change is explained by the rigid-band model (with lattice distortion) the full treatment including potential effects provides a small but significant improvement.

The scattering terms can also be obtained from pseudopotential theory. For a single plane wave scattering off a concentration c of impurities the scattering rate is given, in terms of host and impurity form factors (Harrison 1970) as

$$\tau^{-1} = \frac{m^*\Omega c}{2\pi\hbar^3 k_f}\int_0^{k_f} |\Delta W(\mathbf{q})|^2 \mathbf{q}\, d\mathbf{q}, \tag{41}$$

where $\mathbf{k}' = \mathbf{k} + \mathbf{q}$ and

$$\Delta W(\mathbf{q}) = \langle \mathbf{k}'|W^{\text{i}}|\mathbf{k}\rangle - \langle \mathbf{k}'|W^{\text{h}}|\mathbf{k}\rangle. \tag{42}$$

The form factor for the impurity in the environment of the host can be derived using the same argument as that discussed above, i.e. by rescreening a bare pseudopotential. Although this formula involves only a single plane wave, and therefore does not give any anisotropy, this can be introduced by expanding the wave-function as a sum of plane waves (Ziman 1961b; Fukai 1969).

$$\psi_{\mathbf{k}} = \sum_{\mathbf{g}} c_{\mathbf{g}}(\mathbf{k})|\mathbf{k}+\mathbf{g}\rangle \tag{43}$$

so

$$\tau^{-1}(\mathbf{k}) = \frac{c\Omega}{4\pi^2\hbar}\int_{\mathrm{FS}} |M_{\mathbf{kk}'}|^2 \frac{\mathrm{d}S_{\mathbf{k}'}}{\hbar v_{\mathbf{k}'}}, \tag{44}$$

where the matrix element

$$M_{\mathbf{kk}'} = \sum_{\mathbf{g}}\sum_{\mathbf{g}'} c^*_{\mathbf{g}'}(\mathbf{k}')c_{\mathbf{g}}(\mathbf{k})\Delta W(|\mathbf{q}+\mathbf{g}'-\mathbf{g}|), \tag{45}$$

and where the integral is over the non-spherical Fermi surface. There are, however, very few examples in the literature of such a formula being applied to dHvA data.

The primary limitation of a pseudopotential model is that band-gaps at the Fermi surface must be small so it cannot be used, at least without considerable complication, for transition metals or even noble metals where the d-band resonance is near the Fermi energy. Another restriction comes from the fact that the matrix elements of the potential are calculated in the Born approximation and this is only valid if the difference between host and impurity potentials is small, i.e. essentially only for isovalent impurities (Heine & Weaire 1970).

8.4.3 Phase shift analysis

In this section a partial wave analysis is applied to a single impurity imbedded in a pure host crystal. To render the problem soluble a fundamental approximation is made; both the host and impurity atoms are described by an array of spherically symmetric potentials confined to 'muffin-tins' (MTs) with a flat region (the MT zero, MTZ) between the atoms. As a model for dilute alloys this is sometimes questioned because, as the evidence of the Friedel oscillations shows, the perturbation introduced by an impurity is long-range. We believe that a multiple scattering formalism does treat this long-range perturbation of the wave-functions correctly and would emphasize that it is only the potentials, not the wave-functions, that are assumed to be localized within the MT radius. Even this is less stringent as a constraint than it might appear because, as Andersen (1971a) and Ball (1972) have shown, the Korringa, Kohn and

Rostoker (KKR) method can be extended formally to include the case of overlapping spherical potentials which extend as far as the centre of the nearest neighbour atom and indeed, when the potentials are characterized by their phase shifts, the results of the phase-shift analysis show no dependence on the MT radius.

The scattering by an isolated impurity in the MT model has been discussed by several authors (e.g. Dupree 1961; Beeby 1967; Johnson 1968; Harris 1970; Holzwarth 1975). They conclude that an exact solution of the problem requires the integral over the Brillouin zone of the inverse of the KKR matrix. Simplifying approximations have been used by Lasseter & Soven (1973) and Lehmann (1975) to help in evaluating this integral but as Holzwarth emphasizes the exact calculation is not intractable. It uses standard KKR programs and the results can be expressed in a form independent of impurity so that the integration need only be carried out once for each host metal. Furthermore experimental data can be parametrized in a physically significant fashion without having to evaluate the full Brillouin zone integral, just a simpler Fermi surface integral.

In the analysis outlined below we rely heavily on results obtained by Morgan (1966) and Holzwarth (1975). To clarify the physics the simple example of phase-shift analysis of the scattering of free electrons is discussed first and when possible the more general results are compared with this.

Free-electron scattering

In the standard treatment of scattering from a spherically symmetric potential, $V(r)$, which it will be assumed is zero outside a (MT) radius R_M, a plane wave $\exp(i\mathbf{k}\cdot\mathbf{r})$ is decomposed into partial waves. If $j_l(kr)$ is the spherical Bessel function which is regular at the origin and Y_{lm} is a real normalized spherical harmonic

$$\exp(i\mathbf{k}\cdot\mathbf{r}) = \sum_l \sum_m 4\pi i^l j_l(kr) Y_{lm}(\hat{\mathbf{r}}) Y_{lm}(\hat{\mathbf{k}})$$

$$= \sum_l (2l+1) i^l j_l(kr) P_l(\hat{\mathbf{k}}\cdot\hat{\mathbf{r}}), \tag{46}$$

where P_l is a Legendre polynomial. For $r > R_M$ the total solution consists of the incoming plane wave plus a scattered wave given by

$$\psi_{\text{scatt}} = \sum_l \sum_m 4\pi i^l [i \sin \eta_l \exp(i\eta_l) h_l(kr)] Y_{lm}(\hat{\mathbf{k}}) Y_{lm}(\hat{\mathbf{r}}). \tag{47}$$

The spherical Hankel function $h_l(kr) = j_l + \mathrm{i}n_l$, where n_l is the spherical Neumann function, and the phase shifts η_l are given by

$$\sin \eta_l \exp(\mathrm{i}\eta_l) = -\frac{2mk}{\hbar}\int_0^{R_\mathrm{M}} V(r) j_l(kr) R_l(r) r^2 \,\mathrm{d}r, \tag{48}$$

where R_l is the radial part of the wave-function inside the impurity potential. An attractive (negative) potential implies positive phase shifts and it is conventional to choose $\eta_l \to 0$ as $V \to 0$.

For scattering through an angle θ the scattering amplitude, obtained by matching the wave-functions at $r = R_\mathrm{M}$, is given by

$$f(\theta) = \frac{1}{k}\sum_l (2l+1) \sin \eta_l \exp(\mathrm{i}\eta_l) P_l(\cos\theta) \tag{49}$$

and the total (isotropic) scattering rate is obtained, in terms of the velocity of the incoming particles and the total scattering cross-section as

$$\tau^{-1} = v\int_{4\pi} |f(\theta)|^2 \,\mathrm{d}\Omega = \frac{4\pi\hbar}{mk}\sum_l (2l+1)\sin^2\eta_l. \tag{50}$$

In T-matrix notation the scattering between states $\mathbf{k}$ and $\mathbf{k}'$ (separated by angle θ) is given by

$$T_{\mathbf{k}\mathbf{k}'} = -\frac{2\pi\hbar^2}{m} f(\theta). \tag{51}$$

Inspection of (49)–(51) shows, as an example of the optical theorem (see e.g. Callaway 1974), that

$$\tau^{-1} = -\frac{2}{\hbar}\,\mathrm{Im}\, T_{\mathbf{k}\mathbf{k}}. \tag{52}$$

Anisotropy and other effects of the host atoms

For an impurity in the environment of host atoms it is assumed that the band-structure problem for the pure host can be solved exactly (within the MT approximation) so that in the interstitial region where the potential is taken as zero the Bloch states, energy $\mathscr{E}$, can be written using the KKR formalism (see e.g. Segall & Ham 1968) as

$$\psi_\mathbf{k}(\mathbf{r}) = \sum_{lm} \mathrm{i}^l a_{lm}(\mathbf{k})[j_l(\kappa r) + \mathrm{i}\sin\eta_l^\mathrm{h} \exp(\mathrm{i}\eta_l^\mathrm{h}) h_l(\kappa r)] Y_{lm}(\hat{\mathbf{r}}), \tag{53}$$

where η_l^h are the phase shifts characterizing the host atoms, $\mathbf{r}$ is measured with respect to the position of the atom in a cell and $\kappa^2 = 2m\mathscr{E}/\hbar^2$. In the free-electron case the coefficients $a_{lm}(\mathbf{k})$ just reduce to $4\pi Y_{lm}(\hat{\mathbf{k}})$. Inside the impurity MT the wave-function is of the form

$$\phi_{\mathbf{k}}(\mathbf{r}) = \sum_{lm} \mathrm{i}^l c_{lm}(\mathbf{k}) R_l^{\mathrm{i}}(r) Y_{lm}(\hat{\mathbf{r}}). \tag{54}$$

In the free-electron case the scattered wave $b_{lm}(\mathbf{k})h_l(\kappa r)Y_{lm}(\hat{\mathbf{r}})$ satisfies Schrödinger's equation in free space but here it is not an eigenstate of the host crystal so that $\phi_{\mathbf{k}}(\mathbf{r})$ must be matched not only to the incident wave and a scattered wave but also to a backscattered wave of the form $T^{\mathrm{M}}_{lml'm'}b_{l'm'}(\mathbf{k})j_l(kr)Y_{lm}(\hat{\mathbf{r}})$. The matrix $T^{\mathrm{M}}_{lml'm'}$ (introduced by Morgan 1966) describes the scattering, by the surrounding lattice, of the diverging wave into incoming waves. Matching of all the waves at $r = R_{\mathrm{M}}$ gives

$$\begin{aligned} c_{lm} &= \mathrm{i}a_{lm}\left[1 + \sum_{l'm'} T^{\mathrm{M}}_{lml'm'} b_{l'm'}/a_{lm}\right] \Big/ \kappa R_{\mathrm{M}}^2 [R_l^{\mathrm{i}}, h_l] \\ &= \mathrm{i}\sum_{l'm'} A_{lml'm'} a_{l'm'} / \kappa R_{\mathrm{M}}^2 [R_l^{\mathrm{i}}, h_l], \end{aligned} \tag{55}$$

where $[f, g] = [f\,\mathrm{d}g/\mathrm{d}r - g\,\mathrm{d}f/\mathrm{d}r]_{r=R_{\mathrm{M}}}$ and where the backscattering or renormalization matrix, $A_{lml'm'}$, has been introduced.

If the potential difference between impurity and host is $\delta V(r)$ the T-matrix element, defined between two equal energy states $\mathbf{k}$ and $\mathbf{k}'$, is

$$T_{\mathbf{kk}'} = \int_0^{R_{\mathrm{M}}} \psi^*_{\mathbf{k}'}(r)\delta V(r)\phi_{\mathbf{k}}(r) r^2\,\mathrm{d}r \tag{56}$$

and because $\delta V(r)$ differs from zero only in the impurity MT the volume integral can be replaced by a surface integral which gives, after some manipulation,

$$T_{\mathbf{kk}'} = -\frac{\hbar^2}{2m\kappa} \sum_{lml'm'} a^*_{lm}(\mathbf{k}')a_{l'm'}(\mathbf{k})A_{lml'm'} \sin\Delta\eta_l \exp(\mathrm{i}\Delta\eta_l), \tag{57}$$

where $\Delta\eta_l = \eta_l^{\mathrm{i}} - \eta_l^{\mathrm{h}}$.

The T-matrix, which contains all the scattering information, can therefore be separated into a sum of terms each of which is the product of (*a*) the wave-function amplitudes $a_{lm}(\mathbf{k})$ which exhibit anisotropy over the Fermi surface, and which are a characteristic of the host alone; (*b*) a backscattering or renormalization matrix $A_{lml'm'}$ which reflects the fact that the impurity is not isolated but is in the environment of the surrounding host lattice and which appears to be an extremely complicated function of both host and impurity atoms and (*c*) a phase-shift term dependent only on the difference between the impurity and host phase shifts.

An important simplification occurs in cubic crystals for low values of l. In the environment of the impurity the alloy does not have spherical symmetry but it does have cubic symmetry so that the basis functions should not in general be spherical harmonics but rather the appropriately symmetrized lattice harmonics. For cubic crystals the expansion should therefore be made in Kubic harmonics for which the notation Y_L is used to denote a Kubic harmonic belonging to a representation of the cubic group; $L \equiv l, \Gamma$ where Γ is an irreducible representation and γ denotes a member of that representation. In this case not only is the above analysis valid but for low l values (≤ 2) Blaker & Harris (1971) obtained the very important result that the backscattering matrix $A_{LL'}$ is diagonal and *independent* of $\mathbf{k}$ with elements denoted by A_L. This being so the *anisotropy* of the T-matrix is contained totally in the coefficients $a_{L\gamma}(\mathbf{k})$ (using the new basis set) while the terms $A_L \sin \Delta\eta_L \exp(i\Delta\eta_L)$ are independent of $\mathbf{k}$ and characteristic only of the impurity in the host. As discussed in Chapter 9 this means that dHvA scattering results and also the Fermi surface changes can be deconvoluted to yield phase-shift parameters characterizing the impurity.

In a cubic crystal there is only one irreducible representation corresponding to each of the s- and p-scattering channels (see Table 8.1). For d-scattering there are two representations though there is often only a small error involved in combining them into one $l = 2$ term (Lee, Holzwarth & Coleridge 1976a). When f-scattering is considered, the Γ_{15} representation is now common to both p- and f-scattering so that in an angular momentum representation $A^{\Gamma}_{ll'}$ has off-diagonal terms, and for $l = 4$ the situation is even more complicated. However, if the phase shifts are characterized not by the angular momentum but by the irreducible representation L then A^{Γ}_{L} is always diagonal (Callaway 1974).

TABLE 8.1. *Irreducible representations of the cubic group for angular momentum $l \leq 4$ (from Lee et al. 1976a)*

l				
0	Γ_1			
1	Γ_{15}			
2	Γ_{12}	Γ'_{25}		
3	Γ'_2	Γ_{15}	Γ_{25}	
4	Γ_1	Γ_{12}	Γ'_{15}	Γ'_{25}

Determination of the wave-functions

Analysis of dHvA data in alloys involves fitting anisotropies to a T-matrix of the form

$$T_{\mathbf{kk}} = -\sum_L t_L(\mathbf{k}) A_L \sin \Delta\eta_L \exp(\mathrm{i}\Delta\eta_L), \tag{58}$$

where we have written $t_L(\mathbf{k}) = (\hbar^2/2m\kappa)\sum_\gamma |a_{L\gamma}(\mathbf{k})|^2$. As discussed by Holzwarth (1975) it is possible to determine $t_L(\mathbf{k})$ experimentally by fitting the Fermi surface of the pure host metal with a KKR parametrization. In the appropriate lattice representation the KKR matrix takes the form

$$M_{LL'}^{\gamma\gamma'} = \cot \eta_L^{\mathrm{h}} \delta_{LL'} \delta_{\gamma\gamma'} + \kappa^{-1} B_{LL'}^{\gamma\gamma'}, \tag{59}$$

where $B_{LL'}^{\gamma\gamma'}$ are the structure factors and the Fermi surface is specified by the solution of the KKR equation

$$\sum_{L'\gamma'} M_{LL'}^{\gamma\gamma'} u_{L'\gamma'}^0 = \lambda^0 u_{L\gamma}^0 = 0. \tag{60}$$

If the zero eigenvector is normalized to unity, i.e.

$$\sum_{L\gamma} |u_{L\gamma}^0|^2 = 1, \tag{61}$$

then the wave-functions, normalized in a unit cell, are given by

$$t_L(\mathbf{k}) = \frac{\hbar v_{\mathbf{k}} \sum_\gamma |u_{L\gamma}^0(\mathbf{k}, \mathscr{E}_{\mathrm{F}})|^2}{\sin^2 \eta_L^{\mathrm{h}} |\nabla_{\mathbf{k}} \lambda^0(\mathbf{k}, \mathscr{E}_{\mathrm{F}})|} \tag{62}$$

This form of normalization is particularly convenient because a Fermi surface fit involves a constant energy, variable $\mathbf{k}$, search which automatically yields $\nabla_{\mathbf{k}}\lambda^0$; it is equivalent to an identity (Coleridge 1973b) relating $t_L(\mathbf{k})$ to $\partial\mathbf{k}/\partial\eta_L$.

It is an empirical fact that the anisotropy of $t_L(\mathbf{k})$ is independent, to a good approximation, of the choice of the Fermi energy parameter $\mathscr{E}_{\mathrm{F}}$. Physically this is because the choice of $\mathscr{E}_{\mathrm{F}}$ just corresponds to moving the MTZ up or down and is more an artefact of the KKR formalism than of any fundamental importance. Indeed, if the MTs are allowed to overlap by a moderate amount the KKR formalism remains intact (Ball 1972) and the flat potential region can be totally eliminated. With a reasonable choice of Fermi energy parameter (so that convergence of the KKR matrix is optimized) only 3 phase shifts ($l \leq 2$) are needed to parametrize many metals and the sum in (58) can be quite reasonably truncated at $l = 2$.

Friedel phase shifts

If the anisotropy of the wave-functions is essentially independent of the Fermi energy parameter it is reasonable to form the normalized quantity $t_L(\mathbf{k})/I_L$, where I_L is the Fermi surface average of $t_L(\mathbf{k})$ in the form

$$I_L = \frac{1}{n(L)} \frac{\Omega}{8\pi^2} \int_{\mathrm{FS}} t_L(\mathbf{k}) \frac{\mathrm{d}S_{\mathbf{k}}}{\hbar v_{\mathbf{k}}}, \tag{63}$$

where $n(L)$ is the degeneracy of the Lth representation, i.e. for $l=0$, $n(\Gamma_1)=1$; for $l=1$, $n(\Gamma_{15})=3$; and for $l=2$, $n(\Gamma_{12})=2$ and $n(\Gamma'_{25})=3$. Note that because a phase-shift parametrization of the host gives the quantity $t_L(\mathbf{k})/v_{\mathbf{k}}$ in (62), the Fermi surface average I_L is completely determined by a Fermi surface integral without the need to know the velocities $v_{\mathbf{k}}$.

For a concentration c of impurities and with the wave-functions normalized to the unit cell volume Ω, the scattering rate can be obtained in two ways (cf. (50) and (52)):

$$\tau(\mathbf{k})^{-1} = -\frac{2c}{\hbar} \operatorname{Im} T_{\mathbf{kk}} \tag{64a}$$

$$= c\frac{2\pi}{\hbar} \frac{\Omega}{8\pi^3} \int_{\mathrm{FS}} \frac{|T_{\mathbf{kk'}}|^2 \,\mathrm{d}S_{\mathbf{k'}}}{\hbar v_{\mathbf{k'}}}, \tag{64b}$$

so that

$$\operatorname{Im}[A_L \sin \Delta\eta_l \exp(\mathrm{i}\Delta\eta_l)] = |A_L|^2 \sin^2 \Delta\eta_l I_L. \tag{65}$$

If A_L is written as $|A_L| \exp(\mathrm{i}\theta_L)$ and ϕ_L is defined by

$$\phi_L = \Delta n_l + \theta_L \tag{66}$$

then

$$\sin \phi_L = |A_L| I_L \sin \Delta\eta_l. \tag{67}$$

The Friedel phase shifts, ϕ_L, named this way for reasons that will shortly appear obvious, include both the phase shift from the difference in potential and the phase shift from the backscattering factor. In terms of them the T-matrix takes the simple form,

$$T_{\mathbf{kk}} = -\sum_L I_L^{-1} t_L(\mathbf{k}) \sin \phi_L \exp(\mathrm{i}\phi_L). \tag{68}$$

The T-matrix is derived from experiment and therefore independent of the MTZ. Furthermore $t_L(\mathbf{k})/I_L$ is found empirically to be independent of the MTZ so the Friedel phase shifts, unlike A_L and $\Delta\eta_l$, are also independent of the MTZ. It can be shown that they represent eigenstates of the S-matrix (Holzwarth 1975; Lasseter & Soven 1973; Callaway 1974) and satisfy the Friedel sum rule in the form

$$\Delta Z = \frac{2}{\pi} \sum_L n(L)\phi_L. \tag{69}$$

The Friedel phase shifts are obviously analogous to the phase shifts used in the free-electron model, and are the natural choice for any calculations using the model. They are independent of the choice of MTZ and include the renormalization of the wave-functions by the back-scattering from the host. The correspondence with the free-electron result can be seen rather clearly by calculating an average scattering rate over the Fermi surface (cf. (50))

$$\langle \tau(\mathbf{k})^{-1} \rangle = -\frac{2c}{\hbar} \int_{\mathrm{FS}} \mathrm{Im}\, T_{\mathbf{kk}} \frac{\mathrm{d}S_{\mathbf{k}}}{\hbar v_{\mathbf{k}}} \Big/ \int_{\mathrm{FS}} \frac{\mathrm{d}S_{\mathbf{k}}}{\hbar v_{\mathbf{k}}} \tag{70a}$$

$$= c\frac{4\pi\hbar}{m_{\mathrm{th}} k_{\mathrm{fe}}} \sum_L n(L) \sin^2 \phi_L, \tag{70b}$$

where m_{th} is the thermal mass, as deduced from the electronic specific heat, and k_{fe} is the radius of the free-electron sphere.

The corresponding integral involving Re $T_{\mathbf{kk}}$ gives the total volume change (δV) of the Fermi surface because, using (6), (18) and (32),

$$\delta V = \int_{\mathrm{FS}} \Delta k_r \, \mathrm{d}S_{\mathbf{k}} = c \int_{\mathrm{FS}} \mathrm{Re}\, T_{\mathbf{kk}} \frac{\mathrm{d}S_{\mathbf{k}}}{\hbar v_{\mathbf{k}}} \tag{71}$$

so that

$$\frac{\delta V}{V_{\mathrm{fe}}} = c\frac{2}{\pi} \sum_L n(L) \sin \phi_L \cos \phi_L, \tag{72}$$

where V_{fe} is the volume of the free-electron Fermi surface ($=4\pi^3/\Omega$). For small ϕ_L this is equivalent to the rigid-band result but as the phase shifts increase the contribution to the volume change from each phase shift is significantly smaller than the rigid-band prediction until for phase shifts $\sim\pi/2$ it is approximately zero. For even larger phase shifts δV is negative and then becomes zero again when $\phi_L = \pi$. This effect, which has been observed experimentally (Coleridge 1975), can be understood if the spectral density takes the form proposed by Stern (1968). The incoherent background term (see Fig. 8.4) contains states that do not contribute to the 'coherent' Fermi surface.

For example, when a phase shift is near $\pi/2$ the impurity is in a resonant virtual bound state and any conduction electrons contributed by the impurity resonate and become localized. They do not contribute to the propagated coherent wave and have little net effect on the real part of the Fermi surface. But because the impurity is in a virtual bound state there is a large probability of conduction electrons being captured, exchanging with the localized electrons, and contributing to the large scattering rate. When the phase shift is π it represents a bound state that has been localized and drawn below the Fermi surface. It contributes neither to the volume of the Fermi surface nor to the scattering. For phase shifts between $\pi/2$ and π the impurity is behaving as a contributor of holes, attracting electrons from the conduction band and producing a reduction in the Fermi surface size.

It is apparent that the Friedel phase shifts in the alloy are the equivalent of the impurity phase shifts for an isolated impurity and free electrons and the level shifts $\Delta\mathscr{E}_l$ for cN impurities are therefore proportional to $2cN\sqrt{\mathscr{E}^0_{\mathbf{k}}}\phi_l$ (cf. p. 331). By contrast the self-energy, which is proportional to $\sin\phi_l \exp(\mathrm{i}\phi_l)$, is always less than ϕ_l, because the incoherent states are not counted.

Although the Friedel phase shifts may be considered as experimental parameters they cannot be uniquely determined from experiment. The Fermi surface area changes give $\sin\phi_L \cos\phi_L$, so there is an ambiguity of $\phi_L + n\pi$ or $(n+\frac{1}{2})\pi - \phi_L$ in determining ϕ_L, and the Dingle temperatures give $\sin^2\phi_L$, with an ambiguity $\pm\phi_L + n\pi$. In practice the ambiguities can usually be resolved by reference to the Friedel sum rule, which gives the sign of the phase shifts and the correct order of magnitude. A calculation of the resistivity, e.g. by using the free-electron formula or a more sophisticated approach, can also be useful (Coleridge, Holzwarth & Lee 1974). If the assumptions of the theory are correct the two sets of phase shifts derived from the separate experimental quantities should be the same but in practice this does not always happen (see p. 354). The discrepancies are tentatively attributed to distortion of the lattice which has, of course, been neglected in the calculation.

Calculation of the backscattering

Despite the fundamental nature of the Friedel phase-shift parameters and their convenience for purposes of parametrization it is obviously desirable to evaluate the backscattering coefficients so as to deduce experimental values for the phase shifts $\Delta\eta_l$ or, conversely, to deduce Friedel phase shifts from *ab initio* calculations (Lee *et al.* 1976a). The relevant quantity is the complex matrix $\chi_{L\gamma,L'\gamma'}$ (we use the notation of Holzwarth) whose elements are those of the Brillouin zone integral of the inverse of the KKR secular matrix (M) at energy $\mathscr{E}_F$, i.e.

$$\chi_{L\gamma L'\gamma'}(\mathscr{E}_F)=\frac{\Omega}{(2\pi)^3}\int_{BZ}[M(\mathbf{q},\mathscr{E}_F)]^{-1}\,d^3\mathbf{q}. \tag{73}$$

The integrand is singular on the Fermi surface, i.e. the constant energy surface defined by det $M=0$ and the singularity introduces an imaginary term that must be evaluated as a surface integral with diagonal elements given by†

$$\operatorname{Im}\chi_{LL}=\sin^2\eta_l^{\mathrm{h}}I_L, \tag{74}$$

where I_L is defined in (63).

For $l\leq 2$ in a cubic lattice the matrix $\chi_{L\gamma,L'\gamma'}$ is diagonal with elements denoted by χ_L and for each l value, including $l=2$, the Friedel phase shifts are related to χ_L by a single parameter

$$\xi_l=\operatorname{Re}\chi_L+\operatorname{Im}\chi_L\cot\phi_L \tag{75a}$$

$$=(\cot\eta_l^{\mathrm{h}}-\cot\eta_l^{\mathrm{i}})^{-1}. \tag{75b}$$

In the noble metals χ_L has been evaluated by Holzwarth. She has also discussed in detail both the general non-relativistic case (i.e. $l>2$) and the relativistic generalization, which is not trivial.

For $l\leq 2$ the backscattering factor can be written explicitly as

$$A_L=\frac{\exp(-\mathrm{i}\Delta\eta_l)\sin^2\eta_l^{\mathrm{h}}}{\sin\eta_l^{\mathrm{i}}\sin\eta_l^{\mathrm{h}}-\chi_L\sin\Delta\eta_l} \tag{76a}$$

$$=\frac{1-\mathrm{i}T_{LL}^{\mathrm{M}}t_l^{\mathrm{h}}}{1-\mathrm{i}T_{LL}^{\mathrm{M}}t_l^{\mathrm{i}}}, \tag{76b}$$

where

$$T_{LL}^{\mathrm{M}}=\mathrm{i}(\chi_L)^{-1}-\mathrm{i}(t_l^{\mathrm{h}})^{-1} \tag{77}$$

is Morgan's backscattering matrix (equation (55)) and t_l^{h}, not to be confused with the wave-function intensities $t_L(\mathbf{k})$, is a partial T-matrix element defined by

$$t_l^{\mathrm{h}}=\sin\eta_l^{\mathrm{h}}\exp(\mathrm{i}\eta_l^{\mathrm{h}}). \tag{78}$$

† When generalized, I_L is just the density matrix coefficient $T_{ll'}^{\mathrm{i}}$ introduced by Butler, Olson, Faulkner & Gyorffy (1976) in connection with a rigid MT calculation of the electron–phonon interaction.

The dilute limit of the coherent potential model

The coherent potential approximation is usually employed for concentrated alloys but in the dilute limit it should reduce to the single impurity result. In this approximation the random alloy is replaced by periodic array of coherent potentials chosen so as to have a Green's function that is equal to the true Green's function averaged over all possible configurations of the alloy. Soven (1970) treats the coherent MT potential model using a set of (complex) phase shifts η_L^c defined by

$$\cot \eta_L^c = \cot \eta_l^{av} + (\cot \eta_l^i - \cot \eta_L^c)\chi_L^c(\cot \eta_l^h - \cot \eta_L^c), \qquad (79)$$

where

$$\cot \eta_l^{av} = c \cot \eta_l^i + (1-c) \cot \eta_l^h, \qquad (80)$$

and where χ_L^c is just the Brillouin zone integral as defined in (73) but with η_L^c replacing η_l^h. The solution of the coherent potential problem is obtained by reiterating (79).

A similar approach that is sometimes used is the averaged T-matrix approximation, ATA (see e.g. Ehrenreich & Schwartz 1976), in which the phase shifts in the KKR equation are replaced by the complex phase shifts η_L^{ATA} defined by

$$t_L^{ATA} = ct_L^i + (1-c)t_L^h. \qquad (81)$$

The dilute limit of the single impurity problem can be expressed in the form of a partial T-matrix, $\bar{t}_L$ (Coleridge 1975), where

$$\bar{t}_L = cA_L t_L^i + (1-cA_L)t_L^h, \qquad (82)$$

which shows that the ATA differs from the single impurity result to the extent that the backscattering of the surrounding medium is not included.

Using (75), (76) and (82),

$$(\bar{t}_L)^{-1} = (t_l^h)^{-1} - c(\xi_l - \chi_L)^{-1} + O(c^2) \qquad (83)$$

and, following Soven, χ_L^c can be expanded as a function of concentration

$$\chi_L^c = \chi_L + c\chi_L^{(1)} + c^2\chi_L^{(2)} + \ldots \qquad (84)$$

so that equation (61) of Soven can be written as

$$(t_L^c)^{-1} = (t_l^h)^{-1} - c(\xi_l - \chi_L)^{-1} + O(c^2). \qquad (85)$$

Therefore, in the dilute limit $\bar{t}_L = t_L^c$ and the coherent potential approximation reproduces the single impurity result. This is not true of the average T-matrix approximation.

8.4.4 Hybrid phase shift/plane-wave approach

In Morgan's (1966) treatment of scattering an alternative expression for the T-matrix is given in terms of an effective matrix element for scattering between plane-wave states $\mathbf{k}+\mathbf{g}$, $\mathbf{k}'+\mathbf{g}'$.

$$T_{\text{eff}}(\mathbf{k}+\mathbf{g}, \mathbf{k}'+\mathbf{g}') = \sum_{lm} \beta_{lm}(\mathbf{k}+\mathbf{g})\beta_{lm}(\mathbf{k}'+\mathbf{g}')A_L \sin \Delta\eta_l \exp(\mathrm{i}\Delta\eta_l), \quad (86)$$

where $\beta_{lm}(\mathbf{k}+\mathbf{g})$ is a term in the plane-wave expansion (equation (46)). The T-matrix can therefore be evaluated using phase shifts to describe the potentials but with the wave-functions determined by a plane wave expansion (equation (43)). Although, in general, a large number of plane waves must be considered, in the spirit of the pseudopotential treatment, only a small number is required for free electron-like metals. One of the simplest expansions is to describe a noble metal by one plane wave on the belly and two, equal amplitude, plane waves on the neck (Ziman 1961b; Brown & Morgan 1971). More elaborate schemes have been used, e.g. a 4 plane wave, 5 d-state interpolation scheme (Coleridge & Templeton 1971a), a KKR–Ziman expansion (Coleridge 1972) or a pseudopotential calculation (Sorbello 1974b, 1977). These schemes have the particular advantage that only a simple calculation is needed to obtain wave-functions when calculating the self-energy or parametrizing experimental data in the form of phase shifts.

As argued above the backscattering from the host lattice can be absorbed into Friedel phase shifts, and this occurs automatically when fitting experimental data, but when the host metal has a weak potential the backscattering is expected to be small, $A_L \approx 1$, so that the scattering phase shifts can then be approximated by the effect of an isolated impurity. Sorbello argues that in this way the pseudopotential model can be justified for alloys even when the Born approximation is not valid.

8.4.5 Lattice distortion

Changes of lattice constant affect the Fermi surface size and the scattering in different ways. The usual experimental approach for size changes is to scale all measured Fermi surface areas to the alloy Brillouin zone dimensions which, it is usually assumed, are just those measured by X-rays. If $\Delta a/a$ is the fractional lattice constant change on alloying, all Fermi surface areas in the alloy are therefore corrected by a factor $(1+2\Delta a/a)$ before comparing them with the pure metal values. Within the framework of the phase-shift analysis the Friedel sum of the phase shifts deduced from the corrected areas should then give ΔZ but this will include a contribution of order $(3/c)(\Delta a/a)$ from the lattice constant correction.

It must be emphasized that it is incorrect to deduce phase shifts by parametrizing Fermi surface data uncorrected for lattice distortion. When the effect on both electron and hole orbits is considered, this becomes obvious. As illustrated in Fig. 8.7, a decrease in lattice constant increases both hole and electron areas by the same fraction as the Brillouin zone dimensions increase but the effect of increasing one of the phase shifts is to increase the size of electron orbits and decrease the size of hole orbits. Therefore a uniform change of lattice constant cannot be simulated by just modifying the phase shifts.

Although the uniform part of lattice constant change might not affect the scattering there is also a localized portion that does contribute an extra scattering term. Within a continuum model (Eshelby 1954; Blatt 1957) the scattering phase shifts then obey a modified Friedel sum rule with an effective valence difference

$$\Delta Z^* = \Delta Z - Z\gamma\delta\Omega/\Omega_0, \tag{87}$$

where $\gamma\delta\Omega/\Omega_0$ is that fraction of the volume change that is localized round the impurity, Z is the valence of the host and γ is of order $\frac{2}{3}$. An impurity which is larger than the surrounding atoms expands the lattice, removes to infinity part of the total charge, reduces the amount of

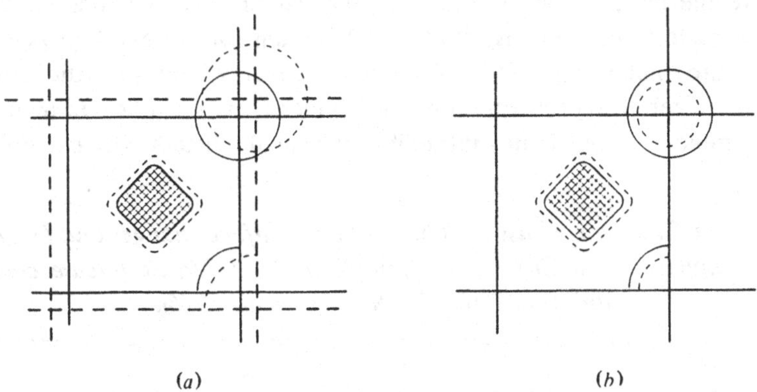

Fig. 8.7. Effect on the Fermi surface of a lattice constant change compared with a potential change. (*a*) The lattice constant is reduced so the Brillouin zone dimensions are increased. Both the electron surface (shown hatched) and the hole surface are uniformly expanded by the same amount. (*b*) If the potential is changed, by increasing a phase shift, for example, the electron surface expands but the hole surface contracts.

screening required and therefore reduces the effect of the impurity potential. It might be expected therefore that the Friedel sum phase shifts determined from Dingle temperature data should differ from those determined from the (corrected) Fermi surface size by an amount of order $Z\gamma(\delta\Omega/\Omega_0)$ and it should not be possible to find one set of phase shifts to explain both sets of experimental results if the lattice distortion is important.

Table 8.2 shows the results of phase-shift analyses in Cu(Ni) and Cu(Zn) alloys. Although it is a little difficult to estimate reliably the errors in the Friedel sums it is clear that there are differences between the two sets of phase shifts and it is the Friedel sum for the Fermi surface data that is correct. Further evidence of this can be seen in the scattering data analysed by Coleridge *et al.* (1974). What is not clear, however, is whether the discrepancy is quantitatively of the amount predicted by the Eshelby–Blatt theory.

A consideration of the scattering process suggests that the effect of lattice distortion should appear, not in the impurity phase shifts but rather in the backscattering factors A_L. Indeed, a treatment by Lodder (1976) shows that for a cluster (j) of MT potentials on a distorted lattice the T-matrix is given by

$$T_{\mathbf{kk}'} = -\frac{\hbar^2}{2m\kappa}\sum_j\sum_L a_L^{j*}(\mathbf{k}')\sum_{L'} a_{L'}^j(\mathbf{k})A_{LL'}^{\mathrm{D}} \times \sin(\eta_{l'}^j - \eta_l^{\mathrm{h}})\exp(\eta_{l'}^j - \eta_l^{\mathrm{h}}), \quad (88)$$

where the new, distorted, backscattering factor $A_{LL'}^{\mathrm{D}}$ depends on the lattice distortion, the phase shifts and Brillouin zone integrals $\chi_{LL'}(R_{jj'})$ which are similar to χ_L but involve also the distance between lattice sites in the cluster. The problem is obviously much more complicated than in the simple case but is not intractible for small clusters. For example,

TABLE 8.2. *Friedel phase shift fits for Fermi surface changes and Dingle temperatures in Cu(Zn) (I. M. Templeton & P. Vasek private communication) and Cu(Ni) (Lee et al. 1976a)*

	$l=$	0	1	2	Friedel sum
Cu(Zn)	ϕ_l^{FS}	0.29 ± 0.07	$0.18_6\pm0.02$	$0.14_3\pm0.02$	1.02
	ϕ_l^{DT}	$0.15_4\pm0.04$	$0.19_0\pm0.006$	$0.13_6\pm0.01$	0.89
Cu(Ni)	ϕ_l^{FS}	$-0.27_3\pm0.2$	$-0.05_3\pm0.04$	$-0.23_9\pm0.07$	−1.03
	ϕ_l^{DT}	-0.07 ± 0.2	$-0.03_8\pm0.03_5$	$-0.25_8\pm0.06$	−0.94

evaluation of the integrals $\chi_{LL'}(R_{jj'})$ needs little more calculation than that involved in determining χ_L.

Equation (88) implies that the separation into s, p and d partial waves will no longer be as simple as in the single substitutional impurity case but from Table 8.2 it is apparent that only the s phase shift is appreciably different. This can be understood in a qualitative fashion if all off-diagonal terms in $A^{\mathrm{D}}_{LL'}$ are ignored and if the main effect of the lattice distortion is just to change the phase of A_L by a small amount δ_L. If

$$A^{\mathrm{D}}_{LL'} \approx |A_L| \exp\,[\mathrm{i}(\theta_L + \delta_L)] \tag{89}$$

then

$$A^{\mathrm{D}}_{LL'} \sin \Delta\eta_l \exp\,(\mathrm{i}\Delta\eta_l) \approx I_L^{-1} \sin \phi_L \exp\,[\mathrm{i}(\phi_L + \delta_L)], \tag{90}$$

and for small phase shifts $\cos\,(\phi_L + \delta_L)$ is approximately unity so that values of ϕ_L deduced from Re $T_{\mathbf{kk}}$ will be approximately correct. In Im $T_{\mathbf{kk}}$, however, the term $\sin\,(\phi_L + \delta_L)$ may differ appreciably from ϕ_L, so phase shifts deduced from the scattering might be wrong. To a first approximation the lattice distortion is isotropic so, not unexpectedly, it is the s phase shift that is the most affected by the lattice distortion. This tentative explanation must await confirmation from detailed calculations using Lodder's formalism.

A simpler approach to lattice distortion is the pseudopotential approach of Benedek & Baratoff (1973; see also Popovic, Carbotte & Piercy 1973). Scattering amplitudes are determined from a plane-wave matrix element

$$M(\mathbf{q}) = N^{-1}[W^{\mathrm{i}}(q) - W^{\mathrm{h}}(q) + W^{\mathrm{h}}(q)S^{\mathrm{d}}(\mathbf{q})], \tag{91}$$

where $W(q)$ is an OPW form factor (for host or impurity) and where S^{d} is a structure factor appropriate to the distorted lattice. Unfortunately this approach does not appear capable of explaining the scattering results in K(Rb) and K(Na) alloys, as measured by Llewellyn, Paul, Randles & Springford (1977). An alternative approach is that of Béal-Monod & Kohn (1968), where the effects of the distorted lattice are derived by considering the electrostatic potential of the displaced atoms. They are expressed in terms of phase shifts and structure factors which can be particularly large for the ⟨111⟩ direction in noble metals. Unfortunately, in copper, the theory predicts the p phase shift to be dominant, which is in disagreement with experiment.

8.4.6 Interstitial impurities

The example of an interstitial impurity, to be specific hydrogen in copper, is a special case of a cluster calculation but it has been discussed by Holzwarth & Lee (1975) as an extension of the substitional impurity case. The hydrogen occupies an octahedral interstitial site which is characterized by the cubic point group symmetry so that a decomposition into s-, p- and d-waves is still possible but the wave-functions must now be evaluated at the impurity site. This is not a lattice site of the pure host crystal so the KKR matrix involved in calculating the wave-functions and the integral χ_L must involve two structure factors, one for the host lattice sites and one for the interstitial sites. The potential on interstitial sites is zero except for the impurity atom. The wave-function intensities, i.e. the coefficients $t_L(\mathbf{k})$, are changed markedly from the substitutional case; e.g. on the neck there is strong s- and d-scattering which is almost the reverse of the substitutional case.

The comparison with experiment in Holzwarth & Lee's paper is not valid because the experimental results then available were in error. An analysis of the correct experimental results is given by Wampler & Lengeler (1977) and, as expected, they find strong s-scattering and smaller amount of p-scattering. The Friedel sum is 0.88 which is consistent with the amount of lattice distortion that would be expected in this system.

8.5 Summary of experimental results

An attempt has been made to list all dHvA experiments in dilute alloys published before July 1978. In the interests of brevity preliminary reports of experiments are not included if a detailed account has been published later and papers that discuss experimental work that is published elsewhere have also been excluded.

Aluminium

Shepherd & Gordon (1968) have measured frequency shifts, Dingle temperatures and cyclotron masses for various frequencies less than 1 MG in Al(Zn), Al(Si), Al(Ge), Al(Mg) and Al(Ag). They found that frequencies obeyed a rigid-band model, that the Dingle temperatures were approximately isotropic and that the cyclotron masses were unchanged. Abele & Blatt (1970) studied the low frequencies (less than 5 MG) in Al(Mg), using the magnetothermal oscillations, and found some deviation from rigid-band behaviour. In Al(Mn), Paton (1971) has measured frequency differences, Dingle temperatures and cyclotron masses and interprets the results in terms of localized spin fluctuations.

Antimony

The Sb(Sn) alloy system has been studied by Ishizawa & Tanuma (1965), Dunsworth & Datars (1973), and Harte, Priestley & Vuillemin (1978). The Sb(Te) system has also been studied by the latter authors and by Altounian & Datars (1975). Frequency changes of over 50 per cent were observed and cyclotron mass changes of 30 per cent. The agreement with a rigid-band model is good when the changes of effective mass are included.

Beryllium

The only beryllium-based alloy that appears to have been measured is Be(Cu). Goldstein, Sellmyer & Averbach (1970) have measured the frequency shifts, Dingle temperatures and some cyclotron masses and Tripp, Everett, Fiske & Gordon (1970) have measured frequencies. Frequencies less than 15 MG were studied and in both cases there was a fair agreement with the rigid-band model but with most of the change being attributable to lattice constant effects.

Bismuth

Using the de Haas–Shubnikov oscillations Bhargava (1967) has studied the frequency shifts in Bi(Pb) and summarizes the dHvA data then available in bismuth alloys including the early work of Shoenberg & Uddin (1936).

Chu & Kao (1970) and Chao, Lin & Kao (1973) summarize early work in the Bi(Sb) system and use the de Haas–Shubnikov oscillations to determine frequency shifts, Dingle temperatures and effective masses.

Rode & Lowndes (1977) have studied Bi(Sb) and Bi(Si) alloys in the quantum limit and pay particular attention to the effect of the scattering on the line-shape.

Cadmium

V. Sukhaparov & I. M. Templeton (private communication), while investigating the effect of pressure on the Fermi surface of cadmium, also measured two Cd(Mg) alloys arguing that the lattice constant change is in the same sense as that produced by pressure.

Carbon

Bender & Young (1971) have reported the observation of new Shubnikov–de Haas oscillations in graphite intercalated with bromine. They suggest that a sequence of intermetallic compounds C_nBr are produce rather than a dilute alloy. Batallan, Bok, Rosenman & Melin (1978) report the observation of magnetothermal oscillations in $C_{12n}(SbCl_5)$ and discuss their results in terms of a two-phase model.

The effect on the de Haas–Shubnikov oscillations of neutron irradiation of graphite has been studied by Cooper, Smith, Woore & Young (1971), and by Dillon & Spain (1977). Frequency shifts, Dingle temperatures and cyclotron masses were studied.

Chromium

When more than about 1 per cent of Mn is added to chromium the spin-density wave becomes commensurate with the lattice. This effect has been seen by Gutman & Stanford (1971) and by Graebner (1971). In more dilute Cr(Mn) and in Cr(V) the changes of frequency seen by Gutman and Stanford were interpreted in terms of a change in **q**-vector rather than a change of band structure.

Copper

Chollet & Templeton (1968) have measured frequency shifts and Dingle temperatures for neck and belly oscillations in Cu(Zn), Cu(Cd), Cu(Al), Cu(Pd) and Cu(Ni) alloys. Coleridge & Templeton (1971a) repeated some of these measurements and also measured Cu(Ge), Cu(Si), Cu(Co), Cu(Fe), Cu(Mn) and Cu(Cr). In general the results did not agree with the rigid-band model and were discussed using a hybrid plane-wave/phase-shift approach.

Dingle temperatures for all symmetry directions have been measured by Coleridge (1972) in Cu(Fe), by Poulsen, Randles & Springford (1974) in Cu(Ni), Cu(Ge) and Cu(Au) and by I. M. Templeton & P. Vasek (private communication) in Cu(Zn). The frequency shifts at symmetry directions have been obtained by Templeton & Coleridge (1975a) in Cu(Ni) and Cu(Al). In all cases the results were fitted by phase-shift parametrizations.

In the magnetic alloys Cu(Cr), Cu(Mn), Cu(Fe) and Cu(Co) the spin-zero technique was used by Coleridge, Scott & Templeton (1972) to study the exchange splitting and the differential scattering, and in Cu(Fe) Alles & Higgins (1974) used the wave-shape analysis technique on the neck oscillations.

The Dingle temperatures at many orientations have been measured by Wampler & Lengeler (1977) in copper with interstitial hydrogen and the results parametrized with a phase-shift fit.

The effect of dislocations in copper has been investigated by Terwilliger & Higgins (1970, 1973), Coleridge & Watts (1971) and Chang & Higgins (1975). Chang *et al.* (1975a) have also looked at the scattering produced by dislocation loops that result from neutron irradiation.

Gold

In early work on gold alloys King-Smith (1965) measured the ⟨111⟩ frequencies and Dingle temperatures in Au(Ag) and Au(Pt) and Springford, Stockton & Templeton (1969) measured Dingle temperatures in Au(Ag).

A detailed analysis of Dingle temperatures, both on- and off-symmetry axes, has been made by Lowndes, Miller, Poulsen & Springford (1973b) in Au(Ag), Au(Cu), Au(Zn) and Au(Fe). Dingle temperatures at symmetry directions have been measured and parametrized by phase shifts in Au(Ga) by Dye *et al.* (1977) and in Au(Co) by Chung & Lowndes (1976).

Frequency shifts at the symmetry directions have been measured and fitted with phase-shift parametrizations in Au(Ag) by Templeton & Coleridge (1975b) and in Au(Ga) by Coleridge†.

Magnetic impurity states have been studied in Au(Fe) by Lowndes, Crabtree, Ketterson & Windmiller (1973a) and Chung *et al.* (1976) and in Au(Co) Chung & Lowndes (1977) have seen magnetic effects from *pairs* of impurity atoms.

Dingle temperatures in quenched gold have been studied by Lengeler (1977), Lengeler & Uelhoff (1975), Chang *et al.* (1975b), Chang, Crabtree & Ketterson (1977). The quenching, and any subsequent annealing, can produce vacancies, stacking-fault tetrahedra and dislocations each of which has a different scattering anisotropy. With care the contribution of each can be separated.

Indium

Higgins, Kaehn & Condon (1969) have studied the frequency shifts of the β oscillations in In(Cd) for frequencies less than 5 MG and Holtham & Parsons (1976) have studied the βs in In(Pb), In(Tl). In both cases the behaviour was predominantly rigid-band-like but with small deviations produced by changes in potential.

† See Coleridge, P. T. (1979) *J. Phys. F* **9**, 473.

Lead

Anderson & Hines (1970) have measured the frequency shift and Dingle temperature for the β frequency (50 MG) in Pb(Bi) alloys and Anderson, Lee & Stone (1975) have made the same measurements in Pb(Bi), Pb(Tl) and the isoelectronic Pb(BiTl) alloy for several frequencies up to 160 MG. The results are mainly rigid-band-like. Tobin, Sellmyer & Averbach (1971) have studied the γs (18 MG) in Pb(In) and find a new frequency is produced by alloying.

Magnesium

Hornbeck, Fung & Gordon (1977) have measured the frequency shifts and some Dingle temperatures for the low frequencies (less than 3 MG) in Mg(Li) and Mg(In) alloys.

Brown & Friedberg (1979) have studied the effects of edge dislocations on Dingle temperatures for ultrapure magnesium and very dilute Mg(Cd), Mg(Zn) alloys.

Molybdenum

Arko & Mueller (1975) have studied the Dingle temperatures for the third-band hole octahedron in Mb(W), Mo(Ru) and Mo(Fe) alloys and invert the data using cubic harmonics. In Mo(U) they made a detailed analysis for the whole Fermi surface.

Palladium

Hörnfeldt, Ketterson & Windmiller (1969) have looked at the exchange splitting in Pd(Co) for two orbits using the spin-zero technique and Hörnfeldt, Dronjak & Nordborg (1976) have made a similar study in Pd(Ni).

Griessen *et al.* (1977b) have introduced up to 1.5 per cent of hydrogen interstitially into Pd and measured the frequency shifts for three small orbits. The changes are much less than a rigid-band model predicts, even when lattice constant effects are considered†.

Potassium

Llewellyn *et al.* (1977) have measured Dingle temperatures in K(Rb) and K(Na) alloys and have fitted their results with a Kubic harmonic expansion.

† See also: Bansil, A *et al.* (1979). *Solid State Commun.* **32**, 1115.

Rhodium

Cheng, Higgins, Graebner & Rubin (1979) have studied the Dingle temperatures for several orbits in the magnetic alloy Rh(Fe). On the smallest sheet of Fermi surface the Dingle temperature is found to vary linearly with temperature and there is a shift in the *g*-factor.

Silver

For the ⟨111⟩ neck and belly a measurement of frequency shifts and Dingle temperatures in Ag(Au) and Ag(Cu) alloys was made by King-Smith (1965) and Springford *et al.* (1969) measured Dingle temperatures in Ag(Au), Ag(Cu) and Ag(Cd).

Dingle temperature measurements at symmetry directions and phase-shift parametrizations have been made by Brown & Myers (1972) in Ag(Au), Ag(Sn), Ag(Cd) and Ag(Ge) and by Sang & Myers (1976) with the transition metal solutes Pd, Y, Gd, Mn and Fe.

In Ag(Au), frequency shifts at the symmetry directions have been measured and parametrized by Templeton & Coleridge (1975b).

Zinc

Higgins & Marcus (1966) have measured frequency shifts and effective masses in Zn(Cu) and Zn(Al) alloys for some of the frequencies less than 10 MG and find the dominant effect to be lattice constant changes. Dingle temperatures on the needles were measured by Hedgecock & Muir (1963) and Paton, Hedgecock & Muir (1970). The relationship between Dingle temperatures and magnetic breakdown has been studied by Higgins & Marcus (1967) in Zn(Cu), Zn(Al) and Zu(Mn) and by Buot, Li & Ström-Olsen (1976) in Zn(Mn), Zn(Cr) and Zn(Fe).

9

Electron quasi-particle lifetimes in metals

M. SPRINGFORD

9.1 Introduction

The concepts of electron lifetime and mean free path are central to our visualization of the physical processes in terms of which we describe the electronic properties of metals. We shall be concerned in this chapter with the application of the dHvA effect to study the scattering of conduction electrons by the relatively simple scattering centres encountered in a dilute random alloy. The dHvA effect proves to be a particularly powerful tool in this regard, as experiments relate to a localized and identifiable group of electrons at the Fermi surface, as a result of which, by the use of a suitable inversion procedure, the pattern of variation of the electron lifetime over the Fermi surface may be deduced. Operationally, electron scattering is investigated in the dHvA effect through measurements of the Dingle temperature x. In order to gain some insight into the nature of this parameter, we shall use the concept of the self-energy to present in a unified way the manner in which both the electron–impurity and the electron–phonon interaction, enter the dHvA effect. Finally, we shall review the most comprehensive experimental results to date pertaining to the anisotropy of electron scattering over the Fermi surface, which refer to the alkali and noble metals. A related set of problems, an example of which is the scattering of conduction electrons by impurities which carry a local magnetic moment, lend themselves to an investigation by the technique of waveshape analysis and this field is reserved for the following chapter.

9.2 The Dingle temperature

The most striking consequence of the addition of impurities to a metal on the dHvA effect is a reduction in the signal amplitude. Historically this problem was first considered by Dingle & Shoenberg (1950), Robinson (1950) and Dingle (1952) who argued that the effect of collisions was to broaden the otherwise well-defined Landau levels and hence, because the dHvA effect arises from the movement of the Landau levels through the chemical potential μ, to cause a reduction in the amplitude of the oscillations. On the basis of the assumption that the level broadening is Lorentzian of half-width Γ and is independent of energy, the amplitude of the rth harmonic of the dHvA amplitude is reduced by the factor

$$\exp(-2\pi r\Gamma/\hbar\omega_c). \tag{1}$$

Parametrizing this result in terms of a Dingle temperature,

$$x = \Gamma/\pi k_B, \tag{2}$$

then the amplitude reduction factor may be written in a form which shows that x is similar, but not exactly equivalent, to a change in temperature

$$\exp(-rKm_c x/H), \tag{3}$$

K being a constant, 146.9 kG K^{-1}, and m_c the cyclotron effective mass. Noting, however, from the work of Brailsford (1966) that $\Gamma = \hbar/2\tau$, it follows that x is related to an average around the extremal orbit on the Fermi surface of the reciprocal lifetime

$$x = \frac{\hbar}{2\pi k_B}\left\langle\frac{1}{\tau}\right\rangle, \tag{4}$$

in which a factor of 2 has been changed from Dingle's original treatment so that τ corresponds to the mean time between collisions (see the discussion of this point by Chambers, Chapter 3, pp. 106–7). Conceptually, this is an easily understood process. The orbiting electrons can be considered as a flux of particles scattered by a localized impurity potential which, if known, can be used to obtain a mean collision time. Whilst Dingle's theory is essentially phenomenological in character, its main conclusion, expressed in (1), has been confirmed by more elaborate methods.

Before proceeding to a discussion of the methods and results of some lifetime measurements in metals, we wish to gain some insight into the nature of the Dingle temperature and in particular to discuss in a unified way the scattering of electron quasi-particles by both impurities and phonons. The influence of the electron–phonon interaction on dHvA effect amplitudes was first considered by Fowler & Prange (1965) and then, in rather more detail, by Engelsberg & Simpson (1970). It is evident that the influence of such interactions may be incorporated into the theory of the dHvA effect by adding the full self-energy to the non-interacting single particle energies in the oscillating part of the thermodynamic potential. In carrying out the calculation the oscillatory part of the self-energy in the presence of a magnetic field is neglected, so that terms of order $(\hbar\omega_c/\mu)^{\frac{1}{2}}$ are dropped, but one is left with an electron self-energy, $\Sigma(\tilde{\mathscr{E}}, T)$, which has rather strongly temperature-dependent real and imaginary parts at the Fermi energy. This formulation of the problem has a rather wider validity and we shall use it here to describe also the influence of impurities. This should be valid provided that we can neglect the oscillatory part of the self-energy characterizing the electron–impurity interaction which, as shown in Chapter 8 and by Strěda (1974), entails avoiding the region of low quantum numbers. With the assumption of an isotropic parabolic dispersion relation for the non-interacting single-particle spectrum, we have, for the oscillatory magnetization (equation (2.15) of Engelsberg & Simpson (1970)),

$$\frac{\tilde{M}}{V} = -\frac{m_c^{\frac{3}{2}}(\hbar\omega_c)^{\frac{1}{2}}}{\pi^2}\frac{\mu}{H}\operatorname{Re}\sum_{r=1}^{\infty}\frac{(-1)^r}{r^{\frac{1}{2}}}\exp\left(\frac{2\pi i r\mu}{\hbar\omega_c}-\frac{\pi i}{4}\right)$$
$$\times\cos\left(\frac{\pi r g}{2}\frac{m_c}{m}\right)J(H, T), \tag{5}$$

in which $\tilde{\mathscr{E}}$ measures the energy from the chemical potential, μ, i.e. $\tilde{\mathscr{E}} = \mathscr{E} - \mu$, and $J(H, T)$ is given by

$$J(H, T) = \int_{-\infty}^{\infty} \mathrm{d}\tilde{\mathscr{E}}\, f_0 \exp\left\{\frac{2\pi i r}{\hbar\omega_c}[\tilde{\mathscr{E}} + \Sigma(\tilde{\mathscr{E}}, T)]\right\}, \tag{6}$$

where f_0 is the Fermi–Dirac function, $(e^{\beta\tilde{\mathscr{E}}}+1)^{-1}$. Thus, in so far as it may be assumed that the self-energies expressing the electron–impurity and electron–phonon interactions are independent of magnetic field, the entire effects of such interactions are contained in the integral expression, $J(H, T)$.

We consider first the case of non-interacting fermions for which $\Sigma(\tilde{\mathscr{E}}) = 0$ and evaluate (6) by contour integration. The integrand has poles when $(e^{\beta\tilde{\mathscr{E}}}+1) = 0$ and hence for $\tilde{\mathscr{E}} = i\omega_n$, such that

$$\omega_n = (2n+1)\pi k_B T. \tag{7}$$

It then follows by summing the residues in the standard way that

$$J_0(H, T) = -2\pi i k_B T \sum_{n=0}^{\infty} \exp\left(-\frac{2\pi r\omega_n}{\hbar\omega_c}\right), \tag{8}$$

which reduces to the thermal damping term first given by Lifshitz & Kosevich (1955)

$$J_0(H, T) = -\left(\frac{i\hbar\omega_c}{2\pi r}\right)\frac{X_r}{\sinh X_r}, \tag{9}$$

where $X_r = 2\pi^2 r k_B T/\hbar\omega_c$.

9.2.1 Impurity scattering

For the case of impurity scattering we shall make the approximation, discussed in Chapter 8, that the self-energy is energy-independent and may be approximated by its zero field value,

$$\Sigma(\tilde{\mathscr{E}}) = \Sigma^0 = \Delta + i\Gamma. \tag{10}$$

We may then take Σ^0 outside the integral in (6) to yield, in addition to (9), the factor

$$\exp\left[\frac{2\pi r}{\hbar\omega_c}(i\Delta + \Gamma)\right], \tag{11}$$

which embodies both the frequency shift and damping term referred to in equation (3) of Chapter 8. For the damping induced by impurity scattering we have thus reproduced (1) but with Γ now interpreted as the imaginary part of the self-energy. Furthermore, in the spirit of the preceding chapter (see equation (32) of Chapter 8) we may write for the self-energy the forward scattering part of the T-matrix, so that both the frequency shift and the damping term are readily amenable to calculation for specific impurities. We could also have elaborated the above argument to include a slow dependence of self-energy on energy by expanding $\Sigma(\tilde{\mathscr{E}})$ in a Taylor series. Dropping second-order and higher terms, the above argument is modified only to the extent that (9) is now renormalized by the electron–impurity interaction, exactly as found by Soven (1972) (see also equation (33) of Chapter 8),

$$\frac{(\omega_c)_{\text{eff}}}{\omega_c} = \left\{1 - \text{Re}\left[\frac{\partial\Sigma(\tilde{\mathscr{E}})}{\partial\tilde{\mathscr{E}}}\right]_\mu\right\}^{-1}. \tag{12}$$

9.2.2 Electron–phonon interaction

We turn now to a consideration of the electron–phonon interaction whose influence on the dHvA effect is rather subtle. Essentially the problem is to evaluate (6), but with $\Sigma(\tilde{\mathscr{E}}, T)$ now having both energy and temperature-dependent real and imaginary parts. This has been done most elegantly by Engelsberg & Simpson (1970), whose approach for completeness we shall indicate briefly. Rather than integrate (6) by parts, which in view of the presence of $\Sigma(\tilde{\mathscr{E}}, T)$ is difficult to handle analytically, we return to a contour integration and, on the assumption that $\Sigma(\tilde{\mathscr{E}}, T)$ is analytic, define $\zeta(\omega_n)$ as the full self-energy evaluated at the imaginary frequencies $(\mathrm{i}\omega_n)$ where the poles ω_n of the Fermi function are already defined in (7). Thus

$$\zeta(\omega_n) \equiv \mathrm{i}\Sigma(\mathrm{i}\omega_n). \tag{13}$$

As before, (6) may be evaluated to yield

$$J(H, T) = -2\pi \mathrm{i} k_B T \sum_{n=0}^{\infty} \exp\left\{\frac{2\pi r}{\hbar\omega_c}[\omega_n + \zeta(\omega_n, T)]\right\} \tag{14}$$

but with $\zeta(\omega_n, T)$ now given from the theory of the electron–phonon interaction by

$$\zeta(\omega_n, T) = \pi k_B T \int_0^{\infty} \frac{\mathrm{d}\nu\, 2\alpha^2(\nu)F(\nu)}{\nu}\left\{1 + 2\sum_{j=1}^{n}\left[1 + \left(\frac{2\pi j k_B T}{\nu}\right)^2\right]^{-1}\right\}. \tag{15}$$

We notice first that, unlike the impurity scattering case, the real and imaginary parts of the self-energy operator are *not* now separable in their influence on the dHvA effect. Indeed, as a consequence of $\zeta(\omega_n, T)$ being a wholly real quantity, the dHvA frequency is unchanged in accordance with our expectation since $\mathrm{Re}\,[\Sigma(\tilde{\mathscr{E}}, T)]_\mu = 0$. The behaviour of (14) in various limits has been discussed by Mueller & Myron (1976) and Engelsberg (1978). Both at high temperatures and as $\omega_c \to 0$ only the first pole, ω_0, contributes effectively to the summation in (14). Under these conditions, where $\zeta(\omega_0, T) = \lambda_0 \pi k_B T$, (14) reduces to

$$J(H, T) = -\frac{\mathrm{i}\pi k_B T}{\sinh\,[2\pi^2 r k_B T(1+\lambda_0)/\hbar\omega_c]} \tag{16}$$

so that effective masses, deduced experimentally from the temperature dependence of the dHvA effect amplitude will be renormalized by the electron–phonon interaction, a result first given by Wilkins & Woo (1965). Here, λ_0 is the zero field temperature renormalization constant

$$\lambda_0 = \frac{m_c^*}{m_c} - 1 = \int_0^{\infty} \frac{\mathrm{d}\nu\, 2\alpha^2(\nu)F(\nu)}{\nu}. \tag{17}$$

The limiting behaviour at high fields and low temperatures is less easy to discern. However, as Mueller & Myron (1976) have indicated, whilst the self-energy itself is not field-dependent, with increasing field an increasingly large number of terms are effective in the summation in (14). Thus, for a metal having sufficiently low phonon frequencies that the term $(2\pi jk_B T/\nu)^2$ in (15) is significant at the highest effective values of j $(=n)$ which contribute to the summation in (14), the departures from (16) and hence from 'quasi-particle behaviour' will become observable. For a favourable situation, such as the β-orbit in mercury, such an effect is predicted (Mueller & Myron 1976) to amount to a change in the dHvA amplitude of ~20 per cent for $T = 1$ K and $H = 100$ kG, although the experiments of Palin (1972) at rather lower fields failed to reveal the presence of any significant departures. However, we note that the experiments of Palin were not performed at the optimum (i.e. lowest effective mass) orientation in mercury, but, in view of the fact that measurements were made in part with a torque magnetometer, a non-symmetry orientation, for which $m_c^* = 0.183$ m rather than the minimum value of $m_c^* \sim 0.15$ m, was selected.

Recent measurements by Elliott, Ellis & Springford (1978) under more favourable experimental conditions have revealed the deviations from quasi-particle behaviour predicted by Engelsberg & Simpson (1970). In these measurements the optimum crystallographic orientation was selected for which $m_c^* = 0.152$ m and, bearing in mind the uncertainty in $\alpha^2(\nu)F(\nu)$ appropriate to the β-orbit, experiment and theory, as shown in Fig. 9.1, are seen to be in reasonable quantitative agreement.

One way of accounting for the surprising smallness of the influence of the electron–phonon interaction in the dHvA effect is to argue that there is a measure of cancellation between the real and imaginary parts of $\Sigma(\tilde{\mathscr{E}}, T)$ in (6). However, in such an argument, the energy and temperature dependences need to be carefully handled and it is difficult to find a simple physical model to explain the effect which is not seriously misleading. Nevertheless, it is clear that, for the majority of metals, such effects will be quite negligible under normal experimental conditions and that the sole effect of the electron–phonon interaction will be expressed by (16).

We note also that, in so far as we can make a separation of the self-energies arising from impurity and phonon interactions (Grimvall 1974), the cyclotron frequency in the Dingle term (equation (11)) is unrenormalized. However, in evaluating Dingle temperatures from experimental measurements one uses measured cyclotron masses, m_c^*,

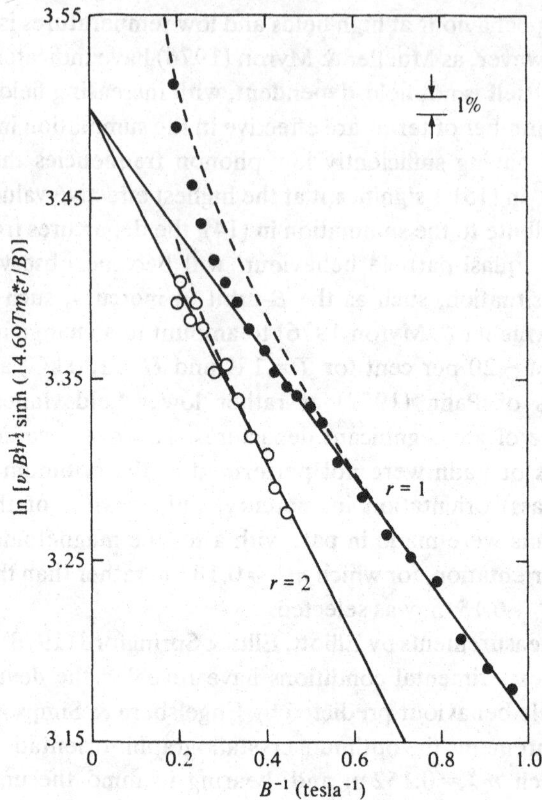

Fig. 9.1. Dingle plots showing the field dependence of the fundamental ● ($r = 1$) and second harmonic ○ ($r = 2$) of the dHvA effect amplitude for the β-orbit in mercury. Experiments were performed at $T = 2.1$ K over the field range ~10–50 kG. At the lower fields ($B \lessapprox 20$ kG) the results for the fundamental show quasi-particle (linear) behaviour, the solid line corresponding to the theory of Lifshitz & Kosevich (1956) for a Dingle temperature of 0.15 K and $m_c^* = 0.152$ m. At higher magnetic fields the dHvA amplitude is seen to diverge in accord with the theory of Engelsberg & Simpson (1970) (broken lines) (Elliott, Ellis & Springford 1978).

which are, as we have seen, renormalized by the electron–phonon interaction. Dingle temperatures determined in this way must therefore be interpreted accordingly as the renormalized values, x^*, such that $m_c^* x^* \equiv m_c x$.

9.3 The inversion of experimental data

The central problem in studying electron scattering by means of the dHvA effect is to deduce and understand the pattern of variation of the reciprocal electron quasi-particle lifetime, $\tau^{-1}(\mathbf{k})$, over the Fermi surface, from measurements of Dingle temperatures $x(\theta, \phi)$ made at different orientations (θ, ϕ) of the magnetic field with respect to the crystal axes. The procedure by which this is accomplished is termed *inversion* and in this section, and prior to discussing some experimental results, we discuss several schemes that have been developed for this purpose.

9.3.1 Basic inversion equations

In forming the orbital average, we assume that each local value $\tau(\mathbf{k})$ contributes in proportion to the time, dt, spent by the wave-vector $\mathbf{k}$ in each region along the Fermi surface orbit (Lowndes, Miller, Poulsen & Springford 1973b),

$$\left\langle\frac{1}{\tau(\mathbf{k})}\right\rangle = \frac{1}{T_0}\oint_0^{T_0}\frac{\mathrm{d}t(\mathbf{k})}{\tau(\mathbf{k})}, \tag{18}$$

where the period $T_0 = 2\pi/\omega_c$, $\omega_c = eH/m_c$ and the cyclotron mass $m_c = \hbar^2(\partial\mathscr{A}/\partial\mathscr{E})/2\pi$. Using the Lorentz force equation and combining (4) and (18) we may express the measured quantity in the dHvA effect, m_cx, in terms of a line integral around the particular orbit

$$m_cx = \frac{\hbar}{2\pi}\oint\frac{x(\mathbf{k})\,\mathrm{d}k_\perp}{v_\perp}, \tag{19}$$

in which $\mathbf{v}_\perp$ is the component of the Fermi velocity in the plane normal to $\mathbf{H}$ at the point $\mathbf{k}$ and $x(\mathbf{k})$ is the *local* Dingle temperature given by

$$x(\mathbf{k}) = \frac{\hbar}{2\pi k_B}\frac{1}{\tau(\mathbf{k})}. \tag{20}$$

For the purposes of computation it is sometimes more convenient to change the variable in (19) to an angle η which specifies that position of the projection $k_\perp$ of the $\mathbf{k}$ vector on to a plane normal to $\mathbf{H}$ for a state on the orbit (cf. Fig. 8.2 of Chapter 8) so that,

$$m_cx = \frac{\hbar}{2\pi}\oint_0^{2\pi} x(\eta)W(\eta)\,\mathrm{d}\eta, \tag{21}$$

where it may be shown (Springford 1971) that the weight function, $W(\eta)$, can be calculated from

$$W(\eta) = \frac{k_\perp^2(\eta)}{\mathbf{v}_\perp(\eta)\cdot\mathbf{k}_\perp(\eta)} \tag{22}$$

and where

$$m_c = \frac{\hbar}{2\pi}\oint_0^{2\pi} W(\eta)\,\mathrm{d}\eta = \frac{\hbar}{2\pi}\oint\frac{\mathrm{d}k_\perp}{v_\perp}. \tag{23}$$

In the noble metals, whilst m_c is now known for arbitrary field orientations, the measured values at specific orientations may be used (Halse 1969; Lengeler *et al.* 1977) to deduce the anisotropy of the Fermi velocity. In such cases it is convenient to eliminate m_c between (19) and (23) to yield

$$x = \oint \frac{x(\mathbf{k})\,\mathrm{d}k}{v_\perp} \bigg/ \oint \frac{\mathrm{d}k}{v_\perp}. \tag{24}$$

A basic requirement of any inversion scheme is that it should conform with the requirements of the crystal symmetry. It is an additional advantage if a representation of $\tau^{-1}(\mathbf{k})$ may be made with only a few adjustable parameters. Three different schemes, based on a Fourier series, Kubic harmonic and phase-shift expansion, have successfully been used. In each case Dingle temperatures are expressed in the form

$$x(\theta, \phi) = \sum_j C_j X_j(\theta, \phi), \tag{25}$$

in which $X_j(\theta, \phi)$ are calculated quantities and the C_j are adjustable parameters determined by a least-squares fitting procedure to the experimental quantities $x(\theta, \phi)$. Each scheme carries particular advantages and disadvantages, which are separately discussed.

9.3.2 Fourier series inversion

In view of the particular topological properties of the Fermi surfaces of the noble metals, namely a fcc lattice in which the Fermi surface approximates spheres that are multiply-connected in the ⟨111⟩ directions, a symmetrized Fourier series inversion scheme has been found to give rapid convergence. As a Fourier series representation we write

$$\begin{aligned}
x(\mathbf{k}) &= \sum_{l,m,n} C_{l,m,n} \cos(\tfrac{1}{2}lak_x)\cos(\tfrac{1}{2}lak_y)\cos(\tfrac{1}{2}lak_z) \\
&= C_{000} + C_{110}(\mathcal{S}\cos(\tfrac{1}{2}ak_x)\cos(\tfrac{1}{2}ak_y)) \\
&\quad + C_{200}(\mathcal{S}\cos(ak_x)) + C_{211}(\mathcal{S}\cos(ak_x)\cos(\tfrac{1}{2}ak_y)\cos(\tfrac{1}{2}ak_z)) \\
&\quad + C_{220}(\mathcal{S}\cos(ak_x)\cos(ak_y)) \\
&\quad + C_{310}(\mathcal{S}\cos(\tfrac{3}{2}ak_x)\cos(ak_y) + \mathcal{S}\cos(\tfrac{3}{2}ak_y)\cos(\tfrac{1}{2}ak_x)) \\
&\quad + C_{222}(\cos(ak_x)\cos(ak_y)\cos(ak_z)) \\
&\quad + \ldots.
\end{aligned} \tag{26}$$

in which $l+m+n$ is even for an fcc lattice and $\mathscr{S}$ denotes a sum over permutations of x, y and z. The lattice parameter is denoted by a and k_x, k_y and k_z are the coordinates in **k**-space of a point on the Fermi surface.

Combining (24) and (26) we may write

$$x(\theta, \phi) = \sum_{l,m,n} C_{l,m,n} x_{l,m,n}(\theta, \phi), \tag{27}$$

in which (θ, ϕ), which specifies a field direction with respect to the crystal axes, is determined by the path of the line integral in (24). The $x_{l,m,n}$ are computed quantities depending on the shape of the Fermi surface and the derived quasi-particle velocity distribution over it. Tabulated values for a number of important orbits in Cu, Ag and Au are given elsewhere (Lowndes *et al.* 1973b; Poulsen, Randles & Springford 1974). Having determined the $C_{l,m,n}$ in (27) by a least-squares fitting procedure to the experimental data $x(\theta, \phi)$, then local values of $x(\mathbf{k})$ are readily found from (26). It is important to note that the representation (26) is entirely empirical and that the fitted coefficients, $C_{l,m,n}$, cannot simply be related to physical properties of the impurities. There remains the problem of selecting the particular Fourier coefficients and of truncating the expansion. In work on the noble metals, Poulsen *et al.* (1974) and Templeton & Coleridge (1975a) have used the procedure of fitting the coefficients in order of increasing anisotropy (corresponding to increasing $l^2+m^2+n^2$) and of selecting as the best fit, that with the minimum number of coefficients for which the rms deviation R_f between measured x^m and fitted x^f values is less than the rms error R_m in the data.

$$\begin{aligned} R_f^2 &= \sum_{i=1}^{N} \frac{[(x_i^m - x_i^f)/\sigma_i]^2}{\sum_{i=1}^{N} (1/\sigma_i)^2}, \\ R_m^2 &= \frac{1}{N} \sum_{i=1}^{N} (\sigma_i^2). \end{aligned} \tag{28}$$

The fit so obtained reflects the minimum anisotropy consistent with the experimental data. The error has been judged by the rms variation in the local value $x(\mathbf{k})$ as the number of coefficients, n, is varied from the number corresponding to the best fit to the maximum number N. Only for $n < N$ is the number of degrees-of-freedom finite and the resultant inversion significant in the statistical sense. Where a comparison has been made (e.g. Coleridge, Holzwarth & Lee 1974; Coleridge 1975) the good agreement between the results for the anisotropy deduced by this method and those which depend on the validity of a phase-shift description of electron scattering, lends support to the physical assumptions of the latter model.

9.3.3 Kubic harmonic inversion

An elegant inversion scheme based upon suitability symmetrized spherical harmonics has been described by Mueller (1966). For application to closed surfaces having inversion symmetry, the method was originally used to relate cross-sectional areas and radius vectors, but also provides an attractively simple scheme for the present application (Arko & Mueller 1975). Restricting our attention to cubic symmetry, a function $y(\theta', \phi')$ may be expressed as the sum of Kubic harmonics,

$$y(\theta', \phi') = \sum_{i,m} {}_iD_m \, {}_iK_m(\theta', \phi'), \tag{29}$$

in which ${}_iK_m$ is the ith Kubic harmonic of order m and ${}_iD_m$ are expansion coefficients. If $z(\theta, \phi)$ denotes the line integral about a closed path in the central plane

$$z(\theta, \phi) = \oint y(\theta', \phi') \, d\eta \tag{30}$$

and $d\eta$ is an angle in the central plane which is normal to (θ, ϕ), then following Mueller (1966) we may write

$$z(\theta, \phi) = 2\pi \sum_{i,m} {}_iD_m \, P_m(0) \, {}_iK_m(\theta, \phi) \tag{31}$$

such that the $P_m(0)$ are Legendre polynomials of order m evaluated at the origin, with m necessarily even because of the inversion symmetry.

Expressing the experimentally measured quantities $m_c x$ and m_c in the form

$$m_c x(\theta, \phi) = \sum_{i,m} {}_ia_m \, {}_iK_m(\theta, \phi), \tag{32}$$

$$m_c(\theta, \phi) = \sum_{i,m} {}_ib_m \, {}_iK_m(\theta, \phi), \tag{33}$$

and recalling that $dk = k_\perp \, d\eta$, it follows from (23), (24) and (31) that

$$\frac{\hbar}{2\pi}\left[\frac{x(\mathbf{k})k}{v_\mathbf{k}}\right]_{\theta', \phi'} = \frac{1}{2\pi} \sum_{i,m} \frac{{}_ia_m}{P_m(0)} \, {}_iK_m(\theta', \phi'), \tag{34}$$

$$\frac{\hbar}{2\pi}\left[\frac{k}{v_\mathbf{k}}\right]_{\theta', \phi'} = \frac{1}{2\pi} \sum_{i,m} \frac{{}_ib_m}{P_m(0)} \, {}_iK_m(\theta', \phi'), \tag{35}$$

whence $\tau^{-1}(\mathbf{k})$ is obtained from the ratio without the need to determine $[k/v_{\mathbf{k}}]_{\theta',\phi'}$.

In the case where both m_c and $m_c x$ are measured in the same experiment at the same orientations, (θ, ϕ), then x may be expanded directly in the general form of (25)

$$x(\theta, \phi) = \sum_{i,m} {}_iC_m \, {}_iK_m(\theta, \phi), \tag{36}$$

when from (20), (30) and (31) it follows that

$$\tau^{-1}(\mathbf{k}) \equiv \tau^{-1}(\theta', \phi') = \frac{2\pi k_B}{\hbar} \sum_{i,m} \frac{{}_iC_m}{P_m(0)} \, {}_iK_m(\theta', \phi') \tag{37a}$$

or

$$\tau^{-1}(\theta', \phi') = \sum_{i,m} {}_iB_m \, {}_iK_m(\theta', \phi'). \tag{37b}$$

It may be seen from these equations that, since $P_m(0) < 1$, the anisotropy in $\tau^{-1}(\theta', \phi')$ will be greater than in the measured quantity $x(\theta, \phi)$. Alternatively, if one seeks to determine the anisotropy in $\tau^{-1}(\theta', \phi')$ to an accuracy of say p per cent, then the anisotropy in $x(\theta, \phi)$ must be determined to q per cent where $q < p$ and p/q is determined by the set of coefficients ${}_iC_m$. A related question is the accuracy, itself a function of $\mathbf{k}$, with which the local values $x(\mathbf{k})$ are known. The point has been discussed by Arko & Mueller (1975). If $\sigma(\theta, \phi)$ are the errors in the quantities $m_c x(\theta, \phi)$ (typically $\sigma/m_c x$ will be in the range 1–10 per cent) and it is assumed that the errors in m_c are negligible in comparison, then we may write for the fractional error in the locally determined values,

$$\frac{\Delta\{x(\mathbf{k})\}}{x(\mathbf{k})} = -\frac{\Delta\{\tau(\theta', \phi')\}}{\tau(\theta', \phi')} = \frac{\left[\sum_{i,m}\sum_{j,k} \frac{{}_iK_m(\theta', \phi')}{P_m(0)} \frac{{}_jK_k(\theta', \phi')}{P_k(0)} M^{-1}_{imjk}\right]}{\sum_{i,m} \frac{{}_ia_m}{P_m(0)} \, {}_iK_m(\theta', \phi')} \tag{38}$$

in which M_{imjk} is a measurement matrix defined as

$$M_{imjk} = \sum_{n=1}^{N} \frac{{}_iK_m(\theta_n, \phi_n) \, {}_jK_k(\theta_n, \phi_n)}{\sigma^2(\theta_n, \phi_n)}, \tag{39}$$

the summation being taken over the N orientations at which measurements are made. That the errors in the inverted quantity as a function of position on the Fermi surface can be cast in this analytical form is a useful feature of the Kubic (spherical) harmonic based scheme.

9.3.4 Phase shift inversion

In contrast to the previous two inversion schemes, the phase shift method is based upon a physical model for the influence of impurities on the dHvA effect. The model, which was developed in detail in Chapter 8, is based on the muffin-tin (MT) approximation for both the host lattice and impurity atoms. A central result is that, consequent on the addition of impurities, the complex change of wave vector, $\Delta\mathbf{k}_c$, is given by the forward part of the T-matrix

$$\Delta\mathbf{k}_c = c\left(\frac{\partial k}{\partial \mathscr{E}}\right)T_{\mathbf{kk}}, \tag{40}$$

which from equation (68) of Chapter 8 we may rewrite

$$\Delta\mathbf{k} = -\frac{c}{\hbar v_{\mathbf{k}}}\sum_L \frac{t_L(\mathbf{k})}{I_L}\sin\phi_L \exp(\mathrm{i}\phi_L). \tag{41}$$

Here the sum is taken over partial waves and the anisotropic host-dependent quantities $t_L(\mathbf{k})$ are normalized by their Fermi surface average, I_L, as defined in equation (63) of Chapter 8. The Friedel phase shifts, ϕ_L, characterize the difference between impurity and host potentials and include both a term which arises from this difference in potential and also a contribution from the backscattering by the host lattice. They are a particularly appropriate choice for a parametrization of the effect of impurities, being independent of the choice of the MT zero and obeying the Friedel sum rule,

$$\Delta z = \frac{2}{\pi}\sum_L n(L)\phi_L, \tag{42}$$

in which $n(L)$ is the degeneracy of the Lth representation and Δz is the number of electrons displaced into the impurity cell. Recalling that the electron quasi-particle lifetime is obtained from the optical theorem by

$$\tau^{-1}(\mathbf{k}) = -\frac{2c}{\hbar}\,\mathrm{Im}\,T_{\mathbf{kk}} \tag{43a}$$

$$= \frac{2c}{\hbar}\sum_L \frac{t_L(\mathbf{k})}{I_L}\sin^2\phi_L, \tag{43b}$$

it then follows from (19) that

$$m_c x = \frac{c\hbar}{2\pi^2 k_B}\sum_L \sin^2\phi_L \frac{1}{I_L}\oint \frac{t_L(\mathbf{k})\,\mathrm{d}k_\perp}{v_\perp}. \tag{44}$$

Given the identity for $t_L(\mathbf{k})$ (Coleridge 1972) that

$$t_L(\mathbf{k}) \equiv \hbar v_{\mathbf{k}}(\mathbf{k})\left(\frac{\partial k}{\partial \eta_L}\right)_\mu, \tag{45}$$

and defining anisotropic host-dependent functions

$$H_L(\theta, \phi) = \frac{1}{I_L}\left(\frac{\partial \mathscr{A}}{\partial \eta_L}\right)_\mu, \tag{46}$$

then we may write (44) in the form

$$m_c x = \frac{c\hbar^2}{2\pi^2 k_B} \sum_L \sin^2 \phi_L H_L(\theta, \phi), \tag{47}$$

being of the form of the general inversion equation (equation (25)).

As before, knowing $H_L(\theta, \phi)$, the coefficients $\sin^2 \phi_L$ are determined by a least-squares fitting procedure to the experimental data $m_c^* x^*$ ($= m_c x$). In the present case, however, the fitting parameters are physically significant and an estimate of the errors that should be attached to them is obviously of importance in subsequent discussions. The $H_L(\theta, \phi)$ do not represent a set of orthogonal functions and, to take account of this degeneracy, Templeton & Coleridge (1975a) have used a procedure in which the uncertainty in ϕ_L is taken to be the amount by which it may be varied such that the error of the new fit to the data, made by optimizing the remaining parameters, was $\sqrt{2}$ times that for the best fit. Such an uncertainty represents the bounds of an acceptable fit to the data. It should also be noted that ϕ_L will only be obtained with some ambiguity, e.g. Dingle temperature data do not determine the sign of ϕ_L and yield its magnitude modulo π.

9.4 The intercept technique

The methods and precautions for the accurate measurement of Dingle temperatures have been discussed elsewhere (see e.g. Shoenberg 1969; Springford 1971; Chang & Higgins 1975) and we shall not repeat them here. We shall, however, draw attention to a technique which enables very small *changes* in Dingle temperature to be revealed. Such changes might exist, for example, between different crystal orientations owing to the presence of a scattering anisotropy or, even at a fixed orientation, as the result of a change of some external agency. This *intercept technique* has been used to determine changes in Dingle temperature with an accuracy of only a few millikelvin (Paul & Springford 1977).

The principle of the method may be appreciated with reference to Fig. 9.2 which shows schematically the field dependence of the dHvA amplitude, in the conventional form of a Dingle plot, for two cases where only the Dingle temperature is imagined to have changed. The theoretical variation is shown by the broken lines, $Y = PH^{-1} + Q$, and experimental measurements are supposed to have been made over an experimentally accessible range of H^{-1}. The ordinate Y is the logarithm of the appropriately normalized dHvA amplitude and the intercept, Q, is a function of both apparatus and Fermi surface constants. For the case that Fig. 9.2 refers to the low-frequency field modulation method of detection, these symbols would be translated as follows

$$Y = \ln\left[\frac{v_0 H^{\frac{1}{2}} \sinh X}{2TJ_2(\Lambda)}\right], \tag{48a}$$

$$P = -Km_c^* x^*, \tag{48b}$$

$$Q = \ln\left[C\Theta\Omega 2fF\cos(\pi S)\mathscr{A}''^{-\frac{1}{2}}\right], \tag{48c}$$

in which v_0 is the rms voltage at twice the modulation frequency f induced in a pick-up coil of coupling constant Θ, C is a numerical constant, Ω the sample volume, S the spin factor, $\mathscr{A}''$ the curvature factor and $J_2(\Lambda)$ is a second-order Bessel function whose argument is related to the amplitude

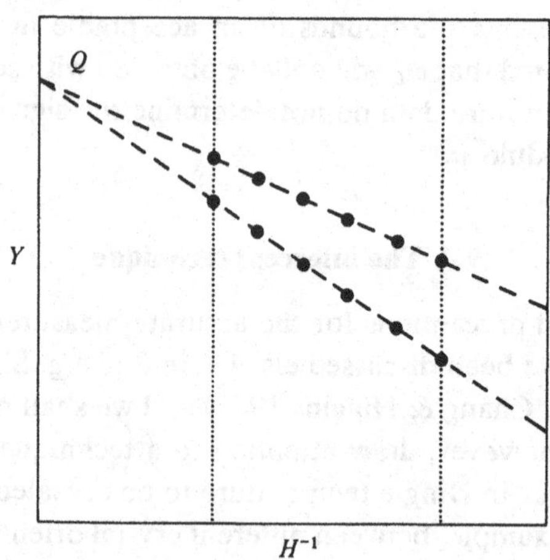

Fig. 9.2. Variation of the amplitude of the dHvA effect with reciprocal magnetic field depicted schematically in the form of a *Dingle plot*, for two cases where only the Dingle temperature has changed. Experimental measurements are restricted to the accessible range of H^{-1}. The broken lines, $Y = PH^{-1} + Q$ depict the theoretical variation with Y, P and Q defined in equation (48).

of field modulation h_0 by $\Lambda = 2\pi F h_0/H^2$. For simplicity we include only the fundamental dHvA harmonic. In practice, where x^* is independent of H, v_0 is measured at a constant temperature for a series of values of H^{-1}, and $m_c^* x^*$ is determined by equating p to the gradient of the line $Y = pH^{-1} + q$, that best fits the experimental data, as judged by the method of least squares.

Now it is evident that if the common intercept Q, in Fig. 9.2, were known, then p and consequently $m_c^* x^*$ could be determined with less uncertainty. Q, however, refers to the dHvA amplitude at infinite field and is experimentally inaccessible. Nor of course can it be calculated with any certainty from (48). The intercept method is based on the following result. Whilst the magnitude of Q is uncertain, it may be known to be a constant for two experiments such as depicted in Fig. 9.2. It may then be shown that any arbitrary constant value, β, of the intercept may be included in the least-squares fitting procedure above to determine the *difference* between the two gradients with improved accuracy. We stress that the argument applies only to differences in x^* and not to the determination of the absolute value of x^*. If X_i are the set of n values of H_i^{-1} at which the measurements are made, then inclusion of the point $(0, \beta)$ reduces the uncertainty with which the difference between the two Dingle temperatures may be measured by a factor f where for $n \gg 1$,

$$f \approx \left[1 + \frac{\sum_i X_i^2}{n\sum_i X_i^2 - (\sum_i X_i)^2}\right]. \tag{49}$$

Depending on the field range over which measurements are made it can be that $f \gg 1$. As proved elsewhere (Paul & Springford 1977) the method is exact if certain experimental conditions are fulfilled:

(i) experimental errors occur in Y_i but not in H_i;
(ii) the set of n values of H_i^{-1} are identical for each graph;
(iii) the set of relative weights to be attached to the measured values Y_i are the same for each graph.

We note that conditions closely approximating these are readily achieved in practice.

In any real metal, Q will of course be anisotropic and the method would therefore seem to be inappropriate for the investigation of scattering anisotropies. It turns out, however, that, for studies in dilute alloys, we may use the technique to determine more accurately small changes in the coefficients, C_j, in the representation for $x(\theta, \phi)$ (equation (25)) rather than directly in the Dingle temperatures themselves. Suppose, for example, that $x'(\theta, \phi)$ represents the apparent anisotropy in which experiments are analysed on the assumption that the intercept is independent of orientation

$$x'(\theta, \phi) = \sum_j C'_j K_j(\theta, \phi), \tag{50}$$

C'_j are fitted coefficients in which an arbitrary constant value of the intercept, $(0, \beta)$, is included. The variation $x'(\theta, \phi)$ therefore contains a systematic error, $x_s(\theta, \phi)$, which, in view of the symmetry properties of Q, we can also express in terms of the same set of basis functions

$$x_s(\theta, \phi) = \sum_j S_j K_j(\theta, \phi). \tag{51}$$

If $x^*(\theta, \phi)$ expresses the desired anisotropy which arises from a concentration, c, of impurities

$$x^*(\theta, \phi) = c \sum_j C_j K_j(\theta, \phi), \tag{52}$$

and $x_0(\theta, \phi)$ is a general anisotropic background term which we shall assume is independent of c and embraces the effects of residual impurities, etc.,

$$x_0(\theta, \phi) = \sum_j O_j K_j(\theta, \phi), \tag{53}$$

then since

$$x'(\theta, \phi) = x^*(\theta, \phi) + x_s(\theta, \phi) + x_0(\theta, \phi), \tag{54}$$

it follows that

$$x'(\theta, \phi) = \sum_j (cC_j + S_j + O_j) K_j(\theta, \phi) \tag{55}$$

and the coefficients C_j which express the desired variation may then be recovered from measurements of C'_j at two or more concentrations. Implicit in this analysis is that neither Q nor x_0 is a function of solute concentration. Whilst for Q this should be valid in the dilute limit, x_0 could vary through a concentration dependence of the dislocation field, and such a variation could not be distinguished from the true variation of x^*. The same problem of interpretation arises, however, even without using the intercept technique. Finally x_0 has been assumed to possess the symmetry of the lattice, requiring carefully annealed and strain-free single crystals.

9.5 Experimental results

Of the many dHvA effect studies in dilute alloys that are summarized in Chapter 8, only a minority to date have yielded detailed information on the anisotropy of the electron–impurity scattering rate over the whole Fermi surface. These are confined to the monovalent metals, Cu, Ag, Au and K as hosts, in which the Fermi surface is a single sheet and the simple spectrum of dHvA frequencies allows relatively easy measurement of the Dingle temperatures which characterize different extremal orbits. Whilst in principle such measurements can readily be extended to both polyvalent and transition metals, the co-existence of a larger number of dHvA frequencies which derive from the more complex Fermi surface topologies in these metals greatly complicates the experimental problems. In order to resolve nearly degenerate frequencies and allow properly for the temperature and field dependence of the harmonic content, one needs to employ considerable computing power in the performance of such experiments and to use techniques such as digital filtering in their analysis. Thus, for the present, we only have available experimental results for the simplest cases. Nevertheless, the results have shed some new light on some old problems and the interplay between theory and experiment has been especially fruitful.

9.5.1 The noble metals

In the noble metals we have an ideal laboratory for a study of electron lifetimes by the techniques described in this chapter. Their Fermi surface parameters, such as radius vectors, Fermi velocities and effective masses, are known in considerable detail and their metallurgy is well understood. Additionally a wealth of other experiments, including the ordinary transport properties, have been studied in both the pure metals and their alloys and, based on these, Ziman (1961a) has argued that conduction electron scattering should be appreciably anisotropic and linked to the large distortions of the Fermi surfaces from a sphere. The results of experiments on the ordinary transport properties are often interpreted in terms of a two-band model in which, for example, carriers in the 'neck' and 'belly' regions of the Fermi surface are characterized by different lifetimes. Such models often suggest appreciable differences between the scattering rates in the different bands but Springford (1971) has concluded that, with regard to electron–impurity scattering, the assembled results of experiments on the Hall effect, magnetoresistance and deviations from Mattheissen's rule, do not present a coherent picture of the scattering anisotropy. Such *direct* measurements as are provided by dHvA effect studies in dilute alloys are therefore of interest in the clarification of the transport properties. Additionally, a proper understanding of the scattering requires a description of both electron wavefunctions and scattering centres and to date the emphasis has been rather in this area.

Dingle temperature measurements in the noble metals are complicated by the existence of a small mosaic spread of crystal orientation which seems inevitably to exist throughout a single crystal. Because of the anisotropy of the Fermi surface, this leads to phase-smearing of the quantum oscillations and to a spurious contribution to the Dingle temperature, a point to be discussed more fully in Chapter 10. For this reason measurements are confined either to the principal symmetry directions or to a few singular orientations for which the dHvA frequency is stationary with respect to rotation in a given plane. In copper, for example, the belly frequency has such a turning point at 16.4° from ⟨100⟩ in the (100) plane and at 11.8° from ⟨100⟩ in the (110) plane. The influence and characterization of mosaic spread has been discussed in detail by both Miller, Poulsen & Springford (1972) and by Chang & Higgins (1975). Based on the measurements of Dingle temperatures as a function of solute concentration at seven field directions in the crystal, the results for copper and gold hosts are collected in Table 9.1. Fourier series

TABLE 9.1. *Summary of the coefficients, $C_{l,m,n}$, in a Fourier series representation for the renormalized scattering rate $\tau^{*-1}(\mathbf{k})$ according to equation (26), for various impurities in copper and gold hosts. Coefficients are in units of $2\pi k_B/\hbar = 8.225\,88 \times 10^{11}$ per atomic % of solute. Results for copper are from Poulsen et al. (1974) and for gold from Lowndes et al. (1973b)*

	C_{000}	C_{110}	C_{200}	C_{211}
Cu(Ni)	73.406	141.919	16.852	—
Cu(Ge)	−120.821	−720.008	−88.300	—
Cu(Au)	27.624	44.091	5.932	—
Cu(H)				
Au(Ag)	14.6272	1.0288	5.0821	17.9638
Au(Cu)	15.3560	1.2803	4.6109	17.8937
Au(Zn)	25.794	−132.67	8.7548	114.99
Au(Fe)	215.43	418.56	17.935	−170.33

inversion has been used and the tabulated coefficients $C_{l,m,n}$ express the variation of the renormalized scattering rate $\tau^{*-1}(\mathbf{k})$ over the Fermi surface according to (26). We note that three and four coefficients are required in copper and gold respectively to represent adequately the measured Dingle temperature, the inversion procedure having been described in Section 9.3.2. As a pictorial representation of this data, we reproduce in Fig. 9.3 topographic maps of $\tau^{*-1}(\mathbf{k})$ for copper constructed over the basic 1/48th part of the stereographic projection. The smaller inset maps give an indication of the relative uncertainties in the inversion by indicating at a number of discrete points the percentage variation of $\tau^{*-1}(\mathbf{k})$ that results as the number of fitted coefficients is varied from 3 to 7. The most striking features of these results are the relatively large anisotropies (a factor of ~2.5 in Cu(Ni)) and the different pattern of variation of $\tau^{*-1}(\mathbf{k})$ for the scattering by different impurities.

Whilst the Fourier series inversion method has the attraction that it makes no *a priori* assumption about the pattern of variation of $\tau^{*-1}(\mathbf{k})$ and is based entirely on quantities that can be measured in a dHvA experiment, it is not possible to attach any simple physical significance to the coefficients which are summarized in Table 9.1. For this reason, an analysis of the same experimental data in terms of partial waves gives us a new insight into the factors governing the anisotropy. Referring to (43) we note that the anisotropy is determined entirely by that of the host-dependent functions $t_L(\mathbf{k})$ which from equation (58) of Chapter 8 are given by

$$t_L(\mathbf{k}) = \frac{\hbar^2}{2m\kappa} \sum_{\gamma} |a_{L\gamma}(\mathbf{k})|^2. \tag{56}$$

Here, the index $L = l, \Gamma$ is an abbreviated notation in which Γ denotes an irreproducible representation of the cubic group and γ denotes a particular member of that group. For the noble metals in which we can restrict our attention to $l \lesssim 2$ we note that there is only one irreducible representation for $l = 0$ (Γ_1) and $l = 1$ (Γ_{15}) but two for $l = 2$ (Γ_{12} and $\Gamma_{25'}$) (Lee, Holzwarth & Coleridge 1976a). The function $t_L(\mathbf{k})$ is a measure of the changing character or composition of the electron wave-function over the host Fermi surface, when expressed in terms of partial waves. In Fig. 9.4 we depict the variation of $2t_L(\mathbf{k})/\hbar$ for copper for $l = 0$, 1, and 2. The maps were computed by D. L. Randles (private communication) from the data calculated for copper by Coleridge (1972) using the KKR method. For $l = 2$ the results for Γ_{12} and $\Gamma_{25'}$ have been appropriately averaged to yield a single map.

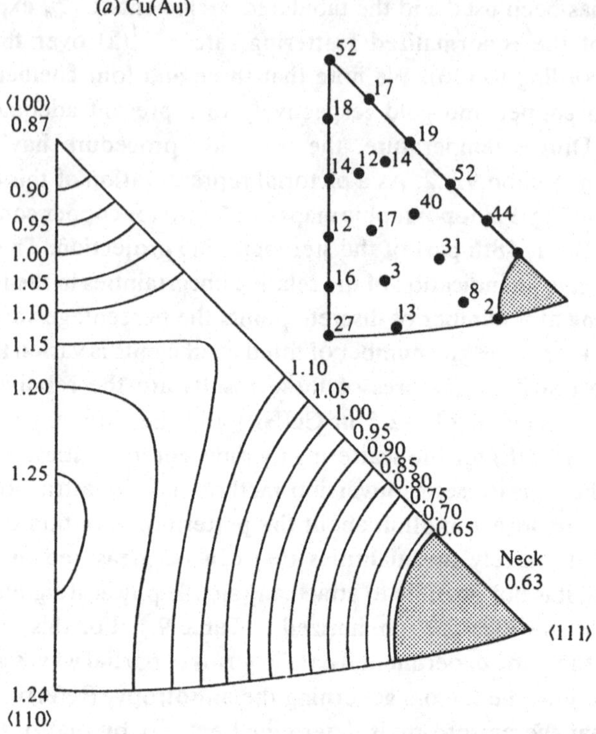

Fig. 9.3. Variation of the renormalized scattering rate, $\tau^{*-1}(\mathbf{k})$, over the Fermi surface of copper in the dilute alloys (*a*) Cu(Au), (*b*) Cu(Ge) and (*c*) Cu(Ni). The units are 10^{13} s^{-1} in (*a*) and (*c*), 10^{14} s^{-1} in (*b*), and the maps are drawn over the irreducible 1/48th segment of the [001] stereographic projection. The smaller inset maps indicate, as a percentage, the relative uncertainty in the experimental results at a number of discrete points (Poulsen *et al.* 1974).

It is apparent from a careful inspection of Figs. 9.3 and 9.4 that the anisotropy of $\tau^{*-1}(\mathbf{k})$ for Cu(Au), Cu(Ge) and Cu(Ni) closely resembles the pattern of variation of $2t_L(\mathbf{k})/\hbar$ for $l = 0$, 1 and 2 respectively. We may then say approximately that the influence of Au, Ge and Ni solutes in a copper host is to give rise to dominantly s-, p- and d-wave scattering respectively. This also accords with our understanding of the nature of the perturbation to which the presence of these impurities in copper will give rise. In the absence of any lattice strain effects the influence of a monovalent impurity such as gold will be particularly strongly felt close to the core region of the impurity and therefore by conduction electrons of predominantly s-character. A germanium impurity with its screening

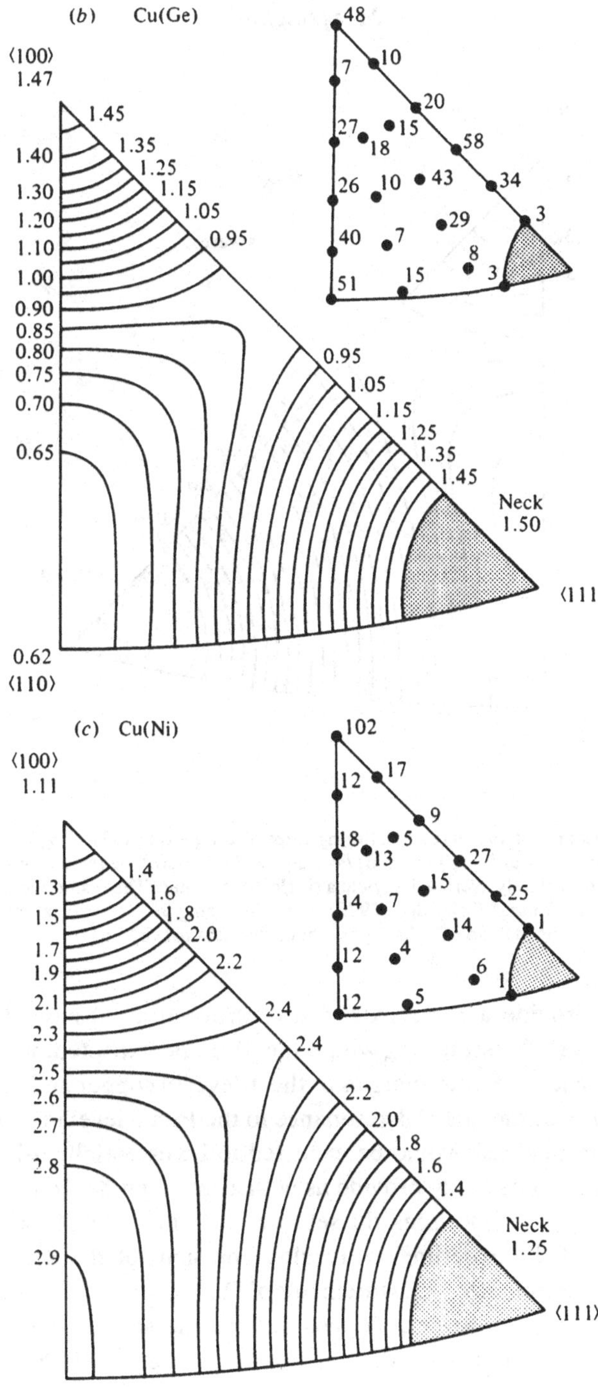
(b) Cu(Ge)
⟨100⟩
1.47
1.45
1.35
1.25
1.15
1.05
0.95
1.40
1.30
1.20
1.10
1.00
0.90
0.85
0.80
0.75
0.70
0.65
0.95
1.05
1.15
1.25
1.35
1.45
Neck
1.50
⟨111⟩
0.62
⟨110⟩
48
10
7
20
27
15
18
58
26
10
43
34
3
29
40
7
8
3
51
15
(c) Cu(Ni)
⟨100⟩
1.11
1.4
1.6
1.8
2.0
2.2
2.4
1.3
1.5
1.7
1.9
2.1
2.3
2.5
2.6
2.7
2.8
2.9
2.4
2.2
2.0
1.8
1.6
1.4
Neck
1.25
⟨111⟩
2.91
⟨110⟩
102
17
12
9
18
5
13
27
15
25
14
7
1
14
12
4
6
1
12
5

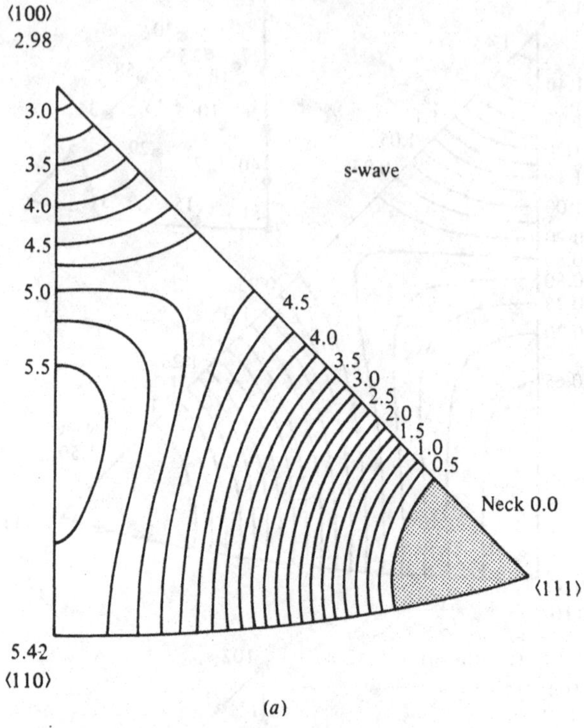

Fig. 9.4. Variation of the partial scattering rates (see equation (43)), $2t_L(\mathbf{k})/\hbar$, over the Fermi surface of copper for (*a*) $l=0$, (*b*) $l=1$ and (*c*) $l=2$. In (*c*) the results for Γ_{12} and $\Gamma_{25'}$ have been appropriately averaged to yield a single d-wave map. The results were computed from the calculations of Coleridge (1972), the units being $10^{13}\,s^{-1}$ per atomic per cent for (*a*) and $10^{-14}\,s^{-1}$ per atomic per cent for (*b*) and (*c*).

cloud will provide a more extended centre and a substantial p-wave contribution to the scattering would therefore be expected, whereas, in view of the characteristic energies of the d-level in copper and nickel and the proximity of the nickel d-resonance to the Fermi level in copper, the scattering in this case would be expected to be essentially d-like. Such arguments of course only provide us with a rough guide. In more detail and referring to (47) we seek the set of values of Friedel phase shifts ϕ_L which best fit the experimental results. The signs of the ϕ_L, however, cannot be determined by measurements of Dingle temperatures alone. If, as an additional constraint, we also require the set of phase shifts to reproduce correctly the residual resistivity (Lee *et al.* 1976a) then the ambiguity in sign can generally be resolved. The results of such a procedure for a number of non-magnetic impurities in the noble metals

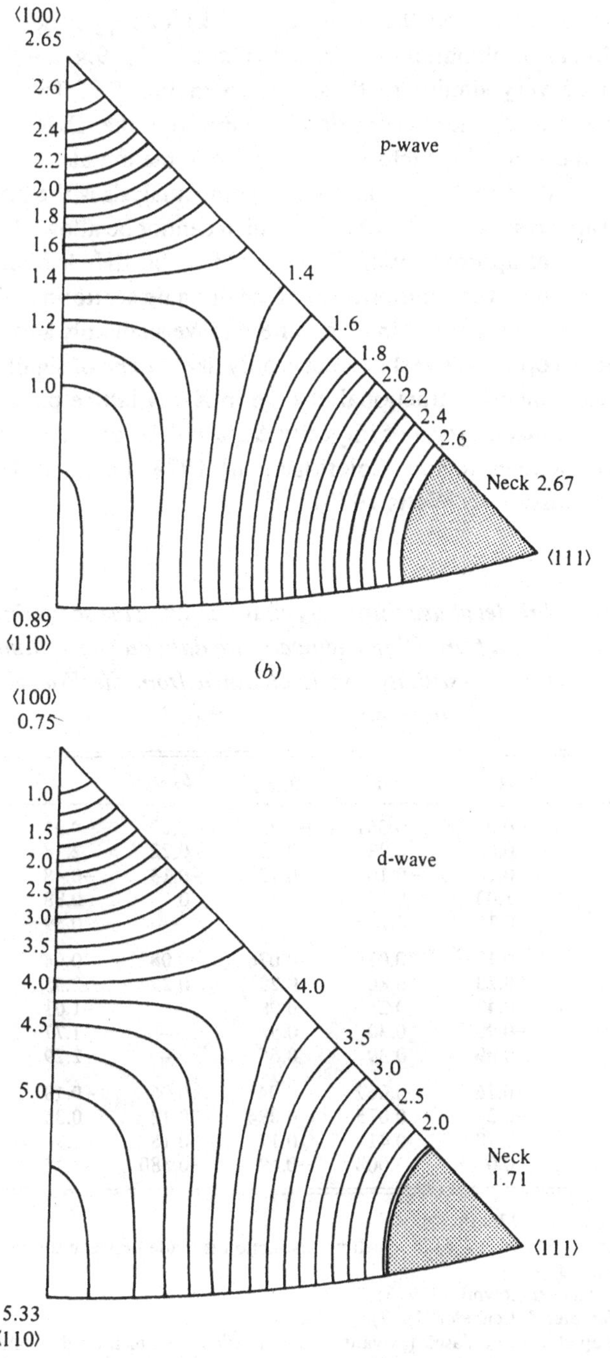
⟨100⟩
2.65
2.6
2.4
2.2
2.0
1.8
1.6
1.4
1.2
1.0
p-wave
1.4
1.6
1.8
2.0
2.2
2.4
2.6
Neck 2.67
⟨111⟩
0.89
⟨110⟩
(*b*)
⟨100⟩
0.75
1.0
1.5
2.0
2.5
3.0
3.5
4.0
4.5
5.0
d-wave
4.0
3.5
3.0
2.5
2.0
Neck
1.71
⟨111⟩
5.33
⟨110⟩

(*c*)

are summarized in Table 9.2. Maps of $\tau^{*-1}(\mathbf{k})$ for copper derived in this way are linear combinations of those depicted in Fig. 9.4, and have been found to be very similar to those shown in Fig. 9.3. This is a most reassuring result and lends considerable support to the KKR calculations of the partial scattering factors, $t_L(\mathbf{k})$, in the noble metals.

A complication that arises in the foregoing analysis is the influence of lattice strain around an impurity site, which cannot be allowed for within the MT crystal approximation, but to which the dHvA effect will be sensitive in view of the importance of small angle scattering. This point has already been discussed in Chapter 8 and we shall only add here that, in the above copper alloys the effect is only likely to be of significance for Cu(Au) for which the fractional change in X-ray lattice parameter per unit atomic concentration of impurity $\Delta a/a = 0.14$ and then only in the vicinity of the neck region (Coleridge *et al.* 1974; Lee *et al.* 1976a; see also the discussion by Watts 1973).

TABLE 9.2. *Friedel phase shifts, ϕ_L, characterizing the scattering in noble metal hosts derived from Dingle temperature data and constrained to give the correct residual resistivity. Δz is obtained from the Friedel sum rule according to equation (42)*

	ϕ_0	ϕ_1	$\phi_{2,\Gamma_{12}}$	$\phi_{2,\Gamma_{25'}}$	Δz	Notes
Cu(Ni)	0.0	−0.061	−0.26	−0.27	−0.96	*a*
Cu(Ge)	0.09	0.73	0.21	0.22	2.14	*a*
Cu(Au)	0.20	−0.10	−0.13	−0.13	−0.48	*a*
Cu(H)	1.03	0.12	0	0	0.88	*d*
Cu(Zn)	0.15	0.19	0.14	—	0.89	*e*
Ag(Au)	0.19	0.088	−0.076	−0.082	0.04	*a*
Ag(Sn)	0.23	0.80	0.20	0.22	2.35	*a*
Ag(Bd)	−0.42	−0.25	−0.08	—	−1.01	*b*
Ag(Y)	−0.90	0.30	0.56	—	1.77	*b*
Ag(Gd)	−0.89	0.39	0.66	—	2.29	*b*
Au(Ag)	−0.26	−0.067	0.04	0.06	−0.13	*a*
Au(Cu)	−0.20	0.075	0.084	0.12	0.35	*a*
Au(Ga)	0.98	0.41	0.12	0.18	1.9	*a*
Au(Co)	0.0	−0.008	−0.49	−0.580	−1.75	*c*

[a] From Lee *et al.* (1976a).
[b] From Sang & Myers (1976) in which no distinction is made between the two representations Γ_{12} and $\Gamma_{25'}$.
[c] Fit 2 of Chung & Lowndes (1976).
[d] From Wampler & Lengeler (1977).
[e] I. M. Templeton & P. Vasek (private communication) in which no distinction is made between Γ_{12} and $\Gamma_{25'}$.

9.5.2 Potassium

In view of the relatively small distortions of the Fermi surfaces of the alkali metals, anisotropic scattering effects are expected to be smaller than those encountered in the noble metals. For potassium this is indeed so but surprisingly, as we shall illustrate, the measured anisotropy of $\tau^{*-1}(\mathbf{k})$ is appreciably more difficult to account for theoretically in this case. The reason for this is interesting. Because of the diminished anisotropy of the wave-function variation over the Fermi surface, the lattice strain field which accompanies the inclusion of an impurity, and to which we referred only briefly in the last section, now assumes a relatively greater importance. Such an effect, however, as is well known (Blatt 1957; Harrison 1958) is not easy to take into account in theories of electron scattering and this applies also to the phase-shift approach (see the discussion of this point in Chapter 8). Using the method of lattice statics, Benedek & Baratoff (1973) have calculated that for the alloys K(Rb) and K(Na) the anisotropy of the lattice distortion field will give rise to an anisotropy in $\tau^{*-1}(\mathbf{k})$ amounting to ~10 per cent and 15 per cent respectively. This, however, as shown in the following, is of the same order as that which derives from the wave-function variation, even though the Fermi surface of potassium is distorted from a sphere by only ~0.1 per cent in radius vector. Thus we have a situation in which the scattering anisotropy is influenced by two effects of different physical origin which are not easily encompassed within a single theoretical model.

The electronic structure of potassium is influenced by the position of this metal at the head of the 3d-transition series in the periodic table and therefore by the presence of low-lying d-bands. In common with the other alkali metals, the Fermi surface is drawn out in the ⟨110⟩ directions towards the Brillouin zone faces, but unlike the others the Fermi wave-vector has its maximum value in the ⟨100⟩ directions. Arising out of considerations of the electron–ion interaction in the alkali metals, M. J. G. Lee (private communication) has calculated the quantities $(\partial k/\partial \eta_L)_\mu$ for the $l = 0$, 1 and 2 partial waves in potassium at a mesh of points over the Fermi surface. These quantities are directly related through (45) to the partial scattering factors, $t_L(\mathbf{k})$, which describe the varying character of the electron wave-functions. Expressing (45) in terms of an expansion in Kubic harmonics

$$\frac{2t_L(\mathbf{k})}{\hbar} = 2v_{\mathbf{k}}(\mathbf{k})\left\{\frac{\partial k(\theta,\phi)}{\partial \eta_L}\right\}_\mu = \sum_{i,m} {}_iD^L_m\, {}_iK_m(\theta,\phi), \tag{57}$$

TABLE 9.3. *Coefficients, ${}_iD^L_m$, in a 3-term Kubic harmonic series representation of the partial scattering rates $2t_L(\mathbf{k})/\hbar$ (see equation (57)) for the $l = 0$, 1 and 2 partial waves in potassium. No distinction has been made between the Γ_{12} and $\Gamma_{25'}$ representations for $l = 2$. Units are 10^{15} s^{-1}. The computed values of I_L (equation (63) of Chapter 8) are also given. The results were calculated from a mesh of 50 values of $(\partial k/\partial \eta_l)_\mu$ over the 1/48th basic segment of the Fermi surface by M. J. G. Lee (private communication). A renormalized quasi-particle velocity of $v = 7.017 \times 10^7$ cm s^{-1} is used.*

	I_L	${}_1D_0$	${}_1D_4$	${}_1D_6$
$l=0$	1.124	2.500	−0.0834	−0.0876
$l=1$	1.123	7.492	0.1648	0.3537
$l=2$	1.020	11.342	0.5033	−0.8321

we have upon substitution into (47)

$$\tau^{-1}(\theta, \phi) = c \sum_L \sum_{i,m} {}_iD^L_m \, {}_iK_m(\theta, \phi) \frac{\sin^2 \phi_L}{I_L}. \tag{58}$$

Using Lee's results, the computed coefficients ${}_iD^L_m$ are given for potassium in Table 9.3. Using these results we show in Fig. 9.5 the variation of $2t_L(\mathbf{k})/\hbar$ over the Fermi surface for the $l = 0$, 1 and 2 partial waves. The overall anisotropy displayed in these maps is ~10 per cent (s-wave), ~20 per cent (p-wave) and ~35 per cent (d-wave) and, for a given impurity which is characterized by a set of ϕ_L and ignoring scattering into f and higher angular momentum states, the anisotropy of $\tau^{-1}(\theta, \phi)$ should be represented by a linear combination of these three maps.

We now turn to the experimental results for potassium of Llewellyn, Paul, Randles & Springford (1977a, 1977b) for the two alloys K(Rb) and K(Na). The intercept technique was used to achieve the necessary resolution of the small changes in Dingle temperature and the experimental results were inverted using the Kubic harmonic scheme described in Section 9.3.3. Summarized in Table 9.4 are the coefficients, ${}_iB_m$, which, according to (37), represent the anisotropy of $\tau^{*-1}(\theta, \phi)$. These results are given pictorially in Fig. 9.6. Thus we conclude that the anisotropy of the scattering rate in potassium is 10.8 per cent and 19.4 per cent for the scattering by rubidium and sodium impurities respectively. The sense of the anisotropy is the same in both cases, having 'warm' spots for scattering in the ⟨110⟩ regions of the Fermi surface.

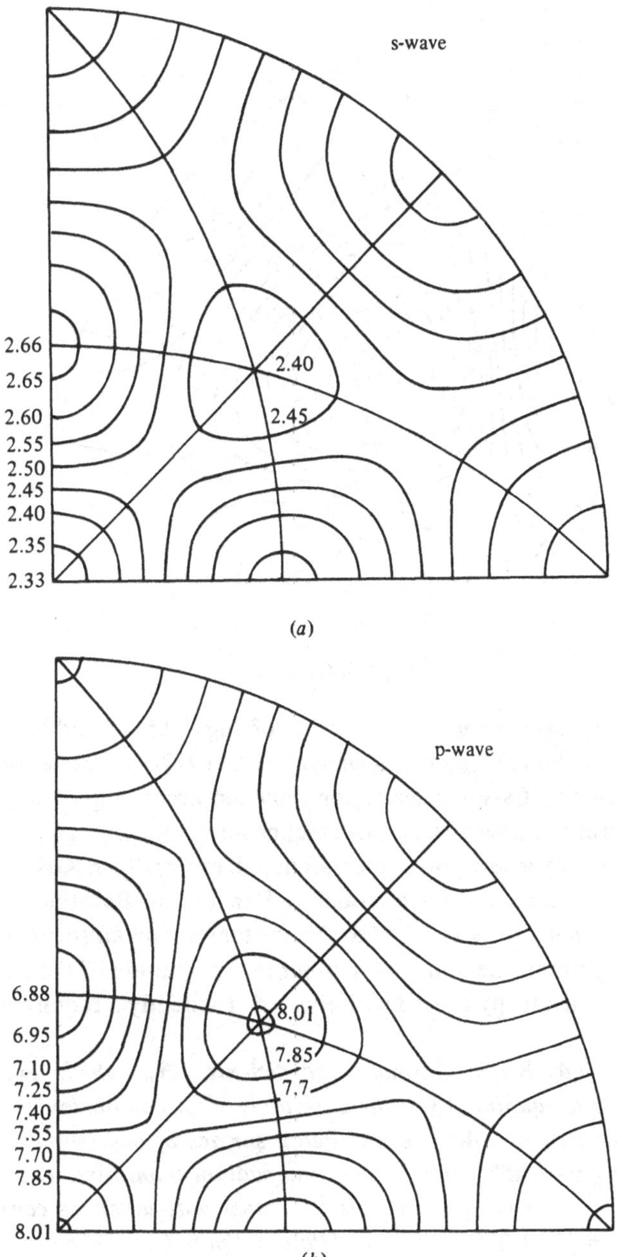

Fig. 9.5. Variation of the partial scattering rates (equation (57)), $2t_L(\mathbf{k})/\hbar$, over the Fermi surface in potassium, computed from calculations by M. J. G. Lee (private communication) for (*a*) $l = 0$, (*b*) $l = 1$ and (*c*) $l = 2$. For $l = 2$ the results for the Γ_{12} and $\Gamma_{25'}$ representations are appropriately averaged to yield a single map. The units are $10^{15}\ s^{-1}$ per atomic per cent.

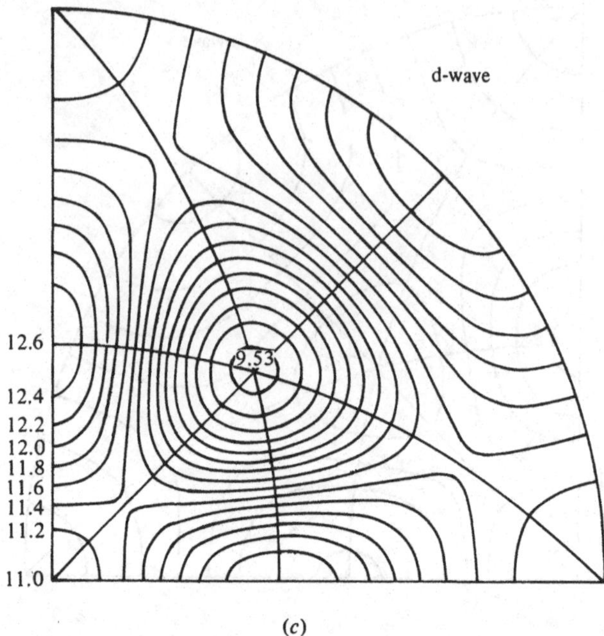

(c)

Fig. 9.5. *Continued.*

As may be seen from a comparison of Figs. 9.5(*a*) and 9.6, and is confirmed by the analysis of Llewellyn *et al.* (1977b), a model based on the assumption of s-wave scattering only can account qualitatively for these experimental results. In this case, however, the overall magnitude of the anisotropy is seriously underestimated, especially in K(Na). On the other hand, according to the model of Benedek & Baratoff (1973) in which the anisotropy is ascribed to the existence of an anisotropic lattice strain field around the impurity, neither the magnitude of the scattering rate nor its anisotropy accord with Fig. 9.6. Evidently a requirement for

TABLE 9.4. *Kubic harmonic coefficients, ${}_iB_m$ which express, according to equation (37), the anisotropy of the conduction electron scattering rate $\tau^{*-1}(\mathbf{k})$ over the Fermi surface of potassium due to scattering by 1 at.% of rubidium and sodium impurities. The units are $10^{12}\ s^{-1}$. The cyclotron effective mass was taken as constant $m_c^* = 1.230 \pm 0.013\ m$. (Paul & Springford 1978)*

	${}_1B_0$	${}_1B_4$	${}_1B_6$
K (Rb)	0.783 ± 0.001	−0.038 ± 0.005	−0.014 ± 0.001
K (Na)	4.092 ± 0.027	−0.320 ± 0.011	−0.150 ± 0.011

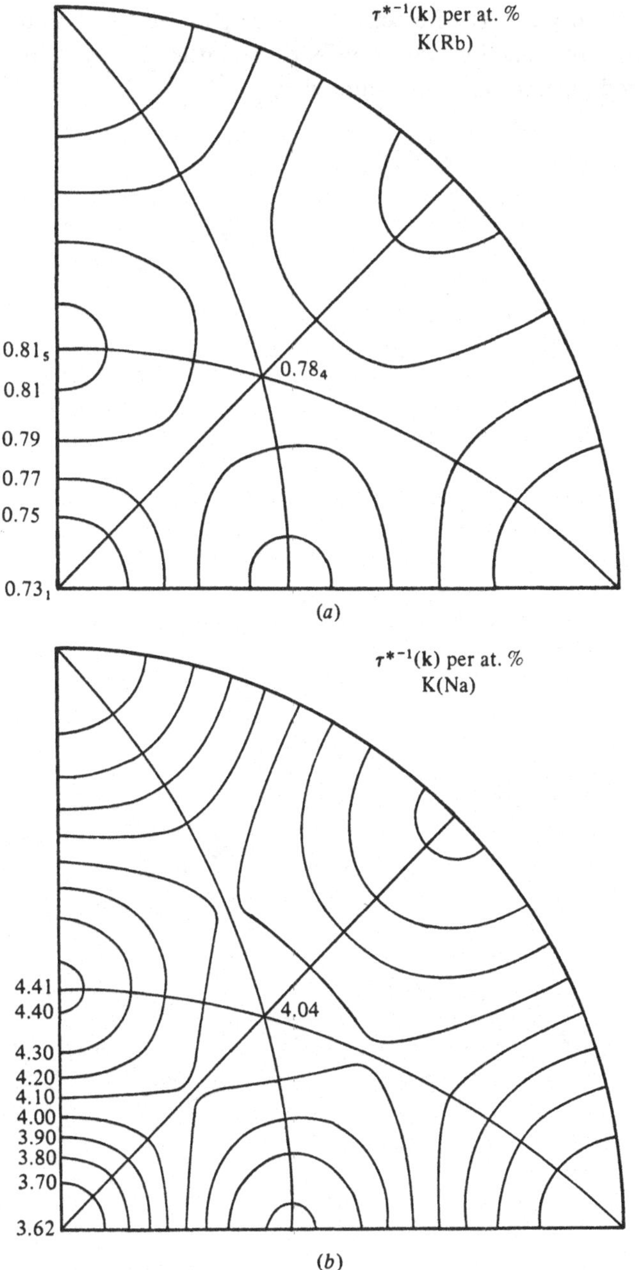

Fig. 9.6. Variation of the renormalized scattering rate $\tau^{*-1}(\mathbf{k})$ over the Fermi surface of potassium for scattering by (a) rubidium and (b) sodium impurities. The maps are constructed in an octant of the [001] stereographic projection with contours drawn at intervals of ~2.5 per cent. The units are $10^{12}\,s^{-1}$ per atomic per cent (Llewellyn *et al.* 1977a).

any satisfactory theory of electron scattering in the alkali metals is that both the wave-function variation and the lattice strain field must be considered together in a consistent way.

10

Waveshape analysis in the de Haas–van Alphen effect†

R. J. HIGGINS AND D. H. LOWNDES

10.1 Information content in the de Haas–van Alphen waveshapes

The de Haas–van Alphen (dHvA) effect remained a laboratory curiosity, observable only in anomalous semimetals like bismuth until a series of careful investigations (Marcus 1949; Shoenberg 1952a, 1957, 1960a) showed that with sufficiently pure metals, and sufficiently high magnetic fields and low temperatures, the phenomenon was observable in any metal, degenerate semiconductor, or metallic compound. These experiments stimulated the theoretical explanation (Lifshitz & Kosevich 1956) that the effect was indeed generally observable in any metal, and contained a wealth of information summarized by the expression

$$\tilde{M} = -\hat{m} \sum_{r=1}^{\infty} C_r D^r \cos(\pi r S) \sin\left[2\pi r\left(\frac{F}{H} - \gamma\right) + p\frac{\pi}{4}\right], \tag{1}$$

in which r is the dHvA harmonic index, and

$$\left.\begin{aligned} C_r &= \frac{\nu T F}{(\mathscr{A}'' r \hbar)^{\frac{1}{2}}} \frac{1}{\sinh(rKm_c T/mH)}, \\ D^r &= \exp(-rKm_c x/mH), \\ S &= m_c g_c/2m, \\ \nu &= [4k_B/\sqrt{(2\pi)}]\left(\frac{e}{hc}\right)^{\frac{3}{2}} = 1.304\times 10^{-5}\ (\mathrm{Oe}^{\frac{1}{2}}/\mathrm{K}), \\ K &= 2\pi^2 k_B mc/e\hbar = 146.9\ (\mathrm{kOe/K}), \end{aligned}\right\} \tag{2a}$$

† Research supported by the US National Science Foundation under grants DMR 76 14878 and DMR 74-07652 A02.

where F is the dHvA frequency ($=\hbar c\mathscr{A}/2\pi e$), $\mathscr{A}$ is the extremal cross-sectional area of the Fermi surface normal to the magnetic field, γ is the Onsager phase factor ($=\frac{1}{2}$ for free electrons), $p=-1(+1)$ if the extremal Fermi surface cross-section is a maximum (minimum) with respect to variation of k_H, $\mathscr{A}''$ is the Fermi surface curvature factor, $|d^2\mathscr{A}/dk_H^2|$, m_c/m is the cyclotron effective mass (in units of the free electron mass), k_B is Boltzmann's constant, $\hat{m}$ specifies the direction of $\tilde{M}$,

$$\hat{m}=\hat{H}-\frac{1}{F}\frac{\delta F}{\delta\theta}\hat{\theta}-\frac{1}{F\sin\theta}\frac{\delta F}{\delta\phi}\hat{\phi}, \tag{2b}$$

g_c is the electronic cyclotron orbitally-averaged g-factor, and x is the Dingle temperature (Dingle 1952), which is proportional to the cyclotron orbital average of the electronic scattering rate, $\tau(\mathbf{k})^{-1}$, at the point $\mathbf{k}$ on the Fermi surface (e.g. see Springford 1971),

$$x=\frac{\hbar}{2\pi k_B}\left\langle\frac{1}{\tau(\mathbf{k})}\right\rangle=\frac{\langle\Gamma\rangle}{\pi k_B}, \tag{3}$$

where $\langle\Gamma\rangle$ is the broadening (assumed to be Lorentzian) of the Landau levels.

The fact that non-vanishing quantum oscillations originate only from orbits which are 'extremal areas' (maximum or minimum cross-sections viewed along the magnetic field direction) is what has given the dHvA effect its power as a microscopic tool. The spectra which may be obtained by varying the magnetic field direction were found to be readily interpretable in terms of surprisingly simple models (Harrison 1960) for the 'nearly free electron' metals, and to provide for the first time a precise caliper against which numerical energy-band calculations could be compared (e.g. Mattheiss 1964b; Segall 1962; Andersen & Mackintosh 1968).

During this decade of Fermiography (summarized, for example, by Cracknell 1971) it also became clear that the dHvA effect contained, in principle, far more information than that provided by the frequencies alone. Measurements of the cyclotron effective mass (through the temperature dependence of amplitudes, C_r in (2a)) proved to be more readily possible in dHvA than in cyclotron resonance, and led to the first direct measurement of the many-body electron–phonon mass enhancement. Measurements of the field dependence of amplitudes (D^r in (2a)) led to unambiguous measurements of conduction electron lifetimes and their anisotropy over the Fermi surface (Springford 1971; and Springford, this volume) which, when interpreted using real metal wavefunctions and partial wave analysis techniques, have led to a detailed understanding of impurity–conduction electron interactions.

Although the Lifshitz–Kosevich (LK) expression (equation (1)) is a harmonic series, the harmonic content is preferentially damped out by thermal and impurity broadening, and under ordinary conditions, the susceptibility oscillations are very nearly simple sine waves. For example

$$A_r \propto \frac{\exp(-rKm_c x/mH)}{2\sinh(Krm_c T/mH)} \approx \exp[-Krm_c(T+x)/mH]. \tag{4}$$

Under typical experimental conditions ($m_c \approx m$, $H \approx 50$ kOe, $(T+x) \approx 2$ K) then the rth harmonic is reduced by $\exp(-6r) = (0.0025)^r$. However, with the development of the field modulation technique (Shoenberg & Stiles 1964; Goldstein, Williamson & Foner 1965; Stark & Windmiller 1968; for a review, see Gold 1968) the added variable of the modulation field amplitude made it possible to enhance the harmonic content by many orders of magnitude relative to the dHvA fundamental amplitude, thus making available the full information content of the dHvA waveshape which is the subject of this article. This extension in experimental technique and the resulting extensions in the basic LK theory are described in Section 10.2.

It was discovered early (Shoenberg 1962) that the reliable measurement of the amplitudes of dHvA oscillations required not only careful control of experimental variables such as magnetic field homogeneity (which had led Landau in his famous 1930 diamagnetism paper to predict that the oscillatory terms in his susceptibility theory would be unobservable) but also a high degree of control over the physical perfection of crystals. The observation of unusual field dependences of amplitudes and of unusual harmonic content was shown to be sensitively related to the degree of crystalline substructure present, such as subgrain boundaries and lattice dislocations. The interpretation of such effects has since then been made using methods which have come to be known as 'dephasing' approaches; these form the subject of Section 10.3. These have led to the use of the dHvA effect not only as a microscopic tool for the study of lattice imperfection scattering but also, using the phase information contained in the waveshapes, as a *predictive* tool for eliminating the errors due to these effects in other measurements.

A variety of non-linear effects alter the dHvA waveshapes from the basic LK expression (equation (1)). When band-gaps are small enough for magnetic breakdown to occur, certain oscillations disappear (Shoenberg 1962) and other new spectral terms appear (Stark & Falicov 1967; Falicov, this volume) as the magnetic field is raised. The exploration of the rich spectral information contained in magnetic breakdown has cast light on some fundamental questions dealing with Bloch electrons in a magnetic field. The fact that a conduction electron sees B rather than H results in a non-linear modulation of (1) which can either lead to appreciable harmonic distortion of the waveshape for a given orbit, or the mixing of several spectral terms resulting in unexpected sum and difference frequencies (Shoenberg 1962; Pippard 1963; see also Pippard, this volume). This effect becomes appreciable whenever the field spacing (period) of oscillations becomes comparable to the amplitude of the magnetization oscillation itself. In practice, the effect produces appreciable harmonic distortion for $F > 10^7$ G, and becomes the dominant contribution to harmonic content for $F > 10^8$ G. This magnetic interaction (MI) or 'Shoenberg effect' is a subject of interest not only in its own right, but is a complication which must be dealt with before the waveshape analysis technique can be generally applied. This is dealt with briefly in Section 10.2, and in detail in Sections 10.4 and 10.5.

The dHvA effect also contains spin-dependent information, since the Zeeman effect splits Landau levels into two sets of spin-(up, down) levels, and the observed oscillations are a superposition of those arising independently from the two spins. Although resolved spin-splitting of Landau levels may be occasionally seen in semimetals such as bismuth (Smith, Baraff & Rowell 1964), or in oscillations arising from very small pieces of Fermi surface such as the needle in zinc (Stark 1964), spin-splitting is not usually resolved with the weak harmonic content which is normally present (for the reasons discussed above). The two spin components interfere to give the factor $\cos(\pi r S)$ in (1). If it happens that $S = (n + \frac{1}{2})$, the amplitude of the fundamental vanishes and, if m_c is known, g may be calculated. However, the occurrence of such 'spin-splitting zeros' of the dHvA fundamental is essentially accidental, occurring only at a few magnetic field directions, which in general will not be symmetry directions. Although occasionally useful (see Windmiller and Ketterson 1968), the direct measurement of spin-splitting zeros does not provide the powerful 'handle' one would like to have on local, orbital g-values, for example in measuring anisotropic relativistic effects such as spin–orbit coupling, or for measurements of many-body enhancement effects such as those arising in Fermi liquid theory. This difficulty may be overcome (as we discuss in Section 10.4) by the measurement of orbital

g-factors at *arbitrary* crystal orientations using the relative harmonic content of the oscillations (Shoenberg & Vuillemin 1966; Randles 1972). The additional harmonic content produced by the MI causes an additional complication in this measurement. However, if the waveshape measurement includes the relative phases of the harmonics, the presence of MI harmonic content can turn into an advantage and the LK and MI contributions can be separated by a projection technique (Randles 1972; Alles, Higgins & Lowndes 1975). An extension of this technique using the amplitudes and phases of the first three harmonics (Gold & Schmor 1976) makes possible g-measurements in the presence of MI *without* a knowledge of the detailed shape of the Fermi surface or of the Dingle temperature under certain circumstances.

Perhaps the most powerful applications of spin-resolved dHvA spectroscopy have been to the study of magnetic impurities. Here, the conduction electron–local moment interaction is inherently spin-dependent, and no other experimental technique resolves spin-dependent information so directly. In addition, since the information obtained is local in **k**-space and the host metal Fermi surfaces are well known, the analysis has led to the most detailed microscopic information yet available on the nature of the magnetic impurity states, via partial wave analysis of impurity–host scattering phase-shift differences. The way in which spin-dependent information is contained in the dHvA waveshape in this case is illustrated schematically in Fig. 10.1. The Landau levels (a) are spin-split (b). The impurity moment couples to the conduction electrons via an exchange energy which shifts the g-value and the spin-splitting (c). This was seen initially (Coleridge & Templeton 1971b) as a spin-splitting zero appearing as a function of magnetic field, or as a shift in the angle at which the spin-splitting zero occurred (Coleridge, Scott & Templeton 1972) from which the orbital g-shift could be unambiguously determined. In some cases (e.g. Cu–Mn) a spin-splitting zero tended to become smeared. This is reasonable because in systems where the scattering is strongly energy-dependent (large impurity thermopower) the scattering is spin-dependent. Two different Landau level broadenings, $\Gamma^{\uparrow}$ and $\Gamma^{\downarrow}$, are required (Fig. 10.1(d)) so that the interference of the two sets of oscillations with unequal amplitudes cannot produce a perfect spin-splitting zero. In a pioneering series of measurements, the McGill group (Hedgecock & Muir 1963; Paton & Muir 1968; Paton, Hedgecock & Muir 1970) detected anomalies in the dHvA effect arising from electrons on the 'needle' orbit in Zn(Mn) dilute alloys, and showed how the energy-dependent relaxation time associated

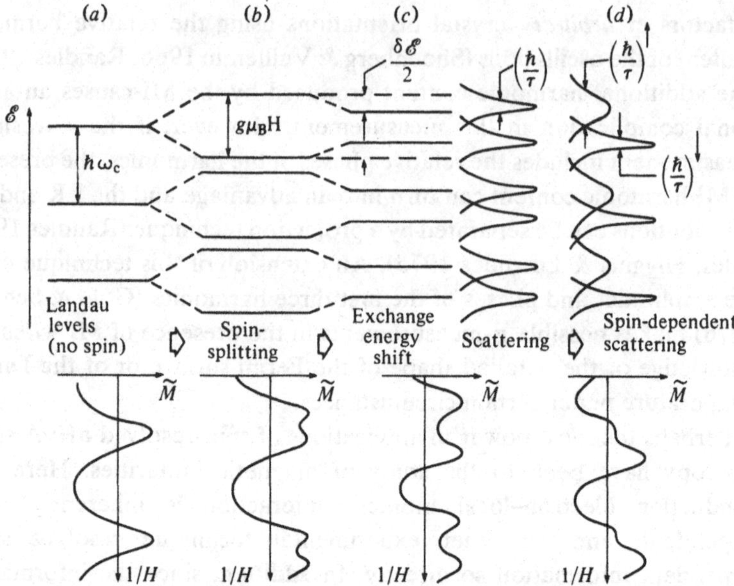

Fig. 10.1. Modification of (*a*) electronic orbital Landau levels due to (*b*) spin-splitting (Zeeman effect), (*c*) exchange interaction with a local magnetic moment (antiferromagnetic case shown) and (*d*) unequal (↑, ↓)-spin level broadening ($\Gamma^{\uparrow} > \Gamma^{\downarrow}$ shown). The corresponding qualitative changes in the waveshape of the oscillatory magnetization are shown below.

with the Kondo effect could be included in the LK expression for the dHvA effect. They later (1970) pointed out the connection between their measurements and the field dependence of the electronic scattering rate, resulting from freeze-out of spin-flip scattering, in the magnetoresistance.

The development of precise waveshape analysis techniques has made it possible to work backwards from measurements of relative harmonic amplitudes and phases to obtain accurate values for both microscopic *g*-shifts and spin-dependent scattering (SDS) rates (Alles, Higgins & Lowndes 1973; Alles & Higgins 1974; C. Hendel & D. H. Lowndes, unpublished data). The general significance of these techniques to the understanding of magnetic impurity problems may be seen by noting that the observables may be related directly to the real and imaginary parts of the impurity self-energy (Shiba 1973). Although the initial measurements were limited to a case where a complication of MI was unimportant, and used only the fundamental and second harmonic information, a considerable advance in the technique becomes possible (Chung, Lowndes & Hendel 1978; D. H. Lowndes & P Reinders, unpublished data) if the full information content of the first three harmonics is used. This makes it possible to project out the MI components, and provides a sufficient number of relative amplitudes and relative phases to determine unambiguously all three of the microscopic parameters (spin-up and spin-down scattering and g-shift) which alter the dHvA waveshape in magnetic impurity systems. Using these techniques it has been possible to use changes in the dHvA waveshape to detect the very onset of magnetism (formation of magnetic pairs, etc.) due to interactions between pairs of non-magnetic solute atoms at very low solute concentrations (e.g. Au(Co), Chung & Lowndes 1977), and to make detailed comparisons of the effectiveness of exchange interactions between conduction electrons and both d-shell and f-shell local moments in producing g shifts and SDS (C. Hendel & D. H. Lowndes, unpublished data). This subject is dealt with in detail in Section 10.5.

10.2 The waveshape of dHvA oscillations

In this section we shall describe the experimental methods and develop the necessary generalizations of the LK theory which provide the foundation and the notation used in later sections. There are just two fundamental modifications. The first, occurring when magnetic impurities are present, modifies both the relative amplitudes and the relative phases of the two spin components of the dHvA signal. The second modification is the modulation of the dHvA waveform (MI). There are other situations in which the dHvA signal is modified from the form of (1), such as those involving magnetic field inhomogeneity and crystalline substructure. These involve computing the net dHvA signal as the coherent superposition of a spatially non-uniform signal, and are reserved for discussion in Section 10.3 on dephasing techniques.

The emphasis in this paper will be on the information contained in the *relative* phases and *relative* amplitudes which can be deduced from a precise Fourier analysis of a discretely sampled data *window* (~10–20 cycles) of dHvA oscillations. The advantages of such measurements are that one needs neither a precise knowledge of the magnetic field, or the quantum index N, or a calibration of the sensitivity of the measuring apparatus. It will be shown that if one Fourier analyses a window of data into sine and cosine components (relative to the *arbitrary* origin, in $1/H$, of the window), then the *relative* phases of the harmonics, $2\theta_1-\theta_2$, $3\theta_3-\theta_1$, and, in general, $m\theta_n-n\theta_m$, are *independent* of the window origin. We shall also show that these relative phases and relative harmonic amplitudes (A_r/A_1) are often sufficient to extract much of the information in (1) (and in its generalizations which follow), including the determination of *absolute* amplitudes from the measured *relative* amplitudes and phases.

10.2.1 Measurable quantities in dHvA waveshape analysis

It will become readily apparent in the following sections that the *vector* magnitudes and phases of the dHvA oscillations are useful quantities. Although a discussion of experimental problems is reserved for Section 10.4 and 10.5, several fundamental points which affect later analysis will be mentioned here.

In measurements with the field modulation technique, one measures an a.c. voltage V which is proportional to the oscillatory magnetization, $\tilde{M}_0$. Discrete samples of V in a sweep of $1/H$ produce a 'window' of dHvA oscillations. This data window is then Fourier analysed into voltage amplitudes, V_r, and phases, θ_r:

$$V(\zeta) = G\tilde{M}(\zeta) = \sum_{r=1}^{\infty} V_r(\zeta) = G \sum_{r=1}^{\infty} \tilde{R}_r(\zeta), \tag{5a}$$

where

$$\tilde{R}_r(\zeta) = |\tilde{M}_r| J_n(r\Lambda) \sin(r\zeta + \theta_r). \tag{5b}$$

Here, the system gain G contains the absolute system sensitivity, amplifier gains, and factors dependent on the amplitude and frequency ω of the modulation signal. The signal is coherently detected at a harmonic frequency $n\omega$. We write $\zeta \equiv 2\pi F/H$; θ_r contains the phase factors of (1) and its later generalizations. J_n is a Bessel function of order n, and $\Lambda = 2\pi Fh/H^2$.

Much of the potential of waveshape analysis technique, particularly in alloys where the harmonic content is exponentially damped, is due to the enhancement of harmonic content possible with the 'Bessel function spectrometer' action implicit in (5). When the modulation field amplitude is set to bring the highest harmonic r of interest near the first Bessel function peak, the amplitude of the rth harmonic is enhanced (relative to the fundamental) by an amount of order $(r)^n$. This enhancement can be several orders of magnitude, bringing harmonic content of less than 1 per cent to be nearly equal in magnitude to the fundamental in the detected signal. An example (Alles & Higgins 1974) is shown in Fig. 10.2.

The observables $|V_r|$ and θ_r are obtained either by direct Fourier transform, by least-squares fitting, or, when the spectrum is complicated, by a successive subtraction procedure (e.g. Alles *et al.* 1975, Appendix C)

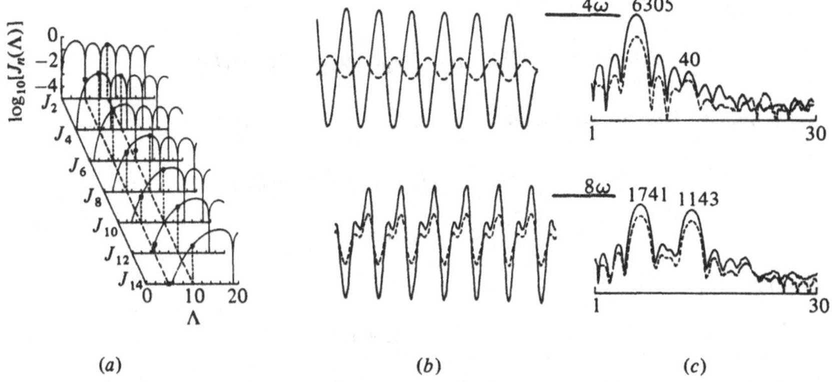

Fig. 10.2. Illustration of the relative enhancement of dHvA harmonic information made possible by use of higher detection harmonics in the field modulation technique. (*a*) Plot on a log scale of the first seven even-order Bessel functions. The dashed lines show typical settings of the modulation argument Λ for 1st and 2nd (■ and ●) dHvA harmonics and the corresponding amplitude factor. (*b*) Example of data showing improvement in dHvA harmonic content between eighth and fourth detection harmonic for the Λ value of (*a*). (*c*) Corresponding Fourier transform (on a log scale of amplitude) for the data of (*b*). The numbers listed with the peaks are amplitudes determined by a successive subtraction fitting method. (From Alles & Higgins 1974.)

which yields sine and cosine Fourier components s_r and c_r, relative to the origin of the data window:

$$\left.\begin{aligned} |V_r| &= (s_r^2 + c_r^2)^{\frac{1}{2}}, \\ \theta_r' &= \tan^{-1}(s_r/c_r). \end{aligned}\right\} \tag{6}$$

There are several complications inherent in the transition from magnetization to voltage oscillations. First, G is not readily measurable. Second, the sign of the measured signal is uncertain in the a.c. method, i.e. one does not know whether a given dHvA peak is a maximum or minimum of $\tilde{M}$. In principle, this sign ambiguity can be resolved, for example by using a calibration sample with known static paramagnetism or ferromagnetism. In practice, however, a reversal of the pickup coil leads reverses the sign and, in addition, there are two choices of lock-in detector phase (of opposite sign) which give a maximum detected signal. Third, the phases which come out of the Fourier analysis are measured with respect to the arbitrary origin in $1/H$ of the data window, not with respect to infinite field, as is implicit in (1).

Since both the magnitude and sign of G are usually not known, the most useful magnitudes are *relative* amplitudes, in which G drops out, for example

$$|V_2|/|V_1| = |R_2|/|R_1|, \ |V_3|/|V_1| = |R_3|/|R_1|. \tag{7}$$

To consider the phase measurement problem, we write:

$$\begin{aligned} V_r(x) &= |R_r|\,|G| \sin[r(\zeta - \zeta_0) + r\zeta_0 + \theta_r + n\pi] \\ &\equiv |R_r|\,|G| \sin[r(\zeta - \zeta_0) + \theta'_r], \end{aligned} \tag{8}$$

where

$$\theta'_r \equiv (r\zeta_0 + \theta_r + n\pi) \tag{9}$$

is the phase of the observed voltage measured with respect to the left edge of the data window and n is 0 or 1 depending on whether an (unknown) accidental sign reversal has taken place. When $V_r(x)$ is resolved into sine and cosine components s_r and c_r, the *measurable* phase angles are, using (6),

$$\theta'_r = (r\zeta_0 + \theta_r + (n + l_r)\pi), \tag{10}$$

where the prime is a reminder of the arbitrary window origin, and where $l_r = (0 \text{ or } 1)$ takes into account the ambiguity (mod π) in determining an angle from the value of its tangent. Thus, quantities of the form

$$(j\theta'_i - i\theta'_j) = (j\theta_i - i\theta_j) + (j - i)n\pi + (jl_i - il_j)\pi \tag{11}$$

(j and i integers) are *independent* of the arbitrary initial phase $r\zeta_0$ of the data window, but do depend on the unknown factors $n\pi$, $l_i\pi$ and $l_j\pi$. For example

$$(2\theta'_1 - \theta'_2) = (2\theta_1 - \theta_2) + n\pi + (2l_1 - l_2)\pi \tag{12}$$

but

$$\begin{aligned} (3\theta'_1 - \theta'_3) &= (3\theta_1 - \theta_3) + 2\pi n + (3l_1 - l_3)\pi \\ &= (3\theta_1 - \theta_3) + (3l_1 - l_3)\pi. \end{aligned} \tag{13}$$

Although n drops out if $(j - i)$ is even, the modulo π ambiguity arising from the Fourier transform phase angle determination remains. However, by considering limiting behaviour of the dHvA waveform (as a function of H and T) in the presence of MI we shall see that it is generally possible to remove even this uncertainty, and to determine *uniquely* the relative phase angles. Thus, the measurable phase *differences* (relative phases) of voltages in a data window do give directly the phase differences of various harmonic dHvA magnetization oscillations, within the cautions just stated.

To summarize, the 'observables' in dHvA waveshape measurement are ratios of harmonic amplitudes (equation (7)) and relative harmonic phase angles (equation (11)). We now show how to use this information.

10.2.2 Modifications due to electron spin: pure metals and orbital g-factors

Under ordinary conditions of temperature and magnetic field, the cusp-like shape (e.g. Gold 1968, Fig. 4*b*) of the $T = 0$ oscillations is not observed, i.e. spin-splitting of the Landau levels does not introduce clearly resolved spin-splitting of the dHvA oscillations. Rather, one observes a decrease in the amplitude due to interference of the spin-up and spin-down oscillations, leading to the $\cos(\pi r S)$ terms in (1). It is useful to review the origin of this spin-splitting term, both as an introduction to methods of measuring the orbitally-averaged g-factor g_c, and in order to introduce a pictorial representation which will be used throughout later sections. Zeeman splitting shifts the Landau levels by an energy $\pm 1/2 g\mu_B H$. The two spin contributions to $\tilde{M}_r$ are therefore phase shifted. For example

$$\tilde{M}_1 \sim \sin 2\pi[F/H + (\tfrac{1}{2}g\mu_B H/\hbar\omega_c)] + \sin 2\pi[F/H - (\tfrac{1}{2}g\mu_B H/\hbar\omega_c)]$$

$$= 2\cos(\pi S)\sin(2\pi F/H), \tag{14}$$

where $\omega_c = eH/m_c$ is the cyclotron frequency. The oscillatory magnetization may be pictured (Fig. 10.3) as the projection on the horizontal axis of a rotating vector, which goes around once each time F/H changes by 1 for the first harmonic $\tilde{M}_1$, twice for the second harmonic $\tilde{M}_2$, etc. Note that the spin-splitting angle

$$\phi_r = r\pi S \equiv r\phi \tag{15}$$

is proportional to the harmonic number, and that the horizontal axis (0°) represents the phase angle of the unsplit oscillation.

When this picture is valid, a measurement of the orbital g-factor may be made in one of three ways:

(*a*) *g from spin-splitting zero.* The effective mass and/or g-factor of an extremal orbit may vary rapidly with magnetic field direction. As a consequence, the quantity πS may pass through an odd multiple of $\pi/2$,

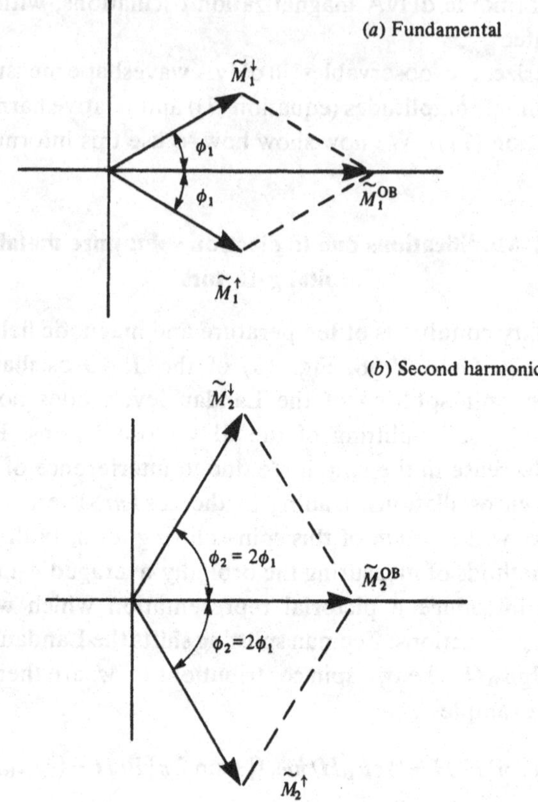

Fig. 10.3. Phasor diagram showing components of the resultant LK rth harmonic amplitude ($r = 1, 2$) from the spin-up and spin-down electrons. *Increasing* dHvA phase is plotted anticlockwise and corresponds to *decreasing* magnetic field. The labels (↑, ↓) refer to electron spin, not magnetic moment.

leading to a vanishing of the first harmonic amplitude (first harmonic spin-splitting zero) as the crystal is rotated through small angular steps, from which (knowing m_c) a set of possible values of g_c may be deduced. A similar technique may be used to determine values of g_c from higher harmonic spin-splitting zeros.

(*b*) *g from absolute amplitudes.* If all other quantities in (2) are known or measured, a measurement of the absolute amplitude $\tilde{M}_1$ yields a set of possible values for g_c through the quantity $\cos \pi S$. This may be done either by an absolute calibration of the apparatus (difficult but possible with the modulation technique and straightforward with absolute susceptibility or torsion methods) or, as developed later in this chapter, by projection techniques which use MI harmonic content to measure absolute amplitudes.

(*c*) *g from relative harmonic amplitudes.* Another measure of g may be obtained from various ratios of dHvA harmonic amplitudes. For example, the ratio of the LK harmonic amplitudes $\tilde{M}_2/\tilde{M}_1$ leads to

$$\frac{\cos(2\pi S)}{\cos(\pi S)} = \left(\frac{\tilde{M}_2}{\tilde{M}_1}\right)\left(\frac{\sqrt{2}\sinh X_2}{\sinh X_1}\right)\exp(Km_c x/mH), \tag{16}$$

where $X_r = rKm_cT/mH$. Other ratios may be used, as discussed in Section 10.4.

Methods (*a*) and (*c*) do not require an absolute calibration of the apparatus and do not require that terms in (2) such as $\mathrm{d}^2\mathscr{A}/\mathrm{d}k_H^2$ be known. Note that since g is contained within a cosine (using any method) it is known only within a certain modularity, and one must appeal to plausibility arguments (e.g. in the case of a nearly free electron-like orbit, $g \approx 2$), to band-structure calculations (for small orbits with important spin–orbit terms, g may differ from 2 by several orders of magnitude), or to theoretical bounds on g, to determine it uniquely. As shown in Fig. 10.4, when S is measured by method (*a*) or (*b*), it is known at best mod 2, $S' = 2n \pm S$, where n = any positive integer. However, if the sign of the signal is unknown, extra modularity occurs (dashed lines in Fig. 10.4), and one only knows the quantity $n \pm S$, where n may be any positive integer. When S is measured by method (*c*), yet another family of roots is possible (Fig. 10.4(*c*)) at half-integer values of n, so that only the quantity $n \pm S$ is known, where n is any positive integer or half integer. Methods of sorting out some of this ambiguity will be discussed in Section 10.4.

10.2.3 Modification due to magnetic impurities: *g*-shifts and spin-dependent scattering

Effective g-shift

There is another, equally interesting, splitting which can occur in a dilute alloy containing magnetic impurities which possess a *local magnetic moment.* For example, rare earth elements and some transition metal elements produce local moments when they are present as dilute solutes in noble metal (Cu, Ag, Au) hosts. The local moment is associated with the d- or f-states of the impurity atom, which are localized in the region of the impurity cell. When these states are embedded in the energy continuum of conduction electron states near $\mathscr{E}_F$, hybridization results in

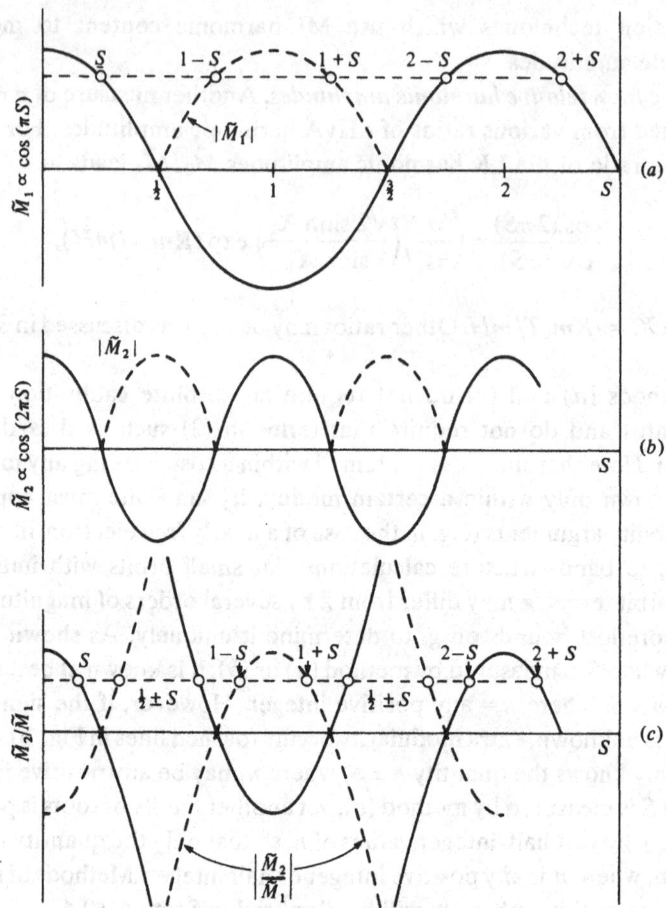

Fig. 10.4. Multiplicity of solutions for the orbital *g*-factor, using measured values of $\tilde{M}_1$, $\tilde{M}_2$, and $\tilde{M}_2/\tilde{M}_1$. One-half of the possible solutions can be eliminated if the algebraic sign of any one of these quantities is known in addition to its magnitude, or if the magnitudes of any two quantities are known. (*a*) $\tilde{M}_1$; (*b*) $\tilde{M}_2$; (*c*) $\tilde{M}_2/\tilde{M}_1$.

a finite lifetime. (For a review, see Kondo 1969; Heeger 1969.) A local magnetic moment can interact with the conduction electron moment via an effective exchange interaction, producing a spin-dependent shift of the electronic energy levels. If we represent this interaction by an isotropic Heisenberg exchange Hamiltonian with exchange constant J, then

$$\mathcal{H}_{\text{ex}} = -c_{\text{m}}JS\sigma. \tag{17}$$

Here, c_{m} is the fractional local moment concentration and $J(<0)>0$ for (anti)ferromagnetic exchange. Assuming that $\mathbf{H}$ tends to polarize $\boldsymbol{\mu}_{\text{imp}}$ parallel to $\mathbf{H}$ (with $\mathbf{S}_{\text{imp}}$ opposite to $\boldsymbol{\mu}_{\text{imp}}$), then the exchange shifts the spin-splitting of energy levels by

$$\delta\mathscr{E} = (\langle\mathscr{H}_{ex}\rangle^{\uparrow} - \langle\mathscr{H}_{ex}\rangle^{\downarrow}) = c_m J\langle S_z\rangle = -c_m|J|\langle S_z\rangle, \tag{18}$$

where we take $\langle S_z\rangle$ to be the component of S along H. Since $\langle S_z\rangle$ is an expectation value, it will be a function of T and H, reflecting the polarization of the local moment by the applied field. Here, the last expression assumes antiferromagnetic exchange. The net effect is to modify the Zeeman splitting, as shown in Fig. 10.1(c). For this reason it is convenient to introduce an 'effective g-factor' g'_c and an 'exchange field', H_{ex}, defined such that the combination of Zeeman energy $g\mu_B H$ and exchange energy shift $\delta\mathscr{E}$ is

$$\Delta\mathscr{E}_\sigma = \frac{\sigma}{2}(g_c\mu_B H + \delta\mathscr{E}) = \sigma g'_c\mu_B H/2, \tag{19}$$

where

$$g'_c = \left(g_c + \frac{\delta\mathscr{E}}{\mu_B H}\right) = \left(g_c + \frac{H_{ex}}{H}\right),\ H_{ex} = \frac{\delta\mathscr{E}}{\mu_B}. \tag{20}$$

Note that $\Delta\mathscr{E}$ is $(<0)>0$ and $g'_c\ (<g_c)>g_c$ for (anti)ferromagnetic exchange.

The effect of this energy shift is to shift the absolute *phase* of the up-spin and down-spin contributions to the dHvA oscillations. The dHvA first harmonic phase is $(2\pi F/H - \gamma + p\,(\pi/4)$, a *decreasing* function of H. On the other hand, the Landau level separations, $\hbar\omega_c$ are *increasing* functions of H. A shift *upward* in Landau level energy by an amount $+\hbar\omega_c$ corresponds to a decrease of 2π in dHvA phase. The net phase shift is

$$\phi_\sigma = -2\pi\left(\frac{\Delta\mathscr{E}_\sigma}{\hbar\omega_c}\right) = -2\pi\sigma\frac{\dfrac{\mu_B H}{2}\left(g_c - \dfrac{\delta\mathscr{E}}{\mu_B H}\right)}{\hbar\omega_c} = -\frac{\pi\sigma}{2}\cdot\frac{m_c}{m}g'_c. \tag{21}$$

For the rth harmonic, the effect of the energy shift $\Delta\mathscr{E}_\sigma$ is to shift the phase r times as fast, i.e.

$$\phi_{r\sigma} = r\phi_\sigma = \pi\sigma r S',\ S' = \frac{g'_c}{2}\left(\frac{m_c}{m}\right). \tag{22}$$

Thus, if the spin-up and spin-down oscillations have the same amplitude but are simply phase-shifted by $\pm\phi_r$, their resultant is

$$\tilde{M}_r = A_r \cos\phi_r \sin[2\pi r(F/H - \gamma)], \tag{23}$$

where A_r is the dHvA amplitude for *spinless* electrons. Thus, in this case, the resultant dHvA amplitude is changed only via the factor $\cos(\pi r S')$. However, as will be shown, the requirement that both spin components have equal amplitudes is a restrictive one, observed only in exceptional cases.

Spin-dependent scattering

Since the local moment-conduction electron exchange interaction is spin-dependent, there is reason to expect the scattering rates for spin-up and spin-down electrons also to be unequal (Beal-Monod & Wiener 1968; Coleridge *et al.* 1972; Alles *et al.* 1973), corresponding to the unequal Landau level broadenings shown in Fig. 10.1(*d*). The amplitudes $A^{\uparrow}$ and $A^{\downarrow}$ also become unequal, and the above analysis must be generalized to allow for a spin-dependent Dingle temperature. Equation (1) then becomes

$$\tilde{M} \equiv \frac{\hat{m}}{2} \sum_{r=1}^{\infty} \sum_{\sigma} C_r D^r E^{\sigma r} \sin\left[2\pi r(F/H - \gamma) + p\frac{\pi}{4} - \sigma\pi r S'\right], \quad (24)$$

where σ is the spin index (+1(−1) for spin-up (-down)), and

$$\left.\begin{aligned} D &= \exp(-Km_c\bar{x}/mH), & E &= \exp(-Km_c\delta x/mH), \\ \bar{x} &= (x^{\uparrow} + x^{\downarrow})/2, & \delta x &= (x^{\uparrow} - x^{\downarrow})/2, \end{aligned}\right\} \quad (25)$$

$x^{\uparrow}$ and $x^{\downarrow}$ are the up-spin and down-spin electron Dingle temperatures in terms of which the mean Dingle temperature, $\bar{x}$, and the difference δx are defined. Note that the *difference* in up- and down-spin scattering rates is $2\delta x$.

In (24) we have separated the spin-independent term, C_r, from the spin-dependent terms DE and DE^{-1}, which describe the effect of SDS on the up-spin and down-spin dHvA amplitudes, respectively. The connection between (24), which is the basis for experimental analysis of the dHvA waveshape, and the theoretical expression (Shiba 1973) for the dHvA effect in the presence of magnetic impurities, is discussed in Section 10.5.

Considering only the first three Fourier components, (24) becomes

$$\begin{aligned} \tilde{M} = -\frac{\hat{m}}{2}\Big\{ & CD\Big[E \sin\Big(\zeta + p\frac{\pi}{4} - \pi S'\Big) + E^{-1} \sin\Big(\zeta + p\frac{\pi}{4} + \pi S'\Big)\Big] \\ & + C_2 D^2\Big[E^2 \sin\Big(2\zeta + p\frac{\pi}{4} - 2\pi S'\Big) + E^{-2} \sin\Big(2\zeta + p\frac{\pi}{4} + 2\pi S'\Big)\Big] \\ & + C_3 D^3\Big[E^3 \sin\Big(3\zeta + p\frac{\pi}{4} - 3\pi S'\Big) + E^{-3} \sin\Big(3\zeta + p\frac{\pi}{4} + 3\pi S'\Big)\Big]\Big\}, \end{aligned} \quad (26)$$

where $\zeta \equiv 2\pi(F/H - \gamma)$.

It is convenient to represent the terms in (26) with the aid of phasor diagrams (Fig. 10.5). The consequence of unequal spin component contributions is to change not only the magnitude but also the absolute phase of the resultant. Note that although the spin-splitting angles, $\phi_r = r\phi$, scale with r, the phase shifts in the *resultant* magnitudes do not, i.e. $\Delta\theta_2 \neq 2\Delta\theta_1$. This leads to a *measurable* quantity, the *relative* phase shift (e.g. $2\theta_1 - \theta_2$) which contains information about the SDS which can be unfolded. Use of the vector diagram (Fig. 10.5) yields the relationships

$$\tilde{M}_r^{\mathrm{LK}} = \frac{C_r D^r}{2}[E^{2r} + E^{-2r} + 2\cos(2r\phi)]^{\frac{1}{2}}, \tag{27}$$

$$\Delta\theta_r^{\mathrm{LK}} = \tan^{-1}\left[\tan(r\phi)\left(\frac{1-E^{2r}}{1+E^{2r}}\right)\right], \tag{28}$$

where LK denotes that we have not so far considered harmonic content due to MI.

The relationship between measurable quantities (e.g. relative harmonic amplitudes and phases) and the quantities D, E and S', from which the scattering rate, its spin dependence and the g-shift may be determined, now follow easily. For example

$$\frac{|\tilde{M}_2^{\mathrm{LK}}|}{|\tilde{M}_1^{\mathrm{LK}}|} = \frac{C_2 D}{C_1}\frac{[E^4 + E^{-4} + 2\cos(4\phi)]^{\frac{1}{2}}}{[E^2 + E^{-2} + 2\cos(2\phi)]^{\frac{1}{2}}}, \tag{29}$$

$$\frac{|\tilde{M}_3^{\mathrm{LK}}|}{|\tilde{M}_1^{\mathrm{LK}}|} = \frac{C_3 D^2}{C_1}\frac{[E^6 + E^{-6} + 2\cos(6\phi)]^{\frac{1}{2}}}{[E^2 + E^{-2} + 2\cos(2\phi)]^{\frac{1}{2}}}, \tag{30}$$

while $(2\theta_1 - \theta_2)^{\mathrm{OB}}$ and $(3\theta_1 - \theta_3)^{\mathrm{OB}}$ (OB = observed) are each shifted in phase by $(2\Delta\theta_1 - \Delta\theta_2)$ and $(3\Delta\theta_1 - \Delta\theta_3)$, respectively, from the values they would taken in the absence of SDS. (We defer until Section 10.5 the discussion of analogous equations for the relative phase angles.) The interrelation of the various quantities above is best seen graphically (Fig. 10.6). A given measured value of the relative phase $(2\theta_1 - \theta_2)$, for example, is compatible with a contour of possible values of ϕ and E (equivalent to S' and δx). The problem of *uniquely* unfolding the quantities of theoretical interest will be discussed in Section 10.5.

Although the form of (26) is convenient for the analysis of experimental data, a slight recasting of form produces an intuitively useful variation.

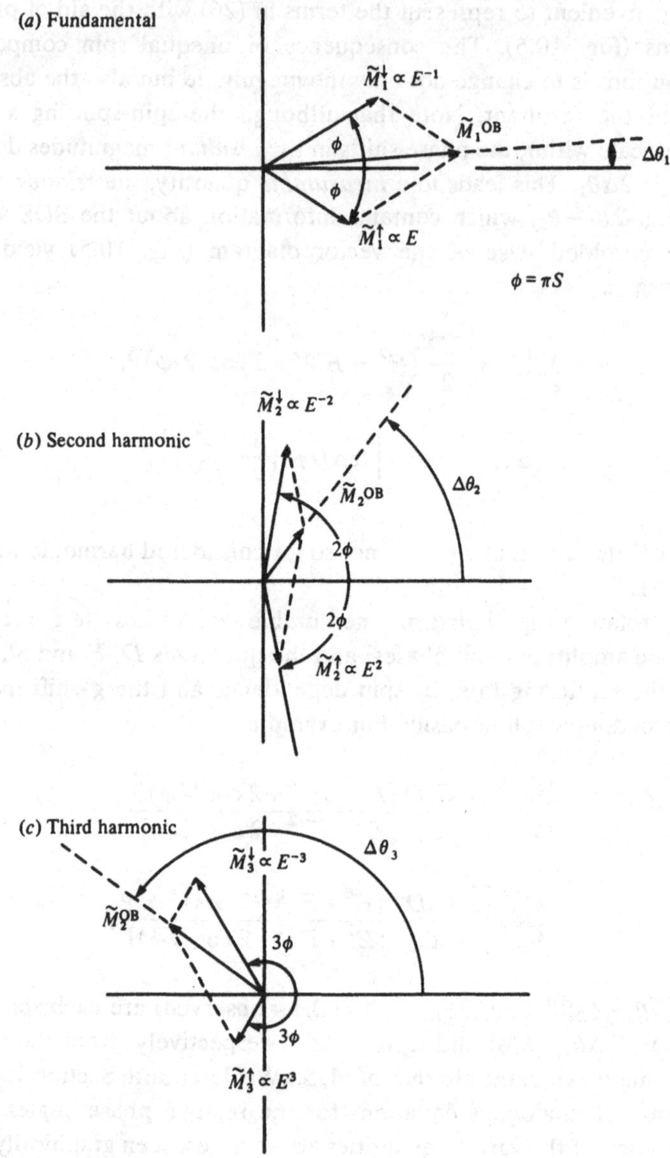

Fig. 10.5. Phasor diagrams for the first three harmonics, showing formation of the resultant dHvA harmonic amplitudes from the spin-up and spin-down components, for the case in which no MI harmonic content is present. Increasing dHvA phase is plotted anticlockwise. $\Delta\theta_r$ is the phase shift induced in the rth harmonic resultant dHvA amplitude by SDS. No SDS ($\delta x = 0$) always produces no resultant phase shift, while for $\phi < \pi/4$ total SDS (only one spin contributing) also produces no phase shift. (The figures correspond to parameter values $E^2 = 0.75$ and $\phi = 40°$.)

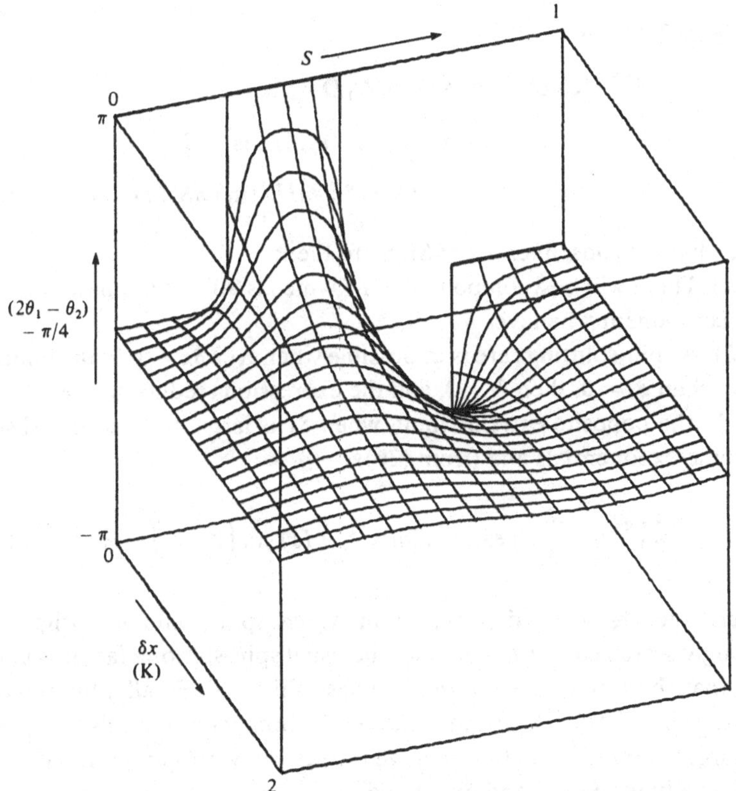

Fig. 10.6. Plot showing the dependence of the observable $(2\theta_1 - \theta_2)$ as a function of δx and S, for the case in which no MI harmonic content is present. (Graphics courtesy of T. G. Matheson.)

Using Fig. 10.5, project $\tilde{M}$ into components. Then

$$\tilde{M}_1^{\mathrm{LK}} = ((\tilde{M}_1^{\uparrow} + \tilde{M}_1^{\downarrow}) \cos \pi S') \sin\left(\zeta + p\frac{\pi}{4}\right) - ((\tilde{M}_1^{\uparrow} - \tilde{M}_1^{\downarrow}) \sin \pi S') \cos\left(\zeta + p\frac{\pi}{4}\right). \tag{31}$$

This is equivalent to

$$\tilde{M}_1^{\mathrm{LK}} = C_1 D\left[\mathscr{X} \sin\left(\zeta + p\frac{\pi}{4}\right) - \mathscr{Y} \cos\left(\zeta + p\frac{\pi}{4}\right)\right], \tag{32}$$

$$\left.\begin{aligned} \mathscr{X} &= \cosh(K m_c \delta x / mH) \cos \pi S', \\ \mathscr{Y} &= \sinh(K m_c \delta x / mH) \sin \pi S', \end{aligned}\right\} \tag{33}$$

whose magnitude is:

$$|\tilde{M}_1^{LK}|/C_1D = (\mathscr{X}^2 + \mathscr{Y}^2)^{\frac{1}{2}}/C_1D$$
$$= [\cosh^2 (Km_c\delta x/mH) \cos^2 \pi S'$$
$$+ \sinh^2 (Km_c\delta x/mH) \sin^2 \pi S']^{\frac{1}{2}}/C_1D. \quad (34)$$

The physical consequences of SDS are then:

(*a*) There will be a component of $\tilde{M}$ which is 90° out of phase with the LK component whenever $\delta x \neq 0$;

(*b*) A spin-splitting zero is impossible whenever $\delta x \neq 0$. In the limiting case when $\delta x = 0$, then $\mathscr{Y} = 0$, and the expressions reduce to the form of (1). In the opposite extreme limit, when δx is large, both cosh and sinh become exponentials and (from (26))

$$\tilde{M}_1^{LK} = -\frac{\hat{m}}{2} C_1 \exp(-Km_c x^{\uparrow\downarrow}/mH) \sin\left(\zeta + p\frac{\pi}{4} \pm \pi S'\right), \quad (35)$$

where the ($\uparrow\downarrow$, +/−) depends upon which spin dominates (the *least* strongly scattered spin). Note that the usual spin-splitting factor is gone, but that there is an extra *absolute* phase shift $\pi S'$. Finally, the relative phase shifts $(2\Delta\theta_1 - \Delta\theta_2)$ and $(3\Delta\theta_1 - \Delta\theta_3)$ both vanish in this limit, as is apparent from (28), so that relative phases carry no spin information in *both* the limits $\delta x = 0$ and $\delta x =$ large.

10.2.4 Magnetic interaction effects

Origin

During experiments involving the use of high magnetic fields to study dHvA in the noble metals, Shoenberg (1962) noticed that the harmonic content of the observed signals was much stronger than is predicted by the LK theory. He proposed that the effective magnetizing field which conduction electrons see is not H but instead is the total magnetic induction,

$$B = H + 4\pi\tilde{M}(1 - \mathscr{D}), \quad (36)$$

where $\tilde{M}$ is the (oscillatory) dHvA magnetization of the sample itself, and $\mathscr{D}$ is the sample's demagnetizing factor ($0 \leq \mathscr{D} \leq 1$). Thus, the conduction electrons *magnetically interact* with themselves, by responding to a B field which includes their own oscillatory magnetization. Shoenberg's conjecture has since been confirmed on thermodynamic and experimental grounds (Pippard 1963; Condon 1966a; Shoenberg & Vuillemin 1966; Phillips & Gold 1969).

The harmonic content generated by MI, though often regarded as a troublesome experimental complication, has recently been turned into an advantage with the development of dHvA waveform analysis techniques, in which the presence of appreciable MI-generated harmonic content (comparable in magnitude to the LK harmonic content) can be helpful in unlocking the information contained in the dHvA harmonic spectrum. The effect on the harmonic content of the dHvA effect, of having the electrons respond to B rather than H, can be seen directly. The first consequence is that $\tilde{M}$ must now be determined self-consistently from an *implicit* equation for $\tilde{M}$

$$\tilde{M} = \sum_{r=1}^{\infty} \tilde{M}_r = \hat{m} \sum_{r=1}^{\infty} A_r \sin\left[2\pi r\left(\frac{F}{H + 4\pi\tilde{M}(1-\mathscr{D})} - \gamma\right) + p\frac{\pi}{4}\right]. \tag{37}$$

The demagnetizing factor $(1-\mathscr{D})$ was not present in the earliest analyses of MI effects, but is essential to the precise analysis of experiments making use of both the harmonic phases and amplitudes. The harmonic amplitude A_r (Phillips & Gold 1969) is simply related to our notation (equation (24)) by

$$A_r = -C_r D' \cos(\pi r S), \tag{38}$$

where we assume for the moment that SDS or exchange energy shifts are absent ($E = 1$, $S' = S$).

Because $|4\pi\tilde{M}| \ll H$ ($|4\pi\tilde{M}| \lesssim 1$ G, $H \gtrsim 50$ kOe, typically), replacing H by B will have negligible effect on the size of $\tilde{M}$. However, $|4\pi\tilde{M}|$ can be an appreciable fraction of the field spacing (H^2/F) of a dHvA oscillation, so that the presence of $4\pi\tilde{M}$, alternately adding or subtracting to small steady changes in the applied field H, can lead to *large* distortions of the *shape* of the waveform, when it is observed and plotted as a function of the *applied* field H. In the limit when $\tilde{M}$ becomes comparable to the field spacing B^2/F between oscillations, thermodynamic instability leads to the formation of diamagnetic domains (Pippard 1963; Condon 1966a). Much of this subject is discussed elsewhere (Pippard, this volume), and we will limit the discussion of MI here to methods projecting out the MI harmonic content in the weak MI limit. It is the development of these projection techniques which have saved the waveshapes technique from being limited to those few systems with negligible MI, and have actually made the MI effect useful, for example in sorting out g-factor ambiguities (Section 10.4), or in uniquely pinning down the signs and magnitudes of g-shifts and scattering lifetimes in magnetic impurity systems (Section 10.5).

Iterative analysis of harmonic content

This discussion follows Phillips & Gold (1969) and introduces the notation needed in later sections. For $|4\pi\tilde{M}/H| \ll 1$, we can rewrite (37) as

$$\tilde{M} = \sum_{r=1}^{\infty} A_r \sin\left[r(\zeta - \kappa'\tilde{M}) + p\frac{\pi}{4}\right], \tag{39}$$

where

$$\zeta \equiv 2\pi(F/H - \gamma), \ \kappa' \equiv (1-\mathscr{D})8\pi^2 F/H^2. \tag{40}$$

Consider the case of *weak* MI, $|\kappa' A_1| < 1$. Equation (39) is an implicit equation for $\tilde{M}$, and a general solution for $\tilde{M}$ as an explicit function of ζ can be found to any desired accuracy by using an iterative scheme of successive approximations, in which the nth approximation to $\tilde{M}$ is given by

$$\tilde{M}^{(n)} = \sum_{r=1}^{\infty} A_r \sin\left[r(\zeta - \kappa'\tilde{M}^{(n-r)}) + p\frac{\pi}{4}\right], \tag{41}$$

with $\tilde{M}^{(0)} \equiv 0$. The right-hand side of (41) can be expanded by trigonometric identities involving powers and cross-products of the amplitudes, A_r. However, the relative magnitudes of the A_r are largely determined by the exponential damping factors,

$$A_r \sim \exp(-rKm_c(T+x)/mH),$$

so that the terms which occur can be classified according to the order $O(i)$ of the resulting negative exponentials. Thus, terms involving A_4, A_1A_2 and A_1A_3 are all $O(4)$. With this classification scheme, the iteration given by (41) is very simple. The result is that any approximation $\tilde{M}^{(n)}$ is exact to $O(n)$. It is convenient to express a given $\tilde{M}^{(n)}$ in terms of coefficients p_r and q_r, which are in phase and out of phase respectively, with the LK magnetization in the absence of MI.

$$\tilde{M}^{(n)} = \sum_{r=1}^{\infty}\left[p_r \sin\left(r\zeta + p\frac{\pi}{4}\right) + q_r \cos\left(r\zeta + p\frac{\pi}{4}\right)\right]. \tag{42}$$

Thus, $q_r \neq 0$ corresponds to an MI-induced *phase shift* of the *observed* rth *harmonic signal*, since some of the signal is now in the 'cosine' part of the resultant waveform. The first 2 orders of coefficients p_r and q_r are easily found, and the results to $O(4)$ are given in Table 10.1 (Phillips & Gold 1969).

Thus, for $|\kappa' A_1| < 1$, the effect of MI on A_1 is small, but for the higher harmonics $r > 2$, the lowest-order MI terms are usually of the same order as the LK harmonic amplitudes. This follows because the *ratio* of MI to LK component in A_2, for example, is proportional to $\kappa A_1^2/A_2$. The Dingle temperature term drops out of this ratio, so that scattering does not affect the *relative* harmonic content $A_r^{\mathrm{MI}}/A_r^{\mathrm{LK}}$.

Vector projection technique for separating MI and LK harmonic content

Although they only used the resultant amplitudes, it is implicit in the analysis of Phillips and Gold (Table 10.1) that the LK and MI components are separable. The key to doing this without measuring absolute phases (references to $H = \infty$) turns out to be the measurement of relative phases. As a consequence, one can, for example, determine orbital g-factors from the relative harmonic amplitudes. The quantity

TABLE 10.1. *Components p_r and q_r of the dHvA harmonics in the presence of MI (Phillips & Gold 1969). Note that Phillips & Gold's amplitudes A_i are defined including a minus sign*

Term	$O(1)$	$O(2)$	$O(3)$	$O(4)$
p_1	A_1		$+\frac{\kappa A_1 A_2}{2\sqrt{2}} - \frac{\kappa^2 A_1^3}{8}$	
q_1			$\pm\frac{\kappa A_1 A_2}{2\sqrt{2}}$	
p_2		$A_2 - \frac{\kappa A_1^2}{2\sqrt{2}}$		$+\frac{\kappa A_1 A_3}{\sqrt{2}} + \frac{\kappa^3 A_1^4}{6\sqrt{2}} - \kappa^2 A_1^2 A_2$
q_2		$\pm\frac{\kappa A_1^2}{2\sqrt{2}}$		$\pm\frac{\kappa A_1 A_3}{\sqrt{2}} + \frac{\kappa^3 A_1^4}{6\sqrt{2}}$
p_3			$A_2 - \frac{3\kappa A_1 A_2}{2\sqrt{2}}$	
q_3			$\pm\frac{3\kappa A_1 A_2}{2\sqrt{2}} \mp \frac{3\kappa^2 A_1^3}{8}$	
p_4				$A_4 + \frac{\kappa^3 A_1^4}{3\sqrt{2}} - \frac{2\kappa A_1 A_3}{\sqrt{2}} - \frac{\kappa A_2^2}{\sqrt{2}}$
q_4				$\pm\frac{\kappa A_2^2}{\sqrt{2}} \pm \frac{\kappa^3 A_1^4}{3\sqrt{2}} \pm \frac{2\kappa A_1 A_3}{\sqrt{2}} \mp 2\kappa^2 A_1^2 A_2$

needed is a *projection*, the LK component. The MI component is in itself also a useful quantity, since it contains the information needed for an *absolute* amplitude measurement without making an absolute gain calibration. It may also be used as an independent check on g (Section 10.4) or as a means to measure x when skin-depth problems are extreme. It will be shown that the information content in the dHvA waveshape is enhanced when MI and LK components are present simultaneously, although the analysis becomes more difficult. The method of analysis will be illustrated in this section using just the second harmonic content for simplicity. Again, we assume here that no magnetic impurity is present, since either a field-dependent g-factor or SDS further complicates the analysis. This restriction will be removed in Section 10.5.

The Phillips and Gold expansion may be written, to second order in negative exponentials, as,

$$\tilde{M} = \hat{m}\left\{C_1 D|\cos \pi S| \sin\left[\zeta + p\frac{\pi}{4} + (1+j_1)\frac{\pi}{2}\right]\right.$$

$$+ C_2 D^2 |\cos 2\pi S| \sin\left[2\zeta + p\frac{\pi}{4} + (1+j_2)\frac{\pi}{2}\right]$$

$$\left. - \frac{\kappa'}{2}(C_1 D \cos \pi S)^2 \sin\left[2\left(\zeta + p\frac{\pi}{4}\right)\right]\right\}. \tag{43}$$

The intrinsic minus sign and the sign of $\cos(r\pi S)$ have both been absorbed in the phase factors $(1+j_r)\pi/2$ where

$$j_r \equiv \begin{cases} +1, & \cos \pi r S > 0, \\ -1, & \cos \pi r S < 0. \end{cases} \tag{44}$$

Rewriting $\tilde{M}$ relative to the left edge of the data window, and absorbing minus signs into phase factors so that all coefficients are positive,

$$\tilde{M} = C_1 D\,|\cos \pi S| \sin\left[(\zeta - \zeta_0) + \zeta_0 + p\frac{\pi}{4} + (1+j_1)\frac{\pi}{2}\right]$$

$$+ C_2 D^2 |\cos 2\pi S| \sin\left[2(\zeta - \zeta_0) + 2\zeta_0 + p\frac{\pi}{4} + (1+j_2)\frac{\pi}{2}\right]$$

$$+ \frac{\kappa'}{2}(C_1 D\,|\cos \pi S|)^2 \sin\left[2(\zeta - \zeta_0) + 2\zeta_0 + 2p\frac{\pi}{4} + \pi\right]. \tag{45}$$

The phase angles of the voltage oscillations, $\tilde{V}$, relative to the left edge of the data window are

$$\theta_1 = \zeta_0 + p\frac{\pi}{4} + (1+j_1)\frac{\pi}{2} + n\pi. \tag{46a}$$

For θ_2, the phase range is

$$\left.\begin{aligned} \theta_2^{\text{LK}} &= 2\zeta_0 + p\frac{\pi}{4} + (1+j_2)\frac{\pi}{2} + n\pi, \\ \theta_2^{\text{MI}} &= 2\zeta_0 + 2p\frac{\pi}{4} + \pi + n\pi. \end{aligned}\right\} \tag{46b}$$

Thus, to modulo (2π), the observable range of relative phase will be

$$(2\theta_1 - \theta_2) = \begin{cases} p\dfrac{\pi}{4} - (1+j_2)\dfrac{\pi}{2} + n\pi & \text{(pure LK)}, \\ -\pi + n\pi & \text{(pure MI)}. \end{cases} \tag{47}$$

Neglecting the factor $n\pi$ for the moment, the calculated range of values of $2\theta_1 - \theta_2$ is displayed graphically in Fig. 10.7, for all four combinations of p (extremal orbit a maximum or minimum) and j_2 ($\cos 2\pi S \lessgtr 0$).

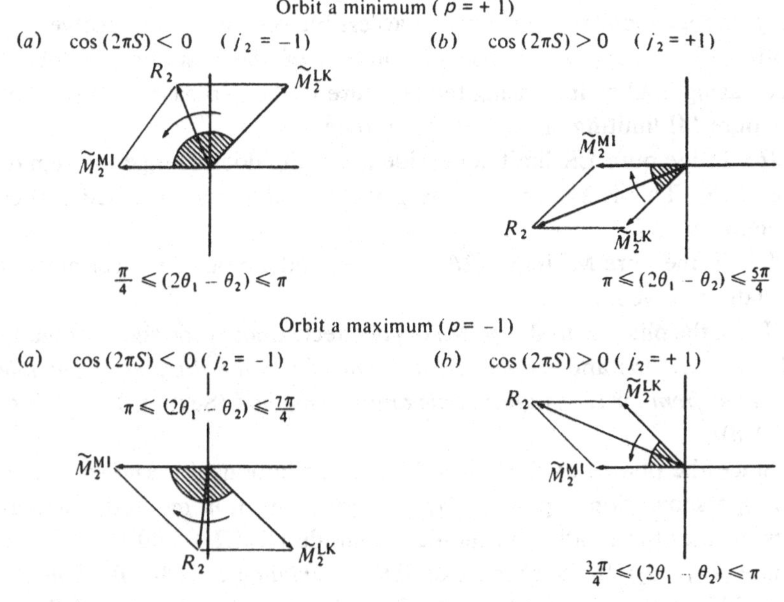

Fig. 10.7. Illustration of the accessible ranges of values for the observable relative phase parameter $2\theta_1 - \theta_2$. For a given orbit ($p = +1$ or $p = -1$), the two possible ranges of values of $2\theta_1 - \theta_2$ are mutually exclusive and depend on the sign of the spin-splitting factor $\cos(2\pi S)$. The value $2\theta_1 - \theta_2 = \pi$ corresponds to the pure MI limit. In general $2\theta_1 - \theta_2$ lies between the pure LK and pure MI limiting values.

Because of the mod (π) uncertainty in determining the phase angles of $\tilde{V}$, any measured value of $2\theta_1-\theta_2$ could equally well have π added to it. However, in practice, there need be *no* such ambiguity. This depends only on (i) the presence of contributions from *both* MI and LK in the observed dHvA second harmonic, and (ii) the fact that the ratio

$$\frac{|\tilde{M}_2^{\rm MI}|}{|\tilde{M}_2^{\rm LK}|}=\frac{\sqrt{2}\nu TF}{(\mathscr{A}''H)^{\frac{1}{2}}}\frac{\sinh(2Km_cT/mH)\,|\cos\pi S|^2}{2\sinh^2(Km_cT/mH)\,|\cos 2\pi S|}\frac{4\pi^2F(1-\mathscr{D})}{H^2} \tag{48}$$

is proportional to $(T/H^{\frac{5}{2}})$, so that the *resultant* goes to the pure LK limit at high H and low T and to the pure MI limit in the opposite case. With reference to Fig. 10.7:

(*a*) If p is known (extremal orbit a maximum or minimum) then there is no ambiguity in $2\theta_1-\theta_2$. For example, for $p=+1$, either $\pi/4\leq(2\theta_1-\theta_2)\leq\pi$ (Case 1) or $\pi\leq(2\theta_1-\theta_2)\leq\frac{5}{4}\pi$ (Case 2). With the addition of a factor π, those two accessible regions remain mutually exclusive. Since p is a simple Fermi surface property, it is either already known for most orbits, or is readily measurable by noting the rate of increase or decrease of F as the crystal is rotated away from the direction of interest. Any doubts about the appropriate accessible region can be resolved by noting (i) the opposite sense of rotation of the resultant, R_2, with decreasing field or increasing temperature in the two cases and (ii) that the pure MI limiting phase is defined to be π.

(*b*) In the pure LK limit, an accidental π flip does change the sign of $\cos(2\pi S)$. This is a source of the g-factor ambiguity discussed earlier (Section 10.2.2).

(*c*) In the pure MI limit, $(2\theta_1-\theta_2)=\pi$, and a mod (π) uncertainty is trivially resolved.

Thus, the phase shift due to MI (*a*) in effect *removes* the sign ambiguity of $\cos(2\pi S)$ (*b*) and *reduces by a factor of two the number of possible g-values from the second harmonic analysis method* (Section 10.2.2, Fig. 10.4(*c*)).

Since the phases of the MI and LK contributions are known (Table 10.1), a separation is possible by a vector projection method, using as observables the relative harmonic amplitude (R_2/R_1), and the relative phase $(2\theta_1-\theta_2)$. In the absence of SDS, a *deviation* of $(2\theta_1-\theta_2)$ *from the pure LK limiting value* can *only be due to MI.* It is convenient to define α as the angle by which θ_2 (and also $(2\theta_1-\theta_2)$ deviates from its LK limiting value:

$$\theta_2=\theta_2^{\rm LK}+\alpha$$

so that

$$(2\theta_1-\theta_2)^{\mathrm{OB}}=(2\theta_1-\theta_2)^{\mathrm{LK}}-\alpha \tag{49}$$

or

$$\alpha=\left[p\frac{\pi}{4}-(1+j_2)\frac{\pi}{2}\right]-(2\theta_1-\theta_2)^{\mathrm{OB}}.$$

We can use the phasor (or vector) diagram idea to reconstruct $\tilde{M}_2^{\mathrm{MI}}$ and $\tilde{M}_2^{\mathrm{LK}}$ in terms of R_2 and α^{OB}. It is convenient to redraw the vector diagrams of Fig. 10.7 with $\alpha=0$ (pure LK case) as the phase reference (Fig. 10.8). Note that increasing dHvA phase (decreasing H) is an anti-clockwise rotation in these diagrams. Note also since

$$\theta_2^{\mathrm{MI}}-\theta_2^{\mathrm{LK}}=p\frac{\pi}{4}+\pi-(1+j_2)\pi/2 \tag{50}$$

then

$$\mathrm{SGN}\,(\alpha)=p*j_2.$$

Thus, a determination of the sign of α is equivalent to removing the ambiguity in the sign of $\cos(2\pi S)$. Separation of the MI and LK

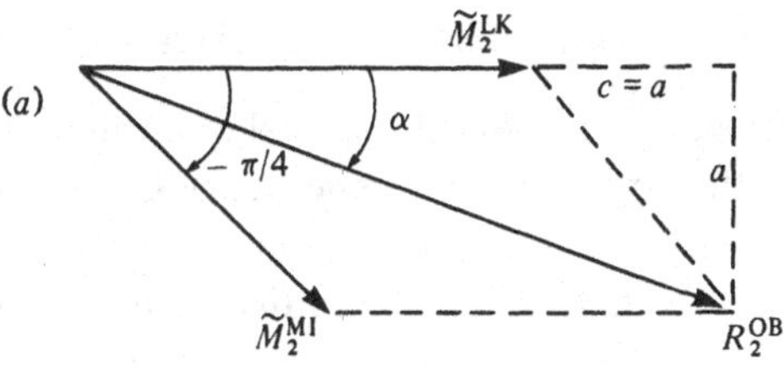

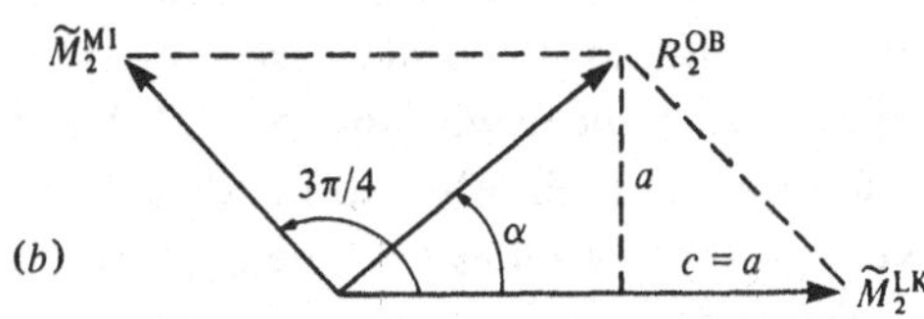

Fig. 10.8. Phasor diagram decomposition of the resultant (observed) second harmonic amplitude into its LK and MI components, for the case of no SDS. (*a*) For $\cos(2\pi S)>0$ ($j_2=+1$); (*b*) for $\cos(2\pi S)<0$ ($j_2=-1$). For either case $\tilde{M}_2^{\mathrm{MI}}=\sqrt{2}\,R_2^{\mathrm{OB}}\sin\alpha=\kappa(1-\mathscr{D})(R_1^{\mathrm{OB}})^2/2$, $\tilde{M}_2^{\mathrm{LK}}=R_2^{\mathrm{OB}}(\cos\alpha-j_2\sin\alpha)$, where α is the angle by which the resultant second harmonic signal, R_2^{OB}, deviates from the pure LK limit (see equation (49)). Both (*a*) and (*b*) are drawn for the case $p=+1$ (extremal orbit an area minimum).

components then follows by geometry. Using Fig. 10.8(a) as an example,

$$\left.\begin{aligned} b &= \tilde{M}_2^{\rm LK} = R_2^{\rm OB} \cos \alpha - c, \\ a &= R_2^{\rm OB} \sin \alpha = c = \tilde{M}_2^{\rm MI}/\sqrt{2}, \end{aligned}\right\} \tag{51a}$$

$$\tilde{M}_2^{\rm LK} = R_2^{\rm OB} (\cos \alpha - \sin \alpha), \tag{51b}$$

$$\begin{aligned} \tilde{M}_2^{\rm MI} &= \sqrt{2} a = \sqrt{2} R_2^{\rm OB} \sin \alpha \\ &= \kappa (1 - \mathscr{D})(R_1^{\rm OB})^2/2. \end{aligned} \tag{51c}$$

The result of other cases (e.g. Fig. 10.8(b)) is identical except for a sign change in the equation for $\tilde{M}_2^{\rm LK}$, since in that case MI subtracts rather than adds to the resultant magnitude. The general result for all cases is:

$$\tilde{M}_2^{\rm LK} = R_2^{\rm OB} (\cos \alpha + p_{j_2} \sin \alpha), \tag{52a}$$

$$\tilde{M}_2^{\rm MI} = \sqrt{2} R_2^{\rm OB} \sin \alpha. \tag{52b}$$

Note that there is no sign ambiguity in projecting $\tilde{M}_2^{\rm LK}$, since the sign of the observable α equals the quantity p_{j_2}.

10.3 Dephasing effects on the dHvA waveshapes

There are many examples where the spatial equivalence of orbits in real space (leading to the large orbital degeneracy of Landau levels) is broken when the environment is spatially non-uniform. We will deal in this section with such effects from crystal lattice defects and magnetic field inhomogeneity. It will turn out that the analysis of waveshapes leads not only to methods of correcting for such undesired background effects, but also to the observation of new phenomena. An electron travelling in such a disturbed environment finds its local phase shifted by the disturbance, so that the energy at which the wave-function will close is shifted slightly. When the contributions of such orbits to the density of states are added up, Landau levels for the crystal as a whole become broadened. Rather than working with the density of states, it is customary to work directly with dHvA oscillations, which are the Fourier transform of the density of states. Such situations are better treated as an interference effect rather than as scattering, and have come to be known as *dephasing* methods (Shoenberg 1962, 1969). If a shift in energy of a given orbit translates to a shift ψ in phase ζ of the dHvA oscillation, then an oscillatory term $\sin \zeta$

becomes

$$\int_{-\infty}^{\infty} \sin(\zeta + \psi) D(\psi \hbar\omega_c/\Gamma)\, d\psi, \tag{53}$$

where $D(Z)$ is the distribution of phase smearing, and $\Gamma/\hbar\omega_c$ characterizes the breadth of the phase smearing. The amplitude is reduced by a factor $|f(\Gamma/\hbar\omega_c)|/f(0)$, where $f(\Gamma/\hbar\omega_c)$ is the Fourier transform of D.

$$f(\Gamma/\hbar\omega_c) = \int_{-\infty}^{\infty} e^{-iZ} D(Z)\, dZ. \tag{54}$$

If it happens that D has Lorentzian form, then the amplitude is reduced by an exponential factor $\exp(-\Gamma/\hbar\omega_c)$ which is indistinguishable in form from the usual Dingle factor, except that the origin of the line broadening is phase smearing rather than electron lifetime. This resembles inhomogeneous broadening in NMR when the problem is simply to average over inequivalent orbits. However, when the disturbance varies *within* a given orbit, the problem is more complicated, and is then analogous to the problem of light scattering in an inhomogeneous medium. Watts (1974a) has used such methods to examine the limits of validity of the intuitive notion (Terwilliger & Higgins 1972) that small angle scattering and dephasing are equivalent. The phase-smeared dHvA amplitude was shown to be the wave amplitude which takes it back to a given starting point, and is thus equivalent to the forward scattering amplitude which determines the conventional scattering lifetime. Watts showed that dephasing methods are valid when:

(*a*) The disturbance changes slowly over a de Broglie wavelength $2\pi/k$;

(*b*) $\Gamma/\hbar\omega_c$ is small compared to the correlation length of the disturbance;

(*c*) The effective refraction index n must be small, where n is a measure of the sensitivity of electron wave-vector to a disturbance u: $n = d \ln k / d \ln u$.

It was shown (Watts 1974b) using these criteria that the dephasing is a valid approach to treat scattering from lattice dislocations, but is only qualitatively useful in the calculation of scattering from impurity strain fields.

Crystalline lattice defects reduce the dHvA amplitude, due to destructive interference of contributions to the observed signal from adjacent regions of crystal. With mosaic structure such as subgrains, adjacent portions of crystal are tipped with respect to the magnetic field direction. Since the extremal orbit cross-section is usually a function of field direction, adjacent portions contribute signals of slightly different frequency, which interfere destructively. Although F varies with angle rather slowly, the typically large value of F/H magnifies the effect, and an angular subgrain structure as small as 0.1° can seriously affect both amplitudes and phases. In the case of lattice dislocations, the local strain alters the local electron density, producing a spatially varying shift in Fermi surface cross-section, with accompanying destructive interference. For typical F/H values a strain of less than 0.1 per cent can seriously reduce the amplitudes. Similar effects occur due to impurity strain fields (Coleridge 1973a) but will be ignored in what follows because they do not alter the waveshape.

In impurity scattering studies, these effects produce an undesirable apparent background 'scattering'. Such effects have been shown (Terwilliger & Higgins 1972) to be responsible in *pure* crystals for Dingle temperatures ranging from 0.1 to several K, several orders of magnitude larger than expected from the residual impurity concentration. Since the Fermi surface sensitivity to strain or tilt varies with angle, these effects produce in addition a spurious *anisotropy* in the apparent scattering rate.

In g-factor and other experiments utilizing the harmonic content, lattice defects can produce even more serious errors, since the phase shift $\delta(F/H)$ is multiplied by the harmonic number r.

Careful studies of these effects, in which the actual crystalline substructure was independently characterized well enough to provide a correlation with dHvA observations, have provided the means of estimating the seriousness of these effects in other experiments. In addition, the extra information provided by the harmonic amplitudes and phases gives a quantitative means of correcting for substructure which has not been metallurgically well characterized. Finally, these tools are a powerful new way to characterize the microscopic disturbance of electronic structure due to lattice defects.

10.3.1 Crystal mosaic structure

The term mosaic structure is often used to indicate that the data looked peculiar and that the crystal was probably not perfect. A metallurgically more precise definition is that dislocations nucleating during crystal growth find a metastable configuration by gliding into arrays which form low-angle grain boundaries, within which the crystallites are often relatively perfect. Mosaic structure is characterized by the mean size of the crystal blocks and the mean tilt or rotation between adjacent blocks, and may be metallurgically characterized, for example, by measurement of the X-ray diffraction line-widths. Strains introduced by handling of specimens may alter this configuration, introducing lattice dislocations in the interior of the blocks, or even macroscopic bending over the length of the specimen. Dislocation effects will be considered in Section 10.3.2. Macroscopic bending can generally be avoided by careful handling and will not be treated here (see Shoenberg 1962). Two useful prototypes of mosaic structure will be considered: a bicrystal, consisting of two blocks tilted by an angle β, and a Gaussian spread of crystallites, with mean square tilt β. The discussion follows that of Shoenberg (1962) but is extended to include harmonic amplitudes and phases (Chang & Higgins 1975). For related examples, see Paton & Slavin (1973).

The resultant magnetization from a crystal with mosaic structure is reduced from the ideal value by

$$R = \iint D(\alpha, \phi) \exp(\mathrm{i}\psi_r) \, \mathrm{d}\alpha \, \mathrm{d}\phi. \tag{55}$$

Here, $D(\alpha, \phi)$ is the probability that a crystallite will be tipped by an angle α, ϕ from the mean direction, and ψ_r is the shift in dHvA phase resulting from that tip. Choosing the axes such that the field rotates in the plane defined by α, the phase shift is:

$$\psi_r = \frac{2\pi r}{H}\left[\frac{\partial F}{\partial \alpha}(\alpha - \alpha_0) + \frac{1}{2}\frac{\partial^2 F}{\partial \alpha^2}(\alpha - \alpha_0)^2 + \ldots + \frac{1}{2}\frac{\partial^2 F}{\partial \phi^2}\phi^2 + \ldots\right]. \tag{56}$$

Near a symmetry direction, the lowest order terms dominate:

$$\left.\begin{aligned} \psi_r &\approx ra(\alpha - \alpha_0) + rb(\alpha - \alpha_0)^2 + rb\phi^2, \\ a &\equiv 2\pi F'(\alpha_0)/H, \\ b &\equiv \pi F''(\alpha_0)/H. \end{aligned}\right\} \tag{57}$$

The net magnetization may be expressed in terms of the amplitude A_r of the dHvA oscillation and its phase shift $\delta\theta_r$ relative to the perfect crystal (for which $A_r = 1$ and $\delta\theta_r = 0$):

$$\left.\begin{aligned} R_r &= A_r \exp(\mathrm{i}\delta\theta_r), \\ A_r &= |R_r| = ((\mathrm{Im}\, R_r)^2 + (\mathrm{Re}\, R_r)^2)^{\frac{1}{2}}, \\ \delta\theta_r &= \tan^{-1}(\mathrm{Im}\, R_r/\mathrm{Re}\, R_r). \end{aligned}\right\} \tag{58}$$

A change in the relative phases $(r\theta_1 - \theta_r)$ occurs when the sine and cosine contributions s_r and c_r do *not* simply scale with harmonic number r, even though the dHvA phase shift ψ_r *does* usually scale with r.

Bicrystal

In this case, the distribution $D(\alpha, \phi)$ is a sum of delta functions separated by an angle 2β. Assuming for simplicity that the normals to the plane of tilt and the plane of H coincide (worst case) and that the crystallites are of equal volume, the amplitude and phase shift are given by:

$$A_r = \cos\left[\frac{r\pi F'(\alpha_0)\beta}{H}\right] \approx \cos\left(\frac{r\pi F''(\alpha_0)\beta}{G}\right). \tag{59}$$

Here, the second approximate form applies when α_0 approaches a symmetry direction; $\delta\theta_r = r\pi F''(\alpha_0)/4\beta^2 H$, and the relative phase shift is $r\delta\theta_1 - \delta\theta_r = 0$. The amplitude is reduced, with a field dependence which can cause an error in scattering temperature measurements. The effect is relatively worse in the higher harmonics ($r > 1$), and in severe cases a beat structure may be observed as a function of field. In noble metals at $H = 50$ kG, for example, with 'typical' values of $F' = 2 \times 10^6$ G deg^{-1}, (59) shows that a beat can occur in the fundamental with a subgrain tilt of only 0.01°! By contrast, when H is very near to a symmetry direction and the second form of (59) applies, the effect is far less serious. For a value of F'' of 4×10^8 G rad^{-2} (typical for the belly orbit in the noble metals), and with $H = 50$ kG, if the field can be aligned to within 0.1°, a substructure of up to 2° can be tolerated without a beat, and up to 0.5° without appreciable (10 per cent) alteration of the amplitude. It is for this reason that dHvA amplitude measurements are conventionally made very close to symmetry directions or turning points, where F' approaches zero.

The absence of a relative phase shift above is misleading, and occurs only in the special case of equal bicrystal volume fractions. The general case is complicated and not particularly illuminating. As an illustrative example, suppose that α is far from a symmetry direction, so that the dHvA phase shift is dominated by F', giving equal and opposite dHvA phase shifts in the two bicrystal contributions to the signal. The phase shift in the resultant signal does not cancel, however, and is

$$\delta\theta_r = \tan^{-1}\left[\frac{(2\eta - 1)\sin(ra\beta/2)}{\cos(ra\beta/2)}\right], \tag{60}$$

where η is the volume fraction of one subgrain. For $\eta = \frac{1}{3}$ and a dHvA phase shift $a\beta/2$ of $\pi/4$, the *relative* phase shift $\delta(2\theta_1 - \theta_2)$ is 54°, visibly altering the waveshape. Thus, the presence of a large relative phase shift when the field is tipped away from a symmetry direction may be used as an indicator of mosaic structure which is far more sensitive than the amplitudes themselves.

In addition, the r-dependence of the dephasing error makes measurements which depend upon harmonic amplitudes (e.g. g-factors) extremely sensitive to mosaic structure at field directions away from symmetry directions. The data themselves, however, can serve as a useful indicator. In the absence of other dephasing errors or of scattering from impurities with a magnetic moment, a conventional Dingle plot for various Fourier components should scale with harmonic number, yielding a harmonic-independent Dingle temperature. An example in a strained crystal is shown in Fig. 10.9. When the field is aligned along the symmetry

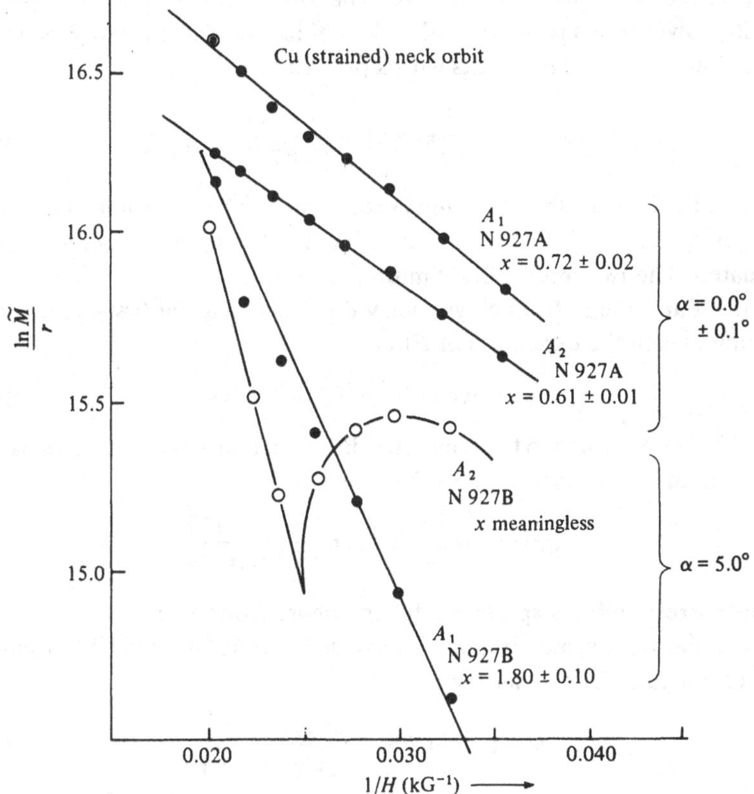

Fig. 10.9. Example of a Dingle plot in a crystal with mosaic structure. If first harmonic information were used alone, the data imply a large angular dependence on the Dingle temperature as H is tipped away from the symmetry direction. However, the lack of scaling of second harmonic Dingle plots, especially in the (lower) tipped example, reveals that the apparent increase in slope of the A_1 plot is a spurious result of mosaic structure dephasing.

direction, the two plots agree within experimental error. When the field is tipped by 5°, the first harmonic component shows a large increase in slope. The second harmonic component, however, shows a beat structure, which translates (using (59)) into a bicrystal mosaic structure of order 0.1°. This lack of agreement between first and second harmonic Dingle plots allows the conclusion that the apparent shift in x with tilt is not an intrinsic scattering effect, but is an error due to mosaic structure dephasing.

Gaussian spread

The case of a random distribution of mosaic blocks is more typical and more difficult to detect than a simple bicrystal. An X-ray Laue spot is blurred, rather than split, and the amplitude plots or rotation diagrams do not display obvious beat structure. The observation of a relative phase shift, however, is a powerful and under-utilized diagnostic tool. Assume a distribution of mosaic blocks with a probability:

$$D(\alpha, \phi) = \frac{1}{\pi\beta^2} \exp\left[-\frac{(\alpha-\alpha_0)^2}{\beta^2} - \frac{\phi^2}{\beta^2}\right], \tag{61}$$

where the line-width β of α and ϕ rotations has been assumed the same for simplicity. Using (55), the resulting amplitude may readily be calculated. The two most useful limiting cases are:

(a) Far enough from a symmetry direction that the first-order term dominates in the expansion of $F(\alpha)$:

$$A_r = \exp(-\tfrac{1}{4}a^2\beta^2) \quad (b\beta^2 \ll 1). \tag{62}$$

(b) Close enough to the symmetry direction that a parabolic expansion of F is adequate, such that $a = 2b\alpha$:

$$A_r = [1 + (b\beta^2)^2]^{-\frac{1}{2}} \exp\left[\frac{-b^2\beta^2\alpha^2}{1 + (b\beta^2)^2}\right]. \tag{63}$$

The corresponding expressions for the phase shifts are:

(a) Far from symmetry, $\delta_r = 0$ provided that $b\beta^2 \ll 1$ (which is usually not fulfilled). More generally,

$$\delta\theta_r = \tan^{-1}(rb\beta^2) - \frac{\frac{1}{4}(ra\beta)^2(rb\beta^2)}{[1 + (rb\beta^2)^2]}. \tag{64}$$

(b) Very near a symmetry direction ($a \approx 2b\alpha$)

$$\delta\theta_r = \tan^{-1}(rb\beta^2) - \frac{r(b\beta^2)^2(b\alpha^2)}{(1 + b\beta^2)}. \tag{65}$$

The *relative* phase shift changes from zero at the symmetry direction to a value which is non-zero (because $\delta\theta_r$ does not scale with r) and which can be quite large.

The consequences of such mosaic structure on scattering measurements were explored by Miller, Poulsen & Springford (1972). They compared the (field-dependent) error of the above case with the constant error of an assumed Lorentzian distribution of mosaic blocks. Both models' fit to experiment yielded roughly the same mosaic spread, with the field dependence of the Gaussian case not discernible over a typical measuring field range. They found that if enough orbits were available which were insensitive to mosaic structure (symmetry directions and turning points), it was possible to generate a Fourier expansion with which the data from other orbits could be corrected for mosaic background using a single parameter (the substructure angular width, which was not measured). Although this provided a self-consistent fit, it is not very direct, and it will now be shown that the added information present in the dHvA waveshapes can provide such a correction during the experiment, for the actual sample being measured.

The angular derivative F' and F'' for the case of Au are given in Fig. 10.10. The relative sensitivity to mosaic structure errors varies in this example by a factor of 10 from neck (least sensitive) to rosette (most sensitive). Since the F' and F'' values also vary greatly with field direction, the error due to mosaic structure can create a spurious anisotropy in any amplitude or phase-related quantity. The most reliable amplitude measurements are made at symmetry directions, where F' values in general pass through zero. This seriously restricts measurement of detailed *anisotropy* in scattering, g-factor, or other amplitude-related quantity. Calculated examples for a 1° tilt in Au are given in Table 10.2.

TABLE 10.2. *Examples of amplitude and phase errors due to a 0.05° Gaussian spread of mosaic structure in Au when $H = 50$ kG and is tipped 1° away from the principal symmetry direction of a given orbit*

Orbit	A_1	A_2	A_2/A_1	$2\theta_1 - \theta_2$ (deg)
Neck_{111}	0.996	0.988	0.991	3×10^{-3}
Belly_{111}	0.65	0.18	0.27	4.1
Belly_{100} (110)	0.74	0.30	0.41	3.1
Belly_{100} (100)	0.65	0.18	0.27	4.1
Rosette_{100} (110)	0.51	0.07	0.13	10.7
Rosette_{100} (100)	0.38	0.02	0.05	15.5

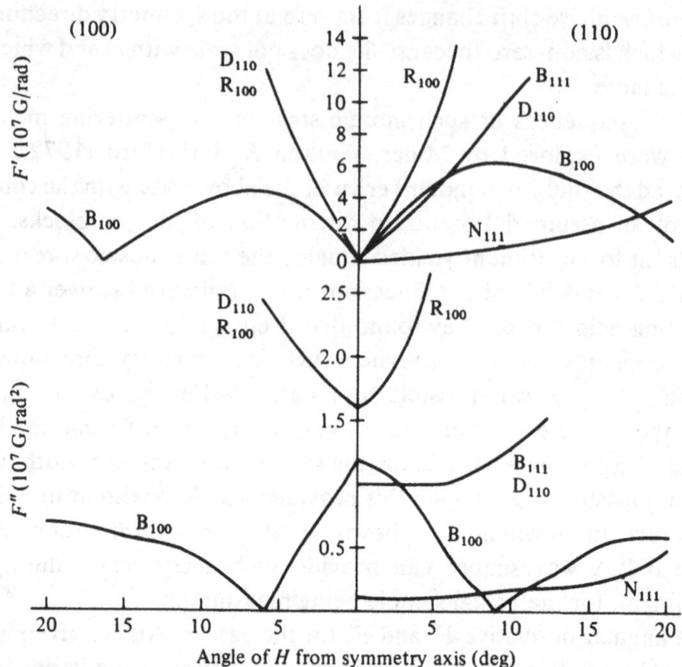

Fig. 10.10. Angular dependence of F' and F'' in Au, illustrating the large angular and orbital dependence of these quantities upon which the sensitivity to mosaic structure dephasing depends (B, R, D, N refer to belly, rosette, dogsbone and neck orbits respectively). (From Miller *et al.* 1972.)

The effect on the neck orbit is negligible, but in the case of the rosette, the fundamental amplitude is reduced by a factor of two and the ratio A_2/A_1 by a factor of 10, making scattering temperature measurements erroneous and g-factor measurements meaningless. Note, however, that in these cases a relative phase shift of order 10° is also present. This readily observable phase shift can be used as an indicator that mosaic structure errors in amplitude are present even in the absence of direct crystalline substructure measurements, which are difficult when the substructure is small (<0.1°).

Calculated examples of amplitude and phase errors as a function of tip are shown in Fig. 10.11 for the example of neck and belly orbits in Au, using (63) and (64) and the data of Fig. 10.9. These errors vary rapidly with angle, scaling with F' and F'' values. Note in the case of the [100] belly orbit an additional off-symmetry angle occurs at 17° at which both amplitude and phase measurements again become reliable. This 'turning point' angle occurs because F' passes through zero (Fig. 10.3). The close

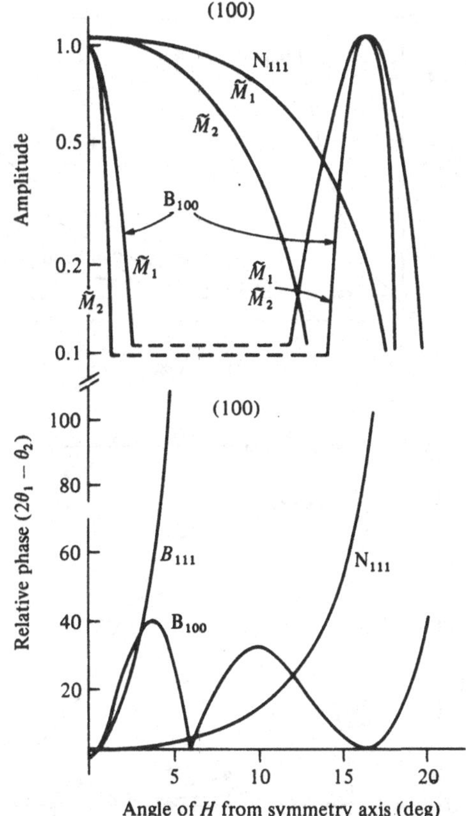

Fig. 10.11. Illustration of the correlation between conditions under which dHvA amplitudes become unreliable, and the (observable) relative phase shift, both due to mosaic structure dephasing. Example calculated for Au using the data of Fig. 10.10, with β (mosaic spread) $= 0.1°$ and $H = 100$ kG.

correlation in Fig. 10.11 between large amplitude reduction and large relative phase shift provides a useful reliability check on amplitude measurements.

We conclude this section with an example (Chang & Higgins 1975) demonstrating that all of these effects are observable and self-consistent. An example of the scattering temperature error in a strained but pure Cu crystal is given in Fig. 10.12(a). Here, with no other scattering centres present, the apparent scattering temperature as a function of angle of tip of H is consistent with a mosaic structure of width 0.1°. Observations of relative phase shifts in that same crystal are shown in Fig. 10.12(b). The values are large, approaching 100°, when the tip is only 5°, in spite of the fact that this orbit is the most *insensitive* to mosaic structure effects of any

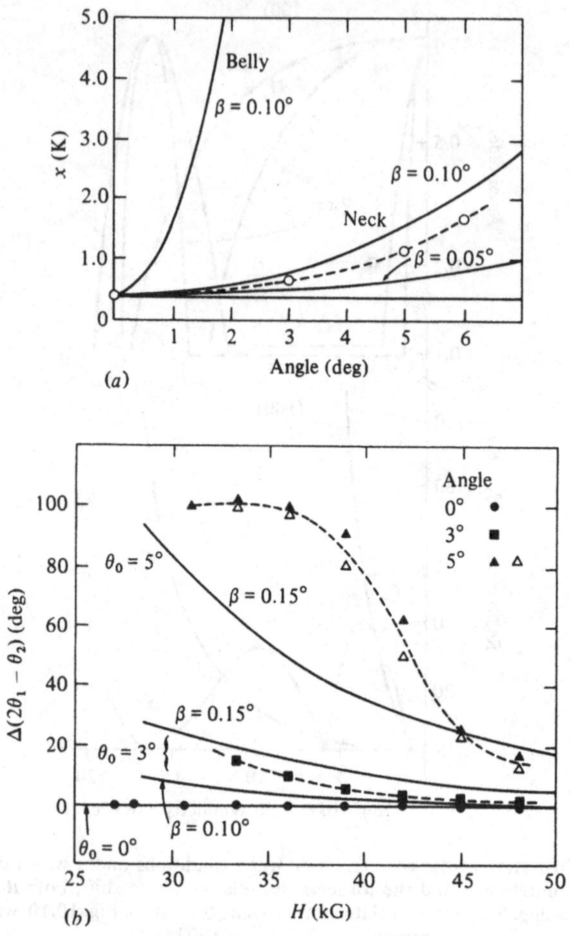

Fig. 10.12. (*a*) Illustration of the observed angular dependence of apparent Dingle temperature in a Cu sample (○), together with calculated values due to mosaic structure of width β. (*b*) Relative phase-shift measurements on the same sample as a function of H, together with calculated values due to mosaic structure. Note the large observable phase shift even when the apparent Dingle temperature errors are moderate. (From Chang & Higgins 1975.)

in the noble metals. The measurements are also consistent (calculated curves) with a mosaic spread of about 0.1°. A direct X-ray diffraction measurement also led to a 0.1° mosaic spread. Thus, the relative phase shift is a strong enough effect to distort the waveshape visibly, in contrast to amplitude measurements, where the effect causes a serious error but carries no characteristic 'signature' which allows it to be distinguished from 'genuine' scattering. Given the self-consistency between mosaic spread estimates from amplitudes, from phases, and from direct X-ray evidence in this carefully controlled example, we conclude that the relative phase-shift method can yield a diagnostic tool and a correction method for the measurement of amplitude-related quantities away from symmetry directions.

10.3.2 Lattice dislocations

An understanding of the interaction of conduction electrons with lattice dislocations is of intrinsic interest, for example, in understanding what limits the conductivity of deformed or irradiated metals, flux trapping in superconductors, and minority carrier lifetime limitations in semiconductors. Dislocations have also been shown (Terwilliger & Higgins 1972, 1973) to provide the practical lower limit to electron lifetimes measured in the dHvA effect. Recent experiments have provided the information needed not only to remove an undesirable background effect in dHvA amplitude measurements, but also to provide a detailed mapping of the **k**-dependent conduction electron–dislocation interaction (Chang & Higgins 1975). In the case of edge dislocations (the only example yet studied systematically) the symmetry of the defect leads to a vanishing dHvA phase shift. In that case, observation of a relative phase shift cannot be used as an indicator of background scattering due to dislocation substructure. Attention focuses instead on the harmonic amplitudes. It will be shown that the Dingle temperatures of the various harmonics need not be the same, and may in addition be field-dependent, because the comparable size of electron orbit and dislocation spacing leads to a geometric scale factor in the problem. As a consequence, the usual method of measuring the amplitude slope as a function of $1/H$ can be in error. The concept of a 'point' Dingle temperature at a single value of H is defined and related to the relative harmonic amplitudes. Such measurements have provided a convincing demonstration that small angle scattering on the quantized Landau level may be treated semiclassically by the 'dephasing' method.

Dislocations are a distinctly different case from mosaic subgrain structure because dislocations can be present in sufficient numbers to cause appreciable amplitude effects even when the crystal is free from metallographically visible subgrain structure. The effects of dislocations occur through the microscopic strain field shifting the quantum phase of an electron as it goes round its orbit. This is physically very different from the mosaic structure case, where the amplitude reduction is merely a consequence of the interference of the oscillatory signal from different crystallites of differing orientation. For examples of metallographic photos of these two cases, see Terwilliger & Higgins (1972), Fig. 10.3 (dislocations) or Chang & Higgins (1975), Fig. 10.2 (mosaic structure). Although it is tempting to make the analogy between homogeneous/inhomogeneous broadening as in NMR, the analogy is not appropriate, since the size of the Landau orbit is comparable to the dislocation spacing. A further complication is that the dislocation strain field is long range ($1/r$, unscreened), so that small angle scattering in fact is dominant, and essentially all scattering events 'count' in determining the dHvA lifetime (unlike the transport lifetime, where only large angle scattering counts). The most useful approach is the method of dephasing (equation (53)). A double average is involved, first *around a given orbit* centred at x_0, y_0,

$$\langle\psi_r\rangle = \frac{1}{2\pi}\int_0^{2\pi} \psi_r(x_0, y_0, \phi)\, \mathrm{d}\phi, \tag{66}$$

and then an average *over orbits* (now inequivalent):

$$\bar{M} = \frac{1}{\Omega}\int_{V_s} M_{0r} \sin(\zeta_r + \langle\psi_r\rangle)\, \mathrm{d}V. \tag{67}$$

The results of such calculations have been shown (Chang & Higgins 1975) to account for the anomalously large dHvA scattering due to dislocations, with the local dephasing given simply by the shift in the Fermi surface due to the local value of the dislocation strain. There are several useful applications of dHvA waveshape methods in this problem. First, this is a case where the Dingle temperature is field-dependent, so that the conventional technique of evaluating x from the slope of $\ln A$ versus $1/H$ is in error. Defining the 'point' Dingle temperature from the amplitude reduction at a single value of $1/H$, the connection with the 'slope' Dingle temperature $\bar{x}$ is:

$$\bar{x} = x + (1/H)\, \mathrm{d}x/\mathrm{d}(1/H), \tag{68}$$

so that the value of $\bar{x}$ may be greater or less than x depending upon whether x is an increasing or decreasing function of $1/H$ (conventional Dingle plots displaying negative or positive curvature, respectively). Since the experimental range of $1/H$ is seldom large enough to resolve the curvature accurately, the apparent ($\bar{x}$) values may be in error. An

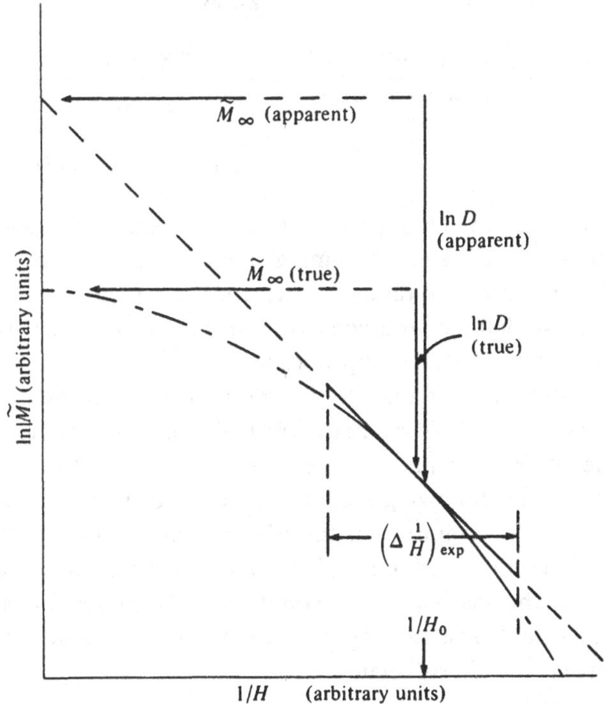

Fig. 10.13. Example where the conventional (slope) Dingle temperature is in error by a factor of two from the 'true' (amplitude reduction) Dingle temperature. The data, assumed known over a finite field range H_{exp}, is fitted to a straight line, yielding $\bar{x}$. If in fact the amplitude varies as a Gaussian function of $1/H$ (as it would, for example, with a Gaussian distribution of strain fields due to dislocations), the log plot in fact varies as a parabola, whose curvature is only barely visible over ΔH_{exp}. Defining the 'true' Dingle temperature in terms of the reduction in amplitude due to this scattering from the infinite field value, the error in using the slope method is a factor of two in this example, chosen such that the slopes of the apparent and true log plots are matched in the experimental field range. (Example chosen so that slopes match at $x_0 = 1/H_0$.)

example is shown in Fig. 10.13. The true scattering is assumed to result in a Gaussian variation of amplitude with $1/H$. Such a case results, for example, when the scattering is due to a Gaussian distribution of internal strain as from dislocations. If the data over the experimental field range are fit in the conventional way by a straight line fit to ln A versus $(1/H)$, an error amounting to a factor of two in x results in the example chosen, where the true and apparent slopes match over the experimental field range! Such an error may be avoided and the correct x value determined by a measurement of x at a *single* value of H using the relative harmonic amplitudes A_2/A_1, since (assuming g is known)

$$\frac{A_2\sqrt{2}\cos(\frac{1}{2}\pi g m_c/m)\sinh 2x}{A_1\cos(\pi g m_c/m)\sinh x}=\frac{\exp[-2K(m_c/m)x/H]}{\exp[-K(m_c/m)x/H]}$$

$$=\exp[-K(m_c/m)x/H]. \qquad (69)$$

In the example shown in Fig. 10.13, the relative harmonic ratios are very different in the two cases. Assuming a true x of 1 K, $m_c=m$, and $H=50$ kG, the harmonic ratios A_2/A_1 are 0.22 (Gaussian) and 0.049 (exponential), so that the observed harmonic ratio can readily distinguish which of these two models is appropriate.

However, this assumes that the Dingle temperature obtained in the first harmonic (x_1) and second harmonic (x_2) amplitudes is identical at a given value of H. This may not be true when the scattering is field-dependent, as in the following example. As pointed out by Watts (1974a), it follows from the definition of the dHvA phase of the rth harmonic $\zeta_r=2\pi rF/H$ that the value of the phase shift of the rth harmonic at H_0 is simply equal to the phase shift of the fundamental at H_0/n. It was inferred from this that the harmonic amplitude reduction (and hence the Dingle temperature) are similarly related:

$$x_r(H_0)=x_1(H_0/r). \qquad (70)$$

This result, which *invalidates* the determination of a 'point' x value from the relative harmonic amplitudes as in (69), is intuitively appealing. Viewed graphically, these are the field values at which the ratio of line-width (assumed field-independent) to field spacing (linear in H) of the various harmonics are equal.

However, this result is not correct when the orbit size is comparable to the defect spacing, and when the scattering centre has a long range. In such a case, there is a geometrical scale factor to the problem: (orbit size/defect spacing). Results of scattering calculations are shown to depend on the 'regime' of this parameter. An example for the case of lattice dislocations is shown in Fig. 10.14. The second harmonic amplitude reduction at H_0 is clearly not equal to the first harmonic amplitude reduction at $H_0/2$. This example, which has a divergence related to the matching of orbit size to defect spacing, rather dramatically demonstrates the failure of (70) when a geometrical scale factor is present. Physically, the *local* dephasing at a point r does scale with harmonic number, but the dephasing *integrated* around the orbit does not scale, since the orbit at

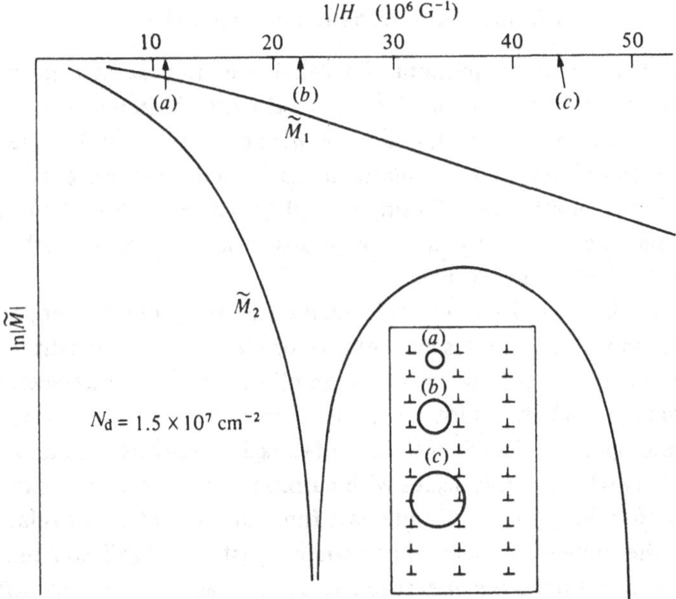

Fig. 10.14. Example of a dephasing situation where the harmonic amplitude reductions do not simply scale to those of the fundamental by equation (70) calculated for a dislocation lattice with density $N_d = 1.5 \times 10^7\ \mathrm{cm}^{-2}$, in a rectangular array whose sides are in the ratio 2 : 1. The second harmonic amplitude displays zeros corresponding to total phase cancellation at values of H which nearly correspond to geometrical matching of orbit diameter and closest dislocation spacing. If equation (70) were appropriate for this case, the fundamental would also vanish at $1/H = 2(1/H_0)$; no dramatic structure is apparent. Note that the values of the slopes do not approach one another until the infinite field limit where the orbit size is significantly smaller than the dislocation spacing.

$H_0/2$ has half the size it had at H_0, and samples a very different range of the spatially varying strain field.

In summary, the application of waveshape analysis methods to defect problems can avoid the errors of the conventional x measurement (when the scattering is field-dependent) by a variety of techniques. When the scattering is by point scatterers (large angle), the relative harmonic amplitude method is appropriate using (69). When the scattering is best viewed as dephasing (small angle) but with orbit size small compared to the defect spacing, the measured amplitudes of the harmonics at a single value of H can yield a map of the field dependence of the scattering, using (70). When there is a geometrical scale factor to the problem such that the orbit samples a range of disturbance to its phase which varies with H, neither equation is directly applicable, but the comparison of conventional $\bar{x}$ values with harmonic amplitude ratios can be used to test a model calculation of the scattering (Fig. 10.14).

10.3.3 Effects of field inhomogeneity

Field homogeneity requirements for reliable amplitude measurements are well known and need not be repeated here. Although the homogeneity measurements for reliable *harmonic* amplitudes simply scale with harmonic frequency, reliable *phase* measurements impose even more demanding requirements. Examples will be given to show how these effects may be used both as a diagnostic tool and as a method of measuring *absolute* amplitudes.

Paton & Slavin (1973) have explored the effects of field inhomogeneity on the harmonic amplitudes and relative phases. They demonstrate that relative phase errors show up before amplitude errors are detectable in the curvature of a Dingle plot. Harmonic amplitudes are more sensitive to field inhomogeneity than the fundamental, as demonstrated in Fig. 10.15, but in a way which simply scales with harmonic number r. In an example (Fig. 10.16) where the amplitude as a function of $1/H$ looks relatively normal (the slope of the second harmonic is just barely shifted from that of the fundamental), the measured relative phases show dramatically large and field-dependent errors. Used as a diagnostic tool, the relative phase has the advantage of being measurable within a single data window, whereas looking for such errors in non-linearities of a Dingle plot requires a substantial field range.

Hörnfeldt, Ketterson & Windmiller (1973a) have treated the amplitude reduction due to field inhomogeneity in a similar way. In addition, they show that a new mixing effect occurs when observing high-frequency oscillations F_2 in the presence of a large-amplitude low-frequency oscillation F_1. If the variation in applied field over the sample is ΔH, the

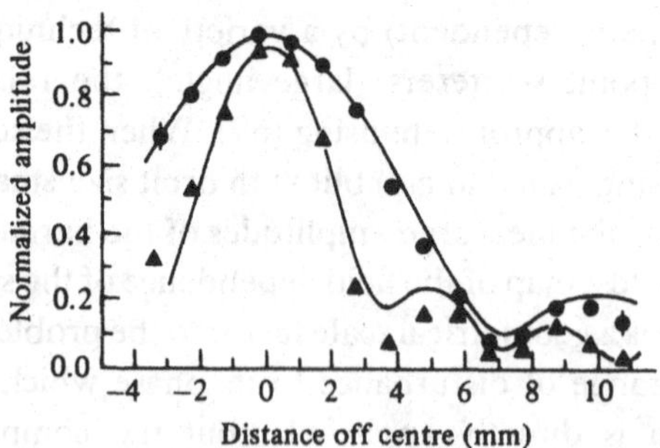

Fig. 10.15. Illustration of amplitude errors due to field inhomogeneity for first (●) and second (▲) harmonic signals, together with calculated values. (From Paton & Slavin 1973.)

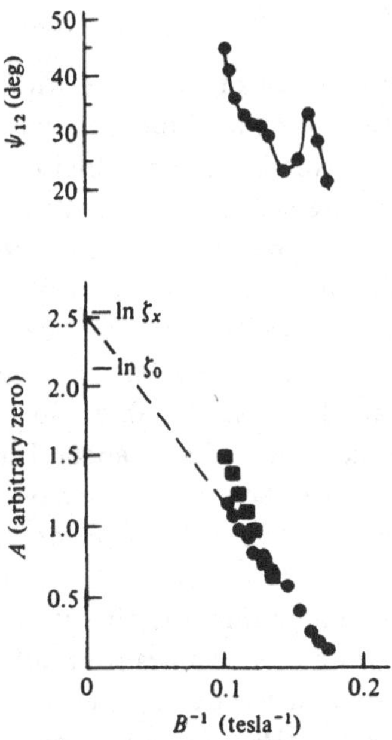

Fig. 10.16. Effect of field inhomogeneity on the Dingle plot (lower curve) and relative phase shift (upper curve) for a [111] neck oscillation in Cu. The second harmonic Dingle plot (■) appears reasonably linear although of different slope from the first (●). The relative phase shift is large, and a sensitive indicator of field inhomogeneity.

variation in magnetic induction is

$$\Delta B = (1 + 4\pi \, \mathrm{d}\tilde{M}_1/\mathrm{d}H)\,\Delta H. \tag{71}$$

As a result, as the dominant oscillation $\tilde{M}_1$ goes through one cycle, the field inhomogeneity seen by electrons generating $\tilde{M}_2$ successively increases ($\mathrm{d}\tilde{M}/\mathrm{d}H > 1$) and decreases ($\mathrm{d}\tilde{M}/\mathrm{d}H < 0$). If the field inhomogeneity is sufficiently strong to cause an amplitude reduction, then the high-frequency oscillations will be modulated in amplitude by the low-frequency term. This occurs in the absence of MI (Section 10.4). In cases where the inhomogeneity can be measured and controlled, the depth of the modulation provides an unambiguous method of absolute amplitude calibration (Alles & Lowndes 1973). This has formal similarities with the same kind of mixing encountered with MI but is useful as a calibration technique even when MI is absent. Such mixing effects seriously alter the measured relative phases, and must be well understood and separated in the spectrum before a relative phase measurement is meaningful.

10.4 g-factors in pure metals

Within the one-electron theory of metals, g-values provide the most direct information on the strength of the (relativistic) spin–orbit interaction. Although such information can be obtained from Fermi surface measurements, the difference between a relativistic and a non-relativistic band-structure calculation is typically small. By contrast, g-shifts of up to 100 have been observed in orbits passing near regions of accidental degeneracy in the band structure, and a factor of 2 shift is not uncommon. In addition, g-measurements provide a fertile testing ground for many-electron effects in metals. Because the observable is the product $m_c g$, electron–phonon renormalization effects cancel (Engelsberg & Simpson 1970), providing a direct handle on other many-body effects such as electron–electron interactions, and electron–exchange enhancement. Because such effects appear differently in other methods of measuring g (for example, exchange enhancement is absent in conduction electron spin resonance) a comparison of g-values from different measurements provides a useful separation of competing effects. Finally, the richness of detail present in the orbital dHvA g-measurements is lost in all other methods of measuring the spin susceptibility. As a result, the comparison of g-values measured by other methods is most clear cut in cases where the distinction between orbital and Fermi surface average g-values is not important, i.e. the alkali metals. (For a comparative review of other methods, see Knecht 1975.)

What is measured by the ratio of the dHvA spin-splitting to Landau level splitting is in fact an orbital average weighted by the Fermi velocity $v_{\mathbf{k}}$

$$g_c = \frac{\oint g(\mathbf{k}, \mathbf{B}) v_{\mathbf{k}}^{-1} \, \mathrm{d}k}{\oint v_{\mathbf{k}}^{-1} \, \mathrm{d}k}. \tag{72}$$

Although this resembles the dHvA orbital relaxation time in that the result is weighted by the time spent at different points along the orbit, it differs in that g is a function both of $\mathbf{k}$ and $\mathbf{B}$, and thus is a tensor quantity. As a result, the geometrical inversion schemes which have been so useful in deducing relaxation time maps fail in the case of g-factors. Relatively little work has been done as yet in metals in exploring the implications of the tensor nature of g.

The discussion which follows begins with a comparison of the various dHvA methods for measuring g-factors. The discussion will focus on methods which make use of the relative amplitudes of cos $(\pi r S)$ in (1). We will ignore those few cases where the harmonic content is so strong that direct spin-splitting may be used to obtain g (for example, Stark (1964) in Zn; Hill & Vanderkooy (1978) in Sb; Takano & Koga (1977) in Bi). A survey is then given of the g-value results to date in the noble metals, alkali metals, polyvalent metals, and transition metals. The results are interpreted briefly in terms of spin–orbit coupling and many-body interactions.

10.4.1 Comparison of dHvA methods of measuring orbital g-factors

dHvA g-factor methods fall into roughly four categories:

1. *Spin-splitting zeros* where the observable is the angle of the magnetic field (relative to the crystal axes) at which the amplitude of the fundamental or of a particular harmonic passes through zero.

2. *Absolute amplitude measurements*, in which g is inferred from the value of $|\cos \pi S|$. The absolute system gain must therefore be calibrated. Although this is relatively easy in the torque method, most g-measurements have made use of the field modulation technique, because of its ability to enhance harmonic content. Calibration of the field modulation method to the desired (1 per cent) accuracy is more difficult, and is made either directly or by an internal calibration using MI.

3. *LK harmonic content methods* utilize a ratio of relative harmonic amplitudes, after a correction is made for MI.

4. *The MI mixing method* makes use of the relative amplitude of sideband terms compared to the fundamental when non-linear mixing occurs.

These methods are compared in Table 10.3. The methods differ not only in what is measured, but also in the additional information needed before g can be inferred. Depending upon the method, the additional information needed may include the Dingle temperature x_D, the curvature factor $d^2\mathscr{A}/dk_H^2$, or the demagnetization factor N. As shown in Table 10.3, the methods differ significantly in their accuracy, ability to make measurements at a general direction of H, independence of system gain calibration, independence of other Fermi surface information, independence of the Dingle factor, independence of skin-effect complications, and independence of MI dominating the signal. Note also that the methods differ in the ambiguity in the value of S determined from the observed quantity. Methods which depend upon a measurement of $\tilde{M}_1$ have four values of S in the basic interval $0<S<2$ consistent with a given measured value of $|\cos \pi S|$. This may be reduced to 2 (S and $2-S$) if the

TABLE 10.3. *Comparison of dHvA g-factor methods*

Method	Measured quantities	Also needed: x_D	Also needed: $\frac{d^2\mathscr{A}}{dk_H^2}$	Also needed: N	Advantages (+)/disadvantages (−): Accur.	Any θ	Indep. of sys. gain	Indep. of Fermi surface info.	Indep. of Dingle factor	Indep. of skin effect	Insens. to MI dominance	Multiplicity $0<S<2$
1. Spin-splitting zeros	$\tilde{M}_1=0$ $(A_r=0)$				+	−	+	+	+		+	4
2. Absolute amplitude												
(i) abs. calib.	$\tilde{M}_1$	x	x		+	+	−	−	−	−	+	4
(ii) MI calib.	$\tilde{M}_2/\tilde{M}_1$, $2\theta_1-\theta_2$			x		+	+	+	+	(+)	−	4
3. Harmonic ratio												
(i) 2nd harm.	$\tilde{M}_2/\tilde{M}_1$, $2\theta_1-\theta_1$	x				+	+	+	−	+	−	8
(ii) 3rd harm.	$\tilde{M}_3/\tilde{M}_1$, $\tilde{M}_2/\tilde{M}_1$ $3\theta_1-\theta_3$ $2\theta_1-\theta_2$					+	+	+	+	+	−	≤12
4. MI mixing sidebands	$\frac{\tilde{M}_a \pm b}{\tilde{M}_a}$				−	+	+	+	+	+	−	4

absolute ($H = \infty$) phase can be determined (Templeton 1972). Measurements which depend upon $\tilde{M}_2$ have an additional multiplicity (Fig. 10.4(*b*)). However, this can be reduced by the use of MI to determine the sign of $\cos 2\pi S$ (Fig. 10.7). As a final qualitative distinction, the methods may be compared as follows:

(*a*) The spin-splitting zero method is potentially the most accurate, but also the least general method.

(*b*) The absolute amplitude method using a calibrated system needs the most additional information, but is not subject to competition from MI, as long as T and H may be adjusted so that the MI contribution to the *fundamental* amplitude is negligible even though MI contributions to the *harmonics* are substantial. Hence, this method is the most general in the sense that all Fermi surface orbits are candidates for g-measurement *only* by this method.

(*c*) Methods which use a ratio of harmonic amplitudes need the least additional information, but are the most subject to being overwhelmed by MI. For example, the ratio $A_2^{\mathrm{MI}}/A_2^{\mathrm{LK}} \propto F^2T/H^{\frac{5}{2}}$. Thus, relatively few orbits ($F \lesssim 10^8$) are candidates for g-measurement by these methods.

Note that it is straightforward to use several of the methods simultaneously in the same experiment. Since experimental errors such as skin effect and phase smearing appear differently in the different techniques, the agreement of results by different methods can be used as a check on the reliability of any one. Similarly, a method which relies upon an imperfectly known quantity (e.g. $\mathrm{d}^2\mathcal{A}/\mathrm{d}k_H^2$ is required for the absolute amplitude method) may be checked using a harmonic ratio method which is independent of that quantity. The characteristic strengths and weaknesses of each of the methods will now be discussed.

Spin-splitting zeros

This method is the most accurate in the sense that it only requires a measurement of the angle at which a zero occurs, and the uncertainty in g is limited only by the accuracy with which the value of m_c is measured (or interpolated) at that same angle. It is also the least general method, since spin-splitting zeros are accidental, occurring at most on a few contour lines on the unit sphere of $\hat{H}$. The potential accuracy may be unrealized, since m_c may be rapidly varying at these arbitrary directions, and is less well known than at symmetry directions. Note that additional spin-splitting zeros occur when $A_r = 0$ ($r \geq 2$) which is quite observable (at least for A_2) using waveshape techniques, although these higher harmonic zeros are made imperfect by MI.

Absolute amplitude via a calibrated system

Here, there are two reliable system calibration methods. The first (Knecht 1975) relies on the fact that the sensitivity of a pickup coil to a sample's magnetization is related to the spatial variation of magnetic field produced by that same pickup coil when it carries a current. The dHvA sample itself is used to calibrate the coil. One measures the current required in the modulation coil to bring the dHvA signal seen in the pickup coil to the first Bessel function zero. Then, the roles are reversed: a current is put in the pickup coil, and one looks for the dHvA signal in the modulation coil, finding the current at which the first Bessel function zero occurs. The required calibration constant can be shown to be (D. Shoenberg, private communication)

$$V/M_0\omega(2\pi F/H_0^2)i \approx \Omega/(i_{01}i_{02}\cos\xi). \tag{73}$$

Here, V is the voltage seen in the small modulation limit for current i in the modulation coil, i_{01} and i_{02} are the currents at which the Bessel zero appears in the normal and reversed configuration, ξ is the angle between $\mathbf{h}$ produced by the pickup coil and the external field $\mathbf{H}$, and Ω is the volume of the sample. The result is accurate to 0.1 per cent for a coil inhomogeneity of up to 10 per cent over the sample's dimensions. The method is conceptually elegant, although it requires a pickup coil–modulation coil arrangement quite different from that normally employed.

A second absolute calibration method utilizes the fact that at a first harmonic spin-splitting zero, the cosine spin-splitting factor in M_2 equals 1. A measurement of the $\tilde{M}_2$ voltage signal, together with calculated values of $d^2\mathcal{A}/dk_B^2$ and the measured Dingle factor, give an absolute system calibration, which may be checked against a calibration at symmetry directions utilizing g-values from harmonic ratios. Note that the vector magnetization (generally not parallel to the pickup coil axis) makes an important correction in the amplitudes away from symmetry directions (Crabtree, Windmiller & Ketterson 1977) and, in addition, uncertainties about possible angular dependence of the coupling between pickup coil and modulation coil must be carefully eliminated before this method becomes reliable. Finally, this method is the only one in which the skin effect can produce appreciable first-order errors. Nonetheless, this method has strong merits as a general technique free from the complications of MI.

Second harmonic amplitude ratio and relative phase: MI and LK projection

These two methods come out of the same observables, the relative harmonic amplitudes $|\tilde{M}_2/\tilde{M}_1|$ and relative phase $2\theta_1 - \theta_2$, together with the projection technique (equation (51) or (52)). The projection of the MI component yields

$$\tilde{M}_2^{\text{MI}}/\tilde{M}_1^{\text{LK}} \propto \kappa (A_1^{\text{LK}})^2/A_1^{\text{LK}} = \kappa A_1^{\text{LK}} \propto \cos \pi S, \tag{74}$$

whereas, the projection of the LK component yields the ratio

$$|\tilde{M}_2^{\text{LK}}/\tilde{M}_1^{\text{LK}}| \propto \cos 2\pi S/|\cos \pi S|, \tag{75}$$

where it is assumed that the sign of the numerator can be determined using the known phase of the MI component (see Fig. 10.7). The two methods provide a complimentary check on one another, since the Dingle factor needed to evaluate g from the LK projection (equation (75)) cancels in the MI component (equation (74)). However, calculation of g from the MI component requires a calculation of the demagnetization factor N, which can be awkward (Crabtree 1977), and is subject to a second-order error due to skin effects (Alles *et al.* 1975).

Third harmonic amplitude ratio

Here, use is made (Gold & Schmor 1976) of the ratio $\tilde{M}_2^2/\tilde{M}_3\tilde{M}_1$ which is unique among the harmonic ratios listed in Table 10.1 in requiring no additional information to determine a g-value. This ratio in the LK limit is given by

$$A_2^2/A_1A_3 = \tfrac{3}{4}\alpha_\infty(1 + \tfrac{1}{3}\tanh^2(Km_cT/m_0H)), \tag{76}$$

where α_∞ is related to S by

$$\alpha_\infty = (\surd 3/2)(1 - \tan^2 \pi S)^2/(1 - 3\tan^2 \pi S). \tag{77}$$

The relationship between α_∞ and S is shown in Fig. 10.17. A given value of α_∞ is consistent with as many as three possible values of S in the intervals $0 < S < 0.5$. It is most convenient to determine α_∞ by some algebraic manipulations to eliminate the explicit dependence upon T

$$\tilde{M}_1/\tilde{M}_3 = \alpha_\infty[(\tilde{M}_1/\tilde{M}_2)^2 - \tfrac{1}{4}(\tilde{M}_1/\tilde{M}_2)_0^2],$$

where

$$(\tilde{M}_1/\tilde{M}_2)_0 = 2\surd 2 \exp(Km_cx/m_0H)\cos \pi S/\cos 2\pi S. \tag{78}$$

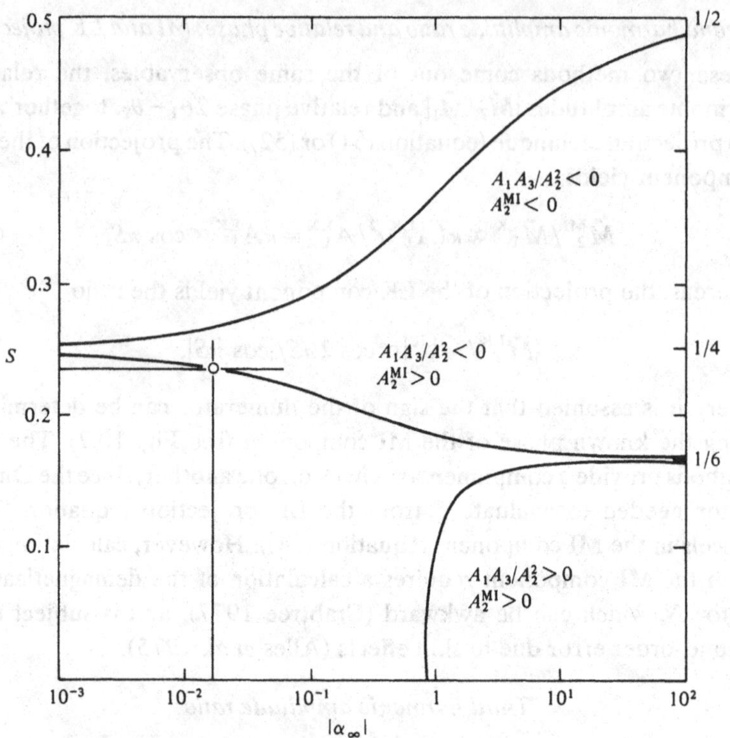

Fig. 10.17. Variation of S as a function of $|\alpha_\infty|$. The solution traverses one of three branches depending upon the sign of A_1A_3/A_2^2 and on the sign of A_2^{MI}. Note that this graph covers only $0 \le S \le 0.5$, with equivalent solutions of the form $\pm p \pm S$, where p is an integer. (From Gold & Schmor 1976.)

In practice, H is held constant and a plot of $\tilde{M}_1/\tilde{M}_3$ as a function of $(\tilde{M}_1/\tilde{M}_2)^2$ yields a straight line of slope α_∞.

The data must first be corrected for MI. In the weak MI limit, an extension of the projection technique (Section 10.2) to include third harmonic information allows the LK components to be calculated from the observed amplitudes and relative phases:

$$\left.\begin{aligned}\tilde{M}_1^{LK} &= \tilde{M}_1^{OB},\\ \tilde{M}_2^{LK} &= \tilde{M}_2^{OB}\,(2)^{-\frac{1}{2}} \sin(2\theta_1 - \theta_2),\\ \tilde{M}_3^{LK} &= \tilde{M}_3^{OB}\,[\sin(3\theta_1 - \theta_3) + 3(2)^{-\frac{1}{2}}(\tilde{M}_2^2/\tilde{M}_1\tilde{M}_3)^{OB} \sin^2(2\theta_1 - \theta_2)].\end{aligned}\right\} \quad (79)$$

Although it suffers from a worse multiplicity than any of the other methods, and is restricted to situations where enough third harmonic content is available to measure, this method appears to have advantages when for some reason the others are excluded by difficulties in measuring x, $d^2\mathscr{A}/dk_H^2$, or $\mathscr{D}$.

MI mixing sidebands

In the weak MI limit, the ratio of sideband amplitude to carrier amplitude in an MI mixing spectrum is a measure of the absolute amplitude of the signal causing the modulation. Crabtree (1977) has shown that when the MI mixing analysis is extended to include demagnetization effects the result is that

$$\frac{\tilde{M}(\text{sideband})}{\tilde{M}(\text{carrier})} = \frac{\tilde{M}_{a\pm b}}{\tilde{M}_a} = \frac{(\kappa'_a \pm \kappa'_b)\tilde{M}_a\tilde{M}_b}{\tilde{M}_a} = (\kappa'_a \pm \kappa'_b)\tilde{M}_b, \tag{80}$$

where κ'_a and κ'_b are individual values of κ' (equation (40)) for frequencies F_a and F_b, which interfere to generate sum and difference components. Thus, if κ'_a and κ'_b can be calculated reliably, the resulting absolute amplitude can be used to infer a g-value without the need for an absolute system calibration.

At first sight, this method appears particularly useful when MI mixing makes the determination of relative phases difficult, eliminating methods 2, 3 and 4 above. However, the analysis summarized by (80) ignores a serious complication which phase smearing may produce in the MI mixing spectrum. Although it is customary to calculate MI harmonic content using an experimentally measured Dingle temperature, Shoenberg (1976) pointed out that this procedure is probably in error in pure crystals which have no inherent scattering. In such a case, the observed Dingle factor is due to some kind of dephasing, and if the scale of this is macroscopic (as in the case of crystalline mosaic structure), then the *micro*scopic magnetization per unit volume seen by an electron is probably considerably larger than that calculated from a measured Dingle factor. Although formidably difficult to deal with in the absence of well-characterized mosaic structure, Shoenberg showed that a new procedure for calculating MI harmonic content is more consistent with the self-modulation harmonic distortion actually observed than the previous method. As one measure of MI mixing, for example the size of the observed frequency modulation due to MI, in a carefully calibrated apparatus, exceeds that predicted by the old method by more than 50 per cent, but agrees to within better than 10 per cent with a calculation using the new method. Shoenberg's conjecture has been independently confirmed recently (Everett & Grenier 1977, 1978) in Pb. Using a ellipsoidal sample such that the size of the MI could be varied in crystallographically equivalent directions, with all other parameters being held constant, they showed that the size of the observed frequency modulation due to MI was consistently at least 25 per cent too large, compared with that calculated, unless this phase smearing correction was properly taken into account.

Although these experiments deal primarily with the modification, due to phase smearing, of the frequency modulation due to MI, it is clear that phase smearing will have similar effects on the MI mixing spectrum so that the use of MI mixing sidebands to determine absolute amplitudes (following (80)) would appear to be unreliable except with nearly perfect samples.

10.4.2 Summary of g-factor results

Alkali metals

The alkali metals, though an experimentalist's nightmare because of their reactivity, are a theorist's dream because they are the closest thing in nature to a free electron metal. Band-structure effects are relatively unimportant, electron–phonon renormalization is small (<10 per cent) and exchange interaction negligible, and thus this family is the ideal testing ground for theories of the electron–electron interaction. The most complete set of measurements are those of Knecht (1975). Three independent measurement methods were used: the LK harmonic ratio, absolute amplitudes by MI projection, and absolute amplitudes by direct calibration. Although differing greatly in their precision (the first being the most precise, and the last the least precise), this resulted in the opportunity for cross-comparison, and therefore increased reliability in view of uncertainties in the crystal perfection in these metallurgically uncharacterized samples. The measurements were extended to Na by Perz & Shoenberg (1976), who were able to cool to low temperatures in such a way that most samples did not undergo a phase transition. The MI harmonic content was large enough so that the expansion to $O(4)$ of Table 10.1 (p.415) was not adequate. Instead, they assumed a value of S and then calculated $\tilde{M}_r$ and θ_r (extending the Phillips and Gold expansion to eighth order) and adjusted S to fit the observed relative phases. A summary of the results for the g-factor and the corresponding spin susceptibility is shown in Table 10.4. The dominant feature is a large (30–40 per cent) positive enhancement, together with a small (5 per cent for Cs) systematic increase with increasing atomic number.

TABLE 10.4. *g-factors and spin susceptibilities for the alkali metals (Knecht 1975; Perz & Shoenberg 1976)*

	Na	K	Rb	Cs
g	2.636 ± 0.024	2.800 ± 0.011	2.83 ± 0.05	$\{2.44 \pm 0.11$ or $3.11 \pm 0.12\}$
χ	1.632 ± 0.007	1.700 ± 0.006	1.724 ± 0.010	$\{1.76 \pm 0.06$ or $2.24 \pm 0.06\}$

Noble metals

An early survey of g-factors in noble metals was done by Randles (1972), using spin-splitting zeros, and (for the neck orbit) LK harmonic ratios, with MI removed by the projection technique. There is evidence of a systematic trend for the neck orbit with increasing atomic number, as shown in Fig. 10.18. (Here, the alternative values $1 \pm S$ have been rejected as a result of the infinite field phase measurements of Coleridge & Templeton (1972).) The values of S and of g_c display a systematic shift

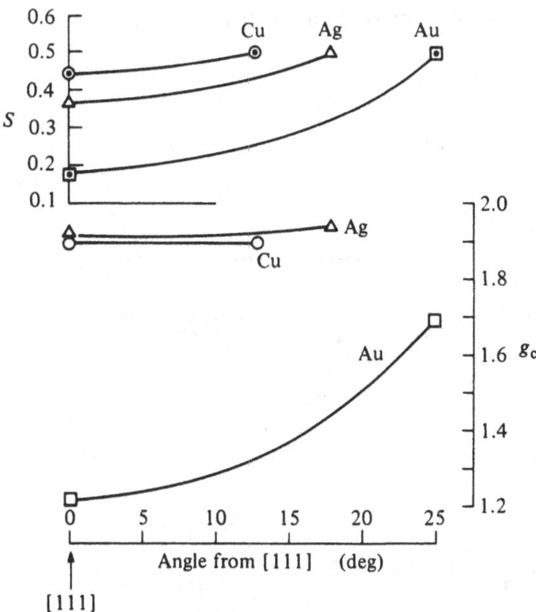

Fig. 10.18. Systematic variation of S and of g for the neck orbit in the noble metals, as a function of magnetic field direction, showing the systematic negative shift and increasing angular dependence with increasing atomic number. Data taken from Randles (1972). The data at [111] are the results of harmonic analysis; other points are spin-splitting zeros.

downwards and an increase in angular dependence with increasing atomic number.

The most complete comparison of results with different methods and in different laboratories is available for Au. The more recent measurements include: Alles *et al.* (1975) (neck by LK harmonic ratio; dogsbone by MI projection, and by spin-splitting zero); and Crabtree *et al.* (1977) (belly, dogsbone and rosette by absolute amplitude). The results are summarized in Fig. 10.19. The values shown for Au are those of Crabtree, Windmiller & Ketterson (1975, 1977). The corresponding values of S are in agreement to within the size of the data points shown with the neck and dogsbone results of Alles *et al.* (1975). Shown for comparison are the results in Cu of Randles (1972). For Au, the nearest alternative value of g on the neck is larger than 13, and is not shown. Alternative choices in Au for the other orbits lead to unphysically large angular dependences. In addition, these orbits represent the major portions of the Fermi surface, and the choice of g shown displays a positive g-shift from 2.00, in agreement with the bulk averaged results of CESR (Dupree, Forwood & Smith 1967; Monot, Chatelain & Borel 1971). The alternative choices (Crabtree *et al.* 1977) are all significantly less than 2.0. These results demonstrate for Au a substantial (15–20 per cent) positive

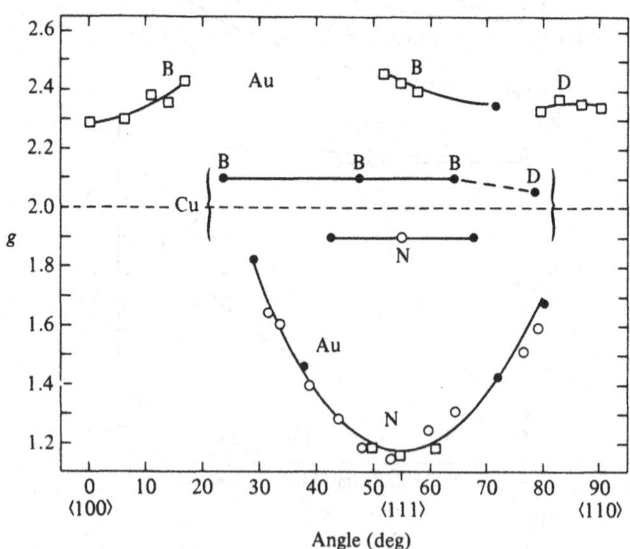

Fig. 10.19. Orbital g-factor map in Au (adapted from Crabtree *et al.* 1977). Data from Randles (1972) for Cu are added for comparison. Points labelled □: absolute amplitude; ○: harmonic ratio; ●: spin-splitting zero.

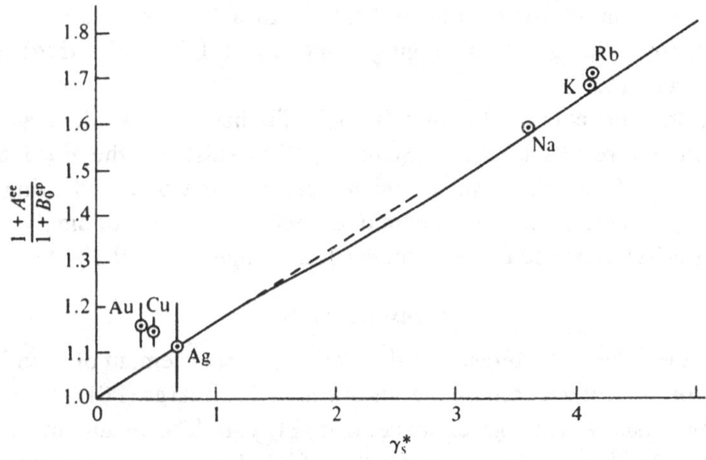

Fig. 10.20. Variation of Fermi liquid parameters $(1 + A_1^{ee})/(1 + B_0^{ep})$ (proportional to χ) as a function of the (renormalized) interelectronic distance r_s^* (in Bohr radii). Curve: theory; points: experimental data in alkali and noble metals. (From Bibby & Shoenberg 1977.)

g-shift over the major portions of the Fermi surface, and an even larger (factor of two) negative g-shift for the neck. Note in Fig. 10.20 that the size of both the positive shift for large orbits, and of the negative shift for the neck orbit, is significantly larger in Au than in Cu.

Polyvalent metals

Here, the data are as yet fragmentary. The principal interest is in the elucidation of spin–orbit coupling strengths. The interpretation of the data is complicated by the fact that no general theory exists, and a given g-value requires for its interpretation an explicit band calculation. However, this is offset by the fact that in this family of metals, explicit pseudopotential band calculations are straightforward and very accurate. The preliminary and fragmentary data indicate that large g-shifts exist, and the field is ripe for further exploration.

The oldest example is that of the needle in zinc, where a g-value of 170 (alternative values 98 or 360) was observed (Stark 1964). The choice of 170 was inferred from the pressure and temperature dependence of the infinite field phase (O'Sullivan & Schirber 1966).

Among the trivalent metals, there appear to be no g-measurements on Al as yet, in spite of the intriguing CESR results (Lubzens, Shanaburger & Schultz 1971) which are most plausibly interpreted as implying a large g-factor anisotropy. Preliminary measurements on In (D. H. Lowndes &

Y. Chung, unpublished data) indicate both a large shift and a large anisotropy, with g-values ranging from 3.2 to 1.7 on the third zone electron surface.

Among the column IV metals, only Pb has been studied as yet. Preliminary results (Gold & Schmor 1976) exist for the third zone electron surface. The results, which were the first obtained using the $\tilde{M}_2^2/\tilde{M}_3\tilde{M}_1$ amplitude ratio technique, include a correspondingly large multiplicity; there are ten possible g-values ranging from 0.384 to 7.746.

Transition metals

Here, the principal interest is in the exchange enhancement of g, and it is therefore useful to compare metals in which this is large (Pd) with others having smaller exchange enhancement (Pt, Rh). The results in Pt and Pd to date (Ketterson & Windmiller 1970; Windmiller, Ketterson & Hörnfeldt 1970) are based upon the spin-splitting zero technique, and indicate a large anisotropy (Table 10.5). However, the interpretation of the Pt and Pd data at that time was based upon the assumption that among the possible values of g, the one closest to 2 should be selected. This is by no means clear in view of the large exchange enhancement in Pd, and it would be interesting to reinterpret those data, together with reducing the multiplicity of possible g-values by using infinite field techniques, wherever possible.

TABLE 10.5

	g_c	
Sheet	Pt	Pd
Γ electrons	1.61–2.13	2.13–2.44
Open hole sheet	1.65–2.22	1.8–1.9

In Rh, measurements have been made (Cheng *et al.* 1979) using the LK projection technique on several of the smaller hole surfaces. The results are given in Table 10.6, where the two smallest possible g-values are listed.

TABLE 10.6

Surface	m_c/m	g_c
L holes (H [111])	0.132 ± 0.002	5.13 ± 0.03, 9.48 ± 0.03
X holes (H [100])	0.453 ± 0.03	1.78 ± 0.01, 2.67 ± 0.01

10.4.3 Interpretation of *g*-factor results

Nature has been kind in providing a useful separation, among the families of metals, of the effects which one hopes to study. The *alkali* metals are the natural candidates for the study of electron–electron interactions, since exchange, electron–phonon, and band-structure effects are small or negligible. The *transition* metals are the natural family in which to study exchange interactions, because of the unfilled d-band with its large $N(\mathscr{E})$ and correspondingly large Stoner enhancement, both of which vary across the d-series. By selecting non-superconducting materials, electron–phonon effects are probably small, though perhaps not negligible (e.g. Pd). A complication for the smaller sheets of Fermi surface in which g is most easily measured, is possible large spin–orbit coupling shifts. The *polyvalent* metals are the natural family in which to study spin–orbit coupling because the d-shell is filled, and exchange enhancement is negligible. The electron–phonon interaction is small in the *noble metals*, though it may be large in polyvalent metals, but requires at most a renormalization with the (observable) cyclotron effective mass m_c, as we shall see. The comparison of orbits which spend most of their time at Brillouin zone boundaries with larger more free electron-like orbits provides a useful separation of spin–orbit effects, and, indeed, several systems with small band-gaps have very large g-shifts.

Many-body effects

Because of the absence as yet of serious attempts to undertake the calculation of exchange enhancement effects on the dHvA g-factors, consideration here will be limited to the electron–electron interaction.

The connection between dHvA g-factors and electron–electron interactions may be seen following the discussion of Knecht (1975). The observed effective mass is enhanced over the band mass m_b by electron–phonon interactions

$$m_c = m_b(1+\lambda). \tag{81}$$

The observed dHvA g-value g_c is altered from the CESR g-value g_s

$$g_c = g_s/(1+B_0), \tag{82}$$

where B_0 is a parameter from Landau–Fermi liquid theory which includes both electron–electron and electron–phonon interactions (see Chapter 2)

$$1+B_0 = (1+B_0^{ee})(1+\lambda). \tag{83}$$

The electron–phonon interaction therefore cancels (Engelsberg & Simpson 1970) in the observables S which goes as the product

$$S=\frac{g}{2}\frac{m_c}{m}=\frac{g_s m_b/m}{1+B_0^{ee}}. \tag{84}$$

With isotropic metals such as the alkalis, Fermi surface and orbital averages are the same, and it has been conventional to express this also in terms of the spin susceptibility χ

$$\frac{\chi}{\chi_0}=\frac{g_s}{2}\frac{g_c m_c}{2m_b}\approx\frac{g_c m_c}{2m_b}, \tag{85}$$

where χ_0 is the spin susceptibility of a free electron gas. The measured g-value computed from S will therefore include the renormalized but measurable m_c. In comparing with theory for χ, one must in addition calculate a value of the band mass m_b. This is a negligible effect in the alkali metals, but is a large correction in the polyvalent metals. The results quoted above (Table 10.4) show both a large enhancement (~70 per cent) in χ and a systematic trend with increasing atomic number. The size of the enhancement agrees qualitatively with theory (for a review, see Knecht 1975) and allows a critical selection between competing theories of the electron–electron interaction in the alkali metals. It can be argued that the systematic trend is not due to spin–orbit coupling, since these orbits are nowhere near Brillouin zone boundaries and band-structure effects are relatively unimportant in the alkalis. The dominant factor in this systematic trend appears to be found within electron–electron theory itself (Bibby & Shoenberg 1977), as shown in Fig. 10.20. The key idea is that the spin susceptibility χ is a function only of a quantity r_s^*, where

$$r_s^*=r_s(m_b/m)/\varepsilon, \tag{86}$$

where r_s is the mean interelectron distance in Bohr radii, and ε is the dielectric constant. The puzzle as to why the electron–electron interaction should be so large in the alkali metals, and only one-fifth as large in the noble metals, is explained by this idea. Referring to Fig. 10.20, the values for the spin susceptibility for *both* the noble metals and the alkali metals clearly fall near the universal curve as a function of r_s^*. Thus, the evidence for electron–electron interactions in the dHvA g-shift appears clear and direct.

Band-structure effects

A qualitative guideline for the magnitude of these effects is due to Cohen & Blount (1960). They showed that in the limit of large spin–orbit coupling, the Landau level splitting and the spin-splitting should be roughly equal, or $S \approx 1$. It is important to realize the restrictions of that calculation. The result is exact only in a low-symmetry situation such as Bi, and when only two energy levels are important and no degeneracy exists. The Cohen and Blount criterion has been applied in many cases where these requirements are not satisfied. Nonetheless, it appears to be a qualitative guideline for understanding the observation of large g-values when m_c is small, i.e. one expects to see $(m_c/m)\ g \cong 2$ when spin–orbit coupling is dominant. The criterion appears to work in the case of Bi, where the requirements are satisfied well, but is violated in Sb and in Zn, for example (Table 10.7).

Another useful criterion for setting bounds on the g-shift due to spin–orbit coupling is that of Dupree & Holland (1967). They show that

$$|\Delta g| < \frac{4l_j^2 \Delta_j}{(2l_j+1)V_{Gj}}, \tag{87}$$

where l_j is the symmetry of the state, Δ_j is the atomic spin–orbit splitting, and V_{Gj} is the direct band gap to the nearest other band. The implications of this criterion are that g-shifts increase with increasing atomic number (Δ_j increases) and in addition increase close to a degeneracy where V_G approaches zero. An example of the application of this criterion was given by Randles for the noble metals, as shown in Table 10.8. It is clear that the p-wave contribution is not enough to account for the observed shifts in the noble metals, but that the d-wave component is of the right order. At that time the neck orbit was thought to be mostly p-like, but more recent partial wave analysis (Coleridge, Holzwarth & Lee 1974) shows that the neck is of mixed symmetry, ~60 per cent p-wave and ~40 per cent d-wave. The Dupree and Holland upper bound offers no way to sum up the contributions from an orbit of mixed symmetry, or, indeed, even of predicting the *sign* of the shift. It would appear possible and useful to extend the partial wave methods developed for explaining impurity scattering anisotropy to deal with g-factor anisotropy as well.

TABLE 10.7

	g	$m_c g$
Bi (electrons)	~225	~2.0 ± 0.2 (most directions)
Sb (electrons)	4–8	0.3–1.0
Zn (needle)	170	1.275

TABLE 10.8. *Spin–orbit contributions to the g-shift in the noble metals, with potassium included for comparison* (*Randles 1972*)

Metal	j	l_j	Δ_j (eV)	$\Delta\mathscr{E}_{so}$ (eV)	δg_{max}
Copper	4p	1	0.03	1.9	0.02
	3d	2	0.25	1.9	0.42
Silver	5p	1	0.11	4.0	0.04
	4d	2	0.55	4.0	0.44
Gold	6p	1	0.47	2.4	0.26
	5d	2	1.52	2.4	2.0
Potassium	4p	1	0.0007	1.2	0.008

In the case of the polyvalent metals, a criterion due to Pippard (1969a) is useful for small orbits whose dynamics are dominated by Bragg reflections. He showed that the g-shift has an upper limit of

$$(g_c - 2) < nm/m_c, \tag{88}$$

where n is the number of Bragg reflections around the orbit.

The connection between Pippard's derivation of this result (using Onsager quantization for the relatively large orbits appropriate to metals) and the essentially atomic-like calculations which were developed for use in semiconductors (Elliott 1954) and semimetals (Cohen & Blount 1960) is not clear. In the latter the assumption is made that the carriers are located essentially at a point in **k**-space, at the bottom of a parabolic band, with no specific reference made to the role of Bragg reflections; a fully quantitative though rather atomic-like calculation results. Pippard's model calculation, while clearly an oversimplification for real metals, contains additional qualitative insight in showing how the orbital g-shift due to spin–orbit interaction may depend on both the number of Bragg reflections and on the relative strength of the spin–orbit interaction potential in comparison with the lattice pseudopotential. Considerable anisotropy of conduction electron g-factors is a natural consequence of this model.

Although (88) has been used to set an upper bound for the g-shift due to spin–orbit interaction (e.g. Gold & Schmor 1976) we are aware of no direct test of Pippard's result. Indeed, even when one uses Pippard's criterion, as in the example of Pb, selection among the remaining multiple g-values which are consistent with the data ultimately has to rely on comparison with an explicit band g-factor calculation. An example is given in Fig. 10.21 (Gold & Schmor 1976) of a calculation at the

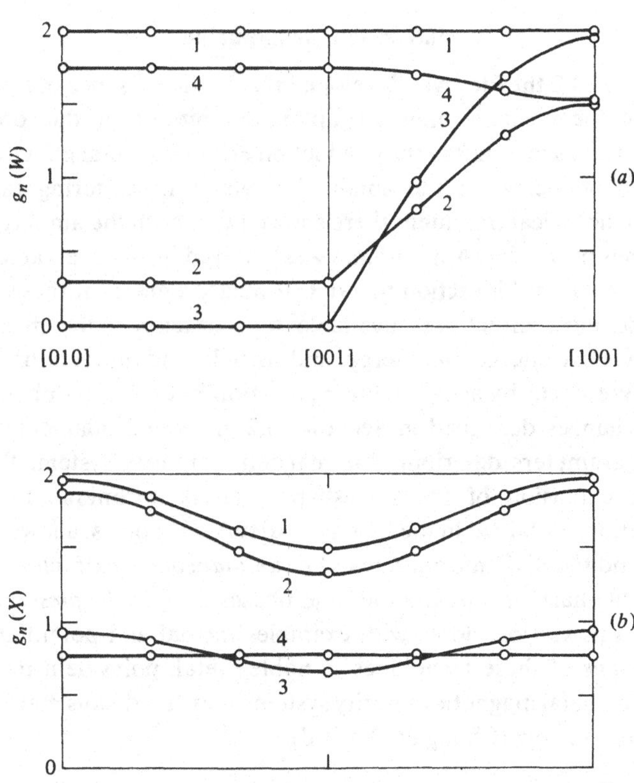

Fig. 10.21. Calculated local $g(\mathbf{k}, \hat{B})$ illustrating the dependence upon field *direction*, for a 4-OPW model calculation in Pb. The index n labels the bands in order of increasing energy. (a) and (b) are for the symmetry points W and X, respectively, $\hat{B}$ denotes magnetic field direction. (From Gold & Schmor 1976.)

symmetry point W in Pb. The g-values vary greatly from band 1 to band 4, and in addition display a large anisotropy as a function of the field *direction*. This is a clear example of the tensor nature of g, making the unfolding of orbital g-values into local $g(\mathbf{k})$ impossible. Gold & Schmor (1976) and Holtham (1976) have pointed out that multiple-OPW methods can lead to serious errors in calculating properties (such as anisotropic g-factors) which depend more critically on wave-function symmetry than does the Fermi surface itself. A full, properly symmetrized set of basis functions need be used throughout the Brillouin zone. Thus, the subject of g-factor anisotropy in polyvalent metals remains open for both experimental and theoretical exploration.

10.5 Magnetic impurities

In Section 10.2 the idea was developed that in the presence of a magnetic impurity, the exchange energy coupling the impurity to the conduction electrons leads to a shift in the conduction electron orbital g-factor and to a spin dependence of the conduction electron scattering rate. The phenomenological treatment there showed that both the amplitudes and the phases of the dHvA harmonics were altered in an observable way. It will be shown in this section that dHvA measurements of these changes offer the most direct way available to characterize the microscopic impurity–host interaction energy, and its field and temperature dependence. We begin by showing the connection between the phenomenological changes described in Section 10.2 and the fundamental microscopic parameters describing the magnetic impurity system. We show that an extension of the waveshape analysis technique to include information contained in the first *three* dHvA harmonics, allows *both* the g-shift and the SDS information to be *simultaneously* extracted from the observable harmonic amplitudes and phases *even in the presence of MI.* Finally, this section closes with examples (actual and potential) of the applications of these techniques to noble metal, polyvalent metal, and transition metal magnetic impurity systems. Further details may be found in a recent review (Chung *et al.* 1978).

10.5.1 Theories of the dHvA effect in the presence of magnetic impurities

The theories for the dHvA effect in the presence of magnetic impurities rest upon the Green's function formalism developed by Engelsberg & Simpson (1970). This approach is particularly powerful because, once the electron self-energy Σ in the presence of the impurity is known, all of the effects on the dHvA signal may be readily calculated. For example, with a magnetic impurity, Σ has the form

$$\Sigma_\sigma = \Delta_\sigma + \mathrm{i}\Gamma_\sigma \tag{89}$$

where both the shift Δ and the width Γ depend upon the spin ($\alpha = \pm 1$). Starting with this expression, Shiba (1973) has shown that the corresponding changes in the dHvA observables are:

$$\left.\begin{aligned} \delta F/H &= -(\Delta_\uparrow + \Delta_\downarrow)/2\hbar\omega_c, \\ \delta g &= (g_c' - g_c) = (\Delta_\uparrow - \Delta_\downarrow)/\mu_B H, \\ \bar{x} &= (\Gamma_\uparrow + \Gamma_\downarrow)/2\pi k_B, \\ \delta x &= (\Gamma_\uparrow - \Gamma_\downarrow)/2\pi k_B. \end{aligned}\right\} \tag{90}$$

Thus, when these four experimental quantities are determined, the *complete* electron self-energy is determined, and may be mapped as a function of H, T, and impurity type. A particularly convenient separation occurs between spin-independent quantities δF and $\bar{x}$ (which depend only upon the average of the two spins) and δg and δx (which are sensitive to the *difference* in behaviour of the spin components of Σ).

The above connection (equation 90) more rigorously involves a sum on the imaginary energy axis of terms in $\Sigma(i\omega_n)$, with $\omega_n = (2n+1)\pi k_B T$, which requires evaluation with some explicit assumptions about the energy dependence of Σ. Thus, the connection of (90) is exact only in the high-temperature limit ($k_B T \gg \hbar\omega_c$; $n = 0$ term only), which is not a useful limit, because the dHvA amplitude goes towards zero there! The correspondence is also correct if it happens that $\Sigma(i\omega_n)$ is energy-independent over a scale of several Landau levels. This is quite plausible, since the Landau level splittings are of the order of 10^{-4}–10^{-3} eV, and magnetic impurity level widths are of the order 10^{-1}–1 eV. Although the consequences of a possible failure of this assumption have not as yet been looked into in detail, the corrections are likely to be unimportant compared to the differences in the results obtained from different models of the magnetic impurity.

A further result has been obtained by Fenton (1975, 1976) who showed that the dominant behaviour of Re Σ is determined by

$$\mathrm{Re}\,\Sigma_\sigma = -\tfrac{1}{2}c_\sigma J\langle S_z\rangle + cV, \tag{91}$$

where the first term is due to exchange scattering and the second term is due to potential scattering. This equation has been proved in two useful limiting cases: the Anderson model or ground state of a Kondo system, and also using the Kondo Hamiltonian in the perturbation regime $T \gg T_K$, $H \gg H_K$. The significance of this result may be seen by substituting (91) into (90), yielding

$$\begin{aligned} \delta g &= -cJ\langle S_z\rangle/\mu_B H, \\ \delta F/H &= -cV/\hbar\omega_c. \end{aligned} \tag{92}$$

Thus, the dHvA frequency shift is a measure of the potential scattering (valence difference) and is independent of the specifically spin-dependent properties of a magnetic impurity. A measurement of the g-shift, however, directly yields the exchange contribution to the electronic self-energy, with J equal to the *bare* exchange coupling constant (uncorrected for Kondo effects) and $\langle S_z\rangle$ containing *both* the ordinary field and temperature dependence of a bare moment *together* with any Kondo screening contribution (reduction) to the moment. This is a particularly useful result, because most other bulk macroscopic properties characterizing magnetic impurity systems involve a mixture of exchange and potential scattering which is not as cleanly separated. In contrast, if Fenton is correct, a dHvA measurement of $\delta g(H, T)$ can yield a direct and complete determination of the field and temperature dependence of an impurity spin through the Kondo transition, including any conduction electron spin polarization screening of the impurity moment (Kondo spin compensation effects). We note that this result for the g-shift is intuitively plausible, since Re Σ may be viewed as a shift in the spin-dependent energy levels of the sort shown in Fig. 10.2, yet differs significantly from earlier expressions for δg (Shiba 1973; Harris, Mulimani & Strom-Olsen 1975a; Harris, Mulimani & Zuckermann 1975b).

Another conceptually general result is that the observation of SDS requires that the potential scattering term V be non-vanishing. Thus, for example, Shiba (1973) finds, using the Friedel–Anderson model, that

$$\delta x \propto \sin(\pi Z_d/5)\sin(2\pi S/5), \tag{93}$$

where Z_d, the number of impurity d-electrons, is a measure of the potential scattering, and S, the impurity moment, is a measure of the exchange scattering. Similarly, Harris *et al.* (1975a) and Mulimani (1976) find, using the s–d exchange model that

$$\delta x \propto \sin(\Delta\eta_L)\,\mathrm{Re}\,(T^J_\uparrow - T^J_\downarrow), \tag{94}$$

where $\Delta\eta_L$ is the difference between impurity and host d-wave potential scattering phase shifts ($L=2$) and T^J_α is the spin-dependent impurity scattering T-matrix. Thus, the observation of SDS is a consequence not only of exchange scattering but requires an admixture of potential scattering. In situations where the Friedel–Anderson model is appropriate, this is a useful indicator of the location of the impurity level with respect to $\mathscr{E}_F$, since the *sign* of δx depends upon whether the impurity resonance is (more than/less than) half-filled. It is consistent, for example, with the observation in Cu–Cr (Coleridge *et al.* 1972) of $\delta x = 0$, with the impurity band presumably half-filled, and is also in one-to-one correspondence with the vanishing of the magnetic impurity thermopower in this system.

As a final qualitative comment, we note that the theories to date are based either on the Friedel–Anderson model or on the s–d exchange model, and have ignored the time scale of the impurity magnetic moment (assuming simply that a moment exists) and are therefore probably inadequate for nearly magnetic spin fluctuation systems such as those based on Al or in transition metal hosts such as Rh or Ir.

A formalism for the calculation of the quantities in (90) which includes real metal band-structure effects has been developed by Harris *et al.* (1975a). Assuming that only the d-wave part of the conduction electron wave-function interacts with the magnetic impurity (d-wave resonance), but including both potential and exchange scattering terms, they showed that all four of the quantities in (90) scale in proportion to the fraction of d-wave charge density for different extremal orbits on the host metal Fermi surface. This proportionality is identical with the orbital scaling factor appearing in earlier treatments using phase-shift analysis of ordinary (non-magnetic) impurity scattering (Blaker & Harris 1971; Coleridge *et al.* 1974) and is in agreement with the few early experimental determinations of $|J|$ for different extremal orbits, using observations of dHvA spin-splitting zeros (Coleridge *et al.* 1972). Although the four observables in (90) are dominated by the self-energy terms listed, there are additional contributions to these observables in the results of Harris *et al.* (1975a), resulting from the interference of potential and exchange scattering interactions, and from the effects of backscattering from surrounding host atoms back onto the impurity site. Even in this case it is clear that dHvA experiments do provide a separation of spin-*independent* and spin-*dependent* contributions to the electronic self-energy (via measurements of $\bar{x}$ and δF versus δx and δg, respectively) but this is no longer the same as completely separating potential scattering effects from exchange scattering effects, due to interference terms. Interference then provides a significant correction to the work of Fenton, in that Harris *et al.* would find that the g-shift contains a contribution due to potential scattering, which is absent in Fenton's result (equation (92)). The results are complicated and beyond the scope of this review, but provide the basis for a method of analysis for real metals which is similar to that provided by the earlier application of phase-shift analysis to describe impurity scattering for non-magnetic impurity systems.

10.5.2 The dHvA waveshape in the presence of magnetic impurities

In the presence of magnetic impurities the dHvA harmonic amplitudes and phases are modified by the development of SDS rates for the conduction electrons and by the exchange energy shift of their Landau levels. As discussed in the previous section these effects are conveniently described using the three microscopic parameters $x^{\uparrow}$, $x^{\downarrow}$, and Δg (or equivalently, $\bar{x}$, δx, and S'). Thus, the observable harmonic amplitude ratios and relative harmonic phases contain information about these microscopic parameters, as was illustrated (for the second harmonic) in Section 10.2. Using just the first three dHvA harmonics one can determine values for *four* independent observables (two relative harmonic phase angles and two harmonic amplitude ratios) so that, in principle, the harmonic analysis of a single 'window' of dHvA data should be sufficient to uniquely determine δx, $\bar{x}$, S' at field H and temperature T. We therefore have a tool for mapping out the combined (H, T) dependences of the spin-dependent impurity self-energy.

In practice, though, the harmonic content of the waveshape contains additional contributions from the MI effect. Chung *et al.* (1978) have shown that the effects of MI and of magnetic impurities are mixed non-linearly in determining the amplitude and phase of each resultant dHvA harmonic, i.e. the resultant waveshape is altered in both amplitude and phase from either the pure LK+magnetic impurity or the pure MI result. One important consequence of this non-linear mixing (which illustrates the danger of observing only a single dHvA harmonic in such a case) is that the observation of a clean rth harmonic spin-splitting zero in a dilute magnetic alloy is *no longer* indicative of equal scattering rates for spin-up and spin-down electrons, in the presence of MI.

Chung *et al.* (1978) have used the iterative procedure of Phillips & Gold (1969) to derive an expression for the first three harmonics of the dHvA waveshape in the case where MI, SDS and exchange energy shifts are all simultaneously present. The nth approximation, $\tilde{M}^{(n)}$, to $\tilde{M}$ is written as

$$\tilde{M}^{(n)} = -\hat{m} \sum_{r=1}^{\infty} A'_r C_r D^r \sin [r(\zeta - \kappa' \tilde{M})^{(n-r)} + p\pi/4 + \Delta\theta_r + (1 - j'_r)\pi/2], \tag{95}$$

with $\tilde{M}^{(0)} = 0$,

$$A'_r = [E^{2r} + E^{-2r} + 2\cos(2\pi r S')]^{\frac{1}{2}}/2,$$

and all other quantities as defined in Section 10.2. Equation (95) is completely analogous with (39), except that the changes in LK phase and amplitude resulting from the presence of magnetic impurities have now been included prior to carrying out the MI expansion. Carrying out this expansion and collecting terms of second and third order in the exponential (scattering and thermal damping) factors, the third-order approximation to $\tilde{M}$ is

$$\tilde{M}^{(3)} = \tilde{M}_1 + \tilde{M}_2 + \tilde{M}_3 + \text{higher harmonics,} \tag{96}$$

where

$$\left.\begin{aligned}
\tilde{M}_1 = -\hat{m}\Big\{ & A_1'C_1D \sin[\zeta + p\pi/4 + \Delta\theta_1 + (1-j_1')\pi/2] \\
& -\frac{(\kappa')^2(A_1'C_1D)^3}{8}\sin[\zeta + p\pi/4 + \Delta\theta_1 + (1-j_1')\pi/2] \\
& -(\kappa'/2)(A_1'C_1D)(A_2'C_2D^2) \\
& \times \sin[\zeta + \Delta\theta_2 - \Delta\theta_1 + (j_1' - j_2')\pi/2]\Big\}, \\
\tilde{M}_2 = -\hat{m}\{ & A_2'C_2D^2 \sin[2\zeta + p\pi/4 + \Delta\theta_2 + (1-j_2')\pi/2] \\
& +(\kappa'/2)(A_1'C_1D)^2 \sin[2(\zeta + p\pi/4 + \Delta\theta_1) + (1-j_1')\pi]\}, \\
\tilde{M}_3 = -\hat{m}\{ & A_3'C_3D^3 \sin[3\zeta + p\pi/4 + \Delta\theta_3 + (1-j_3')\pi/2] \\
& +p(3/8)(A_1'C_1D)^3(\kappa')^2 \cos[3\zeta + p\pi/4 - 3\Delta\theta_1 \\
& +3(1-j_1')\pi/2] + (3\kappa'/2)(A_1'C_1D)(A_2'C_2D^2) \\
& \times \sin[3\zeta + 2p\pi/4 + \Delta\theta_1 + \Delta\theta_2 + (2-j_1'-j_2')\pi/2]\}.
\end{aligned}\right\} \tag{97}$$

The coefficients of the terms in (97) are in error by no more than 2 per cent if $(\kappa' A_1'C_1D) \leq 0.2$. In order to calculate the combined effects of MI and SDS on the resultant second and third harmonic dHvA amplitudes, it is convenient to define all phase angles relative to the left edge of a data window ($\zeta_0 = 2\pi(F/H_0 - \gamma)$). The phase angles can then be written as

$$\left.\begin{aligned}
\theta_1^{\mathrm{LK}} &= (\zeta_0 + p\pi/4 + \Delta\theta_1) - (1 + j_1')\pi/2, \\
\theta_1^{\mathrm{MI}p} &= (\zeta_0 + p\pi/4 + \Delta\theta_1) + (1 - j_1')\pi/2, \\
\theta_1^{\mathrm{MI}q} &= (\zeta_0 + \Delta\theta_2 - \Delta\theta_1) + (j_1' - j_2')\pi/2; \\
\theta_2^{\mathrm{LK}} &= (2\zeta_0 + p\pi/4 + \Delta\theta_2) - (1 + j_2')\pi/2, \\
\theta_2^{\mathrm{MI}} &= (2\zeta_0 + 2p\pi/4 + 2\Delta\theta_1) - j_1'\pi; \\
\theta_3^{\mathrm{LK}} &= (3\zeta_0 + p\pi/4 + \Delta\theta_3) - (1 + j_3')\pi/2, \\
\theta_3^{\mathrm{MI}p} &= (3\zeta_0 + (p - 2q_1')3\pi/4 + 3\Delta\theta_1) + \pi/2, \\
\theta_3^{\mathrm{MI}pq} &= (3\zeta_0 + p\pi/2 + \Delta\theta_1 + \Delta\theta_2) - (j_1' + j_2')\pi/2.
\end{aligned}\right\} \tag{98}$$

All possible minus signs have been incorporated in the phase factors in (98) so that the coefficients in (97) are now treated as being positive.

Equations (97) and (98) demonstrate that MI and magnetic impurity effects mix non-linearly. If SDS (alone) produces phase shifts $\Delta\theta_1$, $\Delta\theta_2$ and $\Delta\theta_3$ in pure LK components, then the second harmonic MI component is shifted additionally in phase by $2\Delta\theta_1$, and the two third harmonic MI components have additional phase shifts $3\Delta\theta_1$, and $(\Delta\theta_1 + \Delta\theta_2)$, respectively, in the presence of SDS. Thus, the MI and LK components at each harmonic no longer differ in phase by a simple fixed amount (such as $p\pi/4$), as in a pure metal, and so the separation into 'p' and 'q' channel signals (Section 10.2) is no longer strictly valid. Instead, the MI and LK phase difference, and the phase of each resultant observable harmonic, depends on both the extent of SDS and the magnitude of the Zeeman and exchange energy shifts, via the quantities $\Delta\theta_r$.

10.5.3 Determination of exchange energy shift and SDS rates from observables of the waveshape

Equations (97) and (98) give the amplitude and phase of each contribution to the first three dHvA harmonics, in terms of the three microscopic parameters which are of main interest in describing the conduction electron–magnetic impurity interaction: δx, $\bar{x}$, and Δg (or S'). Equations expressing the four experimental observables $(\tilde{M}_2/\tilde{M}_1)$, $(\tilde{M}_3/\tilde{M}_1)$, $(2\theta_1 - \theta_2)$ and $(3\theta_1 - \theta_3)$ (using three dHvA harmonics), in terms of δx, $\bar{x}$, and S', can be written down in closed form (Chung *et al.* 1978).

However, the resulting equations are complicated transcendental functions which cannot be analytically inverted to give expressions for the microscopic parameters in terms of measured values of the four observables. Furthermore, the complexity of the equations tends to obscure the degree of functional dependence. It is not clear whether a given observable depends strongly or weakly on any given microscopic parameter; nor is it obvious how best to invert the equations and solve for the microscopic parameters from the harmonic analysis (amplitudes and phases) of a data window.

Two different computer-graphic techniques have been developed to overcome these difficulties. One technique uses a laboratory computer interfaced to a plotter to display the observables as functions of the microscopic parameters, thus developing the needed intuition regarding the strength of functional dependences (Chung *et al.* 1978). The other technique uses the plotter to display graphically the locus of points in the $(\delta x, \bar{x}, S')$ solution space which correspond simultaneously to the measured values of all four observables (Chung *et al.* 1978; D. H. Lowndes & P. Reinders unpublished data).

Functional dependence of observables on δx, $\bar{x}$, and S'

It is helpful first to display the functional relationships between observables and microscopic parameters in the limiting case of *no MI.* In this case, the relationships are relatively simple (though still transcendental) expressions, and both the strength of functional dependences and the most efficient method of solving for the microscopic parameters become clear.

Using the phasor diagram (see Fig. 3 of Chung *et al.* 1978) together with the definitions in Section 10.2 it follows that

$$(2\theta_1-\theta_2)^{\mathrm{OB}}=(2\theta_1-\theta_2)^{\mathrm{LK}}+(2\Delta\theta_1-\Delta\theta_2)+(j_2'-j_2)\pi/2, \qquad (99\mathrm{a})$$

$$(3\theta_1-\theta_3)^{\mathrm{OB}}=(3\theta_1-\theta_3)^{\mathrm{LK}}+(3\Delta\theta_1-\Delta\theta_3)+(j_3'-j_3)\pi/2 -3(j_1'-j_1)\pi/2, \qquad (99\mathrm{b})$$

$$\left[\frac{(\tilde{M}_3/\tilde{M}_1)}{(\tilde{M}_2/\tilde{M}_1)^2}\right]^{\mathrm{OB}}\Big/\left(\frac{C_3C_1}{C_2^2}\right) =\left\{\frac{[E^6+E^{-6}+2\cos(6\pi S')][E^2+E^{-2}+2\cos(2\pi S')]}{(E^4+E^{-4}+2\cos(4\pi S'))^2}\right\}^{\frac{1}{2}}, \qquad (99\mathrm{c})$$

$$(\tilde{M}_2/\tilde{M}_1)^{\mathrm{OB}}/(C_2/C_1)=D\left[\frac{E^4+E^{-4}+2\cos(4\pi S')}{E^2+E^{-2}+2\cos(2\pi S')}\right]^{\frac{1}{2}}, \qquad (99\mathrm{d})$$

$$(\tilde{M}_3/\tilde{M}_1)^{\mathrm{OB}}/(C_3/C_1)=D^2\left[\frac{E^6+E^{-6}+2\cos(6\pi S')}{E^2+E^{-2}+2\cos(2\pi S')}\right]^{\frac{1}{2}}, \qquad (99\mathrm{e})$$

where we have used the definitions

$$(2\theta_1-\theta_2)^{\mathrm{LK}}\equiv p\frac{\pi}{4}+(1+j_2)\frac{\pi}{2}, \qquad (100\mathrm{a})$$

$$(3\theta_1-\theta_3)^{\mathrm{LK}}\equiv p\frac{\pi}{2}+(j_3-j_1)\frac{\pi}{2}, \qquad (100\mathrm{b})$$

where primed quantities refer to a dilute magnetic alloy, unprimed quantities to the host metal and $j_r'\neq j_r$ includes the possibility of an exchange energy shift through an ith harmonic spin-splitting zero in the dilute magnetic alloy. ($j_i=+1(-1)$ if $\cos(\pi rS)>0(<0)$).

All quantities appearing in the factors C_r are either known from the conditions of the experiment (e.g. T, H, m_c) or else cancel out in the ratios. Note in particular that the Fermi surface curvature factor, $\mathscr{A}'' = d^2\mathscr{A}/dk_H^2$, which is usually not accurately known, cancels out in all of the amplitude ratios above.

The first three observables defined in (99a–e) depend only on δx and S', while the last two amplitude ratios depend in addition on $\bar{x}$. A natural procedure is to determine δx and S' from the measured values of $(2\theta_1 - \theta_2)$, $(3\theta_1 - \theta_3)$ and $(\tilde{M}_3\tilde{M}_1/\tilde{M}_2^2)$ and then to determine $\bar{x}$ from the two amplitude ratios $(\tilde{M}_3/\tilde{M}_1)$ and $(\tilde{M}_2/\tilde{M}_1)$, using the now fixed values of δx and S'.

Figs. 10.22 and 10.23 show that the two relative phase angle observables are periodic functions of S' with period equal to 1, are antisymmetric about the value $S' = 0.5$, and are antisymmetric with respect to change of sign of δx. Furthermore, the sensitivity of these two phase angle observables to changes in δx depends critically on the value of S'. For example, for $0 < S' < \frac{1}{6}$, $(3\theta_1 - \theta_3)^{\mathrm{OB}}$ is relatively insensitive to the extremes of *zero* ($\delta x = 0$) and *complete* ($|\delta x| = \bar{x}$) SDS, while for $\frac{1}{6} < S' < \frac{1}{2}$, it is very sensitive to small changes in SDS.

The amplitude ratio $(\tilde{M}_3\tilde{M}_1/\tilde{M}_2^2)$ is mainly useful for determining S' when δx is small, since it responds sensitively to proximity to S' values which produce spin-splitting minima of any of the first three dHvA harmonics. Fig. 10.24 illustrates the influence of small amounts of SDS on the quality of a third harmonic spin-splitting zero. In the presence of MI the zero of the third harmonic amplitude occurs for a *non-zero* value of δx; without MI, one expects a clean spin-splitting zero only if $x^{\uparrow} = x^{\downarrow}$ ($\delta x = 0$). Finally, $\bar{x}$ may be determined from the measured value of either $(\tilde{M}_2/\tilde{M}_1)$ or $(\tilde{M}_3/\tilde{M}_1)$, given the values of δx and S' determined as described above, and that the effect of errors in δx (or in S'), feeding through to produce errors in $\bar{x}$, can also be easily estimated.

Figs. 10.22–10.24 illustrate that even in a case for which MI harmonic content is far from dominant (the ⟨111⟩ neck orbit in Au) its effect on the observables of waveshape analysis is certainly non-negligible, and the effects of MI *must* be included if it is hoped to map out accurately the T and H dependences of δx, $\bar{x}$ and S'. Nevertheless, the three observables $(2\theta_1 - \theta_2)$, $(3\theta_1 - \theta_3)$ and $(\tilde{M}_3\tilde{M}_1/\tilde{M}_2^2)$ remain essentially only functions of the two parameters δx and S'. The only dependence of these three observables on $\bar{x}$, in the presence of MI, is through the weak third-order correction to the fundamental amplitude and phase (Chung *et al.* 1978). Thus, the procedure outlined above, of solving first for δx and S', and

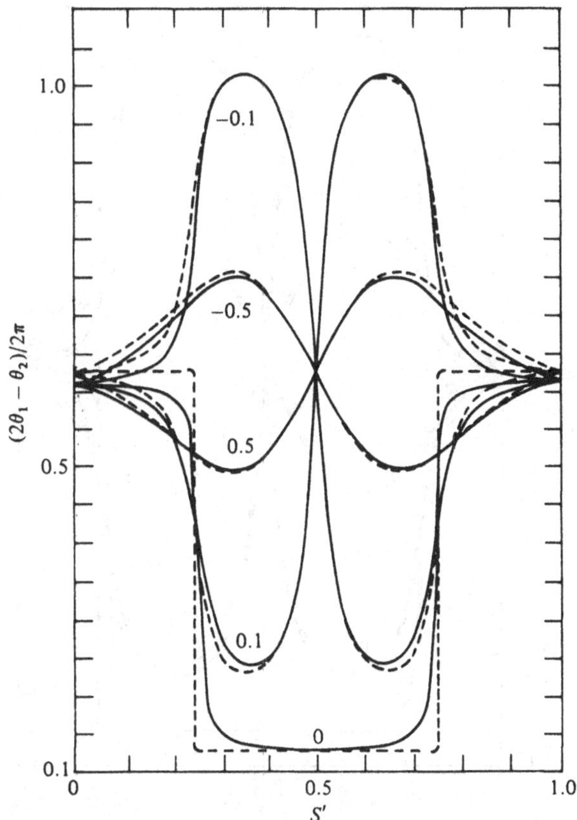

Fig. 10.22. The observable $(2\theta_1 - \theta_2)$ as a function of S', for several fixed values of δx. Broken lines are for the case of *no* MI harmonic content while solid lines give the solutions *including* the effects of MI with the sample's effective demagnetizing factor N taken to be 0.2. Figs. 10.22–10.24 are plotted for $H = 40$ kOe, $T = 1.1$ K and other parameter values (where necessary) for the ⟨111⟩ neck orbit in Au: $F = 1.532 \times 10^4$ T, $m_c/m_0 = 0.28$, $\mathscr{A}''^{-\frac{1}{2}} = 0.26$, $S = 0.16$ and $p = +1$. Fig. 10.22 uses $\bar{x} = 0.5$, and Figs. 10.23 and 10.24 use $\bar{x} = 1.0$, though they depend only very weakly on the value of $\bar{x}$.

then obtaining $\bar{x}$ from measured values of harmonic amplitude ratios, remains valid.

Inversion of observables to obtain SDS rates and exchange energy shifts

A computer-graphic technique provides a convenient solution to the problem of obtaining precise values (together with error estimates) for δx, $\bar{x}$ and S' using the measured values of $(2\theta_1 - \theta_2)$, $(3\theta_1 - \theta_3)$, $(\tilde{M}_2/\tilde{M}_1)$ and $(\tilde{M}_3/\tilde{M}_1)$ for a single short data block at (H, T). The computer is programmed to search the $(\delta x, S')$ space for pairs of values which produce

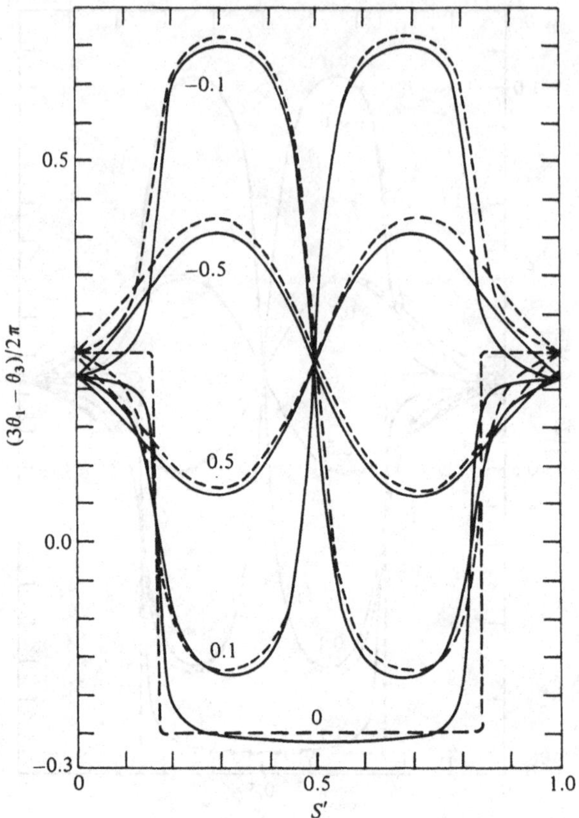

Fig. 10.23. The observable $(3\theta_1 - \theta_3)$ as a function of S', for several fixed values of δx. See also the caption for Fig. 10.22.

values of each of the first three observables in (99a–c) which agree with experiment. Whenever such a pair is found, a point is plotted. Following location of a unique point $(\delta x, S')$ which provides simultaneous solutions for all three observables, the measured values of the two amplitude ratios (equations (99d, e)) may be used to determine $\bar{x}$ (D. H. Lowndes & P. Reinders, unpublished data).

Fig. 10.25 illustrates the loci of solutions $(\delta x, S')$ which correspond to measured values of each of the observables $(2\theta_1 - \theta_2)$, $(3\theta_1 - \theta_3)$ and $(\tilde{M}_3\tilde{M}_1/\tilde{M}_2^2)$. Because of the symmetries present, several sets of solutions are possible on the full interval $-1 < S' < +1$, $-\infty$. With measurements in dilute magnetic alloys this ambiguity is easily resolved by comparing measurements made using data windows at different H and T and

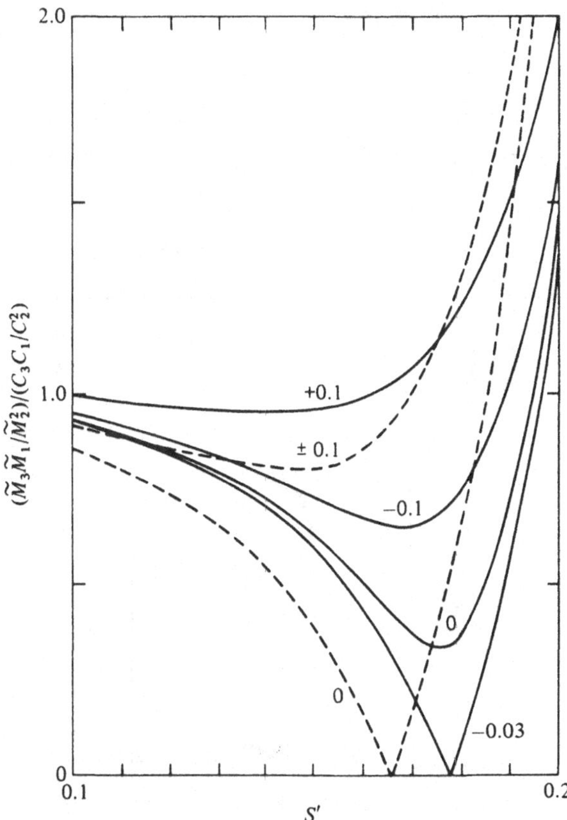

Fig. 10.24. The observable $(\tilde{M}_3\tilde{M}_1/\tilde{M}_2^2)$ (defined in the text) as a function of S', for several values of δx. It has been normalized, for convenience in plotting, by dividing by the known factor (C_3C_1/C_2^2). The figure illustrates the effect of small amounts of SDS on the quality of the third harmonic spin-splitting zero, both with (solid lines) and without (broken lines) MI harmonic content. The lines are labelled by the values of δx. See also the caption to Fig. 10.22.

requiring that the solutions for δx, $\bar{x}$ and H_{ex} should be smoothly varying functions of H and T. Any remaining uncertainty can be eliminated by using measurements on alloys of several different local moment concentrations, as well as on the pure host metal, to trace the evolution of SDS ($\delta x \neq 0$) and exchange energy *shifts* ($H_{ex} \neq 0$) (of course, the ambiguity in determining the absolute host metal g-value for a given orbit remains – see Section 10.4).

Once the region of (δx, S')-space in which the correct solution lies has been determined (for a given sample) in this way, a very precise solution is

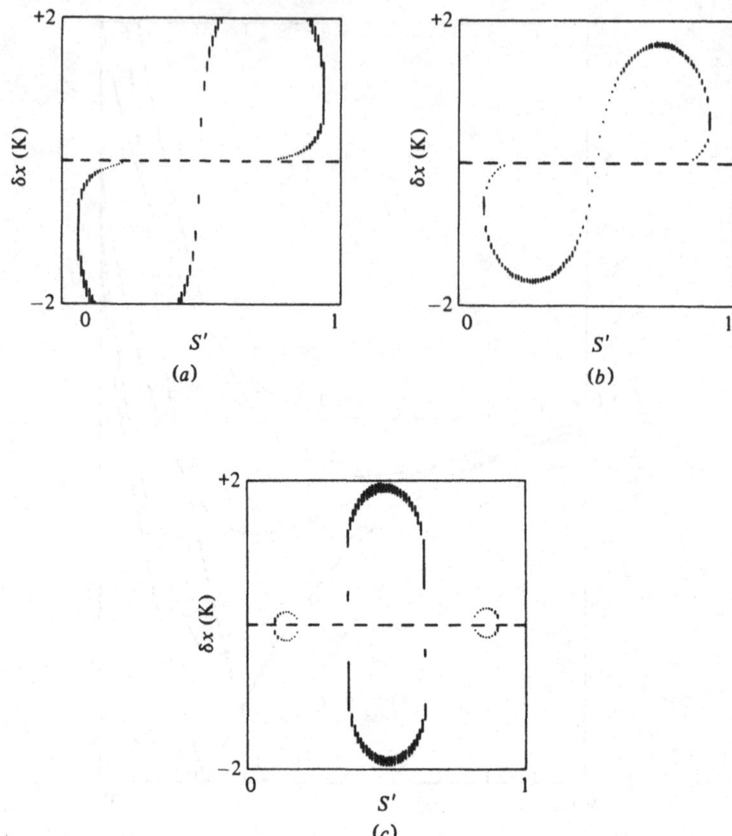

Fig. 10.25. Plot of the loci of solutions (δx, S') corresponding to the measured values (*a*) $(2\theta_1 - \theta_2) = 4.007$ (radians), (*b*) $(3\theta_1 - \theta_3) = 1.932$, and (*c*) $(\tilde{M}_3\tilde{M}_1/\tilde{M}_2^2) = 1.153$. The experimental data were obtained for the ⟨111⟩ neck orbit in an Au(Cr) dilute alloy (D. H. Lowndes & P. Reinders unpublished data). The slight asymmetry is due to MI, while the broken plotted lines are due to discrete stepping of the computer search for solutions.

obtained as illustrated in Fig. 10.26(*a*). A real advantage of this computer-graphic technique is that the error field for the microscopic model parameters, corresponding to estimated experimental errors in the observables, is well defined and conveniently displayed (Fig. 10.26(*b*)). SDS rates are typically determined within a precision of ±10–30 mK, while exchange fields are measured with a precision of ±2 per cent (see Section 10.5.4). Fig. 10.26(*c*) illustrates how the measured values for all five of the observables in (99) may be used simultaneously to display the complete solution (δx, S', $\bar{x}$) in a single graph.

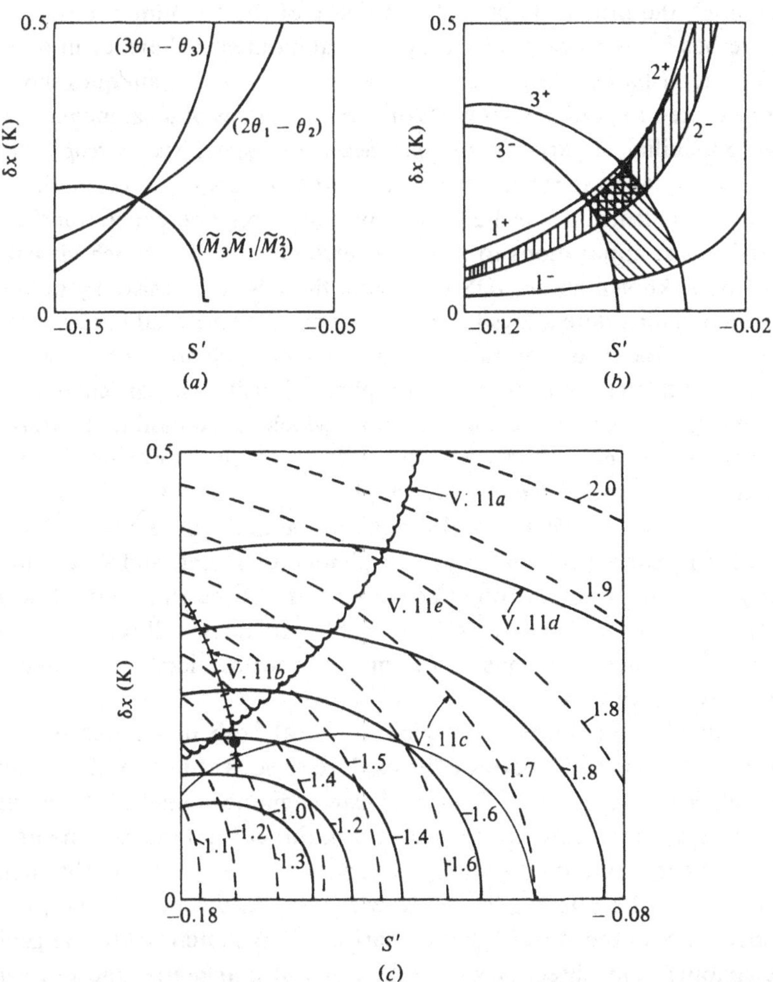

Fig. 10.26. (*a*) Unique solution for $(\delta x, S')$ using measured values of $(2\theta_1 - \theta_2)$, $(3\theta_1 - \theta_3)$ and $(\tilde{M}_3\tilde{M}_1/\tilde{M}_2^2)$ as described in the text, for the $\langle 111 \rangle$ neck orbit in an Au(Cr) dilute alloy at $H = 67.2$ kOe, $T = 1.60$ K. The curves corresponding to these three independent variables intersect in nearly the same point. (*b*) Error field(s) in the $(\delta x, S')$ solution space, corresponding to experimental errors of ± 0.02 radian from the measured value of $(2\theta_1 - \theta_2)$ (curves labelled '1'), $(3\theta_1 - \theta_3)$ (curves labelled '2') and the estimated error in $(\tilde{M}_3\tilde{M}_1/\tilde{M}_2^2)$ (curves labelled '3'). The combined (overlap) error field corresponding to use of all three observables is usually quite small. (*c*) Complete solution $(\delta x, S', \bar{x})$ for a single dHvA data window using all five of the observable parameters defined in equations (99a–e). For this data window all five curves do not pass through a single point, due to experimental errors, but the best solution lies near $S' = -0.16_8$, $\delta x = 0.17_3$ K and $\bar{x} = 1.35$ K.

10.5.4 Applications

Although the principal application to date of the techniques described above has been a systematic study of local moment behaviour in dilute noble metal-based alloys, there is no reason why these techniques should be restricted to noble metal hosts or even to studies of local magnetism. The possibility of carrying out *spin-resolved* **k**-*space spectroscopy* with these techniques suggests several new areas of application: studies of local magnetism and spin-fluctuation or Kondo phenomena in transition metal hosts; applications to itinerant ferromagnets (for which virtually nothing is known about SDS processes, though bulk resistivity data in ferromagnetic dilute alloys have recently been interpreted in terms of a two-band model (up- and down-spin carriers) with strong SDS anisotropy); and, the study of magnetic phase transitions, particularly the paramagnetic–ferromagnetic transition which is accessible to dHvA techniques in one or two very dilute alloy systems. The main requirements for such measurements are simply that the dHvA frequency of the extremal orbits which are used should not be too high (e.g. $\lesssim 10^8$ G), so that MI harmonic content is not completely dominant, and that Dingle temperatures and cyclotron effective masses also be sufficiently low so that third harmonic content is available. In dilute alloys the latter restriction generally means that impurity concentrations are limited to the 0–1000 ppm range.

By making measurements which are local both in **k**-space (orbital averages) and in spin-space we can explore various aspects of theoretical models for the origin and nature of local magnetic moments: the *partial wave nature* of the exchange interaction; direct local measurements of the exchange coupling strength, J_{orbit}; a test of the 'theorem' that non-zero potential scattering is necessary for the occurrence of SDS; field dependence of the local moment polarization (Brillouin versus Langevin behaviour); and, direct access to the *onset of magnetism* due to *interactions* between impurities which are non-magnetic when isolated.

Noble metal hosts

The first application of waveshape analysis techniques to the study of conduction electron–local moment interactions in metals was the study of Cu(Fe) by Alles *et al.* (Alles & Higgins 1974; Alles *et al.* 1973). The inclusion of Mi effects was not necessary in this case, and only the first two dHvA harmonics were used, so that a unique solution for all three of δx, x and H_{ex} was not possible from a single data window. Nevertheless, by assuming linear behaviour of these parameters and combining information contained in a number of data windows, their approximate H- and T-dependences were obtained. This experiment clearly demonstrated that both δx and δg were linearly proportional to $\langle S_z \rangle$ in a Kondo ground-state system where the Friedel–Anderson model is valid.

The most extensive study of magnetic impurity effects using the waveshape method is the broad survey of conduction electron–local moment interactions in Au-based dilute alloys, using both d-shell (Cr, Mn, Fe, Co) and f-shell (Gd, Ho, Yb) local moments, by Lowndes and collaborators at Oregon and Nijmegen (Chung & Lowndes 1976; Chung, Lowndes, Hendel & Rode 1976; Chung *et al.* 1978; D. H. Lowndes & P. Reinders, unpublished data; C. Hendel & D. H. Lowndes, unpublished data). The examples below illustrate the information which can be made available with these techniques. In addition, the examples show how waveshape measurements can be usefully supplemented by observations of harmonic spin-splitting zeros and by conventional Dingle temperature measurements and lifetime mapping.

Fig. 10.27 shows that the spin-resolving capability of the wave-shape technique may be used to view directly the 'freezeout' of spin-flip scattering processes in a magnetic field (due to increasing polarization of a local moment) and the *different rates* at which this freezeout occurs for up-spin and down-spin electrons (D. H. Lowndes & P. Reinders, unpublished data). This effect had been suggested by Beal-Monod & Weiner (1968) as the source of the 'anomalous' negative magnetoresistance, in dilute alloys containing d-shell local moments, but has not before been so directly demonstrated. Furthermore, the result (Fig. 10.27) that $\delta x \neq 0$ for Cr impurities in Au is very different from the situation in Cu(Cr), for which $\delta x = 0$. This is important in relation to the Friedel–Anderson model for localized impurity states, since there exists a theorem that non-zero potential scattering is necessary for SDS to occur (Shiba 1973). Thus, it seems that Cr in Au does *not* satisfy the 'half-filled impurity d-shell' condition, unlike Cr in Cu.

Chung & Lowndes (1977) recently reported the first dHvA observations of *interaction effects* between impurities, via measurements of the SDS of conduction electrons by magnetic *pairs* of Co impurities. The *induced* local moment associated with interacting Co pairs produces the *only SDS* in these dilute alloys, because *isolated* Co impurities (which are also present) carry no local moment at low temperatures. Thus, this technique allows the *onset of magnetism, due to interactions*, to be

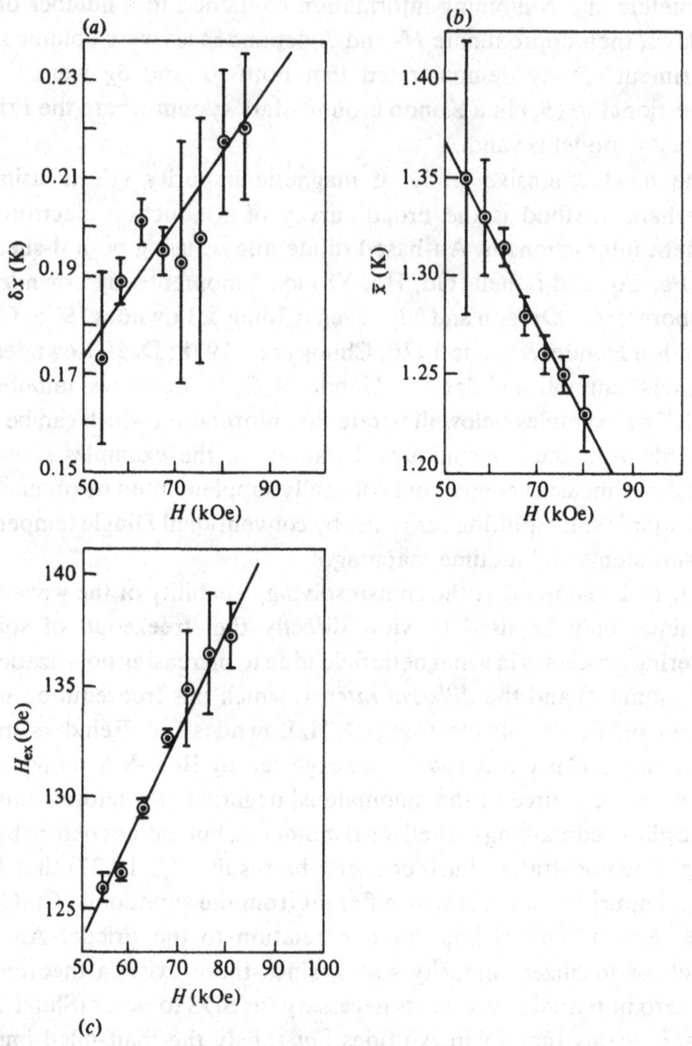

Fig. 10.27. Magnetic field dependence of the quantities $\delta x = (x^{\uparrow} - x^{\downarrow})/2$, $\bar{x} = (x^{\uparrow} + x^{\downarrow})/2$ and $H_{ex} = \Delta E_{ex}/\mu_B$, at $T = 1.5$ K, for electrons on the [111] neck orbit in an Au (230 at. ppm Cr) single crystal. By combining the field dependences in (*a*) and (*b*), one finds that

$$x^{\uparrow}\,(\mathrm{K}) = 1.70 - 0.003_2\,H\,(\mathrm{kOe}),$$

$$x^{\downarrow}\,(\mathrm{K}) = 1.50 - 0.006\,H\,(\mathrm{kOe}),$$

thus directly demonstrating the different rates at which spin-flip scattering processes freeze out. The behaviour of H_{ex} shown in (*c*) is as expected for an impurity moment near saturation in the applied magnetic field (note the suppressed zero for the ordinate).

observed at lower solute concentrations than is possible using measurements of bulk properties in the same alloy system. Phase-shift analysis of the electronic scattering rate anisotropy due to the *isolated* Co impurities shows that the interaction between conduction electrons and Co impurities is a strong, pure d-wave resonance. This is strong evidence that it is the d-*wave part* of the conduction electron wave-functions which is then also responsible for pairing the non-magnetic Co impurities to produce a local moment.

For Au(Fe), a 3:1 ratio of spin-up and spin-down electron scattering rates has been observed for the ⟨111⟩ neck orbit (Chung *et al.* 1976). The magnetic field dependence of the exchange energy was found to be like that of an impurity spin near saturation; from its magnitude the strength of the conduction electron–Fe moment exchange coupling parameter *for the neck orbit* was found to be $J_{\text{neck}} = -0.7$ eV. Using the assumption that the exchange interaction *proceeds through the* d-*wave part* of the conduction electron wave-functions allows one to calculate the Fermi surface-averaged exchange coupling strength, $J_{\text{FS}} = -1.2$ to -1.3 eV, whose value is in excellent agreement with the value needed to explain both bulk magnetization and resistivity measurements in Au(Fe).

Thus, the dHvA effect appears to be a sensitive probe for determining impurity spin behaviour in a magnetic field, and for measuring cyclotron *orbitally-averaged* values of the exchange constant, J_{orbit}, in very dilute local moment systems.

Au(rare earth) dilute alloys

Several striking results have already emerged from these studies (C. Hendel & D. H. Lowndes, unpublished data):

(1) A combination of measurements of harmonic spin-splitting zeros together with third harmonic waveshape measurements for the neck orbit in the (110) plane shows that the *sign* of the exchange interaction changes as the f-shell is filled: Gd (half-filled f-shell) produces *ferromagnetic* exchange, Ho essentially *zero* exchange shift, and Yb produces *anti-ferromagnetic* exchange, with conduction electrons in Au. The Gd result is the first dHvA measurement of a *ferro*magnetic exchange interaction in a noble metal host. Furthermore, the *magnitudes* of the exchange shifts for Gd and Yb are a factor of 5–10 less than those measured for d-shell impurities in Au.

(2) $x^{\uparrow} = x^{\downarrow}$ for scattering by rare earth impurities, in striking contrast to the strongly SDS which was found for the d-shell impurities Cr and Fe in Au.

(3) A preliminary partial wave analysis of the dHvA scattering rate (Dingle temperature) measurements for Au(Gd), Au(Ho) and Au(Yb) appears to have resolved the question of how conduction electrons interact with rare earth impurities: The very large anisotropy observed for different Fermi surface orbits can only be accounted for as scattering of the d-wave part of the conduction electron wave-function. This seems to be very direct evidence of scattering through a localized (virtual bound) d-state resonance.

Transition metal hosts

Here, an extremely wide variety of physical phenomena are accessible. Impurities such as Fe display a magnetic moment in some hosts yet not in others, and display a Kondo effect in some systems, yet in others an 'anti-Kondo' effect is observed, with $dR/dT > 0$ even down to the lowest temperatures (<0.1 K). Because transition metals typically have a number of sheets of Fermi surface with different wave-function symmetry, there is an opportunity to test the generality of the 'd-wave theorem' (coming from the phase-shift analysis of scattering in noble metal hosts), namely, that the scattering rate from a magnetic impurity such as Fe is proportional to the fraction of d wave-function for the orbit in question. Relatively little work has been done as yet with transition metal hosts, in spite of the richness of problems available. The principal complication to be encountered is the difficulty in doing waveshape analysis of the harmonic amplitudes and especially phases (Cheng, Higgins, Graebner & Rubin 1979). We will discuss briefly here two prototype systems: an example of a spin fluctuation system (Rh–Fe), and the phenomena to be expected in strongly exchange enhanced systems based upon hosts such as Pd or Pt.

Spin fluctuation systems

The Rh–Fe system is a prototype of the anti-Kondo effect, in which dR/dT remains positive and non-vanishing, i.e. no resistance minimum and even no residual resistance region (Coles 1964; Graebner *et al.* 1974; Rusby 1974). A number of physically different explanations were proposed to explain why the exchange scattering should be a decreasing function of T in such a system. From the spin-fluctuation point of view (Kaiser & Doniach 1970) the Fe impurity may have no magnetic moment at $T = 0$, but a moment is induced by thermal population of the spin-fluctuation spectrum. Kondo (1968), on the other hand, proposed that if the potential scattering phase shift were large enough to bring the scattering $\sin^2 \Delta\eta_L$ past its peak, additional exchange scattering from a magnetic moment would make the total scattering smaller. Other conceptually different models for this phenomenon include the notion that the screening of the magnetic moment is done by a relatively low-mobility d-band, while most of the electrical current is carried by a high-mobility s-like band (Knapp 1967), and the quite different notion that the relative spin populations of the impurity resonance are a rapidly varying function of temperature (Nagasawa 1972). Although conventional bulk macroscopic measurements are too blunt an instrument with which to distinguish between these physically different models, the fine microscopic detail provided by the dHvA method provides what is needed to sort out the problem.

Experimental results in Rh–Fe (Cheng *et al.* 1979) lead to the following conclusions.

1. The d-wave theorem established for noble metal hosts (Schrieffer 1967; Coleridge *et al.* 1972) is general in the sense that the scattering rate from the purely d-like Fe impurity increases with increasing fraction of d wave-function for different sheets of Fermi surface in the host.

2. The slopes dx/dT as a function of temperature also depend upon wave-function symmetry, and increase with increasing d-content. An unresolved puzzle is that the slope for a purely d-like sheet is much larger than that of the resistivity (see Fig. 10.28), yet since the current is carried

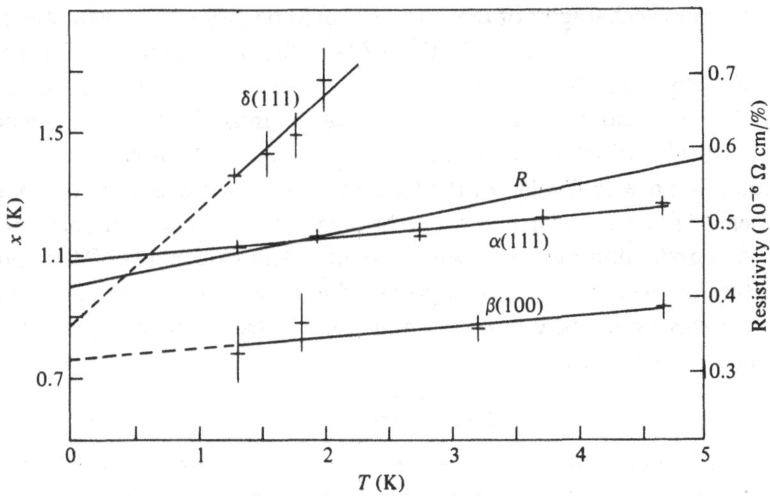

Fig. 10.28. Comparison of the temperature dependence of scattering in Rh (0.05 per cent Fe), for three orbits on the Fermi surface, with the temperature dependence of resistivity. Orbit labels α and β refer to L-centred and X-centred hole pockets, and δ refers to the smaller of two Γ-centred electron sheets (listed in order of increasing d-wave charge density).

mostly by d-like carriers, as established by the APW calculation of Cheng (1977), one would expect these slopes to be similar. Similarly, a comparison of the slopes dx/dT from three different sheets of the Fermi surface shows changes which are in excess of what is expected from the fraction of d wave-function character.

3. The Kondo model is not the explanation for the anti-Kondo $R(T)$ in Rh–Fe because an analysis of $x(T)$ in the limit as $T \to 0$ yields values of the potential scattering phase shifts too small to reach the peak of $\sin^2 \Delta\eta_L$.

4. Knapp's two-band model is also incorrect because the d-like sheets of Fermi surface which couple most strongly to the impurity are also the ones which carry most of the current.

5. Models using phase shift ideas to explain the bulk macroscopic $R(T)$ (Nagasawa 1972) find their natural extension to dHvA measurements in the phase-shift models discussed above in Section 10.5.1. The results are in qualitative agreement with these ideas, except that the experimental slopes dx/dT vary too rapidly to be accounted for with the existing models unless one supposes an (unlikely) orbit-dependent phase shift.

6. An extension of spin-fluctuation ideas into a theory for dHvA observables appears extremely attractive but has not as yet been carried out. Here, the principal new feature to be added to the Kaiser–Doniach–Rivier–Zlatic theory is the dependence of the interaction strength upon the symmetry character of the orbit. It would be particularly valuable to clarify the notion (Rivier & Zlatic 1972a,b) that a *qualitative* difference exists between a d-like host band, in which the interaction proceeds directly by a resonance, and an s–p-like host band, which must first leak onto a d-like orbital before the interaction can occur. However, it is worth noting that previous ideas of the 's–d interaction' in noble metal Kondo systems have been clearly shown to be incorrect, and it is the d-character of the conduction electrons which counts. Any extensions of the spin fluctuation theory must also be prepared to take into account the mixed partial wave symmetry in real conduction electron bands; a 'two-band model' will not do.

Exchange-enhanced hosts

It is well known that the addition of dilute concentrations of the 3d transition elements Mn, Fe, Co and Ni to the 4d host metal Pd results in a sudden magnetic transition from the exchange-enhanced paramagnetic behaviour of the Pd host, to long-range ferromagnetism. Diffuse neutron scattering experiments have verified that the resulting 'giant moments' are in fact spatially extended, ranging over more than 100 Pd atom sites in the vicinity of an Fe or Co impurity. This, and other related evidence, may be interpreted as implying that the *major part* of the ferromagnetism in these dilute alloys arises from the participation of the Pd host matrix.

Theoretical models for *band* magnetism in Pd alloys emphasize the importance of *inter*atomic exchange interactions (involving neighbouring Pd atoms) in addition to the *intra*-atomic exchange of electrons on the same atom. Within such models, the presence of dilute concentrations of Fe, Co or Mn results in the direct polarization (spin-splitting) of the d-bands of the Pd host, which is then responsible for the 'giant moments'. This interpretation of the origin of ferromagnetism in these dilute alloys is attractive because it is very similar to the picture of 'itinerant electron magnetism' which has recently become experimentally well-established for ferromagnetic metals such as Fe and Ni.

Unfortunately, what is missing in the case of Pd dilute alloys is direct, microscopic evidence for the spin-splitting of the Pd d-bands, as a function of solute concentration in the very dilute limit (Windmiller, Ketterson & Hörnfeldt 1969, 1971). There is also a second microscopic aspect of ferromagnetism which has been overlooked in all measurements to date. The possibility of spin-dependent differences in the scattering rates for up-spin and down-spin electrons (majority- and minority-band electrons) and their dependence on solute concentration. Both types of measurements are important because they could directly confirm the applicability of a band picture for the ferromagnetism of Pd 'giant moment' dilute alloys. Both of the required pieces of information can be provided by the dHvA waveshape technique, via measurements of conduction electron exchange energy shifts and SDS rates as functions of solute concentration in the dilute limit. Experiments are currently in preparation for Pd(Fe) (D. H. Lowndes *et al.*, unpublished data) and Pd(Co) (S. Hörnfeldt *et al.*, unpublished data). Related experiments, using the spin-splitting zero technique, are also being carried out for the less strongly exchange-enhanced Pt-based dilute alloys (Hornfeldt, Dronjak & Nordberg, unpublished data).

Polyvalent metal hosts

Waveshape analysis experiments of the type described above have not yet been carried out in dilute alloys based on polyvalent metal hosts. However, both Shiba's theoretical work, and the experimental techniques described above, were preceded by pioneering experimental work in which anomalies in the dHvA effect in Zn-based dilute alloys containing transition metal solute atoms were first detected. In a series of papers, Hedgecock & Muir (1963), Paton & Muir (1968) and Paton *et al.* (1970) showed how the energy-dependent relaxation time associated with the Kondo effect could be included in the expression for the dHvA effect. They later reported a magnetic field and temperature dependence of the Dingle temperature for the electrons on the 'needle' orbit in Zn(Mn) dilute alloys (1970) and pointed out the connection between their measurements and the field dependence of the electronic scattering rate resulting from the freeze-out of spin-flip scattering in the magnetoresistance. This work was extended in the measurements of Li & Ström-Olsen (1973) and the theoretical calculation by Harris, Mulimani & Ström-Olsen (1975a) for Zn(Mn) and Zn(Cr).

Paton (1971) has also shown how localized spin fluctuation effects may be detected using the dHvA effect, via measurements of changes in the cyclotron effective mass and the Dingle temperature in dilute Al(Mn) alloys.

Acknowledgements

We gratefully acknowledge the help of the following individuals in supplying assistance with calculations, graphics, or unpublished data used in this chapter: T. G. Matheson, P. Deimel, Y. K. Chang, Y. Chung and P. Reinders.

11

High-precision measurements of de Haas–van Alphen frequencies

P. T. COLERIDGE AND I. M. TEMPLETON

11.1 Introduction

In the early days of the de Haas–van Alphen (dHvA) effect, any reasonably high-quality single crystal of a 'new' metal was a potential source of Fermi surface information. As many of the frequencies as possible were measured as a function of crystal orientation and the results fitted to some form of model. The accuracy needed for this, typically 1 per cent, made the measurements relatively simple, perhaps even tedious. So long as the magnet calibration was known to about $\frac{1}{2}$ per cent, a count of a hundred or so oscillations yielded the required precision. The most difficult task was the separation of multiple frequencies and the analysis of beat patterns, for example by using resonant circuits in the pulsed field technique. Such measurements proved most valuable in unravelling details of Fermi surface topology and in establishing which computation techniques yielded band structures that were correct, at least at the Fermi energy. Furthermore, they provided sufficiently accurate Fermi surface models to allow a detailed understanding of the amplitude-dependent characteristics of the dHvA effect such as Dingle temperature, g-factors, magnetic breakdown and the Shoenberg effect (magnetic interaction).

The increasing use of band-structure techniques to parametrize pure metal Fermi surfaces and an interest in the measurement of dilute alloy Fermi surfaces have led to a demand for more accurate frequency measurements. In the first case, accurate absolute measurements are required to fit a semi-empirical band structure and, in the second, accurate relative measurements are required to detect the small changes in Fermi surface dimensions that occur for alloy concentrations sufficiently low that scattering of electrons out of the Landau levels does not cause the dHvA signals to vanish.

An example of the relative precision required in investigating alloys is illustrated in Fig. 11.1. The ⟨100⟩ rosette and belly oscillations in a Cu (0.1 at.% Zn) single crystal are compared under similar experimental conditions with the corresponding oscillations in pure copper. The signal strength is a factor of ~200 times lower, near the noise level. The frequency differences are of order 1 in 1000 so a measurement of the *changes* in frequency (and the anisotropy over the Fermi surface of the changes) with a precision of better than 1 per cent requires a relative precision of measurement of a few parts in 10^6. This is typical for large sheets of Fermi surface. For smaller sheets, i.e. lower dHvA frequencies, higher impurity concentrations can often be tolerated because the cyclotron masses are smaller and hence a higher scattering rate ($1/\tau$) gives the limiting value of $\omega_c\tau$ corresponding to the noise figure. In practice, as will be shown below, the precision of frequency measurements is somewhat lower so that no significant improvement in accuracy is obtained by measuring the smaller sheets of Fermi surface.

In contrast to the relative measurements needed in alloys, for pure metals absolute measurements are required. By inverting dHvA data and fitting the results to a band structure it is possible to parametrize the Fermi surface in terms of a set of potential parameters. If a pseudopotential formalism is used a set of pseudopotential coefficients results

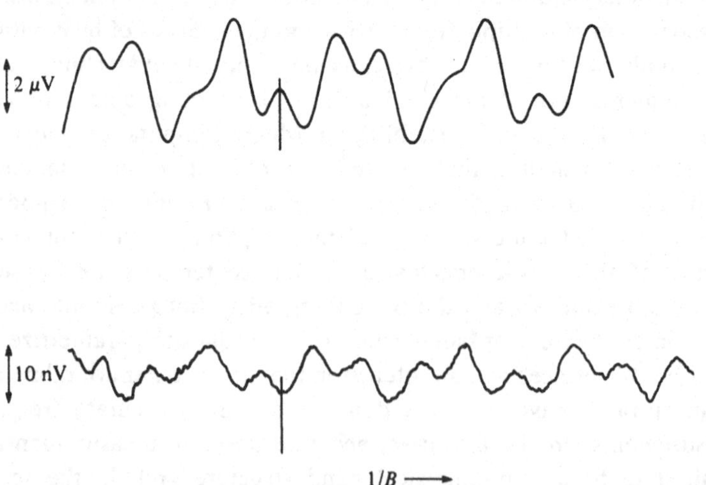

Fig. 11.1. dHvA signals from pure Cu (upper trace) and Cu (0.1 at.% Zn) (lower trace) with the field along ⟨100⟩, taken under similar experimental conditions. The ^{27}Al NMR frequency at the marker is 57.500 00 MHz, corresponding to $B = 51.745\ 86$ kG. Between pure metal and alloy the higher (belly) frequency has increased in phase by 6.45 cycles while the lower (rosette) frequency has decreased in phase by 7.87 cycles.

while 'muffin-tin' methods such as the APW or KKR technique yield phase shifts or equivalently logarithmic derivatives. Band-structure calculations generally obtain accurate energies by making a variational approximation in conjunction with some form of truncation of the basis states to render the calculations feasible. Extending a parametrization to **k**-states of the Fermi surface by using a local approximation or using the parametrization to obtain wave-function amplitudes loses some of the accuracy associated with the variational approach. It is therefore important to determine the Fermi surface, within the fundamental limitations of the model, with the highest possible accuracy so that the experimentally determined potential parameters reflect only errors peculiar to the model used, which can be investigated computationally, not errors in the input data. Such data typically require a precision of measurement of order 0.1 per cent or better.

It is therefore convenient to consider as 'high precision' frequency measurements those with an absolute accuracy significantly better than 0.3 per cent and a relative accuracy (between specimens) of 0.01 to 0.0001 per cent.

11.2 Experimental considerations

Considering only the rth harmonic of a single-frequency component of the dHvA signal the oscillating magnetization can be written as

$$\tilde{M} \sim \cos 2\pi\left(r\frac{F}{B} + \psi_\infty^r\right),$$

where F is the dHvA frequency and B the total field (including the Shoenberg effect). The total phase is conveniently expressed in cycles $(rF/B + \psi_\infty^r)$ with a residual phase number being the fractional part of the total phase number. The 'infinite field phase', ψ_∞^r, is the residual phase number as $B \to \infty$ and is obtained from the Lifshitz–Kosevich formula (1955) as $\gamma r \pm \frac{1}{8} \pm \frac{1}{4}$. Here γ is the Onsager factor ($=\frac{1}{2}$ for all practical purposes, see Gold (1968)), the term $\pm\frac{1}{8}$ is taken according as the extremal area is a minimum or maximum (always provided that the curvature factor $\partial^2\mathcal{A}/\partial k_B^2$ is non-zero) and $\pm\frac{1}{4}$ is taken according as the spin-splitting factor, $\cos(r\pi g m_c/2m)$, is positive or negative. Two additional factors arise in an experimental situation. Use of the frequency modulation technique with detection on the nth harmonic introduces into the *detected voltage* a phase factor of $n/4$ cycles, and the various sign reversals that may occur through the detection system introduce an additional ambiguity of $\frac{1}{2}$ cycle which must be tracked down.

To measure the frequency F with high precision requires certain experimental conditions to be satisfied.

(i) *Orientation.* Even in a metal such as potassium with an almost spherical Fermi surface the variation with orientation of the dHvA frequency is several times larger than the required precision of measurement. This means that not only must the frequency be measured accurately but also it must be characterized with the same precision, i.e. the orientation must also be accurately determined.

(ii) *Magnetic field.* The field B at the *specimen* must be sufficiently homogeneous and must be known with a precision at least equal to the final accuracy required.

(iii) *Phase.* The number of cycles between two or more field values must be known accurately, typically to within a small fraction of a cycle. Both the integral number of cycles and the fractional part of the phase must be known.

(iv) *Spurious responses.* The measured quantity, e.g. $d\tilde{M}/dt$ in the field modulation technique, must be a genuine representation of the single dHvA frequency of interest unadulterated by such effects as other dHvA frequencies, the Shoenberg interaction effect, de Haas–Shubnikov oscillations or other quantum oscillation effects.

We discuss these conditions in sequence, and we also discuss some of the special considerations appropriate to measurements in dilute alloys. Our examples are drawn mainly from the noble metals and superconducting solenoids because of our experience and familiarity with them, but similar considerations apply to all metals and magnet systems.

11.3 Orientation

With care a specimen may be oriented using X-rays and transferred to a cryostat with a total angular error of order $\frac{1}{2}^\circ$. This is frequently unacceptably large; for example in silver an angular error of $\frac{1}{2}^\circ$ at symmetry directions corresponds to frequency errors in both the $\langle 100\rangle$ rosette and the $\langle 111\rangle$ neck of more than 0.02 per cent and the error is worse away from symmetry directions. It is necessary therefore to orient the specimen in the cryostat by using the inherent accuracy obtained from the dHvA oscillations resulting when the crystal is rotated in constant magnetic field.

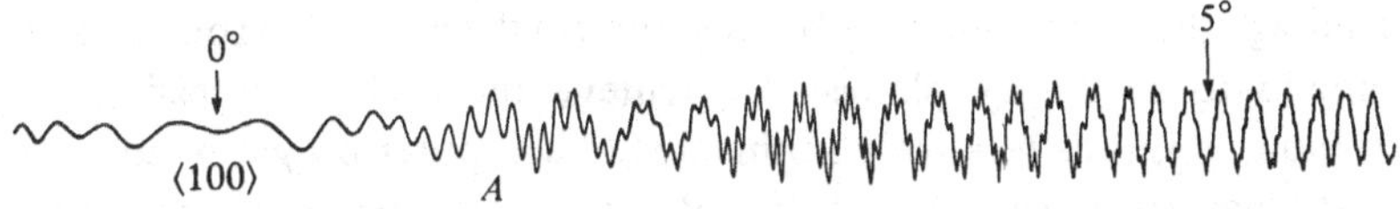

Fig. 11.2. Rotation diagram for pure Cu near ⟨100⟩ in the (110) plane (from Halse 1969). Rosette (the *shorter* period) and belly oscillations are seen clearly at A. Although the rosette is some 2.4 times lower in frequency, it changes in cross-section with angle about 12 times faster than the belly, and hence would be used for a line-up in the ⟨100⟩ direction.

Rotation of the sample in constant field gives a rotation diagram of the type used, for example, by Halse (1969) and illustrated in Fig. 11.2. The angular variation of the dHvA frequency can be mapped out with considerable precision. Indeed if the rotation device has at least 180° of rotation and sufficiently accurate angular calibration the orientation of a completely random specimen can be determined solely from the symmetry of the rotation diagram without recourse to X-rays. This technique was pioneered by Mueller, Windmiller & Ketterson (1970b) in the platinum group metals and has also been employed with success by Gaertner & Templeton (1977) in the alkali metals where the fragility of the specimen and the low density of the metals renders X-ray pictures unreliable as well as imprecise. The technique relies on measuring the angles between the self-intersection of the orbit when mapped into the basic 1/48th of the Brillouin zone (see Fig. 11.3), and has the inherent

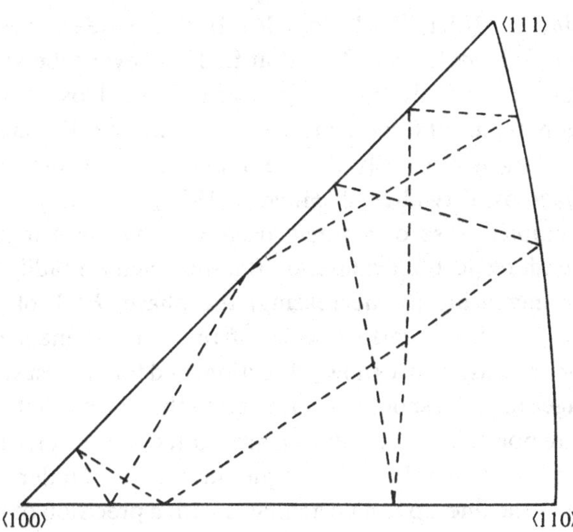

Fig. 11.3. Locus of field directions during rotation of a randomly oriented cubic crystal, remapped into the basic 1/48th of the Brillouin zone (after Gaertner & Templeton 1977).

advantage that more data are obtained than in the case when rotations are confined to symmetry planes. To achieve the highest precision this technique requires relatively sophisticated computer programs to allow for angular errors such as the axis of rotation not being normal to the magnetic field. A prerequisite for the technique to work is the existence of a signal over the complete solid angle, and it is therefore restricted to closed sheets of surface.

To achieve the highest accuracy of orientation in the general case requires that measurements be restricted to symmetry directions. Additional data may be taken at turning-points of the frequency that exist at off-symmetry directions (e.g. near ⟨511⟩ in all the noble metals) but these are not as usefully specified in orientation as the symmetry directions. In fitting pure metal Fermi surfaces the angular position of the turning-points is as crucial a parameter as the actual frequency, and, on alloying, the turning-point is as likely to shift in orientation as to change in frequency. The rotation device needs two axes, preferably orthogonal, with an angular resolution of typically 0.02°. A total angular range of a few degrees is sufficient if the specimen can be cut close to the desired direction. Because of the possibility, discussed below, of a linear transverse variation of magnetic field, any specimen-orienting mechanism should ideally involve as little lateral displacement ($dx/d\theta$) as possible, to avoid an erroneous line-up where the phase change due to tilt happens to balance that due to change of field with displacement, i.e. $(1/F)(dF/d\theta) = -B(d(1/B)/dx)(dx/d\theta)$. In this sense the design shown by Coleridge & Templeton (1972) is at fault: a new probe with the coil system clamped centrally in a ring which is tilted by a similar (but non-orthogonal) pair of drive-rods has proved more satisfactory. A true gimbal-mounting is obviously the ideal solution: one such system has been described by Terwilliger & Higgins (1972).

For the simplest case of a single frequency having a maximum (or minimum) with respect to orientation the specimen is readily aligned by consistently increasing (or decreasing) the phase, F/B, of the dHvA signal. To establish the correct sense of rotation the magnetic field is increased or decreased, in a known direction, and for highest sensitivity in the final stages the field should be adjusted so that the oscillation is near a zero-crossing point. The procedure is illustrated in Fig. 11.4 and details are discussed in Coleridge & Templeton (1972). Under favourable circumstances the line-up can be achieved with a precision of better than 10^{-3} cycles which is usually an order of magnitude smaller than the final precision of measurement. The field at which the line-up is performed is

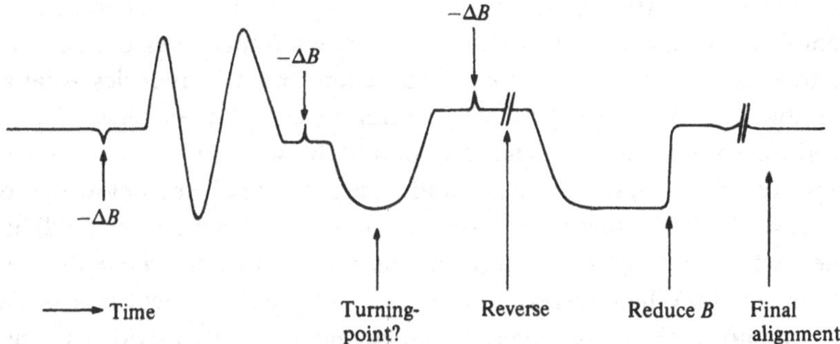

Fig. 11.4. Alignment for minimum cross-section by tilting (or rotation) in constant B. The small momentary field reductions are used to indicate the sense of increasing phase and hence to identify the turning-point (Coleridge & Templeton 1972).

somewhat of a compromise between a low field, to give maximum phase and increased precision, and a high field, to give maximum signal-to-noise ratio. It is a wise precaution to check the line-up as the experiment progresses, since the probe might shift slightly as the dewar is pumped down or the liquid helium level changes. It is even possible in some systems for the direction of the magnetic field relative to the geometric axis of the probe to be a weak function of the field.

If two or more frequencies are present the line-up must concentrate on only one of them, ideally that most sensitive to orientation. If the frequencies are of significantly different periods (e.g. in the noble metals the neck and belly at ⟨111⟩, and the belly and rosette at ⟨100⟩) then it is often possible to adjust the modulation amplitude to null out one frequency, and to line up using the other. If several frequencies are present, which cannot be reduced in number by judicious choice of magnetic field, temperature and modulation amplitude, the procedure becomes much more tedious. An 'on-line' Fourier transform procedure can be helpful in such cases because the amplitude and phase of the several oscillations must be studied repetitively using the same field sweep, and successive adjustments made to the orientation. Alternatively the frequencies can be measured on a grid of points around the symmetry direction and the correct value obtained by suitable interpolation (King-Smith 1965).

At times the symmetry direction is determined not by a turning-point but by the crossing of several branches. In this case, the approach to the exact symmetry direction is characterized by a beat pattern that becomes progressively longer. However, because there are two degrees of freedom at least 3 branches are beating so that the pattern is in general rather complicated, particularly if other, unwanted, frequencies obscure the picture. A general procedure is to reduce unwanted frequencies as far as possible and then to get all the interfering branches in phase, i.e. to produce a beat pattern with no zeros or nulls between infinite field (where they are necessarily all in phase) and a magnetic field well below that of interest. When the first beat zero or null (counting from infinite field) has been established successive adjustments must be made to move the zero to lower field. When the zero has been moved to as low a field as possible, into the noise, the amplitude can then be maximized to provide a precise and final line-up. It is apparent that this technique, illustrated in Fig. 11.5, is extremely tedious because of the necessity of sweeping over a wide field range to see the long beat pattern and to confirm that a particular zero is indeed the 1st or 2nd, etc.; from infinite field. Furthermore, the effect of three or more branches means that the beat pattern is not simple so that only small steps in each of the orthogonal orientations can be made without changing the character of the beat pattern appreciably.

It is emphasized that for accurate frequency measurements it is essential that no beat occurs between the lowest field used and infinite field. Any beat produces a phase modulation which may be too small to be explicitly detected as curvature in a phase plot but which could produce a large error in frequency measurements (see Fig. 11.6). It is, of course, obvious that if the shape of the Fermi surface and the details of the frequency branches are not well understood then any attempt to line up the crystal may well be doomed to failure, particularly in the presence of phenomena such as magnetic breakdown or the field-dependent spin–orbit coupling that occurs in ferromagnetic metals (see Chapter 6).

11.4 Magnetic field

11.4.1 Homogeneity and stability

It goes without saying that the field homogeneity over the dHvA sample should be considerably less than a cycle of the highest frequency to be measured, i.e. better than ~1 in 10^5. Many commercial superconducting solenoids show a marked lack of axial symmetry, so it is not sufficient during the initial evaluation of a solenoid to measure the axial field profile only. While a 'high homogeneity' solenoid may show uniformity better than 1 in 10^5 over 1 cm or more along the axis, the field is often not circularly symmetric and may vary linearly across the bore by ten times

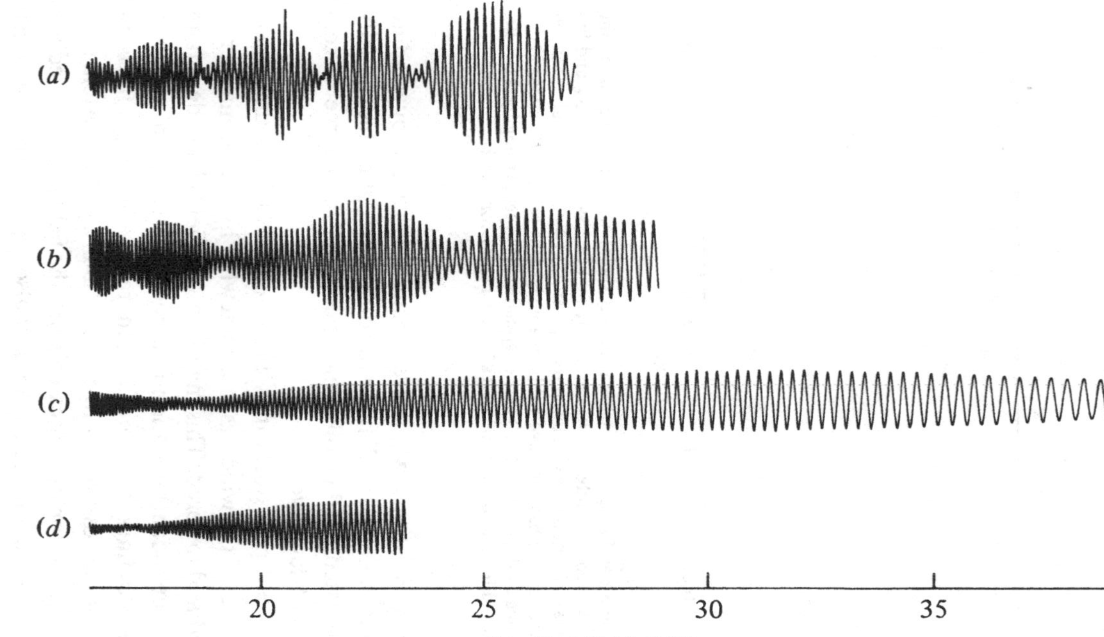

Fig. 11.5. Steps in orienting a ⟨100⟩ aluminium specimen using the γ frequencies (3.91×10^6 G, Coleridge & Holtham, unpublished data). (*a*) Complex beat pattern produced by 4 frequencies beating together. The specimen is 1°–2° from the symmetry axis. (*b*) The various beat periods have lengthened, so the orientation is closer to the symmetry axis. (*c*) The specimen is now sufficiently close to the axis that there is no beat minimum between that at 18 kG and infinite field. The decrease of amplitude above 20 kG, approximately as B^{-4}, is due to the effect of the Bessel function at constant modulation amplitude. (*d*) The 'last zero' is moved to a lower field. Note that the same result could have been obtained by maximizing the amplitude at, say, 20 kG. The specimen is now within 0.1° of the symmetry axis.

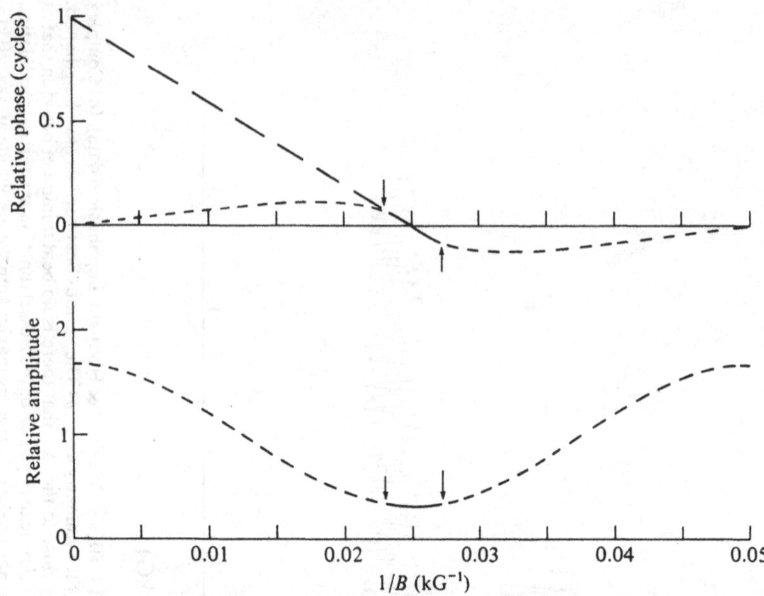

Fig. 11.6. Relative phase and amplitude of the resultant signal in the presence of a long beat. The secondary component differs in frequency by 20 kG, has the same ψ_∞ and has ~0.7 of the main signal amplitude. Over a limited range around 40 kG the resultant shows little amplitude variation, its phase relative to the main signal is reasonably linear and it apparently extrapolates to the correct ψ_∞. The extrapolation is in fact wrong by +1 cycle, and would lead to an error of −40 kG (see Templeton 1972).

this amount. Such variations, which are presumably caused by winding and wire non-uniformities, are difficult to correct except by individually designed and adjusted saddle-coils, and even then the degree of correction may be field-sensitive.

The magnetic field penetrates into and is expelled from the superconductive material of the windings in a somewhat hysteretic manner as the field is raised and lowered. This may lead to a field-dependent (and sweep-direction-dependent) variation in homogeneity. Again, although superconductive windings on a probe can reduce heating effects in modulation coils or improve signal-to-noise ratio in a pickup system they should be used with caution as they can disturb the field in the immediate vicinity of the dHvA sample. The field may also be perturbed by other pieces of superconducting material (e.g. solder) or by magnetic effects which can develop in the stainless steel of a dewar tail. Any miniature coaxial cables which may be used with an NMR system should be chosen with care, since some types have a magnetic core.

While homogeneity requirements become less stringent as the field is increased, the total phase and hence the precision with which the frequency can be defined decreases at the same rate; so, as with orientation, the field range over which measurements are made becomes a compromise between maximum phase and minimum noise.

The question of field stability should also perhaps be discussed here. During field sweeps there may be some non-linear 'events' associated with flux-jumps in the solenoid windings, though these are much less apparent with modern multifilamentary materials. In the stationary but non-persistent state the stability will depend essentially on the power supply, helped by the large inductance of the solenoid and with only second-order effects due to flux redistribution. The latter, however, becomes important in the persistent mode. The observed effect is that the central field changes in a direction *opposite* to the most recent sweep, and this may well, as suggested above, be accompanied by some change in homogeneity, although we have not investigated this. The drift can be made negligibly small by cycling the field in decreasing steps around the final required value. Following such cycling, only when a range of more than 100 G or so is to be covered during one group of measurements does the 'drift-back' tendency become at all noticeable.

11.4.2 Measurement

It is possible in many cases to attain sufficient accuracy by using a transfer standard such as a Hall-effect (Hill & Hwang 1966) or magnetoresistance (Scott, Springford & Stockton 1968) sensor, or even the solenoid current itself (Coleridge & Templeton 1971a), but the primary magnetic field standard for high-precision measurement is ultimately some form of NMR probe. Ideally the NMR sample should be close to, and only a little larger than, the dHvA specimen so that the average field sensed is closely representative of the dHvA field. This is particularly true with electromagnets or in solenoids showing any appreciable field/current hysteresis, since this will be accompanied by changes in flux distribution. For the most precise measurements or in solenoids with questionable homogeneity it is even desirable to mount temporarily a miniature NMR sample in the position normally occupied by the dHvA specimen and compare the two resonance frequencies over the operating field range.

We are assuming here that the dHvA specimen itself, while exhibiting the *internal* effects (magnetic interaction) associated with its own magnetic induction, has no *external* influence on the field-measuring system. In practice such effects should be small and would be eliminated by the averaging procedures discussed below for dealing with phase modulation. For high-precision measurements on ferromagnetic materials, however, it is presumably necessary to measure B at some point remote from the specimen, or to use a calibrated transfer standard. These metals represent a special case and are discussed in Chapter 6.

The choice of material for the NMR resonance is restricted by the requirement for a narrow line-width and strong signal at low temperatures, and for a gyromagnetic ratio such that the resonance frequency in the field to be measured is not so high that the design of the RF system becomes difficult. A convenient nucleus, having a linewidth at 4.2 K of ~9 G, is ^{27}Al (1.1112 MHz/kG). This was originally suggested by Maxfield & Merrill (1965) in the form of coarse filings, but Higgins & Chang (1968) found a fine powder mixed with epoxy resin to be more suitable. This material is readily machinable to any required shape. Other workers have used, for example, sodium dispersion (Hill & Hwang 1966), caesium dispersion (Rupp 1966), sodium caesium and chlorine ions (Anderson, Sandfort & Stone 1972b) and copper powder (Hulbert 1976).

In some cases it is possible to define a magnetic field to a few tenths of a gauss by using the first derivative of a relatively wide, strong resonance so that the absolute precision is in fact as good as with a weak, narrow resonance. The precision under any given conditions can be estimated in practice by a careful examination of the line-shape of the derivative signal, for example with the oscillator locked to a crystal harmonic and the field swept slowly in *both* directions through the resonance. Making allowance for any hysteresis due to 'frequency-pulling', the shape of the dispersion line usually shows some asymmetry which may vary with RF and modulation level. If this is regarded as being due to the presence of a weak absorption (fundamental) component, a simple mathematical treatment of the assumed Lorentzian shape shows that if the observed positive and negative derivative amplitudes are A_1 and A_2 respectively, then the shift is approximately $0.19 \times$ peak separation $\times$ $2(A_1 - A_2)/(A_1 + A_2)$ in the direction of the negative peak. Thus, in ^{27}Al, for a typical asymmetry of 3 per cent, and a peak separation of 9 G, we have a zero error of ~0.05 G or ~1 ppm at 50 kG.

Because of the physical layout of the usual cryostat there is necessarily a relatively long RF path from the NMR oscillator to the probe, and there is an additional problem in the conflicting requirements of low loss in the RF line and low heat influx into the cryostat. Slavin (1972) and Wampler,

Matula, Lengeler & Durkansky (1975) have used tunable NMR oscillators in liquid helium to avoid the long-line problem, but even in this case a coaxial line is necessary to get the signal out of the cryostat. As Wampler *et al.* point out, a coaxial line made from German silver tubes can be much superior in RF performance to a miniature coaxial cable of similarly low thermal conductance. Hill & Hwang (1966) proposed the use of a room-temperature marginal oscillator in which the cable is used as part of the resonant circuit. Variation of the cable length varies the frequency, and frequency-selective feedback is used to force the coil/cable system into higher modes of oscillation. This approach makes changing the frequency difficult because it involves changing cables. Anderson *et al.* (1972b) have described a marginal oscillator (maximum frequency ~45 MHz with a long cable) with switched tuning and narrow resonances that can be used for precise measurement with slow tracking, while Hulbert (1976) used a wideband RF bridge system with frequency modulation and a second harmonic detection for fast tracking.

For our measurements we have found the Hill & Hwang (1966) approach reasonably satisfactory, with Al powder in epoxy resin as the resonance material. Various improvements to the original circuit have been made, notably the addition of a more finely stepped set of tuned circuits to force the oscillator into the correct mode, and a low-temperature coaxial line made from the stainless steel probe support tube with a suitably insulated smaller tube down the centre. The ideal diameter ratio for 50 Ω is 2.3 : 1, but our system of $\frac{1}{2}$-inch and $\frac{3}{16}$-inch o.d. tube performs well in a 50 Ω system with considerably less helium boil-off and better RF performance than with a miniature cable. The strength of the resonance is very dependent on cable matching, and it is particularly important to have good connections between the various cables used.

It is fairly usual to operate a marginal oscillator with the frequency locked to the centre of the first derivative of the resonance. In this condition the highest absolute precision depends on keeping the peak-to-peak modulation level less than the line-width. For a wide resonance such as ^{27}Al this frequently is compatible with the modulation necessary to observe dHvA signals from large pieces of Fermi surface, so it is possible in some cases to use the same modulation field. It is also important for the highest relative precision in certain critical measurements that the operating conditions at any one field (i.e. modulation and RF level) be standardized at some suitable value so that although a magnetic field might not be known absolutely to better than, say, 1 in 10^5 it can be reproduced from day to day and from specimen to specimen to a few parts in 10^7. Again, when the NMR system is operating in this locked mode, the error signal should be reduced to zero before making a frequency measurement: in most cases it is impossible to have a sufficiently high loop gain to achieve this without making the time-constant unacceptably long, so it must usually be done manually or, conveniently, by some form of computer control.

11.5 Measurement of phase

An experimental determination of the total phase number $\psi = (F/B + \psi_\infty)$ can be separated into establishing the integer part of the phase, i.e. the Landau level number, and a measurement of the residual phase number. Provided the frequency is known with sufficient accuracy the integer part can be obtained from the known field B with an error of less than 1 cycle. Even for phase numbers as high as 10 000 the required accuracy can be obtained by the direct counting of a few hundred oscillations between two accurately known field values. In alloy experiments the changes are usually sufficiently small that the frequency will only differ from the pure host metal value by a few cycles, so the host value can be used. Any ambiguity that may result can be resolved in practice by measuring at several field values. Once the integer part of the phase is known, measurement of the residual phase of high frequencies with an accuracy of, say, 0.01 cycles serves to give the total phase and hence the frequency, with a precision of typically 1 in 10^6. For lower frequencies where the total phase may be only a few hundred the precision is of course less but the absolute accuracy for Fermi surface fitting is still satisfactory in most cases.

As mentioned previously, there is in any experimental set-up an ambiguity of $\frac{1}{2}$ cycle in the signal resulting from numerous phase inversions and depending also on the polarity with which the magnet is energized. While it is possible to determine the phase absolutely by placing a piece of iron in the pickup coil at room temperature and magnetizing it in a known sense (thus producing a known sense of hysteresis curvature in the magnetization) and relating this to the sense of the field in the experimental system, a more convenient calibration is desirable following rewiring of the probe or similar experimental changes. Such a calibration may be obtained by studying the relative phases of first and second dHvA harmonics. In some cases, notably in the neck signal in copper where the fundamental is near a spin-zero and the second harmonic is strong, the phase difference $2\psi_1 - \psi_2$ is easily determined by inspection of the characteristic pattern† and will be inverted (changed by $\frac{1}{2}$ cycle) if an odd number of phase reversals is involved. Alternatively, under conditions of strong magnetic interaction, the second harmonic phase is dominated by the Shoenberg effect and this also serves to remove the arbitrary phase factor (see Chapter 10).

The most accurate way of observing the dHvA signal for measuring the phase is to use a static field technique, i.e. the main field is held constant and not swept at all during the actual measurement. With a superconducting solenoid this is conveniently done by adjusting the heater current through the persistent switch so that the field drifts very slowly but can be stopped, essentially instantaneously, at a desired value by switching the current below the critical level. In this way all errors involved in sweeping the field, i.e. the effects of various time constants, etc., are eliminated. As mentioned above, the field may show a tendency to drift slowly back but this can be minimized by first cycling the magnet around the field to be used. A highly stable power supply may ease the problem and simplify the setting of the field but is unlikely to be as stable as a good persistent switch. Drifts of less than 1 in 10^7 during a measurement are desirable although larger but uniform drifts can be tolerated.

For more complicated situations, where a Fourier transform of relatively high dHvA frequencies is required, it can be convenient to stabilize the field with the persistent switch and then step over a *small* range (a few tens of gauss) by means of an auxiliary coil, measuring the NMR frequency and dHvA signal amplitude at each step. The addition of a capacitor and choke allows the modulation coil to serve the dual purpose of carrying both the a.c. modulation and the auxiliary d.c. field but if this is done it is important to consider the homogeneity of the field from the modulation coils and to allow for any differences between the field at the specimen and at the NMR sample. For this reason it is best to step the field equally either side of the zero current value, and take the field at the centre of the range from the mean value of the measured NMR frequencies. This approach cannot be used for lower frequencies as the required d.c. current range becomes unacceptably large.

11.5.1 Graphical techniques for determining phase

For the simplest case of a single uncontaminated frequency the residual phase is most accurately determined near zero-crossing points. Using the

† See e.g. Fig. 11.15.

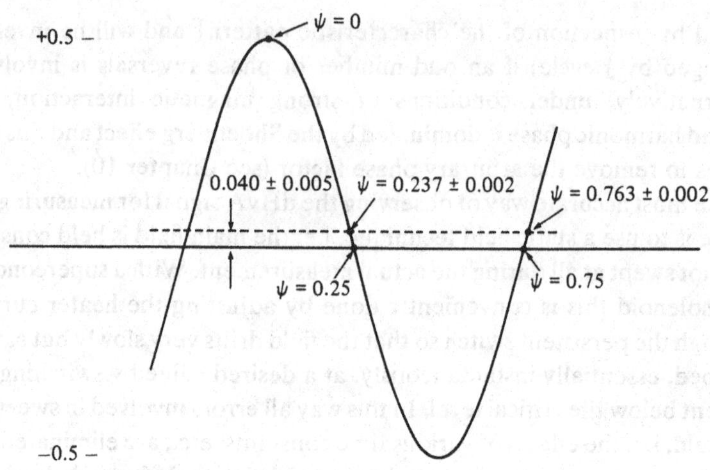

Fig. 11.7. Local phases in a cosine wave at points offset from a zero-crossing.

convention that $\psi = 0$ corresponds to the maximum of a cosine wave the deviation from $\psi = \frac{1}{4}$ or $\frac{3}{4}$ is given for an offset δ by $\sin^{-1}(2\delta/A_{pp})/2\pi$ where A_{pp} is the peak-to-peak amplitude (see Fig. 11.7). For small δs this conveniently reduces to $\delta/\pi A_{pp}$. Errors in defining the base line are eliminated, at least to first order, by averaging pairs of readings at positive- and negative-going crossing points. An advantage of using the zero-crossing points with even harmonic detection in the field modulation method is that these correspond to points where $\tilde{M}$, the magnetization, goes through zero so the Shoenberg effect is smallest at these points and perturbs the measurement least. As illustrated in Fig. 11.8, an accuracy of 0.002 cycles is readily achieved with this technique.

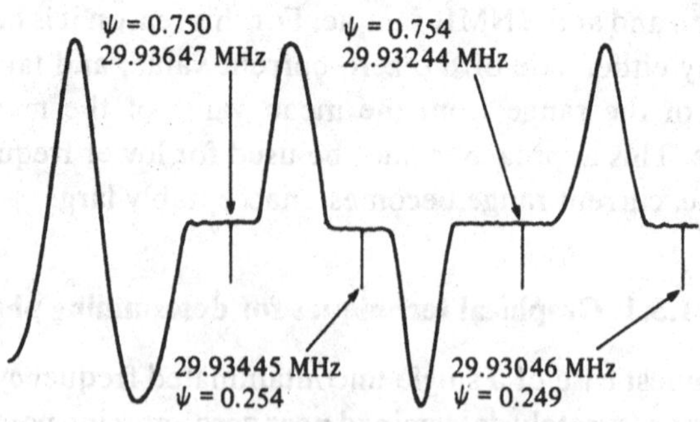

Fig. 11.8. Tracing of experimental record taken during measurement of phase in the ⟨110⟩ dogsbone of Ag. $B \approx 27$ kG, $T \approx 1.1$ K, and the horizontal interval is about 5 min (Coleridge & Templeton 1972).

The data are conveniently presented by using a trial frequency F_t, close to the correct value, and subtracting from the measured phase (plus the Landau level number) the quantity F_t/B, where B is the magnetic field (measured as described above). This 'corrected' phase is then plotted versus B^{-1} and the resultant slope gives a correction frequency to be added to F_t. The plotted phase should extrapolate to ψ_∞ at $B^{-1} = 0$. This is demonstrated in Fig. 11.9. Alternatively in alloys the pure metal frequency is used for F_t and the infinite field phase, which it is assumed changes only by a negligible amount, is also subtracted, giving a corrected phase which is $\Delta F/B$. In principle one field value is then sufficient to determine ΔF, but in practice more points are taken to improve statistics and to check that the integer part of the phase is not in error by one or two cycles.

This straightforward but powerful technique is unfortunately not possible when more than one frequency is present. In this case one approach is to eliminate the spurious frequencies by choice of modulation, field, etc., and then measure the wanted frequency as above. This cannot always be done for one reason or another so it is necessary to modify the technique so as to exploit the property of a sinusoid that the average value of samples taken at equal intervals over an integer number of cycles of the sinusoid is zero. (This is easily verified using a vector representation for the sinusoid.) For example, one procedure is to take

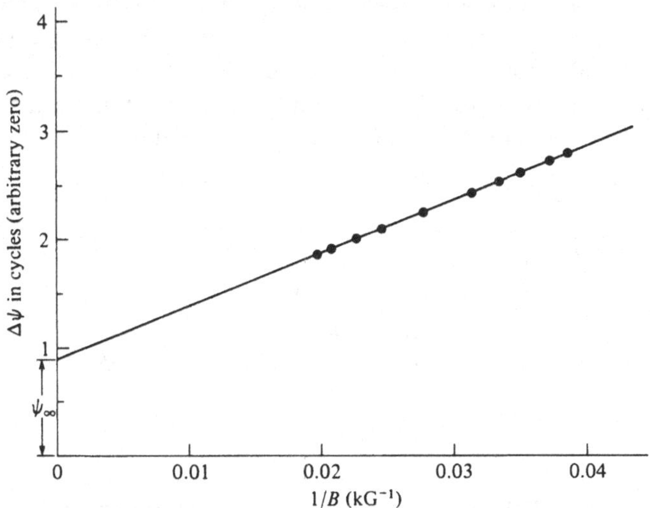

Fig. 11.9. Measured local phase minus fractional part of F_t/B versus B^{-1}, for Ag dogsbone. ψ_∞ has the correct value ~0.875, and the correction to F_t is ~ +50 kG.

sets of readings, always in negative- and positive-going pairs to remove baseline errors, equally spaced over an integer number of cycles of the interfering frequency or frequencies, and this provides a dramatic improvement in accuracy. Even when the number of cycles cannot be chosen over an exact integer an approximate value still gives significant improvement.

An example of this technique is shown in Fig. 11.10. To illustrate an extremely unfavourable case two equal-amplitude signals with a frequency ratio of 5.4 : 2 (chosen deliberately to be incommensurate) have been simulated. At each crossing point the measured phase is taken as $\frac{1}{4}$ or $\frac{3}{4}$ and the value $2\pi F_t/B$ is subtracted, just as described above, to give a 'corrected' phase. When applied to the higher frequency every crossing point is taken, and averaged over 5 cycles the measured phase is 0.868 or 0.883 cycles (depending on which end point is taken) compared with the correct value of 0.875. For the lower frequency the same technique can be used, taking only the four marked crossing points, and here the result obtained is 0.617 cycles, compared with the correct value of 0.676. One would in practice make measurements over more than just two cycles.

Such errors, less than 0.03 cycles for the higher frequency and less than 0.1 cycles for the lower, are typical for such cases. When the spurious frequency is weaker, as can usually be arranged by a suitable adjustment of the modulation, the errors are smaller. Although certainly not as accurate as the more sophisticated techniques described below, this method is quick, easy and, under usual experimental conditions, can regularly yield errors of less than 0.01 cycles. The sinusoid averaging technique need not be restricted to graphical determinations and can be applied very generally. It is, in fact, directly related to the properties of

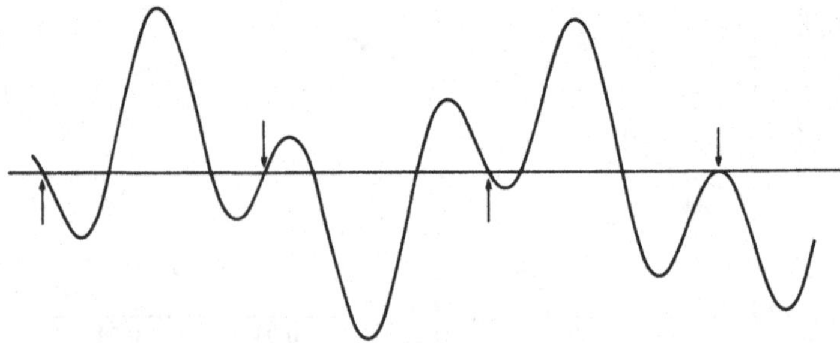

Fig. 11.10. Simulated dHvA signal with two components of frequency ratio 5.4 : 2. All zero-crossing points are used in determining the phase of the higher frequency, and the marked points only are used for the lower frequency.

Fourier transforms and can be applied even with two or three interfering frequencies. It can be used in conjunction with other techniques; for example, if a Fourier transform technique is used to separate two frequencies there might still be a much slower third frequency interfering. Averaging the results of the Fourier transform over a complete cycle of the slow frequency would then essentially eliminate its effect. Even when a low frequency appears weak and unimportant it still may be phase-modulating the higher frequencies through the Shoenberg effect and it is important therefore to use an averaging technique to reduce this interference. This is very obvious in aluminium (Coleridge & Holtham 1977) and indium (P. M. Holtham, unpublished data).

11.5.2 Computer-based techniques

Access to a data acquisition system and suitable computer-processing facilities allows much more accurate measurement of multiple frequencies. In such measurements it is important to distinguish between the problem of obtaining a phase from a short data sample and the different problem of determining frequencies from a long data sample although these two problems are, of course, related.

In determining the phase from a short sample the frequencies present can be considered as known, at least to the precision of the number of cycles sampled (typically less than 10). In addition to the basic frequencies there are possibly harmonics and interaction terms. The obvious procedure is to attempt a discrete Fourier transform of the sampled data and to determine phases from the 'sine' and 'cosine' transforms. We will investigate this approach and discuss some of the pitfalls.

For large data sets a well-established procedure is to use N equally-spaced samples of the data (where N is usually a power of 2) and a Cooley–Tukey (1965) fast Fourier transform (FFT) to obtain the Fourier spectrum. While a conventional transform takes a time $\sim N^2$ for this spectrum, the time for the FFT is $\sim N \log_2 N$, which is at least an order of magnitude faster for N larger than 60 or so. However, for phase measurements where the information in the Fourier transform is only required at a few frequencies there might well be some merit in using the 'brute force' Fourier transform for a small range of frequencies around the few 'known' frequencies. By this means it is possible to obtain a finer frequency resolution, and hence more accurate phase measurements, than is normally the case using the FFT. If the computer has the required memory capacity it is nevertheless just as fast and convenient to use the

FFT and obtain the fine resolution by padding the data with zeros, so we discuss this approach.

The FFT of a data set y_n is conventionally calculated in the form

$$F_j = \sum_{n=0}^{N-1} y_n \exp(-2\pi ijn/N),$$

but when looking for phase information it is advantageous to refer the phase to the centre of the sample rather than the beginning because the phase is then a much slower varying function of j, which will be referred to as the channel number. The shifted transform is

$$F'_j = \sum_{n=-\frac{N}{2}}^{\frac{N}{2}-1} y'_n \exp(-2\pi ijn/N) = (-1)^j F_j.$$

It is apparent that there is a basic asymmetry associated with the end points but this problem can conveniently be solved by adding an extra point $y'_{N/2}$ and weighting the two symmetric end points by $\frac{1}{2}$, i.e.

$$y_0 = \tfrac{1}{2}(y'_{N/2} + y'_{-N/2})$$

then

$$F'_j = \sum_{n=-(\frac{N}{2}-1)}^{\frac{N}{2}-1} y'_n \exp(-2\pi ijn/N) + \tfrac{1}{2}(y'_{N/2} + y'_{-N/2}) = (-1)^j F_j.$$

The phase of F'_j, defined by

$$\psi'_j = -\arctan\left(\frac{\operatorname{Im} F'_j}{\operatorname{Re} F'_j}\right),$$

is now a slow function of j through a peak. This is illustrated in Fig. 11.11 which is the FFT of the sample shown in Fig. 11.10 for a set of 65 data points (end points weighted by $\frac{1}{2}$), padded with zeros to a total of 512 points. An obvious feature is the way the square 'window' on the data has folded into the transform the interference function $\sin x/x$. A decomposition of the transform into orthogonal terms (we use sine and cosine or imaginary and real interchangeably) shows how each term has a positive and negative peak basically because

$$\frac{2}{T}\int_{-\frac{T}{2}}^{\frac{T}{2}} \sin \omega t \sin \Omega t \, dt = \frac{1}{T}\int_{-\frac{T}{2}}^{\frac{T}{2}} \cos(\omega - \Omega)t \, dt + \frac{1}{T}\int_{-\frac{T}{2}}^{\frac{T}{2}} \cos(\omega + \Omega)t \, dt$$

$$= \frac{\sin(\omega - \Omega)T/2}{(\omega - \Omega)T/2} + \frac{\sin(\omega + \Omega)T/2}{(\omega + \Omega)T/2},$$

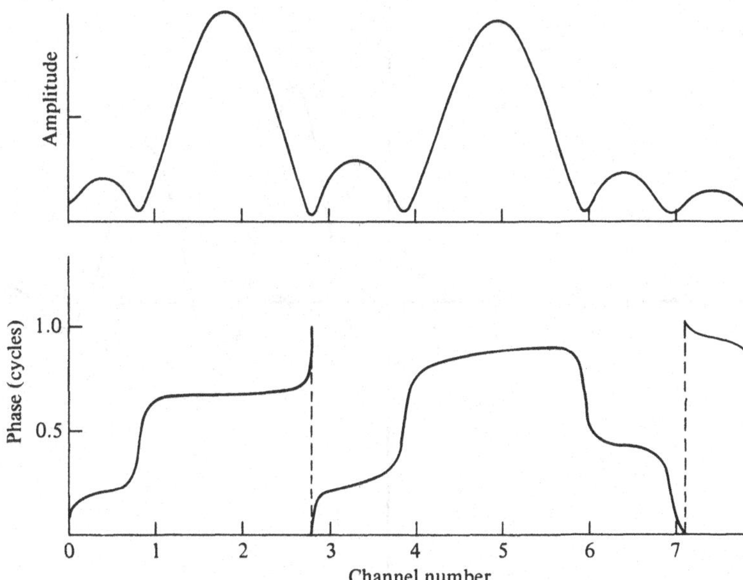

Fig. 11.11. Amplitude and phase of transform for a 65-point sample of the signal of Fig. 11.10. The sample was chosen to include exactly 5 cycles of the higher frequency. The channel number has been normalized to match the number of cycles in the sample.

with a similar form for the cosine integral (Fig. 11.12). It is obvious that the interference between the several peaks and the side-lobes contributes to a complicated variation of the phase with frequency. If the sample length and the frequency channels are chosen carefully so as to position the zeros of the side-lobes favourably, small phase errors (~0.001 cycles) can be obtained, but in general the accuracy is not significantly better with such a transform than the graphical technique discussed in the previous section.

The conventional solution to this problem would be to sample over many more cycles so as to separate the peaks and also to use an alternative form of window to suppress the side-lobes. If, for reasons of field measurement discussed above, we are restricted to short samples it is only possible to try to optimize the shape of the data window so as to suppress the interference effects. Of the several choices possible one of the more accurate for phase measurement is a Gaussian. Nearly all other choices such as a triangular envelope or a cosine bell still retain side-lobes which typically produce phase errors of order 0.01 cycles. The Gaussian still introduces some error because, unless all the phase information is thrown away, there is always some degree of sharp cut-off at the ends.

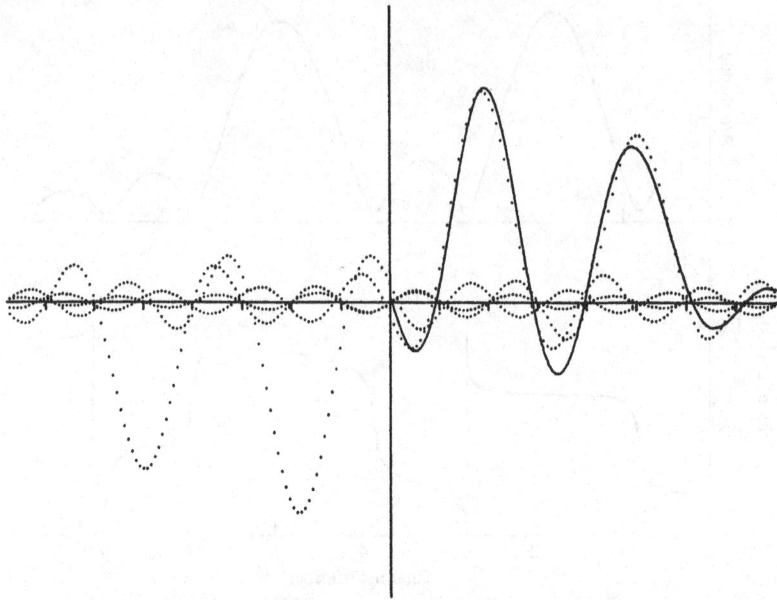

Fig. 11.12. Decomposition of the transform in Fig. 11.11 into its individual components. The combination of the positive side is shown by the solid line.

The compromise of an envelope going to exp (−4) rejects only 50 per cent of the data and gives a very small discontinuity at the ends with a correspondingly small phase error. An example of a transform with such a Gaussian is shown in Fig. 11.13 and demonstrates how well the peaks are separated and the particular insensitivity of phase to frequency channel that results. A convenient rule of thumb is to say that phase errors of less than 0.001 cycles are typical provided the *difference* in the number of cycles for the two frequencies in the data set is at least two: in this case the analysis gave phases which were correct to 0.0009 and 0.0005 cycles for the lower and higher frequencies respectively.

To increase the precision of the phase determination one technique is to use the principle of subtraction (Alles, Higgins & Lowndes 1975). A fitting procedure (typically a Fourier transform) identifies the largest peak in the spectrum and then subtracts from the original data the frequency component corresponding to that peak. This process is repeated until only a noise residue is left.

One disadvantage of using an FFT is the general requirement that the data be equally spaced in reciprocal field. This is sometimes inconvenient, for example if the data points have been taken very accurately using

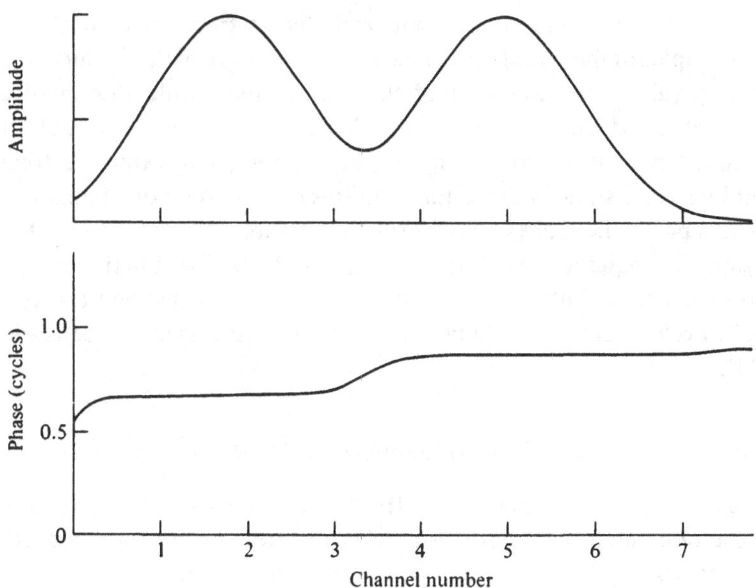

Fig. 11.13. Transform of signal of Fig. 11.10 as in Fig. 11.11 but with an e^{-4} Gaussian window applied to the data before processing.

NMR but at spacings which are only approximately periodic. In this case, knowing the frequencies with fair precision and requiring only the phases, it is possible to make a least-squares fit, for example, to an expression of the form

$$a_0 + a_1 \cos(2\pi F_1/B) + b_1 \sin(2\pi F_1/B) + a_2 \cos(2\pi F_2/B) + \ldots .$$

The normal minimization procedure for a data set y_n at fields B_n leads to expressions of the form

$$a_0 \sum \cos(2\pi F_1/B_n) + a_1 \sum \cos^2(2\pi F_1/B_n) + b_1 \sum \sin(2\pi F_1/B_n)\cos(2\pi F_1/B_n) + \ldots = \sum_n y_n \cos(2\pi F_1/B_n),$$

$$a_0 \sum \sin(2\pi F_1/B_n) + a_1 \sum \cos(2\pi F_1/B_n)\sin(2\pi F_1/B_n) + b_1 \sum \sin^2(2\pi F_1/B_n) + \ldots = \sum_n y_n \sin(2\pi F_1/B_n), \text{ etc.}$$

For equally spaced samples and for frequencies which are an exact sub-multiple of the sample frequency the off-diagonal elements in this matrix vanish (this is just another example of averaging over a sinusoid of equally spaced samples) and the result reduces to the discrete Fourier transform. In the general case, not restricted to frequencies which are sub-multiples of the sample frequency, determination of the coefficients a and b requires the inversion of the matrix and, while this involves well-established computational algorithms, it also carries with it all the associated risks of ill-conditioning. Furthermore, any extensive treatment by a least-squares procedure requires the inclusion of a linear (and maybe a parabolic) background term and consideration of the effects of varying the frequencies slightly. This can rapidly lead to a major computational expense, but in circumstances where this is justified errors of 0.0001 cycles from a short data sample are attainable (see e.g. Lonzarich 1974).

11.5.3 More complicated situations

The simultaneous observation of dHvA oscillations and NMR requires the use of compatible modulation fields and frequencies. In pioneering experiments on the noble metals it was fortunate that the optimum modulation amplitude for the aluminium nuclear resonance (~5–10 G peak to peak) was compatible with the optimum modulation for the high dHvA frequencies at fields of 30–50 kG. Modulation frequencies of the order 40–100 Hz were not only suitable for the NMR but also satisfactorily avoided skin-depth problems in the dHvA specimen. However, a larger modulation amplitude was required for the neck oscillations, and for the other orbits at higher fields. Experience with other metals emphasized the need for more general techniques.

One obvious extension of the static technique is to switch the modulation level to make it separately optimum for the two measurements, dHvA and NMR. For example a modulation field of 100 G or so might be used to observe the neck dHvA oscillations, but with a superconducting magnet in the persistent mode the modulation can be switched to 5 G to make an NMR measurement and then restored to 100 G to confirm that the field has not drifted in the process. One can even use the technique to measure two or more frequencies and NMR at the same field by switching between modulation levels chosen judiciously for suppressing or enhancing particular terms. One rather subtle error must be avoided. At low modulation frequencies (~40 Hz) many audio oscillators are d.c. coupled and may contain a small amount of d.c. offset in the output. If this is passed through the modulation coils the effect of switching the modulation level is to change not only the a.c. modulation field but also to perturb appreciably the d.c. field. The problem is easily cured, either by monitoring the d.c. through the coil and adjusting the offset to zero, or by introducing a capacitor into the circuit.

More sophisticated procedures can be devised. When skin-depth problems become important it is possible to use one frequency for the NMR and another much lower frequency for the dHvA effect. It may even be possible simultaneously to modulate and detect with different levels and frequencies for NMR and dHvA although this may lead to problems associated with non-linearity in the dHvA signals and with saturation in the NMR. Anderson, Lee & Stone (1975) have had some success with switching the modulation levels sufficiently rapidly (up to 5 Hz) that they can make NMR measurements while sweeping the main field.

In some cases the frequency spectrum is so complex that it is not possible to use the static technique. One method of overcoming this problem is to measure one frequency, often the dominant frequency, using the static technique, making as many corrections as possible by averaging to remove the effects of the other frequencies, and then to measure the other frequencies as ratios to the dominant one. Examples are the necks ($\sim 10^7$ G) and the bellies ($\sim 5\times 10^8$ G) in the noble metals or the α and β oscillations ($\sim$2–5×10^5 G) and the γ oscillations ($\sim 3\times 10^6$ G) in aluminium. If the reciprocal field is swept at a uniform rate and the modulation adjusted so that both sets of oscillations are observed a conventional Fourier transform can give the ratio of the frequencies. To achieve a high accuracy it is necessary to take several precautions. A long sample, i.e. wide field range, is necessary and care should be taken to ensure the envelope of the data is symmetrical so that the Fourier transform peak is not distorted. In practice this means some form of compensation for the field dependence of the signal, either instrumentally (e.g. by varying the modulation level) or by preprocessing the data. Gross time-constant errors are eliminated by using a sweep at a uniform rate of B^{-1} but second-order effects may still be present (e.g. magnet hysteresis means that though the magnet current i_B may be swept at a uniform rate of i_B^{-1} the field will not follow precisely). Comparison of sweeps with the field increasing and decreasing serves to check this point. A fairly high precision, ~ 1 in 10^4, can be obtained with this technique, and it can be extended to compare the dHvA frequencies not with another known frequency but with a measurement of magnetic field by magnet current (Coleridge & Templeton 1971a) or magnetoresistive sensor (Scott *et al.* 1968). This field measurement is then calibrated, either by a static comparison with NMR or by the measurement of an accurately known dHvA frequency.

The dynamic measurement can be improved in precision if not only the frequencies but also the phase relationship between them is measured. To convert these phases to frequencies requires, just as in the static measurement, that the frequencies are known with sufficient precision to obtain unambiguously the integer part of the phase. It also requires that the field is known with sufficient precision to identify the phase number of the reference frequency. A simple example of this approach is its use in the noble metals to determine the phase of the neck oscillations relative to those of the belly (Templeton 1974). When the modulation level is set high enough that both signals are observed simultaneously, the NMR still operates with sufficient accuracy that the exact phase numbers of the belly oscillations may be identified during a pause before starting the field sweep. Once the sweep is running at a steady rate, zero-crossing points of the belly, determined either by filtering or digitally, serve to provide rather accurate field points at which to sample the neck oscillations (see Figs. 11.14 and 11.15), though some averaging may be necessary to remove the effects of phase-modulation of the belly oscillations by the neck. Such an approach is, of course, sensitive to time-constant effects and to any time-delay differences between the fast and slow frequencies, so it is essential to establish what these are and correct for them if necessary. Usually averaging of sweeps up and down is sufficient.

A somewhat different technique that should be mentioned is that of linear recursion developed by Anderson *et al.* (1977). Although

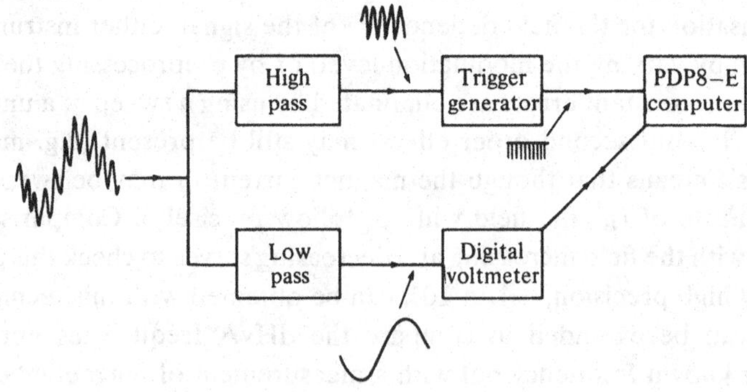

Fig. 11.14. Block diagram of system for sampling low-frequency term once for every zero-crossing of high-frequency term (Templeton 1974).

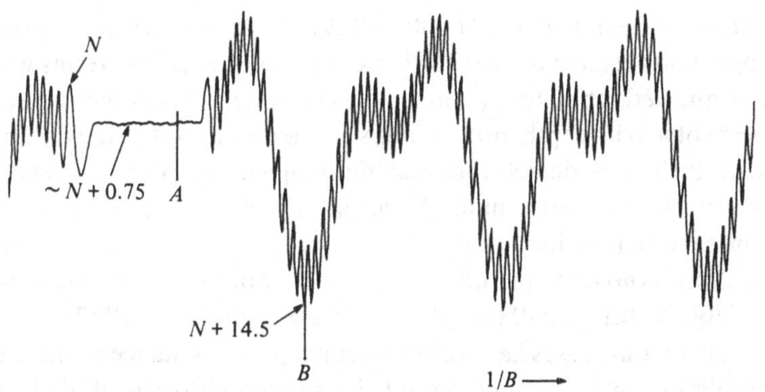

Fig. 11.15. Combined neck and belly oscillations in Cu. The field is determined by NMR at the point A, which defines phase $\sim N+0.75$ for the belly (high frequency). Sampling of the neck (low frequency), twice per belly cycle, starts at the first belly zero-crossing following the marker at point B, i.e. at $N+14.75$, and continues over two neck cycles. The sweep rate gives a belly period of about 0.5 c s^{-1}, thus sampling is at $\sim$1 c s^{-1}. The modulation in this case is set to the 5th maximum of J_2 for the belly.

developed principally for measuring the damping due to temperature and scattering the technique also gives rather precise frequency values. It does, however, require data sampled at equal intervals of B^{-1} with high accuracy and is rather intolerant of frequency modulation from the Shoenberg effect. The process is closely related to the idea of successive subtraction described above in that each time the data is passed through a filter simulated in the computer, a damped sinusoid is removed and then the process is repeated until only a noise signal is left.

It is clearly possible to incorporate in the probe a second, fixed orientation dHvA sample of some material which provides a single accurately known frequency, and to use this as a reference in the ways we have already suggested (see e.g. Coleridge 1966; Harmans & Lassche 1977).

11.6 Spurious effects

To obtain the phase of dHvA oscillations it is of course essential that the measured quantity be a genuine representation of the dHvA magnetization. We discuss here some of the possible errors that may arise.

(i) *Skin depth.* Even the lowest *conveniently* practical modulation frequencies, a few tens of cycles per second, may in some cases be too high for full penetration of the modulation field into the sample. The effects of the eddy currents in this situation have been discussed at some length by

Knecht, Lonzarich, Perz & Shoenberg (1977): for our present discussion the important result is the generation of spurious harmonic terms which must be allowed for in determining a dHvA phase. As they point out, the existence of a skin-depth 'problem' is signalled by a shift in the optimum phase of the lock-in detector for maximum signal, and the observation of an imperfect quadrature null. However, it has been found in several experiments that, so long as the detector is operated at optimum phase and the 'null' is no more than, say, 10 per cent of the maximum, there is no appreciable shift in the dHvA phase (see e.g. Templeton 1972).

We may in some cases have to consider magnetoresistance terms which are oscillating with the same periodicity as the dHvA signal. One such term will be the de Haas–Shubnikov effect, while another could arise from the 'magnetic interferometer' condition, first described by Stark & Friedberg (1971), which may accompany some types of magnetic breakdown. In principle, at least, a reasonable estimate of the error involved can be made by examining the shift of phase (towards its correct value) as the modulation frequency is reduced (Templeton 1972), but in practice this may not always be possible.

(ii) *Magnetic interaction* (Shoenberg effect). We have already discussed some aspects of magnetic interaction, both in the phase-modulation of one dHvA frequency by another, and, in self-interaction, by pointing out that the effect in $d^2\tilde{M}/dB^2$ is at a minimum at zero-crossings. The effect can be quite troublesome, particularly if for any reason it is important to determine the phases of harmonics. An explicit treatment of magnetic interaction in the weak limit has been given by Phillips & Gold (1969), and it is also discussed in Chapters 4 and 10 of this volume. In general the effect should be avoided: a simple reduction of signal amplitude by decreasing the field or increasing the temperature usually suffices for the self-interaction, though for modulation of a high frequency by a low it is sometimes acceptable to 'saturate' the amplitude of the low-frequency term by increasing the field. Fortunately in alloy measurements, because the signals are reduced by scattering, magnetic interaction effects are generally small.

(iii) *Field-amplitude effects*. Because we have, in general, been discussing relatively high dHvA frequencies, we have tended to ignore the fact that dHvA amplitudes vary with field both fundamentally, because of the way the free energy varies, and because of scattering. In most circumstances the resulting phase shift is completely negligible, but for some high-scattering, low-frequency cases (i.e. near the quantum limit) the non-oscillatory field dependences become important, particularly if the detection system involves some form of modulation.

(iv) *Inhomogeneities.* As has already been discussed, or will be discussed in the following section, inhomogeneities of field, of crystal structure or of alloy concentration may give rise to long beats or to phase-smearing which may also vary with field. In general the sample-dependent effects should become less serious at higher fields: unfortunately it is quite possible for the field homogeneity to become worse!

11.7 Alloy measurements: special considerations

As we have seen, it is usually possible to make a dHvA phase-field measurement to ~0.01 cycles or better. Since it is also usually possible to see signals from a major piece of Fermi surface after the addition of sufficient solute to cause a phase shift of at least one cycle, the Fermi surface changes can, in principle, always be measured to better than 1 per cent. If we are to use these figures to determine the effects as a function of solute concentration we must assume that the concentration is known with equal or better precision, that the solute is uniformly distributed and that the crystal structure is essentially perfect; all of which conditions are rarely true.

11.7.1 Solute concentration

The mean solute concentration in relatively large samples of material, cut from regions of an ingot close to that from which the dHvA sample was obtained, can usually be determined by *competent* analysis to 1 or 2 per cent of its actual value, depending on the particular host and solute concerned and on the standards available. In our experience this gives at least as good an estimate of concentration as does a resistance ratio measurement on the actual sample, largely because a dHvA sample (typically 1 mm × 1 mm × 3 mm) is by no means an ideal shape for mounting in any 4-probe resistance-measuring device. Even when the contacts are made on opposing edges of the end-face of the sample, to avoid relative movement during temperature cycling, it is difficult to achieve a measurement consistency much better than about 2 per cent.

11.7.2 Inhomogeneities

Either of these concentration measurements will give no indication of concentration variations within the sample. There may also be local small variations in crystal growth direction ('mosaic structure'). Both of these effects will give rise to a local spread or smearing of dHvA phase which will be field-dependent and will thus tend to give an erroneous phase measurement and Dingle temperature (see e.g. Hornbeck, Fung & Gordon; and Chapter 10 of this volume) and may also contribute to an erroneous, field-dependent line-up. It is thus useful to have measured the Dingle temperatures as well as the frequency changes for various concentrations of a given solute, since a possibly erroneous phase-change figure will often be accompanied by a Dingle temperature which departs noticeably from linearity with concentration.

11.7.3 'Background' effects

A further problem, which serves to emphasize the importance of measuring a series of concentrations, is that certain trace impurities, present in a non-contributing form (e.g. oxide) in the 'pure' host reference material, may have been brought into solution by reduction during the preparation of the alloys. The classic example of this is Fe in Cu, which caused considerable confusion during early thermoelectric power measurements on dilute alloys, when almost every solute appeared to cause the Kondo-effect-related resistance minimum (see Gold *et al.* 1960). Because even pretreatment of the reference material in a reducing atmosphere may not succeed in bringing all the iron into solution, the extrapolation of the alloy results to zero concentration may not agree with the pure host figures, but the *slope* should nevertheless give a valid figure for Fermi surface changes as a function of concentration. Fortunately the presence of any significant quantity of iron or other magnetic impurity is signalled in the dHvA effect by a degradation of spin zeros (Coleridge, Scott & Templeton 1972) or by a relative shift between fundamental and second harmonic phases (Alles & Higgins 1974).

Bibliography

Abel, W. R., Anderson, A. C. & Wheatley, J. C. (1966). *Phys. Rev. Lett.* **17**, 74

Abele, J. C. & Blatt, F. J. (1970). *Phys. Rev.* B **1**, 1298

Abrikosov, A. A., Gor'kov, L. P. & Dzyaloshinskiĭ, I. E. (1963). *Methods of Quantum Field Theory in Statistical Physics.* Englewood Cliffs, N.J.: Prentice-Hall and New York: Dover (1975)

Adams, E. N. & Holstein, T. D. (1959). *J. Phys. Chem. Solids,* **10**, 254

Afanas'ev, A. M. & Kagan, Yu. (1962). *Zh. Eksp. Teor. Fiz.* **43**, 1456

Aigrain, P. (1961). *Proc. 5th Int. Conf. on the Physics of Semiconductors,* Prague 1960, p. 224. New York: Academic

Akhiezer, A. I. (1938). *Zh. Eksp. Teor. Fiz.* **8**, 1330

—, Kaganov, M. I. & Lyubarskiĭ, G. Ya. (1957). *Zh. Eksp. Teor. Fiz.* **32**, 837

Aldred, A. T. (1975). *Phys. Rev.* B **11**, 2597

— & Froehle, P. A. (1972). *Intern. J. Magnetism,* **2**, 195

Alekseevskiĭ, N. E., Brandt, M. E. & Kostina, T. I. (1955). *Dokl. Akad. Nauk. SSSR,* **105**, 46

—, Dubrovin, A. V., Karstens, G. E. & Mikhailov, N. N. (1968). *Soviet Phys.-JETP,* **27**, 188

— & Gaĭdukov, Yu. P. (1959). *Soviet Phys.-JETP,* **10**, 481

—, Gaĭdukov, Yu. P., Lifshitz, I. M. & Peschanskiĭ, V. G. (1960). *Zh. Eksp. Teor. Fiz.* **39**, 120

—, Karstens, G. E. & Mozhaev, V. E. (1964). *Soviet Phys.-JETP,* **19**, 1333

—, Slutskin, A. A. & Egorov, V. S. (1971). *J. Low Temp. Phys.* **5**, 377

Al Khoury, E., *see* Cinti, R. C.

Allen, P. B. (1975). *Phys. Rev.* B **11**, 2693

Alles, H. G. & Higgins, R. J. (1974). *Phys. Rev.* B **9**, 158

—, Higgins, R. J. & Lowndes, D. H. (1973). *Phys. Rev. Lett.* **30**, 705

—, Higgins, R. J. & Lowndes, D. H. (1975). *Phys. Rev.* B **12**, 1304

— & Lowndes, D. H. (1973). *J. Phys.* E **6**, 895

Alpher, R. A. & Rubin, R. J. (1954). *J. Low Temp. Phys.* **9**, 67

Altounian, Z. & Datars, W. R. (1975). *Can. J. Phys.* **53**, 459

Andersen, O. K. (1970). *Phys. Rev.* B **2**, 883

— (1971a). *Computational Methods in Band Theory,* ed. P. M. Marcus, J. F. Janak & A. R. Williams. p. 178. New York: Plenum

— (1971b). *Phys. Rev. Lett.* **27**, 1211

— (1973). *Solid State Commun.* **13**, 133

— (1975). *Phys. Rev.* B **12**, 3060

— & Jepsen, O. (1977). *Physica,* **91**B, 317

—, Klose, W. & Nohl, H. (1978). *Phys. Rev.* B **17**, 1209

— & Mackintosh, A. R. (1968). *Solid State Commun.* **6**, 285

—, Madsen, J., Poulsen, U. K., Jepsen, O. & Kollar, J. (1977). *Physica,* **86–88**B, 249

— *see* Jepsen, O.

— *see* Muller, J. E.

— *see* Poulsen, U. K.

— *see* Skriver, H. L.

Anderson, A. C., *see* Abel, W. R.

Anderson, J. R. & Gold, A. V. (1963). *Phys. Rev. Lett.* **10**, 277

—, Heimann, P., Bauer, W., Schipper, R. & Stone, D. R. (1977). *Bull. Am. Phys. Soc.* **22**, 334

—, Heimann, P., Schirber, J. E. & Stone, D. R. (1976). *Magnetism and Magnetic*

Materials, vol. 29, p. 529. New York: American Institute of Physics
— & Hines, D. C. (1970). *Phys. Rev.* B **2**, 4752
—, Lee, J. Y. M. & Stone, D. R. (1975). *Phys. Rev.* B **11**, 1308
—, O'Sullivan, W. J. & Schirber, J. E. (1967). *Phys. Rev.* **153**, 721
—, O'Sullivan, W. J. & Schirber, J. E. (1972a). *Phys. Rev.* B **5**, 4683
—, Sandfort, R. M. & Stone, D. R. (1972b). *Rev. Sci. Instrum.* **43**, 1129
— *see* Kamm, G. N.
— *see* McCaffrey, J. W.
Anderson, P. W. (1961). *Phys. Rev.* **124**, 41
— & McMillan, W. L. (1967). *Proc. International School of Physics 'Enrico Fermi' Course XXXVII*, ed. W. Marshall. New York: Academic
Andreev, A. F. (1971). *Usp. Fiz. Nauk.* **105**, 113
Andreev, V. V., *see* Kosevich, A. M.
Angadi, M. A. & Fawcett, E. (1973). *Physics in Canada*, **29** (4), 26
Angermüller, H., *see* Becker, H.
Animalu, A. O. E., *see* Vasvari, B.
Appapillai, M. & Heine, V. (1972). *Technical Report*, vol. 5, Solid State Theory Group, Cavendish Laboratory, Cambridge
Argyle, B. E., Charap, S. H. & Pugh, E. W. (1963). *Phys. Rev.* **132**, 2051
Arko, A. J., Crabtree, G. W., Hörnfeldt, S. P., Ketterson, J. B., Kostorz, G. & Windmiller, L. R. (1974). In *Proc. 13th Int. Conf. on Low Temperature Physics*, (*L.T.13*), vol. 4, ed. Timmerhaus, O'Sullivan & Hammel), p. 104. New York: Plenum
—, Marcus, J. A. & Reed, W. A. (1968). *Phys. Rev.* **176**, 671
— & Mueller, F. M. (1975). *Phys. Kondens. Materie*, **19**, 231
— *see* Chang, Y. K.
Aron, P. R. (1972). *J. Low Temp. Phys.* **9**, 67
— *see* Thompson, R. E.
Ashcroft, N. W. & Wilkins, J. W. (1965). *Phys. Lett.* **14**, 285
Au, H.-P., *see* Stanley, D. J.
Avanesyan, G. T., Kaganov, M. I. & Lisovskaya, T. Yu. (1977). *Pisma Zh. Eksp. Teor. Fiz.* **25**, 381
Averbach, B. L., *see* Goldstein, I. S.
— *see* Tobin, P. J.
Azbel', M. Ya. (1960). *Zh. Eksp. Teor. Fiz.* **39**, 1276
— (1963). *Zh. Eksp. Teor. Fiz.* **44**, 983
— & Kaner, É. A. (1956). *Zh. Eksp. Teor. Fiz.* **30**, 811
— & Kaner, É. A. (1957). *Zh. Eksp. Teor. Fiz.* **32**, 896
— & Kaner, É. A. (1958). *J. Phys. Chem. Solids*, **6**, 113
— & Peschanskiĭ, V. G. (1965). *Zh. Eksp. Teor. Fiz.* **49**, 572
— & Peschanskiĭ, V. G. (1968). *Zh. Eksp. Teor. Fiz.* **55**, 1980
— *see* Kaganov, M. I.
— *see* Lifshitz, I. M.
Ball, M. A. (1972). *J. Phys.* C **5**, L23
Baraff, D. R. (1973). *Phys. Rev.* **8**, 3439
Baraff, G. A , *see* Smith, G. E.
Baratoff, A., *see* Benedek, R.
Bass, J., *see* Opsal, J. L.
Batallan, F., Bok, J., Rosenman, I. & Melin, J. (1978). *Phys. Rev. Lett.* **41**, 330
Bauer, G. & Kahlert, H. (1972). *Phys. Rev.* B **5**, 566
Bauer, W., *see* Anderson, J. R.
Baym, G. & Pethick, C. (1978). *The Physics of Liquid and Solid Helium*, Part II, ed. K. H. Bennemann & J. B. Ketterson. New York: Wiley
Béal-Monod, M. T. (1969). *Phys. Rev.* **178**, 874
— & Kohn, W. (1968). *J. Phys. Chem. Solids*, **29**, 1877
— & Weiner, R. A. (1968). *Phys. Rev.* **170**, 552
Becker, H., Dietz, E., Gerhardt, U. & Angermüller, H. (1975). *Phys. Rev.* B **12**, 2084
Beeby, J. L. (1967). *Proc. Roy. Soc.* A **302**, 113
Bender, A. S. & Young, D. A. (1971). *Phys. Stat. Sol.* B **47**, K95
Benedek, R. & Baratoff, A. (1973). *Solid State Commun.* **13**, 385
Bennemann, K. H. & Garland, J. W. (1972). *AIP Conference Proceedings*, **4**, 103
Berglund, C. N., *see* Spicer, W. E.
Bethe, H. A. & Jackiw, R. (1968). *Intermediate Quantum Mechanics*, pp. 56–63. New York: Benjamin
— *see* Sommerfeld, A.
Bhargava, R. N. (1967). *Phys. Rev.* **156**, 785
Bibby, W. M. & Shoenberg, D. (1977). *Phys. Lett.* **60**A, 235
Blaker, J. W. & Harris, R. (1971). *J. Phys.* C **4**, 569

Blatt, F. J. (1957). *Phys. Rev.* **108**, 285
— *see* Abele, J. C.
Bloch, D. & Pavlovic, A. S. (1969). In *Advances in High Pressure Research*, vol. 3, ed. R. S. Bradley, p. 41. New York: Academic
Blount, E. I. (1962). *Phys. Rev.* **126**, 1636
— *see* Cohen, M. H.
Blum, N. A., *see* Foner, S.
Bok, J., *see* Batallan, F.
Borel, J. P., *see* Monot, R.
Bosacchi, B., Ketterson, J. B. & Windmiller, L. R. (1970). *Phys. Rev.* B **2**, 3025
Bourassa, R. R., *see* Lengeler, B.
Boyer, L. L., Papaconstantopoulos, D. A. & Klein, B. M. (1977). *Phys. Rev.* B **15**, 3685
Boyle, D. J. & Gold, A. V. (1969). *Phys. Rev. Lett.* **22**, 461
Brailsford, A. D. (1966). *Phys. Rev.* **149**, 456
Brändli, G. & Griessen, R. (1973). *Cryogenics*, **13**, 299
Brandt, N. B., Itskevich, E. S. & Minina, N. Ya. (1971). *Usp. Fiz. Nauk.* **104**, 459 (*Soviet Phys. Uspekhi*, **14**, 438 (1972))
— *see* Alekseevskiĭ, N. E.
Brillouin, L. (1976). *Wave Propagation in Periodic Structures.* New York: Dover
Brinkman, W. F., Platzman, P. M. & Rice, T. M. (1968). *Phys. Rev.* **174**, 495
Brooker, G. A. & Sykes, J. (1968). *Phys. Rev. Lett.* **21**, 279
— (1970). *Ann. Phys.* **56**, 1
Brown, C. R., Kalejs, J. P., Manchester, F. D. & Perz, J. M. (1972). *Phys. Rev.* B **6**, 4458
Brown, H. R. & Morgan, G. J. (1971). *J. Phys.* F **1**, 132
— & Myers, A. (1972). *J. Phys.* F **2**, 683
Brown, M., Peierls, R. E. & Stern, E. (1977). *Phys. Rev.* B **15**, 738
Brown, R. L. & Friedberg, C. B. (1979). *Phys. Rev.* **19**, 5123
Brueckner, K. A. (1959). *The Many Body Problem*, ed. C. Dewitt, p. 59. New York: Wiley
Bryan, A. C., *see* Thompson, E. D.
Buchsbaum, S. J. & Galt, J. K. (1961). *Phys. Fluids*, **4**, 1514
Budd, H. (1963). *J. Phys. Soc. Japan*, **18**, 142
— (1967). *Phys. Rev.* **158**, 798
Buot, F. A., Li, P. L. & Ström-Olsen, J. O. (1976). *J. Low Temp. Phys.* **22**, 535
Burdick, G. A. (1961). *Phys. Rev. Lett.* **7**, 156
Butler, W. H., Olson, J. J., Faulkner, J. S. & Gyorffy, B. L. (1976). *Phys. Rev.* B **14**, 3823
Bychkov, Yu. A. (1960). *Zh. Eksp. Teor. Fiz.* **39**, 1401 (Eng. trans: *Soviet Phys.–JETP*, **12**, 977 (1961))
— & Gor'kov, L. P. (1962). *Soviet Phys.–JETP*, **14**, 1132 (*Zh. Eksp. Teor. Fiz.* **41**, 1592 (1961))

Callaway, J. (1974). *Quantum Theory of the Solid State.* New York: Academic Press
— & Wang, C. S. (1977a). *Physica*, **91**B, 337
— & Wang, C. S. (1977b). *Phys. Rev.* B **16**, 2095
— *see* Rath, J.
— *see* Singh, M.
— *see* Wang, C. S.
Carbotte, J. P., *see* Popovic, Z.
Carrander, K., Dronjak, M. & Hörnfeldt, S. P. (1977). *J. Phys. Chem. Solids*, **38**, 289
Cashion, J. K., *see* Eastman, D. E.
Castaing, P., *see* Goy, P.
Chakraverty, B. K., *see* Cinti, R. C.
Chambers, R. G. (1950). *Proc. Roy. Soc.* A **202**, 378
— (1952). *Proc. Phys. Soc.* **65**, 458
— (1966). *Proc. Phys. Soc. Lond.*, **88**, 701
— (1968). In *Solid State Physics, vol. 1: Electrons in Metals*, ed. J. F. Cochran & R. R. Haering, p. 283. New York: Gordon & Breach
— (1969). *The Physics of Metals, vol. 1: Electrons*, ed. J. M. Ziman, p. 175. Cambridge: C.U.P.
— (1974). In *Proc. 13th Int. Conf. Low Temperature Physics* (*L.T.13*), vol. 4, ed. Timmerhaus, O'Sullivan & Hammel, p. 3. New York: Plenum
Chandrasekhar, B. S. & Fawcett, E. (1971). *Adv. in Phys.* **20**, 775
—, Fawcett, E., Sparlin, D. M. & White, G. K. (1966). In *Proc. 10th Int. Conf. on Low Temperature Physics* (*LT.10*), vol. 3, p. 328. Moscow
— *see* Green, B. A.
— *see* Thompson, R. E.
Chang, Y. K., Arko, A. J., Crabtree, G. W., Ketterson, J. B., Windmiller, L. R., Higgins, R. J. & Young, Jr, F. W. (1975a). *Int. Conf. on Fundamental Aspects of Radiation Damage in Metals*, Gaithersburg
—, Crabtree, G. W. & Ketterson, J. B. (1977). *Bull. Am. Phys. Soc.* **22**, 23

—, Crabtree, G. W., Ketterson, J. B. & Windmiller, L. R. (1975b). *Bull. Am. Phys. Soc.* **20**, 353
— & Higgins, R. J. (1975). *Phys. Rev.* B **12**, 4261
— *see* Higgins, R. J.
Chao, P. W., Lin, S. Y. & Kao, Y. H. (1973). *Bull. Am. Phys. Soc.* **18**, 93
Charap, S. H., *see* Argyle, B. E.
Chatelain, A., *see* Monot, R.
Chen, A. B. & Segall, B. (1975). *Phys. Rev.* B **12**, 600
Cheng, L. S. (1977). PhD Thesis, U. of Oregon
—, Higgins, R. J., Graebner, J. E. & Rubin, J. J. (1979). *Phys. Rev.* B **19**, 3722
— *see* Higgins, R. J.
Chester, G. V. (1963). *Rep. Prog. Phys.* **26**, 411
Chodorow, M. I. (1939). PhD Thesis, MIT
Chollet, L.-F. & Templeton, I. M. (1968). *Phys. Rev.* **170**, 656
Christensen, N. E. (1972). *Phys. Stat. Sol.* **54**, 551
— (1976). *Phys. Rev.* B **14**, 3446
— (1978). *J. Phys.* F **8**, L51
— & Feuerbacher, B. (1974). *Phys. Rev.* B **10**, 2349
— & Seraphin, B. O. (1971). *Phys. Rev.* B **4**, 3321
— & Willis, R. F. (1978). *Phys. Rev.* (in press)
— *see* Cinti, R. C.
Christy, R. W., *see* Johnson, P. B.
Chu, H. T. & Kao, Y. H. (1970). *Phys. Rev.* B **1**, 2369
Chung, Y. & Lowndes, D. H. (1976). *J. Phys.* F **6**, 199
— & Lowndes, D. H. (1977). *Solid State Commun.* **21**, 647
—, Lowndes, D. H., Hendel, C. & Lin, S. Y. (1978). *J. Low Temp. Phys.* (in press)
—, Lowndes, D. H., Hendel, C., Lin, S. Y. & Rode, J. P. (1976). *Solid State Commun.* **20**, 101
Cinti, R. C., Al Khoury, E., Chakraverty, B. K. & Christensen, N. E. (1976). *Phys. Rev.* B **14**, 3296
Cohen, M. H. & Blount, E. I. (1960). *Phil. Mag.* **5**, 115
Cohen, M. H. & Falicov, L. M. (1961). *Phys. Rev. Lett.* **7**, 231
Cohen, M. L. & Heine, V. (1970). In *Solid State Physics*, vol. 24, ed. F. Seitz & D. Turnbull, p. 38. New York: Academic
Colavita, E., Modesti, S. & Rosei, R. (1976). *Phys. Rev.* B **14**, 3415
Coleman, R. V., Morris, R. C. & Sellmyer, D. J. (1973). *Phys. Rev.* B **8**, 317
Coleridge, P. T. (1966). *Proc. Roy. Soc.* A **295**, 458
— (1969). *J. Low Temp. Phys.* **1**, 577
— (1972). *J. Phys.* F **2**, 1016
— (1973a). *Phil. Mag.* **27**, 1495
— (1973b). *Phys. Rev.* B **7**, 3508
— (1975). *J. Phys.* F **5**, 1317
— & Holtham, P. M. (1977). *J. Phys.* F **7**, 1891
—, Holzwarth, N. A. W. & Lee, M. J. G. (1974). *Phys. Rev.* B **10**, 1213
—, Scott, G. B. & Templeton, I. M. (1972). *Can. J. Phys.* **50**, 1999
— & Templeton, I. M. (1971a). *Can. J. Phys.* **49**, 2449
— & Templeton, I. M. (1971b). *Phys. Rev. Lett.* **27**, 507
— Templeton, I. M. (1972). *J. Phys.* F **2**, 643
— & Watts, B. R. (1971). *Phil. Mag.* **24**, 1163
— *see* Templeton, I. M.
— *see* Lee, M. J. G.
Coles, B. R. (1964). *Phys. Lett.* **8**, 243
Condon, J. H. (1966a). *Phys. Rev.* **145**, 526
— (1966b). In *Proc. 10th Int. Conf. Low Temperature Physics* (*L.T.10*), vol. 3, p. 289. Moscow
— & Walstedt, R. E. (1968). *Phys. Rev. Lett.* **21**, 612
— *see* Halloran, M. H.
— *see* Higgins, R. J.
— *see* Reed, W. A.
— *see* Testardi, L. R.
Cooke, J. F. (1973). *Phys. Rev.* B **7**, 1108
— (1976). In *Proc. Conf. on Neutron Scattering*, vol. 2, Gatlinburg, Tennessee, ed. R. M. Moon, p. 723. Nat. Technical Information Service, U.S. Dept. of Commerce, Springfield, Virginia 22161
Cooke, J. F., Lynn, J. W. & Davis, H. L. (1974). *AIP Conf. Proc.* **24**, 329
Cooley, J. W. & Tukey, J. W. (1965). *Math. of Comput.* **19**, 297
Cooper, J. D., Smith, J. P., Woore, J. & Young, D. A. (1971). *J. Phys.* C **4**, 442
Cornwell, J. F. (1965). *Proc. Roy. Soc.* A **284**, 423
Crabtree, G. W. (1977). *Phys. Rev.* B **16**, 1117
—, Windmiller, L. R. & Ketterson, J. B. (1975). *J. Low Temp. Phys.* **20**, 655
—, Windmiller, L. R. & Ketterson, J. B. (1977). *J. Low Temp. Phys.* **26**, 755
— *see* Arko, A. J.

— *see* Chang, Y. K.
— *see* Dye, D. M.
— *see* Karim, D. P.
— *see* Lowndes, D. H.
Cracknell, A. P. (1971). *Adv. in Phys.* **20**, 1
Cromer, D. T., *see* Liberman, D.
Culp, C. H., *see* Weaver, J. H.

Datars, W. R., *see* Dunsworth, A. E.
— *see* Jenkins, R. M.
— *see* Tanuma, S.
— *see* Vanderkooy, J.
Davis, H. L., *see* Cooke, J. F.
Davison, B. (1957). *Neutron transport theory.* Oxford: Clarendon Press
Davydov, V. N. & Kaganov, M. I. (1972). *Pisma Zh. Eksp. Teor. Fiz.* **16**, 133
de Haas, W. J., *see* Shubnikov, A. V.
de Ribaupierre, Y., *see* Griessen, R.
de Wilde, J. & Meredith, D. J. (1976). *J. Phys.* E **9**, 62
— (1978). PhD Dissertation, Free University of Amsterdam
Dietz, E., *see* Becker, H.
Dillon, R. O. & Spain, I. L. (1977). *Bull. Am. Phys. Soc.* **22**, 381
Dimmock, J. O. & Freeman, A. J. (1964). *Phys. Rev. Lett.* **13**, 750
—, *see* Mueller, F. M.
Dingle, R. B. (1952). *Proc. Roy. Soc.* A **211**, 517
— & Shoenberg, D. (1950). *Nature,* **166**, 652
Dmitrenko, I. M., Verkin, B. I. & Lazarev, B. G. (1958). *Zh. Eksp. Teor. Fiz.* **35**, 328 (*Soviet Phys.-JETP,* **8**, 229 (1959))
— *see* Verkin, B. I.
Dodo, R., *see* Foner, S.
Doi, H., *see* Tanuma, S.
Doniach, S., *see* Kaiser, A. B.
— *see* Murata, K. K.
Dronjak, M., *see* Hörnfeldt, S. P.
— *see* Carrander, K.
Dubrovin, A. V., *see* Alekseevskiĭ, N. E.
Dunifer, G. L., Pinkel, D., & Schultz, S. (1974). *Phys. Rev.* B **10**, 3159
— *see* Schultz, S.
Dunsworth, A. E. & Datars, W. R. (1973). *Phys. Rev.* B **7**, 3435
— *see* Tanuma, S.
Dupree, R., Forwood, C. T. & Smith, M. J. A. (1967). *Phys. Stat. Sol.* **24**, 525
— & Holland, B. W. (1967). *Phys. Stat. Sol.* **24**, 275
Dupree, T. H. (1961). *Ann. Phys.*, **15**, 63
Durkansky, G., *see* Wampler, W. R.

Dye, D. H., Karim, D. P., Ketterson, J. B. & Crabtree, G. W. (1978a). In *Proc. Int. Conf. on the Physics of Transition Metals,* Toronto, 1977, ed. M. J. G. Lee, J. M. Perz & E. Fawcett (Institute of Physics Conference Series No. 39). London: Institute of Physics
—, Ketterson, J. B. & Crabtree, G. W. (1978b). *J. Low Temp. Phys.* **30**
—, Ketterson, J. B., Lowndes, D. H., Crabtree, G. W. & Windmiller, L. R. (1977). *J. Low Temp. Phys.* **26**, 945
Dzyaloshinskiĭ, I. E. (1964). *Zh. Eksp. Teor. Fiz.* **47**, 336
— *see* Abrikosov, A. A.

Eastman, D. E. & Cashion, J. K. (1970). *Phys. Rev. Lett.* **24**, 310
Edel'man, V. S., *see* Khaĭkin, M. S.
— *see* Volodin, A. P.
Edwards, D. M. (1962). *Proc. Roy. Soc.* A **269**, 338
— (1970). *Phys. Lett.* **33**A, 183
— (1974). *Can. J. Phys.* **52**, 704
— (1977). *Physica,* **91**B, 3
— & Wohlfarth, E. P. (1968). *Proc. Roy. Soc.* A **303**, 127
Egorov, V. S., *see* Alekseevskiĭ, N. E.
Ehnson, M. I., *see* Lutskii, V. N.
— *see* Ogrin, Yu. F.
Ehrenreich, H. & Hodges, L. (1968). In *Methods in Computational Physics,* vol. 8, ed. B. Alder, S. Fernbach & M. Rotenberg, p. 149. New York: Academic
— & Schwartz, L. M. (1976). In *Solid State Physics,* vol. 31, ed. F. Seitz, D. Turnbull & H. Ehrenreich, p. 149, New York: Academic
— *see* Gelatt, C. D.
— *see* Hodges, L.
Elbaum, C., *see* Gerstein, I. I.
Elliott, M., Ellis, T. & Springford, M. (1978). *Phys. Rev. Lett.* **41**, 709
Elliott, R. J. (1954). *Phys. Rev.* **96**, 266
Ellis, T., *see* Elliott, M.
Elyashar, N. & Koelling, D. D. (1977). *Phys. Rev.* B **15**, 3620
Engelsberg, S. (1978). *Phys. Rev.* B **18**, 966
— & Simpson, G. (1970). *Phys. Rev.* B **2**, 1657
Enz, C. P. (1968). *Theory of Condensed Matter, Lectures at ICTP, Trieste,* 1967, p. 729. Vienna: IAEA
Eshelby, J. D. (1954). *J. Appl. Phys.* **25**, 255
Evenson, W. E., *see* Wang, S. Q.

Everett, P. M. (1972). *Phys. Rev.* B **6**, 3553
— & Grenier, C. G. (1977). *Phys. Rev.* B **15**, 3826
— & Grenier, C. G. (1978). *Phys. Rev.* B **18**, 4477
— *see* Tripp, J. H.

Fadley, C. S. & Wohlfarth, E. P. (1972). *Comments in Solid State Phys.* **4**, 48
Fairchild, R. W., *see* Kostroun, V. O.
Falicov, L. M., Pippard, A. B. & Sievert, P. R. (1966). *Phys. Rev.* **151**, 498
— & Zuckermann, M. J. (1967). *Phys. Rev.* **160**, 372
— *see* Cohen, M. H.
— *see* Ruvalds, J.
— *see* Stark, R. W.
Fal'kovskiĭ, L. A., *see* Khaĭkin, M. S.
Faulkner, J. S. (1977). *Phys. Rev.* B **16**, 736
— *see* Butler, W. H.
— *see* Painter, G. S.
Fawcett, E. (1956). *Phys. Rev.* **103**, 1582
— (1961). *Phys. Rev. Lett.* **7**, 370
— Griessen, R. & Stanley, D. J. (1976). *J. Low Temp. Phys.* **25**, 771
— & Reed, W. A. (1962). *Phys. Rev. Lett.* **9**, 336
— & Reed, W. A. (1963). *Phys. Rev.* **131**, 2463
— *see* Angadi, M. A.
— *see* Chandrasekhar, B. S.
— *see* Holroyd, F.
— *see* Lee, M. J. G.
Featherston, F. H. & Neighbours, J. R. (1963). *Phys. Rev.* **130**, 1324
Fenton, E. W. (1975). *Solid State Commun.* **16**, 1325
— (1976). *J. Phys.* F **6**, 363
— (1977). *Solid State Commun.* **22**, 63
Feuerbacher, B., *see* Christensen, N. E.
Finkelstein, M. M. (1974). *J. Low Temp. Phys.* **14**, 287
Fiske, J. M., *see* Tripp, J. H.
Flesner, L. D. & Schultz, S. (1976). *Phys. Rev.* B **14**, 4759
Foner, S. (ed.) (1976). *Magnetism, Selected Topics.* New York: Gordon & Breach
—, Dodo, R. & McNiff, E. J. Jr (1968). *J. Appl. Phys.* **39**, 551
—, Freeman, A. J., Blum, N. A., Frankel, R. B., McNiff, E. J. & Praddaude, H. C. (1969). *Phys. Rev.* **181**, 863
— & McNiff, E. J. (1967). *Phys. Rev. Lett.* **19**, 1438
— *see* Goldstein, A.
Forssell, G., *see* Nilsson, P. O.
Forwood, C. T., *see* Dupree, R.
Fowler, M. & Prange, R. E. (1965). *Physics,* **1**, 315
Frankel, R. B., *see* Foner, S.
Freeman, A. J., *see* Dimmock, J. O.
— *see* Foner, S.
— *see* Koelling, D. D.
— *see* Mueller, F. M.
Friedberg, C. B., *see* Brown, R. L.
— *see* Stark, R. W.
Friedel, J. (1954). *Adv. in Phys.* **3**, 446
—, Leman, G. & Olszewski, S. (1961). *J. Appl. Phys. Suppl.* **32**, 3255
Froehle, P. A., *see* Aldred, A. T.
Fröhlich, H. (1950). *Phys. Rev.* **79**, 845
Fuchs, K. (1938). *Proc. Camb. Phil. Soc.* **34**, 100
Fujiwara, Okamoto & Tatsumoto (1965). In *Physics of Solids at High Pressure*, ed. C. T. Tomizuka & A. M. Emrick, p. 261. New York: Academic
Fukai, Y. (1969). *Phys. Rev.* **186**, 697
Fung, W. K. & Gordon, W. L. (1977). *Phys. Rev.* B **15**, 4762
— *see* Hornbeck, L. J.
Furdyna, A. M., *see* Mueller, F. M.

Gaertner, A. A. (1973). *J. Low Temp. Phys.* **10**, 503
— & Templeton, I. M. (1977). *J. Low Temp. Phys.* **29**, 205
Gaffney, J., *see* Overton, W. C.
Gaĭdukov, Yu. P. & Golyamina, E. M. (1976). *Pisma Zh. Eksp. Teor. Fiz.* **23**, 336
— *see* Alekseevskiĭ, N. E.
Galt, J. K., *see* Buchsbaum, S. J.
Gamble, D. & Watts, B. R. (1972). *Phys. Lett.* **40**A, 22
— & Watts, B. R. (1973). *J. Phys.* F **3**, 98
Gantmakher, V. F., *see* Kaner, É. A.
Gaputchenko, A. G., *see* Vinokurova, L. I.
Garland, J. W., *see* Benneman, K. H.
Gaspar, R. (1954). *Acta Phys. Ac. Sci. Hung.* **3**, 263
Gelatt, C. D., Ehrenreich, H. & Watson, R. E. (1977). *Phys. Rev.* B **15**, 1613
Gerhardt, U. (1968). *Phys. Rev.* **172**, 651
— *see* Becker, H.
Gersdorf, R. (1978). *Phys. Rev. Lett.* **40**, 344
Gerstein, I. I. & Elbaum, C. (1973). *Phys. Stat. Sol.* **57**B, 157
Gilat, G. & Raubenheimer, L. J. (1966). *Phys. Rev.* **144**, 390
Girvan, R. F., Gold, A. V. & Phillips, R. A. (1968). *J. Phys. Chem. Solids,* **29**, 1485

Giuliano, E. S., *see* Stocks, G. M.

Gold, A. V. (1958). *Phil. Trans. Roy. Soc. Lond.* A **251**, 85

— (1965). *Proc. Int. Conf. on Magnetism*, p. 124. London: Institute of Physics and The Physical Society

— (1968). In *Solid State Physics, vol. 1: Electrons in Metals*, ed. J. F. Cochran & R. R. Haering, New York: Gordon & Breach

— (1974). *J. Low Temp. Phys.* **16**, 3

—, Hodges, L., Panousis, P. T. & Stone, D. R. (1971). *Intern. J. Magnetism*, **2**, 357

—, MacDonald, D. K. C., Pearson, W. B. & Templeton, I. M. (1960). *Phil. Mag.* **5**, 765

— & Schmor, P. W. (1976). *Can. J. Phys.* **54**, 2445

— *see* Anderson, J. R.

— *see* Boyle, D. J.

— *see* Girvan, R. F.

— *see* Lonzarich, G.

— *see* Phillips, R. A.

Goldstein, A., Williamson, S. J. & Foner, S. (1965). *Rev. Sci. Instr.* **36**, 1356

Goldstein, I. S., Sellmyer, D. J. & Averbach, B. L. (1970). *Phys. Rev.* B **2**, 1442

Golyamina, E. M., *see* Gaĭdukov, Yu. P.

Gordon, W. L., *see* Fund, W. K.

— *see* Hornbeck, L. J.

— *see* Shepherd, J. G.

— *see* Tripp, J. H.

Gor'kov, L. P., *see* Abrikosov, A. A.

— *see* Bychkov, Yu. A.

Goy, P. & Castaing, P. (1973). *Phys. Rev.* B **7**, 4409

— & Castaing, P. (1975). *Phys. Rev.* B **11**, 2696

— & Grimes, C. C. (1973). *Phys. Rev.* B **7**, 299

Graebner, J. E. (1971). In *Proc. 12th Int. Conf. on Low Temperature Physics (L.T.12)*, ed. E. Kanda, p. 601. Japan: Academic

— & Marcus, J. A. (1968). *Phys. Rev.* **175**, 659

—, Rubin, J. J., Shutz, R. J., Hsu, F. S. L., Reed, W. A. & Higgins, R. J. (1974). In *Magnetism and Magnetic Materials*, ed. C. D. Graham, Jr *et al.*, p. 445. New York: American Institute of Physics, 1975

— *see* Cheng, L. S.

— *see* Halloran, M. H.

— *see* Higgins, R. J.

Gray, A. M., *see* Gray, D. M.

Gray, D. M. & Gray, A. M. (1976). *Phys. Rev.* B **14**, 669

Green, B. A. & Chandrasekhar, B. S. (1963). *Phys. Rev. Lett.* **11**, 331

Grenier, C. G., *see* Everett, P. M.

Griessen, R. (1973). *Cryogenics*, **13**, 375

— (1978). In *Proc. Int. Conf. on the Physics of Transition Metals*, Toronto, 1977, ed. M. J. G. Lee, J. M. Perz & E. Fawcett, p. 51 (Institute of Physics Conference Series No. 39). London: Institute of Physics

—, Krugmann, H. & Ott, H. R. (1974). *Phys. Rev.* B **10**, 1160

—, Lee, M. J. G. & Stanley, D. J. (1977a). *Phys. Rev.* B **16**, 4385

— & Olsen, J. L. (1971). *Solid State Commun.* **9**, 1655

— & Sorbello, R. S. (1974). *J. Low Temp. Phys.* **16**, 237

—, Venema, W. J., Jacobs, J. K., Manchester, F. D. & de Ribaupierre, Y. (1977b). *J. Phys.* F **7**, L133

— *see* Brandli, G.

— *see* Fawcett, E.

— *see* Posternak, M.

— *see* Skriver, H. L.

— *see* Sorbello, R. S.

— *see* Stanley, D. J.

Grimes, C. C., *see* Goy, P.

— *see* Walsh, W. M.

Grimvall, G. (1969). *Phys. Kondens. Materie*, **9**, 283

— (1974). *Phys. Kondens. Materie*, **18**, 161

— (1976). *Physica Scripta*, **14**, 63

— (1980). *The Electron–Phonon Interaction in Metals.* Amsterdam: North-Holland

Gschneidner, K. A. (1964). In *Solid State Physics*, vol. 16, ed. F. Seitz & D. Turnbull, p. 295. New York: Academic

Guerrisi, M., Rosei, R. & Winsemius, O. (1975). *Phys. Rev.* B **12**, 557

Gunnarsson, O. (1976). *J. Phys.* F **6**, 587

— (1977). *Physica*, **91B**, 329

— & Lundqvist, B. I. (1976). *Phys. Rev.* B **13**, 4274

Gunnersen, E. M. (1956). *Phil. Trans. Roy. Soc. Lond.* A **249**, 299

Gurevich, V. L. (1959). *Zh. Eksp. Teor. Fiz.* **37**, 71

Gurzhi, R. N. & Kopeliovich, A. I. (1976). *Zh. Eksp. Teor. Fiz.* **71**, 635

Gutman, E. J. & Stanford, J. L. (1971). *Phys. Rev.* B **4**, 4026

Gyorffy, B. L., *see* Butler, W. H.

— *see* Stocks, G. M.

Halloran, M. H., Condon, J. H., Graebner, J. E., Kunzler, J. E. & Hsu, F. S. L. (1970). *Phys. Rev.* B **1**, 366
— *see* Parker, R. D.
Halse, M. R. (1969). *Phil. Trans. Roy. Soc. Lond.* A **265**, 507
Ham, F. S., *see* Segall, B.
Harmans, C. J. P. M. & Lassche, L. (1977). *J. Phys.* E **10**, 155
Harmon, B. N., Schirber, J. & Koelling, D. D. (1978). In *Proc. Int. Conf. on the Physics of Transition Metals*, Toronto, 1977, ed. M. J. G. Lee, J. M. Perz & E. Fawcett (Institute of Physics Conference Series No. 39). London: Institute of Physics
Harris, R. (1970). *J. Phys.* C **3**, 172
—, Mulimani, B. G. & Ström-Olsen, J. O. (1975a). *J. Phys.* F **5**, 1910
—, Mulimani, B. G. & Zuckermann, M. J. (1975b). *Phys. Cond. Mat.* **19**, 269
— *see* Blaker, J. W.
Harrison, W. A. (1958). *Phys. Rev.* **110**, 14
— (1960). *Phys. Rev.* **118**, 1190
— (1966). *Pseudopotentials in the theory of metals.* New York: Benjamin
— (1970). *Solid State Theory.* New York: McGraw-Hill
— & Webb, M. B. (eds.) (1960). *The Fermi Surface.* London: Wiley
Harte, G. A., Priestley, M. G. & Vuillemin, J. J. (1978). *J. Low Temp. Phys.* **31**, 897
Hedgecock, F. T. & Muir, W. B. (1963). *Phys. Rev.* **129**, 2045
— *see* Paton, B. E.
Hedin, L. & Lundqvist, B. I. (1971). *J. Phys.* C **4**, 2064
— & Lundqvist, S. (1969). In *Solid State Physics*, vol. 23, ed. H. Ehrenreich, F. Seitz & D. Turnbull, p. 1. New York: Academic
— *see* Barth, U.
Heeger, A. J. (1969). In *Solid State Physics*, vol. 23, ed. F. Seitz & D. Turnbull, p. 283. New York: Academic
Heimann, P., *see* Anderson, J. R.
Heine, V. (1956). *Proc. Phys. Soc.* A **69**, 505
— (1967). *Phys. Rev.* **153**, 673
— (1970). In *Solid State Physics*, vol. 24, ed. H. Ehrenreich, F. Seitz & D. Turnbull, p. 1. New York: Academic
— & Weaire, D. (1970). In *Solid State Physics*, vol. 24, ed. H. Ehrenreich, F. Seitz & D. Turnbull, p. 249. New York: Academic
— *see* Appapillai, M.
— *see* Cohen, M. H.
— *see* Vasvari, B.
Hendel, C., *see* Chung, Y.
Hennion, M. & Hennion, B. (1978). *J. Phys.* F **8**, 287
Hennion, B., *see* Hennion, M.
Herman, F. & Skillman, S. (1963). *Atomic Structure Calculations*, Englewood Cliffs, N.J.: Prentice-Hall
Herring, C. & Vogt, E. (1956). *Phys. Rev.* **101**, 944
— (1966). In *Magnetism*, vol. 4, ed. G. T. Rado & H. Suhl. New York: Academic
— (1978). In *Proc. Int. Conf. on the Physics of Transition Metals*, Toronto, 1977, ed. M. J. G. Lee, J. M. Perz & E. Fawcett, p. 719 (Institute of Physics Conference Series No. 39). London: Institute of Physics
— & Kittel, C. (1951). *Phys. Rev.* **81**, 869
Hertz, J. A. & Klenin, M. A. (1977). *Physica*, **91**B, 49
Higgins, R. J. & Chang, Y. K. (1968). *Rev. Sci. Instrum.* **39**, 522
—, Cheng, L. S., Graebner, J. E. & Rubin, J. J. (1975). In *Proc. 14th Int. Conf. on Low Temperature Physics*, (*L.T.14*), ed. M. Krusius & M. Vuorio, p. 146. Amsterdam: North-Holland
—, Kaehn, H. D. & Condon, J. H. (1969). *Phys. Rev.* **181**, 1059
— & Marcus, J. A. (1966). *Phys. Rev.* **141**, 553
— & Marcus, J. A. (1967). *Phys. Rev.* **161**, 589
— *see* Alles, H. G.
— *see* Chang, Y. K.
— *see* Cheng, L. S.
— *see* Graebner, J. E.
— *see* Terwilliger, D. W.
Hill, D. A. & Hwang, C. (1966). *J. Sci. Instrum.* **43**, 581
Hill, P. H. & Vanderkooy, J. (1978). *Phys. Rev.* B
Hines, D. C., *see* Anderson, J. R.
Hodges, C. H. (1977). *J. Phys.* F **7**, L249
— *see* Nieminen, R. M.
Hodges, L. & Ehrenreich, H. (1965). *Phys. Lett.* **16**, 203
— Ehrenreich, H. & Lang, N. D. (1966). *Phys. Rev.* **152**, 505
— Ehrenreich, H. & Watson, R. E. (1972). *Phys. Rev.* B **5**, 3953
— *see* Ehrenreich, H.
— *see* Gold, A. V.
Hoekstra, J. A. & Phillips, R. A. (1971). *Phys. Rev.* B **4**, 4184

— & Stanford, J. L. (1973). *Phys. Rev.* B **8**, 1416
Hohenberg, P. C. & Kohn, W. (1964). *Phys. Rev.* B **136**, 864
Højgaard Jensen, H., Smith, H. & Wilkins, J. W. (1968). *Phys. Lett.* **27**A, 532
— (1969). *Phys. Rev.* **185**, 323
Holland, B. W., *see* Dupree, R.
Holroyd, F., Fawcett, E. & Perz, J. M. (1978). In *Proc. Int. Conf. on the Physics of Transition Metals*, Toronto, 1977, ed. M. J. G. Lee, J. M. Perz & E. Fawcett, p. 63 (Institute of Physics Conference Series No. 39). London: Institute of Physics
Holstein, T. (1964). *Ann. Phys.* **29**, 410
—, Norton, R. E. & Pincus, P. (1973). *Phys. Rev.* B **8**, 2649
Holtham, P. M. (1976). *J. Phys.* F **6**, 1457
— & Parsons, D. (1976). *J. Phys.* F **6**, 1481
— *see* Coleridge, P. T.
Holzwarth, N. A. W. (1975). *Phys. Rev.* B **11**, 3718
— & Lee, M. J. G. (1975). *Phys. Kondens. Materie*, **19**, 161
— *see* Coleridge, P. T.
— *see* Lee, M. J. G.
Hornbeck, L. J., Fung, W. K. & Gordon, W. L. (1977). *Phys. Rev.* B **15**, 4750
Hörnfeldt, S. P., Dronjak, M. & Nordborg, L. (1976). *Solid State Comm.* **20**, 1085
—, Ketterson, J. B. & Windmiller, L. R. (1969). *Phys. Rev. Lett.* **23**, 1292
—, Ketterson, J. B. & Windmiller, L. R. (1973a). *J. Phys.* E **6**, 265
—, Windmiller, L. R. & Ketterson, J. B. (1973b). *Phys. Rev.* B **7**, 4349
— *see* Arko, A. J.
— *see* Carrander, K.
— *see* Windmiller, L. R.
Hsu, F. S. L., *see* Graebner, J. E.
— *see* Halloran, M. H.
Hubbard, J. (1963). *Proc. Roy. Soc.* A **276**, 238
— (1967). *Proc. Phys. Soc.* **92**, 921
Hulbert, J. K. (1976). *J. Phys.* E **9**, 283
Hum, R. H., *see* Perz, J. M.
Hwang, C., *see* Hill, D. A.

Ishizawa, Y. & Tanuma, S. (1965). *J. Phys. Soc. Japan*, **20**, 1278
Isihara, A. & Tsai, J. T. (1974). *Physica*, **77**, 247
Itskevich, E. S., *see* Brandt, N. B.
— *see* Vinokurova, L. I.
Izuyama, T., Kim, D. & Kubo, R. (1963). *J. Phys. Soc. Japan*, **18**, 1025
— & Kubo, R. (1964). *J. Appl. Phys.* **35**, 1074

Jackiw, R., *see* Bethe, H. A.
Jacobs, J. K., *see* Griessen, R.
Janak, J. F. (1977). *Phys. Rev.* **16**, 255
— & Williams, A. R. (1976). *Phys. Rev.* B **14**, 4199
— Williams, A. R. & Moruzzi, V. L. (1975). *Phys. Rev.* B **11**, 1522
— *see* Moruzzi, V. L.
Jeans, J. H. (1921). *The Dynamical Theory of Gases*, 3rd edn, p. 371. Cambridge: C.U.P.
Jenkins, R. M. & Datars, W. R. (1973). *Phys. Rev.* B **7**, 2269
Jepsen, O. (1975). *Phys. Rev.* B **12**, 2988
— & Andersen, O. K. (1971). *Solid State Commun.* **9**, 1763
—, Andersen, O. K. & Mackintosh, A. R. (1975). *Phys. Rev.* B **12**, 3084
— *see* Andersen, O. K.
— *see* Muller, J. E.
Johansen, G. (1969). *Solid State Commun.* **7**, 731
— & Mackintosh, A. R. (1970). *Solid State Commun.* **8**, 121
Johansson, B., *see* Skriver, H. L.
Johnson, K. Y. (1968). *Phys. Lett.* **27**A, 138
Johnson, P. B. & Christy, R. W. (1972). *Phys. Rev.* B **6**, 4370
Jones, C. K., *see* Rayne, J. A.
Jones, D. W., *see* Young, R. C.
Jones, W. & March, N. H. (1973). *Theoretical Solid State Physics*, p. 990. London: Wiley
Jordan, R. G., *see* Young, R. C.
Joseph, A. S. & Thorsen, A. C. (1963). *Phys. Rev. Lett.* **11**, 554
— & Thorsen, A. C. (1965). *Phys. Rev.* **138**, A 1159
— *see* Thorsen, A. C.
Joss, W., *see* Posternak, M.
Justi, E. & Scheffers, H. (1938). *Metallwirtschaft*, **17**, 1359

Kadanoff, L. P., *see* Prange, R. E.
Kadigrobov, A. M., *see* Kaganov, M. I.
Kaehn, H. D., *see* Higgins, R. J.
Kagan, Yu., *see* Afanas'ev, A. M.
Kaganov, M. I. & Azbel', M. Ya. (1955). *Dokl. Akad. Nauk SSSR*, **102**, 49
—, Kadigrobov, A. M. & Slutskin, A. A. (1967). *Zh. Eksp. Teor. Fiz.* **53**, 1135
— & Peschanskiĭ, V. G. (1958). *Zh. Eksp. Teor. Fiz.* **35**, 1052
— & Semenenko, A. I. (1966). *Zh. Eksp. Teor. Fiz.* **50**, 630
— *see* Akhiezer, A. I.
— *see* Avanesyan, G. T.

— *see* Davydov, V. N.
— *see* Lifshitz, I. M.
Kahlert, H., *see* Bauer, G.
Kaiser, A. B. & Doniach, S. (1970). *Int. J. Magnetism* **1**, 11
Kalejs, J. P., *see* Brown, C. R.
Kamm, G. N. & Anderson, J. R. (1970). *Phys. Rev.* B **2**, 2944
— & Anderson, J. R. (1974). In *Proc. 13th Int. Conf. on Low Temperature Physics (L.T.13)*, vol. 4, ed. Timmerhaus, O'Sullivan & Hammel, p. 114. New York: Plenum
Kanamori, J. (1963). *Progr. Theor. Phys.* **30**, 275
Kaner, É. A. & Gantmakher, V. F. (1968). *Usp. Fiz. Nauk.* **94**, 123
— & Skobov, V. G. (1963). *Zh. Eksp. Teor. Fiz.* **45**, 610
— & Skobov, V. G. (1966). *Usp. Fiz. Nauk.* **89**, 367
— & Skobov, V. G. (1968). *Adv. in Phys.* **17**, 69
— *see* Azbel', M. Ya.
Kao, Y. H., *see* Chao, P. W.
— *see* Chu, H. T.
Karim, D. P., Ketterson, J. B. & Crabtree, G. W. (1978). *J. Low Temp. Phys.* **30**, 389
— *see* Dye, D. H.
Karstens, G. E., *see* Alekseevskiĭ, N. E.
Kazantsev, V. A., *see* Menshikov, A. Z.
Keeton, S. C. & Loucks, T. L. (1968). *Phys. Rev.* **168**, 672
Ketterson, J. B., Koelling, D. D., Shaw, J. C. & Windmiller, L. R. (1975). *Phys. Rev.* B **11**, 1447
— & Windmiller, L. R. (1970). *Phys. Rev.* B **2**, 4813
— *see* Arko, A. J.
— *see* Bosacchi, B.
— *see* Chang, Y. K.
— *see* Crabtree, G. W.
— *see* Dye, D. H.
— *see* Hörnfeldt, S. P.
— *see* Karim, D. P.
— *see* Lowndes, D. H.
— *see* Mueller, F. M.
— *see* Shaw, J. C.
— *see* Windmiller, L. R.
Khaĭkin, M. S. (1960). *Zh. Teor. Fiz.* **39**, 212
— (1962). *Zh. Teor. Fiz.* **42**, 27
— (1968a). *Zh. Teor. Fiz.* **55**, 1696
— (1968b). *Usp. Fiz. Nauk.* **96**, 409
— & Edel'man, V. S. (1964). *Zh. Eksp. Teor. Fiz.* **47**, 878
—, Fal'kovskiĭ, L. A., Edel'man, V. S. & Mina, R. T. (1963). *Zh. Eksp. Teor. Fiz.* **45**, 1704
— *see* Pudalov, V. M.
— *see* Volodin, A. P.
Kim, D., *see* Izuyama, T.
King-Smith, P. E. (1965). *Phil. Mag.* **12**, 1123
Kip, A. F. (1960). In *The Fermi Surface*, ed. W. A. Harrison & M. B. Webb, p. 146. London: Wiley
Kirichenko, O. V., Lur'e, M. A. & Peschanskiĭ, V. G. (1976a). *Zh. Eksp. Teor. Fiz.* **70**, 337
—, Lur'e, M. A. & Peschanskiĭ, V. G. (1976b). *Fiz. Nizk. Temp.* **2**, 858
Kittel, C. (1963). *Quantum Theory of Solids*, ch. 16, eq. (44). New York: Wiley
— *see* Herring, C.
Klein, B. M., *see* Boyer, L. L.
Kleinman, L. (1978). *Phys. Rev.* B **17**, 3666
Klenin, M. A., *see* Hertz, J. A.
Klose, W., *see* Andersen, O. K.
Knapp, G. S. (1967). *Phys. Lett.* **25**A, 114
Knecht, B. (1975). *J. Low Temp. Phys.* **21**, 619
—, Lonzarich, G. G., Perz, J. M. & Shoenberg, D. (1977). *J. Low Temp. Phys.* **29**, 499
Koelling, D. D. & Freeman, A. J. (1975). *Phys. Rev.* B **12**, 5622
—, Freeman, A. J. & Mueller, F. M. (1970). *Phys. Rev.* B **1**, 1318
—, Mueller, F. M. & Veal, B. W. (1974). *Phys. Rev.* B **10**, 1290
— *see* Elyashar, N.
— *see* Harmon, B. N.
— *see* Ketterson, J. B.
— *see* Schirber, J. E.
Koga, M., *see* Takano, S.
Kohn, W. (1959). *Phys. Rev. Lett.* **2**, 393
— & Rostocker, J. (1954). *Phys. Rev.* **94**, 1111
— & Sham, L. J. (1965). *Phys. Rev.* **140**, A 1133
— *see* Béal-Monod, M. T.
— *see* Hohenberg, P. C.
Kollar, J., *see* Andersen, O. K.
— *see* Poulsen, U. K.
Kondo, J. (1968). *Phys. Rev.* **169**, 437
— (1969). In *Solid State Physics*, vol. 23, ed. F. Seitz & D. Turnbull. New York: Academic
Kondratenko, P. S. (1965). *Soviet Phys.–JETP*, **20**, 1032 (*Zh. Eksp. Teor. Fiz.* **47**, 1536 (1964))

Konstantinov, O. V. & Perel', V. Y. (1960). *Zh. Eksp. Teor. Fiz.* **38**, 161
Kontorovich, V. M. & Stepanova, N. A. (1973). *Pisma Zh. Eksp. Teor. Fiz.* **18**, 381
— & Stepanova, N. A. (1978). *Fiz. Tverd. Tela* (Leningrad) **20**, 245
Kontrym-Sznajd, G., Stachowiak, H., Wierzchowski, W., Petersen, K., Thrane, N. & Trumpy, G. (1975). *J. Appl. Phys.* **8**, 151
Kopeliovich, A. I., *see* Gurzhi, R. N.
Korenman, V., Murray, J. L. & Prange, R. E. (1977). *Phys. Rev.* B **16**, 4032
— & Prange, R. E. (1972). *Phys. Rev.* **6**, 2769
— *see* Prange, R. E.
Korneev, D. N., *see* Lutskii, V. N.
Korringa, J. (1947). *Physica*, **13**, 392
Kosevich, A. M. & Andreev, V. V. (1960). *Zh. Eksp. Teor. Fiz.* **38**, 882
— *see* Lifshitz, I. M.
Koster, G. F., *see* Slater, J. C.
Kostina, T. I., *see* Alekseevskiĭ, N. E.
Kostorz, G., *see* Arko, A. J.
Kostroun, V. O., Fairchild, R. W., Kukkonen, C. A. & Wilkins, J. W. (1976). *Phys. Rev.* B **13**, 3268
Kress, K. A. & Lapeyre, G. J. (1971). *Solid State Commun.* **9**, 827
Krugmann, H., *see* Griessen, R.
Kubo, R. (1973). In *The Boltzmann Equation*, ed. E. G. Cohen & W. Thirring, p. 301. Vienna: Springer
— *see* Izuyama, T.
Kukkonen, C. A., *see* Kostroun, V. O.
Kunzler, J. E., *see* Halloran, M. H.
Kuzmin, N. N., *see* Menshikov, A. Z.

Landau, L. D. (1930). *Z. Physik*, **64**, 629
— (1957a). *Soviet Phys.-JETP*, **3**, 920
— (1957b). *Soviet Phys.-JETP*, **5**, 101
— (1959). *Soviet Phys.-JETP*, **8**, 70
— (1969). *Collected Works in Russian*, vol. 2. Nauka
— & Lifshitz, E. M. (1958). *Statistical Physics*, p. 155. London: Pergamon
— & Lifshitz, E. M. (1959). *Fluid Mechanics*, p. 300. London: Pergamon
— & Luttinger, J. M. (1963), *see* Abrikosov *et al.*
Lang, N. D., *see* Hodges, L.
Langenberg, D. N., *see* Thompson, R. E.
Lapeyre, G. J., *see* Kress, K. A.
Lassche, L., *see* Harmans, C. J. P. M.
Lasseter, N. H. & Soven, P. (1973). *Phys. Rev.* B **8**, 2476
Lax, M. (1958). *Phys. Rev.* **109**, 1921
Lazarev, B. G., Nakhimovich, N. M. & Parfenova, E. A. (1939). *Zh. Eksp. Teor. Fiz.* **9**, 1169
— *see* Dmitrenko, I. M.
— *see* Verkin, B. I.
Lazarus, D., *see* Spurgeon, W. A.
Lee, J. Y. M., *see* Anderson, J. R.
Lee, M. J. G. (1969). *Phys. Rev.* **187**, 901
— (1970). *Phys. Rev.* B **2**, 250
—, Holzwarth, N. A. W. & Coleridge, P. T. (1976a). *Phys. Rev.* B **13**, 3249
—, Perz, J. M. & Fawcett, E. (eds.) (1978). In *Proc. Int. Conf. on the Physics of Transition Metals*, Toronto 1977 (The Institute of Physics Conference Series No. 39). London: Institute of Physics
—, Perz, J. M. & Stanley, D. J. (1976b). *Phys. Rev. Lett.* **37**, 537
— *see* Coleridge, P. T.
— *see* Griessen, R.
— *see* Holzwarth, N. A. W.
— *see* Stanley, D. J.
Leggett, A. J. (1968). *Ann. Phys.* **46**, 76
Legkostupov, M. S. (1972). *Soviet Phys.-JETP*, **34**, 136 (*Zh. Eksp. Teor. Fiz.* **61**, 262 (1971))
Lehmann, G. (1975). *Phys. Stat. Sol.* (b) **70**, 737
Leman, G., *see* Friedel, J.
Lengeler, B. (1977). *Phys. Rev.* B **15**, 5504
— & Uelhoff, W. (1975). *Phys. Lett.* **53**A, 139
—, Wampler, W. R., Bourassa, R. R., Mika, K., Wingerath, K. & Uelhoff, W. (1977). *Phys. Rev.* B **15**, 5493
— *see* Wampler, W. R.
Li, P. L. & Ström-Olsen, J. O. (1973). *J. Low Temp. Phys.* **12**, 255
—, Ström-Olsen, J. O. & Paton, B. E. (1975). *Phys. Cond. Materie*, **19**, 277
— *see* Buot, F. A.
Liberman, D., Waber, J. T. & Cromer, D. T. (1965). *Phys. Rev.* **137**, A27
Lifshitz, I. M. (1948). *Zh. Eksp. Teor. Fiz.* **18**, 298
— (1950). *Report to a session of the Academy of Sciences of the Ukrainian SSR*, Kiev
— (1954). *Zh. Eksp. Teor. Fiz.* **26**, 551
— (1957). *Zh. Eksp. Teor. Fiz.* **32**, 1509
— (1960). *Zh. Eksp. Teor. Fiz.* **38**, 1569
— Azbel', M. Ya. & Kaganov, M. I. (1956a). *Soviet Phys.-JETP*, **3**, 143
— Azbel', M. Ya. & Kaganov, M. I. (1956b). *Zh. Eksp. Teor. Fiz.* **31**, 63
— Azbel', M. Ya. & Kaganov, M. I. (1973). *Electron Theory of Metals* New York:

Consultants Bureau
— & Kaganov, M. I. (1959). *Usp. Fiz. Nauk.* **69**, 419
— & Kosevich, A. M. (1953). *Dokl. Akad. Nauk. SSSR*, **91**, 795
— & Kosevich, A. M. (1954). *Dokl. Akad. Nauk. SSSR*, **96**, 963
— & Kosevich, A. M. (1955). *Zh. Eksp. Teor. Fiz.* **29**, 730 (English trans: *Soviet Phys.-JETP*, **2**, 636 (1956))
— & Kosevich, A. M. (1957). *Zh. Eksp. Teor. Fiz.* **33**, 88
— & Peschanskiĭ, V. G. (1958). *Zh. Eksp. Teor. Fiz.* **35**, 1251
— & Pogorelov, A. V. (1954). *Dokl. Akad. Nauk. SSSR*, **96**, 1143
— *see* Alekseevskiĭ, N.E.
Lifshitz, E. M., *see* Landau, L. D.
Lin, S. Y., *see* Chao, P. W.
— *see* Chung, Y.
Lisovskaya, T. Yu., *see* Avanesyan, G. T.
Liu, S. H. (1976). *Phys. Rev.* B **13**, 3962
— (1978). *Phys. Rev.* B **17**, 3629
— & Toxen, A. M. (1965). *Phys. Rev.* **138**, A487
Llewellyn, B., Paul, D. McK., Randles, D. L. & Springford, M. (1977a). *J. Phys.* F **7**, 2531
— (1977b). *J. Phys.* F **7**, 2545
Lodder, A. (1976). *J. Phys.* F **6**, 1885
Lomer, W. M. (1962). *Proc. Phys. Soc.* **80**, 489
— (1964). *Proc. Phys. Soc.* **84**, 327
London, H. (1940). *Proc. Roy. Soc. Lond.* A **176**, 522
Lonzarich, G. G. (1974). PhD Dissertation, The University of British Columbia (Nat. Library of Canada, Ottawa)
— (1978). In *Proc. Int. Conf. on the Physics of Transition Metals*, Toronto, 1977, ed. M. J. G. Lee, J. M. Perz & E. Fawcett, p. 518 (Institute of Physics Conference Series No. 39). London: Institute of Physics
— Gold, A. V. (1974a). *Can. J. Phys.* **52**, 694
— Gold, A. V. (1974b). *Phys. in Canada*, **30**(4), 56
— *see* Knecht, B.
Loucks, T. L. (1965a). *Phys. Rev.* **139**, A1333
— (1965b). *Phys. Rev.* **139**, A1181
— *see* Keeton, S. C.
Lowde, R. D. & Wohlfarth, E. P. (eds.) (1977). *Itinerant-Electron Magnetism.* Amsterdam: North-Holland (*Physica* **91**B)
Lowndes, D. H., Crabtree, G., Ketterson, J. B. & Windmiller, L. R. (1973a). *Solid State Commun.* **13**, 1855
—, Miller, K. M., Poulsen, R. G. & Springford, M. (1973b). *Proc. Roy. Soc. Lond.* A **331**, 497
— *see* Chung, Y.
— *see* Dye, D. H.
— *see* Rode, J. P.
Lubzens, D., Shanaburger, M. R. & Schultz, S. (1971). *Phys. Rev. Lett.* **29**, 1387 (1971 Erratum: *PRL* 29, 1768)
Lundqvist, B. I., *see* Gunnarsson, O.
— *see* Hedin, L.
Lur'e, M. A. & Peschanskiĭ, V. G. (1972). Abstracts of contributions to: *Proc. All Union Conf. on Low Temp. Phys.*, Donetsk
— *see* Kirichenko, O. V.
Lutskiĭ, V. N., Korneev, D. N. & Elinson, M. I. (1966). *Pisma Zh. Eksp. Teor. Fiz.* **4**, 267
— *see* Ogrin, Yu. F.
Luttinger, J. M. (1960). *Phys. Rev.* **119**, 1153
— (1961). *Phys. Rev.* **121**, 1251
— & Nozières, P. (1962). *Phys. Rev.* **127**, 1423, 1431
Lynch, D. W., *see* Olson, C. G.
— *see* Weaver, J. H.
Lynn, J. W. (1975). *Phys. Rev.* B **11**, 2624
— *see* Cooke, J. F.
— *see* Mook, H. A.
Lyo, S. K. (1977). *Phys. Rev. Lett.* **39**, 363
— (1978). *Phys. Rev.* **17**, 2545
Lyubarskiĭ, G. Ya., *see* Akhiezer, A. I.

McCaffrey, J. W., Anderson, J. R. & Papaconstantopoulos, D. A. (1973). *Phys. Rev.* B **7**, 674
McEwen, K. A. (1971). *Proc. Roy. Soc.* A **322**, 509
Mackintosh, A. R. (1968). *Theory of Condensed Matter*, p. 783. Vienna: IAEA
— *see* Andersen, O. K.
— *see* Johansen, G.
— *see* Skriver, H. L.
— *see* Williams, R. W.
McMillan, W. L. (1968). *Phys. Rev.* **167**, 331
— & Rowell, J. M. (1969). *Superconductivity*, ed. R. D. Parks, pp. 561–613. New York: Marcel Dekker
— *see* Anderson, P. W.
McNiff, E. J., *see* Foner, S.
Madsen, J., *see* Andersen, O. K.
Maglic, R. & Mueller, F. M. (1971). *Int. J. Magnetism*, **1**, 289

Makoshi, K. & Moriya, T. (1975). *J. Phys. Soc. Japan*, **38**, 10
Manchester, F. D., *see* Brown, C. R.
— *see* Griessen, R.
Mann, E. (1971). *Phys. Kondens. Materie*, **12**, 210
— & Schmidt, H. (1975). *Phys. Kondens. Materie*, **19**, 33
March, N. H., *see* Jones, W.
Marcus, J. A. (1949). *Phys. Rev.* **76**, 413
— *see* Arko, A. J.
— *see* Graebner, J. E.
— *see* Higgins, R. J.
Martin, D. L. (1973). *Phys. Rev.* B **8**, 5357
Mathon, J. (1972). *J. Phys.* F **2**, 159
Matsumoto, H., Umezawa, H., Seki, S. & Tachiki, J. (1978). *Phys. Rev.* B **17**, 2276
Mattheiss, L. (1964a). *Phys. Rev.* **133**, A1399
— (1964b). *Phys. Rev.* **134**, A970
— (1966). *Phys. Rev.* **151**, 450
— (1970). *Phys. Rev.* B **1**, 373
Mattocks, P. G. & Young, R. C. (1977a). *J. Phys.* F **7**, L19
— & Young, R. C. (1977b). *J. Phys.* F **7**, 1219
— & Young, R. C. (1978). *J. Phys.* F (in press)
Matula, S., *see* Wampler, W. R.
Maxfield, B. W. & Merrill, J. R. (1965). *Rev. Sci. Instrum.* **36**, 1083
Melin, J., *see* Batallan, F.
Menshikov, A. Z., Kazantsev, V. A., Kuzmin, N. N. & Sidorov, S. K. (1975). *J. Magnetism & Magnetic Materials*, **1**, 91
Meredith, D. J., *see* de Wilde, J.
Merrill, J. R., *see* Maxfield, B. W.
Miedema, A. R. (1976). *J. Less Common Metals*, **46**, 67
Migdal, A. B. (1957). *Soviet Phys.-JETP*, **5**, 333
— (1958a). *Soviet Phys.-JETP*, **7**, 996
— (1958b). *Zh. Eksp. Teor. Fiz.* **34**, 1438
Mijnarends, P. E. (1973). *Physica*, **63**, 235, 248
Mika, K., *see* Lengeler, B.
Mikhailov, N. N., *see* Alekseevskiĭ, N. E.
Miller, K. M., Poulsen, R. G. & Springford, M. (1972). *J. Low Temp. Phys.* **6**, 411
— *see* Lowndes, D. H.
Minina, N. Ya., *see* Brandt, N. B.
Minkiewicz, V. J., *see* Shirane, G.
Mints, R. G. (1969). *Pisma Zh. Eksp. Teor. Fiz.* **9**, 629
Modesti, S., *see* Colavita, E.
Monot, R., Chatelain, A. & Borel, J. P. (1971). *Phys. Lett.* **34**, 57
Mook, H. A., Lynn, J. W. & Nicklow, R. M. (1973). *Phys. Rev. Lett.* **30**, 556
Morgan, G. J. (1966). *Proc. Phys. Soc.* **89**, 365
— *see* Brown, H. R.
Mori, N., *see* Vuillemin, J. J.
Moriya, T. (1977). *Physica*, **91B**, 235
— *see* Makoshi, K.
Morris, R. C., *see* Coleman, R. V.
Morse, R. W. (1960). In *The Fermi Surface*, ed. W. A. Harrison & M. B. Webb, p. 214. London: Wiley
Moruzzi, V. L., Janak, J. F. & Williams, A. R. (1978). *Calculated Electronic Properties of Metals*. New York: Pergamon
—, Williams, A. R. & Janak, J. F. (1977). *Phys. Rev.* B **15**, 2854
— *see* Janek, J. F.
Mozhaev, V. E., *see* Alekseevskiĭ, N. E.
Mueller, F. M. (1966). *Phys. Rev.* **148**, 636
—, Freeman, A. J., Dimmock, J. O. & Furdyna, A. M. (1970a). *Phys. Rev.* B **1**, 4617
— & Myron, H. W. (1976). *Commun. Phys.* **1**, 99
— & Phillips, J. C. (1967). *Phys. Rev.* **157**, 600
— Windmiller, L. R. & Ketterson, J. B. (1970b). *J. Appl. Phys.* **41**, 2312
— *see* Arko, A. J.
— *see* Koelling, D. D.
— *see* Maglic, R.
Muir, W. B., *see* Hedgecock, F. T.
— *see* Paton, B. E.
Mulimani, B. G. (1976). PhD Thesis, McGill University
— *see* Harris, R.
Muller, J. E., Jepsen, O., Andersen, O. K. & Wilkins, J. W. (1978). *Phys. Rev. Lett.* **40**, 720
Murata, K. K. & Doniach, S. (1972). *Phys. Rev. Lett.* **29**, 285
Murray, J. L., *see* Korenman, V.
Myers, A., *see* Brown, H. R.
— *see* Sang, D.
Myron, H. W., *see* Mueller, F. M.

Nagasawa, H. (1972). *Solid State Commun.* **10**, 33
Nakhimovich, N. M., *see* Lazarev, B. G.
Nathans, R., *see* Shirane, G.
Nedorezov, S. S. (1966). *Zh. Eksp. Teor. Fiz.* **51**, 868
Nee, T. W. & Prange, R. E. (1967). *Phys. Lett.* **25A**, 582
Neighbours, J. R., *see* Featherston, F. H.
Nicklow, R. M., *see* Mook, H. A.

Nieminen, R. M. & Hodges, C. H. (1976). *J. Phys.* F **6**, 573
Nilsson, P. O. & Forssell, G. (1977). *Phys. Rev.* B **16**, 3352
— & Sandell, B. (1970). *Solid State Commun.* **8**, 721
Nohe, H., *see* Andersen, O. K.
Norborg, L., *see* Hörnfeldt, S. P.
Norton, R. E., *see* Holstein, T.
Nozières, P., *see* Luttinger, J. M.
— *see* Pines, D.

Ogrin, Yu. F., Lutskiĭ, V. N. & Elinson, M. I. (1966). *Pisma Zh. Eksp. Teor. Fiz.* **3**, 114
Okamoto, *see* Fujiwara
Olsen, J. L., *see* Griessen, R.
Olson, C. G., Piacentini, M. & Lynch, D. W. (1974). *Phys. Rev. Lett.* **33**, 644
— *see* Weaver, J. H.
Olson, J. J., *see* Butler, W. H.
Olszewski, S., *see* Friedel, J.
Onsager, L. (1952). *Phil. Mag.* **43**, 1006
Opsal, J. L., Thaler, B. J. & Bass, J. (1976). *Phys. Rev. Lett.* **36**, 1211
O'Sullivan, W. J. & Schirber, J. E. (1966). *Phys. Rev. Lett.* **16**, 691
—, Switendick, A. C. & Schirber, J. E. (1970). *Phys. Rev.* B **1**, 1443
— *see* Anderson, J. R.
— *see* Schirber, J. E.
Ott, H. R., *see* Griessen, R.
Overton, W. C. & Gaffney, J. (1955). *Phys. Rev.* **98**, 969

Painter, G. S., Faulkner, J. S. & Stocks, G. M. (1974). *Phys. Rev.* B **9**, 2448
Palin, C. J. (1972). *Proc. Roy. Soc. Lond.* A **329**, 17
—, Randles, D. L., Shoenberg, D. & Vanderkooy, J. (1971). In *Proc. 12th Int. Conf. Low Temperature Physics* (*L.T.12*), ed. E. Kanda, p. 513. Japan: Academic
Panfilov, A. S., *see* Svechkarev, I. V.
Panousis, P. T., *see* Gold, A. V.
Papaconstantopoulos, D. A., *see* Boyer, L. L.
— *see* McCaffrey, J. W.
Parfenova, E. A., *see* Lazarev, B. G.
Parker, R. D. & Halloran, M. H. (1974). *Phys. Rev.* B **9**, 4130
Parsons, D., *see* Holtham, P. M.
Paton, B. E. (1971). *Can. J. Phys.* **49**, 1813
—, Hedgecock, F. T. & Muir, W. B. (1970). *Phys. Rev.* B **2**, 4549
— & Muir, W. B. (1968). *Phys. Rev. Lett.* **20**, 732
— & Slavin, A. J. (1973). *Rev. Sci. Instrum.* **44**, 1357
— *see* Li, P. L.
Paul, D. McK. & Springford, M. (1977). *J. Low Temp. Phys.* **27**, 561
— & Springford, M. (1978). *J. Phys.* F **8**, 1713
— *see* Llewellyn, B.
Pavlovic, A. S., *see* Bloch, D.
Pearson, W. B. (1964). *Handbook of Lattice Spacings and Structures of Metals and Alloys*. Oxford: Pergamon
Peierls, R. E. (1955). *Quantum Theory of Solids*. Oxford: O.U.P.
— *see* Brown, M.
Perdew, J. P., *see* Vosko, S. H.
Perel', V. I., *see* Konstantinov, O. V.
Perz, J. M. & Hum, R. H. (1971). *Can. J. Phys.* **49**, 1
— & Shoenberg, D. (1976). *J. Low Temp. Phys.* **25**, 275
— *see* Brown, C. R.
— *see* Holroyd, F.
— *see* Knecht, B.
— *see* Lee, M. J. G.
— *see* Stanley, D. J.
Peschanskiĭ, V. G. (1968). *Pisma, Zh. Eksp. Teor. Fiz.* **7**, 489
— *see* Alekseevskiĭ, N. E.
— *see* Azbel', M. Ya.
— *see* Kaganov, M. I.
— *see* Kirichenko, O. V.
— *see* Lifshitz, I. M.
— *see* Lur'e, M. A.
Petersen, K., *see* Kontrym-Sznajd, G.
Pethick, C., *see* Baym, G.
Pettifor, D. G. (1970). *J. Phys.* C **3**, 367
— (1976). *Commun. Phys.* **1**, 141
— (1977). *J. Phys.* F **4**, 613
— (1978). *J. Phys.* F **8**, 219
Phillips, J. C., *see* Mueller, F. M.
Phillips, R. A. & Gold, A. V. (1969). *Phys. Rev.* **178**, 932
— *see* Girvan, R. F.
— *see* Hoekstra, J. A.
Piacentini, M., *see* Olson, C. G.
Piercy, G. R., *see* Popovic, Z.
Pincus, P., *see* Holstein, T.
Pines, D. & Nozières, P. (1966). *The Theory of Quantum Liquids*, New York: Benjamin
Pippard, A. B. (1947). *Proc. Roy. Soc. Lond.* A **191**, 385
— (1949). *Proc. Roy. Soc. Lond.* A **195**, 336
— (1954). *Proc. Roy. Soc.* **224**, 272
— (1955). *Phil. Mag.* **46**, 1104
— (1957a). *Phil. Mag.* **2**, 1147

— (1957b). *Phil. Trans. Roy. Soc.* A **250**, 325
— (1960). *Proc. Roy. Soc.* A **257**, 165
— (1963). *Proc. Roy. Soc.* A **272**, 192
— (1964). *Proc. Roy. Soc.* A **282**, 464
— (1965). *Proc. Roy. Soc.* A **287**, 165
— (1968). *Proc. Roy. Soc. Lond.* A **308**, 291
— (1969a). In *The Physics of Metals, vol. 1: Electrons*, ed. J. M. Ziman, p. 113. Cambridge: C.U.P.
— (1969b). In *Modern Solid State Physics, vol. 2: Phonons and their Interactions*, ed. J. F. Cochran & R. R. Haering, p. 3. New York: Gordon & Breach
— *see* Falicov, L. M.
Pitaevskii, Y. A. (1959). *Soviet Phys.–JETP*, **10**, 1267
Platzman, P. M. & Wolff, P. A. (1967). *Phys. Rev. Lett.* **18**, 280
— & Wolff, P. A. (1973). *Waves and Interactions in Solid State Plasmas, Solid State Physics Supplement*, vol. 13, ed. H. Ehrenreich, F. Seitz & D. Turnbull. New York: Academic
— *see* Brinkman, W. F.
Pluzhnikov, V. B., *see* Svechkarev, I. V.
Pogorelov, A. V., *see* Lifshitz, I. M.
Popovic, Z., Carbotte, J. P. & Piercy, G. R. (1973). *J. Phys.* F **3**, 1008
Posternak, M., Waeber, W. B., Griessen, R., Joss, W., van der Mark, W. & Wejgaard, W. (1975). *J. Low Temp. Phys.* **21**, 47
Pötzl, H., *see* Seeger, K.
Poulsen, R. G., Randles, D. L. & Springford, M. (1974). *J. Phys.* F **4**, 981
— *see* Lowndes, D. H.
— *see* Miller, K. M.
Poulsen, U. K., Kollár, J. & Andersen, O. K. (1976). *J. Phys.* F **6**, L241
— *see* Andersen, O. K.
Praddaude, H. C., *see* Foner, S.
Prange, R. E. & Kadanoff, L. P. (1964). *Phys. Rev.* A **134**, 566
— & Korenman, V. (1975). *AIP Conf. Proc.* **24**, 325
— & Sachs, A. (1967). *Phys. Rev.* **158**, 672
— *see* Fowler, M.
— *see* Korenman, V.
— *see* Nee, T. W.
Price, P. J. (1957). *I.B.M.J. Res. Dev.* **1**, 147, 239
— (1958). *I.B.M.J. Res. Dev.* **2**, 200
— (1978). *Solid State Electronics*, **21**, 9
Priestley, M. G. (1963). *Proc. Roy. Soc. Lond.* A **276**, 258
— *see* Harte, G. A.
— *see* Vuillemin, J. J.
Privorotskii, I. A. (1967). *Zh. Eksp. Teor. Fiz.* **52**, 1755
Pudalov, V. M. & Khaïkin, M. S. (1974). *Zh. Eksp. Teor. Fiz.* **67**, 2260 (*Soviet Phys.-JETP*, **40**, 1121 (1975))
Pugh, E. W., *see* Argyle, B. E.

Randles, D. L. (1972). *Proc. Roy. Soc.* A **331**, 85
— *see* Llewellyn, B.
— *see* Palin, C. J.
— *see* Poulsen, R. G.
Rasolt, M. & Vosko, S. H. (1974). *Phys. Rev.* B **10**, 4195
Rath, J. & Callaway, J. (1973). *Phys. Rev.* B **8**, 5398
Raubenheimer, L. J., *see* Gilat, G.
Rayne, J. A. & Jones, C. K. (1970). *Physical Acoustics*, vol. 7, ed. W. P. Mason & R. N. Thurston, p. 149. New York: Academic
Reed, W. A. & Condon, J. H. (1970). *Phys. Rev.* B **1**, 3504
— *see* Arko, A. J.
— *see* Fawcett, E.
— *see* Graebner, J. E.
Rees, H. D. (1968). *Phys. Lett.* **26**A, 416
Rennie, R. (1977). *Adv. in Phys.* **26**, 285
Reuter, G. E. & Sondheimer, E. H. (1948). *Proc. Roy. Soc. Lond.* A **195**, 336
Rice, T. M. (1965). *Ann. Phys.* **31**, 100
— *see* Brinkman, W. F.
Richards, F. T. (1974). *Phys. Rev.* B **10**, 3126
Riedi, P. C. (1977). *Phys. Rev.* B **15**, 5197
Rivier, N. & Zlatic, V. (1972a). *J. Phys.* F **2**, L87
— & Zlatic, V. (1972b). *J. Phys.* F **2**, 199
Roaf, D. J. (1962). *Phil. Trans. Roy. Soc.* A **255**, 135
Robinson, J. E. (1950). Thesis, Yale University
Rode, J. P. & Lowndes, D. H. (1977). *Phys. Rev.* B **16**, 2792
— *see* Chung, Y.
Roman, P. (1965). *Advanced Quantum Theory*. Reading: Addison-Wesley
Rosei, R., *see* Colavita, E.
— *see* Guerrisi, M.
— *see* Weaver, J. H.
Rosenman, I., *see* Batallan, F.
Rostocker, J., *see* Kohn, W.
Rowell, J. M., *see* McMillan, W. L.
— *see* Smith, G. E.
Rubin, J. J., *see* Higgins, R. J.
— *see* Graebner, J. E.
— *see* Cheng, L. S.
— *see* Alpher, R. A.

Ruggeri, R., *see* Stocks, G. M.
Rupp, L. W. (1966). *Rev. Sci. Instrum.* **37**, 1039
Rusby, R. L. (1974). *J. Phys.* F **4**, 1265
Ruvalds, J. & Falicov, L. M. (1968). *Phys. Rev.* **172**, 508

Sachs, A., *see* Prange, R. E.
Sandell, B., *see* Nilsson, P. O.
Sandfort, R. M., *see* Anderson, J. R.
Sang, D. & Myers, A. (1976). *J. Phys.* F **6**, 545
Scalapino, D. J., Schrieffer, J. R. & Wilkins, J. W. (1966). *Phys. Rev.* **148**, 263
Scheffers, H., *see* Justi, E.
Schipper, R., *see* Anderson, J. R.
Schirber, J. E. (1971). *Phys. Lett.* **35**A, 194
— (1974). *Honda Memorial Series on Materials Science*, p. 141. Tokyo: Maruzen Co.
— (1978). *Proc. Int. High Pressure Conf.*, Boulder 1977
— & O'Sullivan, W. J. (1970a). *Colloque International du C.N.R.S., Sur les Proprietes Physiques des Solides sous Pression*, Grenoble, **188**, 113
— & O'Sullivan, W. J. (1970b). *Phys. Rev.* B **1**, 1443
— Schmidt, F. A. & Koelling, D. D. (1977). *Phys. Rev.* B **16**, 4235
— *see* Anderson, J. R.
— *see* Harmon, B. N.
— *see* O'Sullivan, W. J.
Schmidt, F. A., *see* Schirber, J. E.
Schmidt, H., *see* Mann, E.
Schmor, P. W., *see* Gold, A. V.
Schrieffer, J. R. (1964). *Theory of Superconductivity*, pp. 158–63. New York: Benjamin
— (1967). *J. Appl. Phys.* **38**, 1143
— *see* Scalapino, D. J.
— *see* Wang, S. Q.
Schultz, S. & Dunifer, G. (1967). *Phys. Rev. Lett.* **18**, 283
— *see* Lubzens, D.
— *see* Pinkel, D.
Schwartz, L. M., *see* Ehrenreich, H.
Scott, G. B., Springford, M. & Stockton, J. R. (1968). *J. Phys.* E **1**, 925
— *see* Coleridge, P. T.
Seeger, K. & Pötzl, H. (1973). In *The Boltzmann Equation*, ed. E. G. Cohen & W. Thirring, p. 341. Vienna: Springer
Segall, B. (1961). *Phys. Rev.* **125**, 109
— & Ham, F. S. (1968). *Methods in Computational Physics*, vol. 8, ch. 7. New York: Academic
— *see* Chen, A. B.
Seitz, F. (1940). *The Modern Theory of Solids*, pp. 234–46. New York: McGraw-Hill
— *see* Wigner, E. P.
Seki, S., *see* Matsumoto, H.
Sellmyer, D. J., *see* Coleman, R. V.
— *see* Goldstein, I. S.
— *see* Tobin, P. J.
Semenenko, A. I., *see* Kaganov, M. I.
Seraphim, B. O., *see* Christensen, N. E.
Sham, L. J., *see* Kohn, W.
Shanaburger, M. R., *see* Lubzens, D.
Shapira, Y. (1968). *Physical Acoustics*, vol. 5, p. 1. New York: Academic
Shaw, J. C., Ketterson, J. B. & Windmiller, L. R. (1972). *Phys. Rev.* B **5**, 3894
— *see* Ketterson, J. B.
Shepherd, J. G. & Gordon, W. L. (1968). *Phys. Rev.* **169**, 541
Shiba, H. (1973). *Prog. Theor. Phys.* **50**, 1797
Shimuzu, M. (1977). *Physica*, **91**B, 14
Shirane, G., Minkiewicz, V. J. & Nathans, R. (1968). *J. Appl. Phys.* **39**, 383
Shoenberg, D. (1939). *Proc. Roy. Soc. Lond.* A **170**, 341
— (1952a). *Phil. Trans. Roy. Soc. Lond.* A **245**, 1
— (1952b). *Superconductivity*, 2nd edn, p. 95. Cambridge: C.U.P.
— (1957). *Progress in Low Temperature Physics*, vol. 2, p. 264. Amsterdam: North-Holland
— (1960a). In *The Fermi Surface*, ed. W. A. Harrison & M. B. Webb, p. 74. New York: Wiley
— (1960b). *Phil. Mag.* **5**, 105
— (1962). *Phil. Trans. Roy. Soc.* A **255**, 85
— (1968). *Can. J. Phys.* **46**, 1915
— (1969). *Phys. Kondens. Materie*, **9**, 1
— (1976). *J. Low Temp. Phys.* **25**, 755
— & Stiles, P. J. (1964). *Proc. Roy. Soc. Lond.* **281**, 62
— & Templeton, I. M. (1973). *Physica*, **69**, 293
— & Uddin, M. Z. (1936). *Proc. Roy. Soc. Lond.* A **156**, 687
— & Vuillemin, J. J. (1966). In *Proc. 10th Int. Conf. on Low Temperature Physics (L.T.10)*. Moscow
— & Watts, B. R. (1965). In *Proc. 9th Int. Conf. on Low Temperature Physics (L.T.9)*, p. 831. New York: Plenum
— & Watts, B. R. (1967). *Phil. Mag.* **15**, 1275
— *see* Bibby, W. M.
— *see* Dingle, R. B.

— *see* Knecht, B.
— *see* Palin, C. J.
— *see* Perz, J. M.
Shubnikov, A. V. & de Haas, W. J. (1930). *Commun. Kamerling Onnes Lab. Univ. Leiden*, **19**, 207a
Shutz, R. J., *see* Graebner, J. E.
Sidorov, S. K., *see* Menshikov, A. Z.
Siegmann, H. C. (1978). In *Proc. Int. Conf. on the Physics of Transition Metals*, Toronto, 1977, ed. M. J. G. Lee, J. M. Perz & E. Fawcett, p. 276 (Institute of Physics Conference Series No. 39). London: Institute of Physics
Sievert, P. R., *see* Falicov, L. M.
Silin, V. P. (1958). *Soviet Phys.-JETP*, **6**, 387
— (1959). *Soviet Phys.-JETP*, **8**, 870
Simpson, G., *see* Engelsberg, S.
Singh, M., Wang, C. S. & Callaway, J. (1975). *Phys. Rev.* B **11**, 287
Skillman, S., *see* Herman, F.
Skobov, V. G. (1959). *Zh. Eksp. Teor. Fiz.* **37**, 1467
— *see* Kaner, E. A.
Skriver, H. L., Andersen, O. K. & Johansson, B. (1978a). *Phys. Rev. Lett.* **41**, 42
—, Venema, W., Walker, E. & Griessen, R. (1978b). *J. Phys.* F **8**, 2313
Slater, J. C. (1937). *Phys. Rev.* **51**, 151
— (1951). *Phys. Rev.* **81**, 385
— (1974). *Quantum Theory of Molecules & Solids*, vol. 4. New York: McGraw-Hill
— & Koster, G. F. (1954). *Phys. Rev.* **94**, 1498
Slavin, A. J. (1972). *Cryogenics*, **12**, 121
— (1973). *Phil. Mag.* **27**, 65
— *see* Paton, B. E.
Slutskin, A. A. (1967). *Zh. Eksp. Teor. Fiz.* **53**, 767
— (1970). *Zh. Eksp. Teor. Fiz.* **58**, 1098
— (1973). *Zh. Eksp. Teor. Fiz.* **65**, 2114
— *see* Alekseevskiĭ, N.E.
— *see* Kaganov, M. I.
Smith, G. E., Baraff, G. A. & Rowell, J. M. (1964). *Phys. Rev.* **135**, A1118
Smith, H., *see* Højgaard, Jensen H.
Smith, J. P., *see* Cooper, J. D.
Smith, M. J. A., *see* Dupree, R.
Smith, N. V. (1974). *Phys. Rev.* B **9**, 1365
— *see* Traum, M. M.
Soden, R. R., *see* Testardi, L. R.
Sokoloff, J. B. (1978). *Phys. Rev.* B **17**, 2380
Sommerfeld, A. & Bethe, H. A. (1933). Elektronentheorie der Metalle, *Handbuch d. Physik*, **24**, 333–622.
Sondheimer, E. H. (1950). *Phys. Rev.* **80**, 401
— *see* Reuter, G. E.
Sorbello, R. S. (1974a). *J. Phys.* F **4**, 503
— (1974b). *J. Phys.* F **4**, 1665
— (1977). *Phys. Rev.* B **15**, 3045
— & Griessen, R. (1973). *Solid State Commun.* **12**, 689
Soven, P. (1970). *Phys. Rev.* B **2**, 4715
— (1972). *Phys. Rev.* B **5**, 260
— *see* Lasseter, N. H.
Spain, I. L., *see* Dillon, R. O.
Sparlin, D. M., *see* Chandrasekhar, B. S.
Spicer, W. E. & Berglund, C. N. (1964). *Phys. Rev. Lett.* **12**, 9
Springford, M. (1971). *Adv. in Phys.* **20**, 493
—, Stockton, J. R. & Templeton, I. M. (1969). *Phys. Kondens. Materie*, **9**, 15
— *see* Elliott, M.
— *see* Llewellyn, B.
— *see* Lowndes, D. H.
— *see* Miller, K. M.
— *see* Paul, D.McK.
— *see* Poulsen, R. G.
— *see* Scott, G. B.
Spurgeon, W. A. & Lazarus, D. (1972). *Phys. Rev.* B **6**, 4396
Stachowiak, H. (1976). In *Magnetism in Metals & Magnetic Compounds*, ed. J. T. Lopuszanski, A. Pekalski & J. Przystawa. New York: Plenum
— *see* Kontrym-Szajd, G.
Stanford, J. L., *see* Gutman, E. J.
— *see* Hoekstra, J. A.
Stanley, D. J., Perz, J. M. & Au, H-P. (1976). *Can. J. Phys.* **54**, 1234
— Perz, J. M., Lee, M. J. G. & Griessen, R. (1977). *Can. J. Phys.* **55**, 344
— *see* Fawcett, E.
— *see* Griessen, R.
— *see* Lee, M. J. G.
Stark, R. W. (1964). *Phys. Rev.* A **135**, 1698
— & Falicov, L. M. (1967). In *Progress in Low Temperature Physics*, vol. 5, ed. C. J. Gorter, p. 235. Amsterdam: North-Holland
— & Friedberg, C. B. (1971). *Phys. Rev. Lett.* **26**, 556
— & Friedberg, C. B. (1972). *Phys. Rev.* B **5**, 2844
— & Windmiller, L. R. (1968). *Cryogenics*, **8**, 272
Stearns, M. B. (1977). *Physica*, **91B**, 37
Stepanova, N. A., *see* Kontorovich, V. M.

Stern, E. A. (1968). *Phys. Rev.* **168**, 730
— (1969). *Phys. Rev.* **188**, 1163
— (1972). *Phys. Rev.* B **5**, 366
— *see* Brown, M.
Stiles, P. J., *see* Shoenberg, D.
Stocks, G. M., Gyorffy, B. L., Guiliano, E. S. & Ruggeri, R. (1977). *J. Phys.* F **7**, 1859
— *see* Painter, G. S.
Stockton, J. R., *see* Scott, G. B.
— *see* Springford, M.
Stone, D. R., *see* Anderson, J. R.
— *see* Gold, A. V.
Strěda, P. (1974). *Czeckoslovak J. Phys.* B **24**, 794
Stringfellow, M. W. (1968). *J. Phys.* C **1**, 950
Ström-Olsen, J. O., *see* Buot, F. A.
— *see* Harris, R.
— *see* Li, P. L.
Supek, I. (1940). *Z. Phys.* **117**, 125
Svechkarev, I. V. & Pluzhnikov, V. B. (1973). *Phys. Stat. Sol.* **55**B, 315
— & Panfilov, A. S. (1974). *Phys. Stat. Sol.* **63**, 11
Switendick, A. C., *see* O'Sullivan, W. J.
Sykes, J., *see* Brooker, G. A.

Tachiki, J., *see* Matsumoto, H.
Takano, S. & Koga, M. (1977). *J. Phys. Soc. Japan*, **42**, 853
Tange, A. & Tokunaga, T. (1969). *J. Phys. Soc. Japan*, **27**, 554
Tanuma, S., Datars, W. R., Doi, H. & Dunsworth, A. (1970). *Solid State Commun.* **8**, 1107
— *see* Ishizawa, Y.
Tatsumoto, *see* Fujiwara
Taylor, P. L. (1963). *Proc. Roy. Soc.* A **275**, 200, 209
— (1970). *A Quantum Approach to the Solid State.* Englewood Cliffs, N.J.: Prentice-Hall
Templeton, I. M. (1966). *Proc. Roy. Soc.* A **292**, 413
— (1972). *Phys. Rev.* B **5**, 3819
— (1974). *Can. J. Phys.* **52**, 1628
— & Coleridge, P. T. (1975a). *J. Phys.* F **5**, 1307
— & Coleridge, P. T. (1975b). In *Proc. 14th Int. Conf. on Low Temperature Physics* (*L.T.14*), ed. M. Krusius & M. Vuorio, p. 143. Amsterdam: North-Holland
— *see* Chollet, L. F.
— *see* Coleridge, P. T.
— *see* Gaertner, A. A.
— *see* Shoenberg, D.
— *see* Springford, M.
Terwilliger, D. W. & Higgins, R. J. (1970). *Phys. Lett.* A **31**, 316
— & Higgins, R. J. (1972). *J. Appl. Phys.* **43**, 3346
— & Higgins, R. J. (1973). *Phys. Rev.* B **7**, 667
Testardi, L. R. & Condon, J. H. (1970). *Phys. Rev.* B **1**, 3928
— & Soden, R. R. (1967). *Phys. Rev.* **158**, 581
Thaler, B. J., *see* Opsal, J. L.
Thompson, E. D., Wohlfarth, E. P. & Bryan, A. C. (1964). *Proc. Phys. Soc. Lond.* **83**, 59
Thompson, R. E., Aron, P. R., Chandrasekhar, B. S. & Langenberg, D. N. (1971). *Phys. Rev.* B **4**, 518
Thorsen, A. C., Joseph, A. S. & Valby, L. E. (1966). *Phys. Rev.* **150**, 523
—, Joseph, A. S. & Valby, L. E. (1967). *Phys. Rev.* **162**, 574
— *see* Joseph, A. S.
Thrane, N., *see* Kontrym-Sznajd, G.
Titeica, S. (1935). *Ann. Phys.* **22**, 128
Tobin, P. J., Sellmyer, D. J. & Averbach, B. L. (1971). *J. Phys. Chem. Solids*, **32**, 1721
Tokunga, T., *see* Tange, A.
Toxen, A. M., *see* Liu, S. H.
Traum, M. M. & Smith, N. V. (1974). *Phys. Rev.* B **9**, 1353
Tripp, J. H., Everett, P. M., Fiske, J. M. & Gordon, W. L. (1970). *Phys. Rev.* B **2**, 1556
Trumpy, G., *see* Kontrym-Sznajd, G.
Tsai, J. T., *see* Isihara, A.
Tsarev, V. A. & Zavaritskii, N. V. (1965). *Soviet Phys.-JETP*, **21**, 85 (*Zh. Eksp. Teor. Fiz.* **48**, 125 (1965))
Tsui, D. C. (1967). *Phys. Rev.* **164**, 669
Tukey, J. W., *see* Cooley, J. W.

Uddin, M. Z., *see* Shoenberg, D.
Uelhoff, W., *see* Lengeler, B.
Umezawa, H., *see* Matsumoto, H.

Valby, L. E., *see* Thorsen, A. C.
Vanderkooy, J. & Datars, W. R. (1967). *Phys. Lett.* **25**A, 258
— *see* Hill, P. H.
— *see* Palin, C. J.
van der Mark, W., *see* Posternak, M.
Van Dyke, J. P. (1973). *Phys. Rev.* B **7**, 2358
Van Hove, L. (1953). *Phys. Rev.* **89**, 1189
Van Vleck, J. H. (1932). *The Theory of Electric and Magnetic Susceptibilities*, p. 100. Oxford: O.U.P.

Vasvari, B., Animalu, A. O. E. & Heine, V. (1967). *Phys. Rev.* **154**, 535
Veal, B. W., *see* Koelling, D. D.
Venema, W. J., *see* Griessen, K.
— *see* Skriver, H. L.
Verkin, B. I., Dmitrenko, I. M. & Lazarev, B. G. (1956). *Zh. Eksp. Teor. Fiz.* **31**, 538 (*Soviet Phys.-JETP*, **4**, 432 (1957))
— *see* Dmitrenko, I. M.
Vilenkin, A. & Taylor, P. L. (1979). *Phys. Rev.* B **20**, 576
Vinokurova, L. I., Gaputchenko, A. G. & Itskevich, E. S. (1977). *JETP Lett.* **26**, 317 (*Pisma Zh. Eksp. Teor. Fiz.* **26**, 443 (1977))
Vogt, E., *see* Herring, C.
Volodin, A. P., Khaĭkin, M. S. & Edel'man, V. S. (1973a). *Pisma Zh. Eksp. Teor. Fiz.* **17**, 491
—, Khaĭkin, M. S. & Edel'man, V. S. (1973b). *Zh. Eksp. Teor. Fiz.* **65**, 2105
von Barth, U. & Hedin, L. (1972). *J. Phys.* C **5**, 1629
Vosko, S. H. & Perdew, J. P. (1975). *Can. J. Phys.* **53**, 1385
— *see* Rasolt, M.
Vuillemin, J. J. & Mori, N. (1977). *Solid State Commun.* **23**, 719
— & Priestley, M. G. (1965). *Phys. Rev. Lett.* **14**, 307
— *see* Harte, G. A.
— *see* Shoenberg, D.

Waber, J. T., *see* Liberman, D.
Waeber, W. B., *see* Posternak, M.
Wagner, D. K. (1979). *Adv. in Phys.* (to appear)
Walker, E., *see* Skriver, H. L.
Walsh, W. M. & Grimes, C. C. (1964). *Phys. Rev. Lett.* **13**, 523
Wampler, W. R. & Lengeler, B. (1977). *Phys. Rev.* B **15**, 4614
—, Matula, S., Lengeler, B. & Durkansky, G. (1975). *Rev. Sci. Instrum.* **46**, 58
— *see* Lengeler, B.
Wang, C. S. & Callaway, J. (1974). *Phys. Rev.* B **9**, 4897
— & Callaway, J. (1977). *Phys. Rev.* B **15**, 298
— *see* Callaway, J.
— *see* Singh, M.
Wang, S. Q., Evenson, W. E. & Schrieffer, J. R. (1969). *Phys. Rev. Lett.* **23**, 92
Watson, R. E., *see* Gelatt, C. D.
— *see* Hodges, L.
Watts, B. R. (1973). *J. Phys. F.* **3**, 1345
— (1974a). *J. Phys.* F **4**, 1371
— (1974b). *J. Phys.* F **4**, 1387
— *see* Coleridge, P. T.
— *see* Gamble, D.
— *see* Shoenberg, D.
Weaire, D., *see* Heine, V.
Weaver, J. H. (1978). In *Proc. Int. Conf. on the Physics of Transition Metals*, Toronto, 1977, ed. M. J. G. Lee, J. M. Perz & E. Fawcett (Institute of Physics Conference Series No. 39). London: Institute of Physics
—, Lynch, D. W., Culp, C. H. & Rosei, R. (1976). *Phys. Rev.* B **14**, 459
— & Olson, C. G. (1977). *Phys. Rev.* B **15**, 4602
—, Olson, C. G. & Lynch, D. W. (1975). *Phys. Rev.* B **12**, 1293
—, Olson, C. G. & Lynch, D. W. (1977). *Phys. Rev.* B **15**, 4115
Webb, M. B., *see* Harrison, W. A.
Weiner, R. A., *see* Beal-Monod, M. T.
Wejgaard, W., *see* Posternak, M.
Wheatley, J. C. (1975). *Rev. Mod. Phys.* **47**, 467
— *see* Abel, W. R.
White, R. M. (1970). *Quantum Theory of Magnetism*, ch. 8. New York: McGraw-Hill
White, G. K., *see* Chandrasekhar, B. S.
Wierzchowski, W., *see* Kontrym-Sznajd, G.
Wigner, E. P. & Seitz, F. (1955). *Solid State Physics*, **1**, 96
Wilkins, J. W. (1968). *Observable Many-Body Effects in Metals.* Copenhagen: Nordita
— & Woo, J. W. F. (1965). *Phys. Lett.* **17**, 89
— *see* Ashcroft, N. W.
— *see* Højgaard Jensen H.
— *see* Kostroun, V. O.
— *see* Muller, J. E.
— *see* Scalapino, D. J.
Williams, A. R., *see* Janak, J. F.
Williams, A. R., *see* Moruzzi, V. L.
Williams, R. W. & Mackintosh, A. R. (1968). *Phys. Rev.* **168**, 679
Williamson, S. J., *see* Goldstein, A.
Willis, R. F., *see* Christensen, N. E.
Wilson, A. H. (1953). *The Theory of Metals*, 2nd edn (1st edn 1936). Cambridge: C.U.P.
Windmiller, L. R. & Ketterson, J. B. (1968). *Rev. Sci. Instrum.* **39**, 1672
—, Ketterson, J. B. & Hörnfeldt, S. (1969). *J. Appl. Phys.* **40**, 1291
—, Ketterson, J. B. & Hörnfeldt, S. (1970).

J. Appl. Phys. **41**, 1232
—, Ketterson, J. B. & Hörnfeldt, S. (1971). *Phys. Rev.* B **3**, 4213
— *see* Arko, A. J.
— *see* Bosacchi, B.
— *see* Chang, Y. K.
— *see* Crabtree, G. W.
— *see* Dye, D. H.
— *see* Hornfeldt, S. P.
— *see* Ketterson, J. B.
— *see* Lowndes, D. H.
— *see* Mueller, F. M.
— *see* Shaw, J. C.
— *see* Stark, R. W.
Wingerath, K., *see* Lengeler, B.
Winsemius, O., *see* Guerrisi, M.
Wohlfarth, E. P. (1975). *Comments Solid State Phys.* **6**, 123
— (1976). In *Magnetism, Selected Topics*, ed. S. Foner. New York: Gordon & Breach
— *see* Edwards, D. M.
— *see* Fadley, C. S.
— *see* Lowde, R. D.
— *see* Thompson, E. D.
Wolff, P. A., *see* Platzmann, P. M.
Woo, J. W. F., *see* Wilkins, J. W.
Wood, J. H. (1962). *Phys. Rev.* **126**, 517
Woore, J., *see* Cooper, J. D.

Yamada, H. (1974). *J. Phys.* F **4**, 1819
Young, D. A., *see* Bender, A. S.
— *see* Cooper, J. D.
Young, R. A. (1968). *Phys. Rev.* **175**, 813
Young, R. C., Jordan, R. G. & Jones, D. W. (1973). *Phys. Rev. Lett.* **31**, 1473
— *see* Mattocks, P. G.

Zavaritskii, N. V., *see* Tsarev, V. A.
Zil'berman, G. E. (1955). *Zh. Eksp. Teor. Fiz.* **29**, 762
Ziman, J. M. (1961a). *Adv. in Phys.* **10**, 1
— (1961b). *Phys. Rev.* **121**, 1320
— (1965). *Proc. Phys. Soc.* **86**, 337
— (1969). *Elements of Advanced Quantum Theory*. Cambridge: C.U.P.
Zlatic, V., *see* Rivier, N.
Zornberg, E. I. (1970). *Phys. Rev.* B **1**, 244
Zuckermann, M. J., *see* Falicov, L. M.
— *see* Harris, R.

Index